Introduction to Electromobility

Martin Doppelbauer

Introduction to Electromobility

Technology, Best Practice, Energy and Environment

Springer

Martin Doppelbauer
Elektrotechnisches Institut
KIT Karlsruher Institut für Technologie
Karlsruhe, Baden-Württemberg, Germany

ISBN 978-3-658-45481-4 ISBN 978-3-658-45482-1 (eBook)
https://doi.org/10.1007/978-3-658-45482-1

Original Published by Springer Vieweg, Springer Fachmedien Wiesbaden

This Springer imprint is published by the registered company Springer Fachmedien Wiesbaden GmbH, part of Springer Nature.
The registered company address is: Abraham-Lincoln-Str. 46, 65189 Wiesbaden, Germany

Preface

This book on electromobility covers a wide range of topics, from the beginnings and fundamentals of electrical engineering to hybrid and electric vehicle drives, battery storage, and power electronics, as well as energy, environmental, and infrastructure aspects.

It, thus, provides a broad and technically sound overview of the wide fields of physical relationships, materials, and technologies relevant to hybrid and electric vehicles. The technical focus is on the areas of electric drive technology, energy storage, and charging technology. A separate chapter is devoted to the many different hybrid drive structures, including a description of power-splitting transmissions. The essential aspects of vehicle technology, especially about power requirements and driving cycles, are also briefly presented.

We live in a time of radical change. To stop man-made global warming, a variety of efforts are needed to reduce anthropogenic CO_2 emissions. The final chapters of the book examine the contribution that electrically powered passenger cars can make in this regard. In detail, the book deals with issues of energy generation, environmental protection, and infrastructure. The book is not intended to compete with the many studies in these fields. Rather, it is intended to provide a comprehensive overview, thus enabling a deeper understanding of the interrelationships.

The book is suitable for readers who have a general knowledge of technology and mathematics, but no special knowledge of electric circuits, electromagnetic motor design, chemical engineering, automotive engineering, or energy technology.

The focus will be on passenger cars, although the basic procedures and principles are applicable to all types of electric traction drives—from Pedelecs to commercial vehicles such as buses, trains, trucks, and more.

Naturally, with such a broad field, abridgments and omissions must be made. Nevertheless, the author is confident that the most relevant topics have been covered in a reasonably complete overview.

The author is often asked about reducing the number of cars and strengthening local and long-distance public transport. Without a doubt, more developed mobility concepts are a key lever in improving CO_2 emissions from the transport sector.

However, this is much more a political than a technical issue. The field of public transportation has been neglected in Germany and many other countries for decades. In this book, I can only highlight the quite remarkable environmental potential of electromobility.

Even if we just replace all of today's cars with electric cars, we will have already achieved a great deal in climate protection. In addition, it is important to reduce individual passenger transport in favor of a new, multimodal mobility. However, this is not the subject of this book.

The book was compiled from four lectures for the module "Electric and Hybrid Drives" in the bachelor's degree program in automotive engineering, which I wrote for the Wilhelm Büchner University of Applied Sciences in Darmstadt. I would like to thank my colleagues from Darmstadt for kindly releasing the texts and graphics for use in this project.

The content is based on my interdisciplinary course "Hybrids and Electric Vehicles", which I have been holding annually at the Karlsruhe Institute of Technology (KIT) since the winter semester 2011/12. The lecture is open to students of all engineering courses and industrial engineering. It is also part of the "Studium Generale" at KIT and enjoys great popularity.

I would like to thank my colleague Prof. Marc Hiller for his support, especially in writing the chapters on power electronics, and Dr.-Ing. Jan Richter from Batemo GmbH for his support in the sections on modeling lithium-ion batteries and for kindly providing the illustrations.

Second Edition

The landscape of electromobility is rapidly evolving. When the first edition of this book was released, electric cars were a rare sight in many European countries. Fast charging stations were scarce, and embarking on a long-distance journey with an electric car was deemed an adventurous endeavor, particularly in southern parts of Europe.

A profound transformation took place in recent years. Charging infrastructure has proliferated, with stations now available at all major service points, especially along highways. Most automotive manufacturers offer multiple all-electric models, with many signaling their intent to phase out internal combustion engines entirely in the coming years.

Due to increased production volumes and improved manufacturing processes, costs are expected to reach parity with internal combustion engine vehicles within the next decade. This situation is already visible on the Chinese market, where the prices of electric cars are sometimes even lower than those of equivalent combustion engine cars. Simultaneously, technology is advancing rapidly: electric motors are becoming smaller and lighter, power electronics are transitioning to novel semiconductor materials, charging technologies for commercial vehicles with megawatt-level capacities are being developed, and entirely new technologies are being introduced in batteries.

What was once the subject of ridicule and apprehension among conservative enthusiasts of internal combustion engines has now firmly established itself in the mainstream market.

There is a growing resistance to electric mobility in some circles, which is not objectively justified but is often loudly expressed, particularly on social media. Similar behavior could be observed towards vaccinations during the COVID-19 pandemic. The transition to a new technology is perceived as a personal threat by some individuals. They are opposing on principle because the issue is driven by politics and science.

There is no doubt that electromobility is not perfect today. The mining conditions for some raw materials need to be improved, we need more electricity generated from renewable sources and we need to increase recycling rates. Nevertheless, electric cars are overall much cleaner than cars with combustion engines. The fact that they are also our only realistic chance to significantly reduce the environmental footprint of the transportation sector in the future and make our environment more livable, while maintaining current individual mobility, is often not understood. Environmental improvements are no longer apparent in cars with internal combustion engines, including hybrids, for the past 15 years. Fuel consumption is stagnating at a high level. I consider all promises of achieving further significant savings with new combustion engines to be misleading.

Furthermore, electromobility also signifies a shift in revenues and profits. There are several financially strong protagonists, such as Russia and the oil-producing countries of the Gulf region, and certainly some large automotive companies that will suffer as a result. It can be assumed that targeted disinformation campaigns are being run to delay or prevent electromobility.

It seems sensible to put forward a few theses, which will be examined and substantiated later in the book:

- Electric cars are nowadays fully practical. You can comfortably travel with them all across Europe, even for long distances and professional purposes. Only in Southern Europe, the expansion of the fast charging network lags behind a bit, but it is developing rapidly.
- Electric cars are significantly more environmentally friendly than vehicles with combustion engines. In Germany, grid electricity currently (as of 2024) saves around half the CO_2 emissions over the typical lifetime of an electric car compared to similar diesel and petrol vehicles, including hybrids, and even two thirds in the average European electricity mix. In addition, particulate matter and noise emissions are lower and all other pollutant emissions are avoided altogether. The CO_2 footprint of electric cars will continue to improve in the future as the electricity mix becomes ever cleaner.
- From today's perspective, there is no alternative to battery-powered electric cars for largely climate-neutral individual mobility, even in the long term. Synthetic fuels or hydrogen are too complex, inefficient, and therefore expensive to produce. They are not suited to support a large fleet of many million vehicles.

- Driving an electric car is much safer than driving a combustion engine vehicle. Fires occur 5–10 times less frequently and there are no spontaneous explosions immediately following severe accidents. In general, occupant accident protection is better for technological reasons.
- The raw materials required for electromobility are available in sufficient quantities and can be mined in an environmentally friendly manner under acceptable working conditions. Where this is not yet the case, the focus is on the pursuit of higher profits rather than technical necessity. We can influence this. All components in electric cars, especially the batteries, can be largely recycled. The raw materials invested in production are therefore not lost, but can be reused. A conventional car burns around 3 to 4 times its own weight in crude oil.
- The additional electric energy required for electromobility is manageable and can be generated entirely from renewable and environmentally friendly sources in Europe.
- The infrastructure for electromobility will undoubtedly cost a significant amount of money, but it can be rolled out over two to three decades, making it affordable. Just 100 years ago, hardly anyone could imagine that we would have electricity, hot water, and a toilet in every house.
- There are currently no recognizable technical obstacles that would fundamentally stand in the way of the widespread introduction of electric cars.
- Mobility is diverse, ranging from bicycles and cars to light and heavy trucks, buses, agricultural vehicles, ships, and aircraft. Batteries are advantageous when relatively little energy is consumed relatively quickly. It remains to be seen which energy form will prevail in the long term for heavy commercial vehicles, ships, and aircraft. Synthetic fuels are likely to play a role in this context.

I am pleased to present an updated and significantly expanded edition of my book on electromobility, where these future topics have been given ample space. This edition is being published in English for the first time, aiming to reach a broader audience and contribute to the ongoing discourse on the present and future of electric vehicles.

Karlsruhe, Spring 2024 Martin Doppelbauer

Contents

About the Author

Martin Doppelbauer studied electrical engineering at the Technical University of Dortmund, Germany, where he received his doctorate in 1995 with a thesis on the analytical field-harmonic calculation of single-phase AC commutator motors. From 1995 to 2010, he held various positions in the electrical industry in Baden-Württemberg, Germany, most recently as head of electric motor development at SEW Eurodrive GmbH & Co KG in Bruchsal.

Since the beginning of 2011, Professor Doppelbauer has held the professorship for Hybrid Electric Vehicles (HEV) at the Institute of Electrical Engineering (Elektrotechnisches Institut—ETI) of the Karlsruhe Institute of Technology (KIT).

With about 70 employees, the ETI conducts research on topics of electric drive and energy technology, ranging from power electronics, control engineering, and electric motors to applications of battery storage and fuel cells in these fields. The ETI is headed by the two professors Marc Hiller with a focus on power electronics and applications in the field of renewable energy and Martin Doppelbauer. His research focus is on increasing the power density of electric traction drives for passenger cars, commercial vehicles and Pedelecs through innovative motor concepts, cooling, and mechanical design.

Prof. Doppelbauer has been active in the field of international standardization of electric machines for many years and is, among other roles, the chairman of technical committee 2 (TC2 rotating electric machinery) at the International Electrotechnical Commission (IEC) in Geneva.

In addition to lecturing on hybrid and electric vehicles, which forms the basis for this book, Prof. Doppelbauer lectures on the design and industrial use of electric drives and on theoretical electrical engineering. He has also been Dean of Studies for the Mechatronics and Information Technology program at the Karlsruhe Institute of Technology since 2012.

Abbreviations

AC	Alternating current
ADAC	Allgemeiner Deutscher Automobil-Club e.V.
AFC	Alcaline fuel cell
AFIM	Axial-flux induction machine
AFM	Axial-flux machine
AFPM	Axial-flux permanent magnet synchronous machine
B4B	Batteries-for-Buildings
BEV	Battery electric vehicle
BJT	Bipolar junction transistor (also BT = bipolar transistor)
BLDC	Brushless DC motor (the wrong term for a PM synchronous motor with integrated power electronics—correct term EC motor)
BMS	Battery management system
BRICS	Brasil, Russia, India, China, South Africa
BSG	Belt starter generator
BtL	Biomass-to-liquid (process to produce liquid fuel from biomass (rapeseed, straw, etc.))
CADC	Common Artemis Driving Cycle (obsolete)
CAN	Controller Area Network (Real-time network system)
CAPEX	Capital expenditures
CCGT	Combined cycle gas turbine
CcH2	Cryo-compressed hydrogen
CCS	Carbon capture and storage
CCS	Combined charging system
CCCV	Constant current constant voltage (battery charging procedure)
CEE	European plug (blue = single-phase, red = three-phase)
CH2	Compressed hydrogen (gaseous hydrogen with 350–700 bar pressure)
CHAdeMO	CHArge de MOve (Japanese standard for DC charging)
ChaoJi	Chinese/Japanese standard for high-power DC fast charging
CID	Current interruption device (thermal fuse)

CIS	Commonwealth of independent states (states of the former Soviet Union)
CNG	Compressed natural gas
CoFe	Cobalt-iron (soft magnetic material for laminated electrical sheets)
CP	Control pilot
CP	Charging point
CPV	Concentrated photovoltaik
CS	Charging station
CSMS	Charging station management system
CSP	Concentrated solar power (solar thermal power plant)
CtL	Coal-to-liquid (process for the production of liquid fuel from coal)
CTP	Cell to pack design (of batteries)
CVT	Continuous velocity transmission (continuously variable transmission)
DAC	Direct air capture (of CO_2)
DC	Direct current
DE	Drive-end (the side of an electric motor that is connected to the gearbox)
DKE	Deutsche Kommission Elektrotechnik Elektronik Informationstechnik in DIN und VDE (national German standards organization for electrotechnical topics)
DOD	Depth-of-discharge (0 to 100%)
DWV	Deutscher Wasserstoff-Verband e.V.
E-Fuels	Designation for synthetic fuels obtained from electrical energy
EBIT	Earnings before interest and taxes
EC-Motor	Electronically (brushless) commutated motor
ECE	Urban driving cycle in the NEDC
EEG	Erneuerbare-Energien-Gesetz (German renewable energies act)
EESM	Electrically excited synchronous machine
EIM	External identification means (registration procedure at a charging station)
ELV	Extra low voltage
EM	Electric motor
EPA	Environmental protection agency
EUDC	Extra Urban Driving Cycle
EV	Electric vehicle
EVSE	Electric vehicle supply equipment (charging point)
EVT	Electric continuously variable transmission
FC	Fuel cell
FCEV	Fuel cell electric vehicle
FET	Field-effect transistor
FHEV	Full-hybrid electric vehicle
FOM	Figure of merit
FTP	Federal Test Procedure (US driving cycle)

GaN	Gallium Nitride (wide bandgap semiconductor material)
GEM	Gearbox-integrated electric motor
GHG	Greenhouse gas (CO_2, CH_4, …)
GDP	Gross domestic product
GtL	Gas-to-liquid (process to produce liquid fuel from gas)
GWP	Global warming potential
HDV	Heavy-duty vehicle
HEV	Hybrid electric vehicle
HGI	Hydrogen gas injector
HPC	High power charging
HREE	Heavy rare earth elements (for example dysprosium)
HV	High voltage
HVDC	High voltage direct current
ICCB	In-cable control box
ICCT	International council on clean transportation
ICE	Internal combustion engine
IEA	International energy agency
IEC	International electrotechnical commission (international standards organization for electrical engineering)
IEEE	Institute of electrical and electronics engineers (US-American professional association of electrical engineers)
IGBT	Insulated-gate bipolar transistor
IM	Induction machine
IMD	Insulation monitoring device
IMG	Integrated motor-generator
IPM	Permanent magnet synchronous machine with internal permanent magnets and hence increased reluctance torque, also IPM-PMSM
ISG	Integrated starter-generator
ISO	International standards organisation (international standards organization for non-electrical topics)
LCO	Lithium cobalt oxide (cathode material of a lithium-ion battery)
LCV	Light commercial vehicle
LFP	Lithium iron phosphate (cathode material of a lithium-ion battery)
lge	Liters of gasoline (petrol) equivalent
LH2	Liquified H_2 (liquid hydrogen at a temperature of 20 K or -253 °C)
LIB	Lithium-ion battery
LiPo	Lithium polymer battery (lithium-ion battery with a gel-like polymer-based separator)
LMO	Lithium manganese oxide (cathode material of a lithium-ion battery)
LNG	Liquified natural gas
LOHC	Liquid organic hydrogen carrier
LPG	Liquified petroleum gas (autogas)
LREE	Light rare earth elements (for example neodymium)
LTO	Lithium titanium oxide (anode material of a lithium-ion battery)

LV	Low-voltage
MCHEV	Micro hybrid electric vehicle
MCS	Megawatt charging system
MCV	Medium-size commercial vehicle
MENA	Middle east and north Africa
MESC	Multifunctional energy storage composite
MHEV	Mild hybrid electric vehicle
MMF	Magnetomotive force (also ampere turns or current linkage)
MOSFET	Metal-oxide field-effect transistor
MTPA	Maximum torque per ampere (torque control with minimum current)
MTPL	Maximum torque per losses (torque control with minimum total losses)
MTPV	Maximum torque per volt (torque control with minimum voltage)
NACS	North American charging system (Tesla plug)
NCA	Nickel cobalt aluminum oxide (cathode material of a lithium-ion battery)
NDE	Non-drive end (opposite to the side of an electric motor that is connected to the gearbox)
NdFeB	Neodymium-iron-bor (rare earth permanent magnet composite material)
NEDC	New European Driving Cylce
NiMH	Nickel metal hydride battery
Nm^3	Cubic meter gas volume at standard conditions (1,013 bar and 0 °C)
NMC	Nickel manganese cobalt (cathode material of a lithium-ion battery)
NPN	Three-layer semiconductors with positive, negative and positive doping
OBC	On-board charger
OCPP	Open charge point protocol
OCV	Open circuit voltage
OME	Oxymethylene ether (synthetic fuel)
OPEX	Operational expenditures
PAFC	Phosphoric acid fuel cell
P&C	Plug-and-charge
PE	Protective earth
PEM	Polymer electrolyte membrane (type of fuel cell)
PEMS	Portable emission measuring system
PFAS	Per- and polyfluoroalkyl substances
PHEV	Plug-in hybrid electric vehicle
PLC	Powerline communication
PMSM	Permanent magnet synchronous machine (also PSM)
PP	Proximity pilot/plug present
PTC	Positive temperature coefficient (heating resistor, thermal fuse)
PtG	Power-to-gas (process for producing hydrogen from electricity—electrolysis)

PtX	Power-to-X (general term for processes in which other energy sources (always hydrogen in the first step) are produced from electricity)
PV	Photovoltaik (electricity generation with solar cells)
PWM	Pulse width modulation
RCD	Residual current detector
RDE	Real driving emissions
REO	Rare-earth oxides
RFM	Radial flux machine
RMS	Root mean square
SAE	Society of automotive engineers (US-American professional association of automotive engineers)
SAIDI	System average interruption duration index (average grid downtime per customer and year in minutes)
SCHUKO	German household plug
SEI	Solid electrolyte interphase
SESM	Seperately excited synchronous machine, see EESM
SiC	Silicon carbide (wide bandgap semiconductor material)
SLAC	Signale level attenuation characterization
SmCo	Samarium-cobalt (rare earth permanent magnet composite material)
SMG	Separate motor-generator
SOC	State of charge (0 to 100%)
SOH	State of health (0 to 100% or kWh remaining capacity)
SOP	State of power (momentarily available battery power)
SPM	Permanent magnet synchronous machine with surface magnets and hence no reluctance torque, also PMSM-SPM
SRM	Switched reluctance machine
SSB	Solid-state battery (also semi solid-state battery)
synRM	Synchronous (sinusoidal current) reluctance machine
TEM	Traction electric motor
TTR	Through the road hybrid (hybrid system with different motor technologies per axle)
TtW	Tank-to-wheel (energy chain from the vehicle tank to the wheel-track contact)
UDC	Urban driving cycle (like ECE)
UDDS	Urban dynamometer driving schedule (US-American driving cycle, see also FTP)
V2B	Vehicle-to-building (like V2H but for apartment building)
V2G	Vehicle-to-grid (energy recovery from the vehicle battery into the supply network)
V2H	Vehicle-to-home (energy recovery from the vehicle battery into the home network)
V2L	Vehicle-to-load (electric car with household AC socket)
V2x	Vehicle-to-everything

VDE	Verband der Elektrotechnik, Elektronik und Informationstechnik (German professional association of electrical engineers), see IEEE
WBG	Wide bandgap (type of semiconductor)
WGSR	Water-gas shift reaction
WHO	World health organization
WLTP	Worldwide harmonized light-duty vehicles test procedure
WtT	Well-to-tank (energy chain from the energy source to the vehicle tank)
WtW	Well-to-wheel (energy chain from the energy source to the wheel/road contact)
XtL	X-to-liquid (general term for all processes to produce liquid fuel from different (X) energy sources)

List of Formula Symbols

Symbol	Unit	
α		Pole form coefficient
α	rad	Pitch angle
α	1/°C	Temperature coefficient
η		Efficiency
κ	1/(Ωm)	Electric conductivity
ω	1/s	Angular velocity (angular frequency)
φ		Phase angle
ϕ	Vs or Wb	Magnetic flux
ψ	Vs or Wb	Linked magnetic flux
ρ	Ωm	Specific resistance
σ	W/kg	Specific losses
ϑ	°C	Ambient temperature
ϑ		Angular displacement
Θ	A	Current linkage
Ψ	Vs	Magnetic flux
ρ	kg/m^3	Density
μ_0	N/A^2	Permeability of free space = $4\pi \cdot 10^{-7}$ N/A^2
τ_p	m	Pole pitch = Distance between two magnetic poles in arc length
μ_r		Relative permeability
A	A/m	Current loading
A	m^2	Area
B	Vs/m^2 or T	Magnetic flux density
C	F	Capacity
C	N/m^2	Esson's utilization factor
D	m	Range (Distance)

Symbol	Unit	
E	V/m	Electric field strength
F	N	Force
f_1	1/s	Frequency of stator voltage and current
f_S	N/m^2	Surface force
G	1/Ohm	Magnetic conductance
g	m/s^2	Gravitational acceleration
H	A/m	Magnetic field strength
h_{PM}	m	Height of a permanent magnet
I	A	Current
i		Transmission ratio
j		Complex constant
J	A/m^2	Current density
J	kgm^2	Moment of inertia
k		General machine constant
l	m	Length of the laminated sheets in axial direction (iron length)
L	H	Inductance
m	kg	Mass
M	g/km	Specific consumption
n	1/s	Rotor speed
n_1	1/s	Synchronous speed
p		Number of pole pairs
p	W/m^3	Power density
P	W	Active power
q		Hole number (number of slots per pole and phase)
Q	var	Reactive power
Q	Ah	Capacity (battery)
Q		Coil quality
R	Ohm	Ohmic resistance
r	m	Radius
s	%	Slip
S	m^2	Area
S	VA	Apparent power
T	s	Period duration
T	Nm	Torque
t	s	Time
U	V	Voltage

Symbol	Unit	
v	m/s	Speed
V	A	Magnetic voltage
V	kWh	Energy/Consumption
w		Number of turns of a coil
W	J or Ws or Wh	Work/Energy
X	Ω	Reactance
ξ		Winding factor
z		Total number of conductors in the stator or rotor
Z	Ω	Complex resistance (impedance)
z		Number of teeth on a gearwheel

Orders of magnitude

Hundred = 100 = 10^2
Thousand = 1000 = 10^3
Million = 1,000,000 = 10^6
Billion = 1,000,000,000 = 10^9
Trillion = 1,000,000,000,000 = 10^{12}

List of Figures

List of Tables

1 Introduction

1.1 DC Motors (Direct Current Motors)

Systematic research into electrical and magnetic phenomena became possible after **Alessandro Volta** (Italian) invented the electric battery around 1800. He succeeded in producing a continuous current flow (direct current/DC) from a stack of silver and zinc plates suspended in an acid bath.

The first generation of a mechanical movement due to electromagnetic fields was achieved by **Michael Faraday** (English) in 1821. He designed two apparatuses to prove the electromagnetic rotation. A vertically suspended current-carrying wire moved in a circular path around a permanent magnet. The lower end of the wire was immersed in a bath of liquid mercury and thus electrically contacted.

One year later, **Christian Oested** (Dane) found the generation of a magnetic field by electric current. He observed the deflection of a compass needle near a current-carrying wire. From this, **André-Marie Ampère** (French) shortly thereafter developed the cylindrical air coil and **William Sturgeon** (English) the electromagnet with iron core for field amplification (Fig. 1.1).

The first known example of a rotating electric motor was published in 1822 by **Peter Barlow** (Englishman): He sent an electric current through a rotatably mounted disc suspended in a transverse magnetic field. The moving electrons in the disk are deflected by the Lorentz force. This force results from the interaction of electric current and magnetic field. It generates a rotary motion in the apparatus. However, this Unipolarmotor was a mere demonstrator and much too weak to do anything useful with it.

Subsequently, dozens of inventors in Europe and the United States began to work on electric machines. Electromagnetism was a fashionable topic for many years, much like computers or the smartphone today. Therefore, it is not possible to name a single person as the inventor of the rotating electric machines we know

M. Doppelbauer, *Introduction to Electromobility*,
https://doi.org/10.1007/978-3-658-45482-1_1

Fig. 1.1 Barlow wheel[1]

today. Often, the inventors knew nothing about each other and independently developed similar solutions.

In 1827 or 1828, **Istvan Ányos Jedlik** (Hungarian) built a rotary apparatus with electromagnets and a commutator. However, Jedlik did not publicly report on his machine until decades later, and the actual date of invention is thus uncertain. A working sample of the apparatus is displayed in the Museum of Decorative Arts in Budapest.

Sometime before 1830, **Johann Michael Ekling**, a mechanic in Vienna, Austria, built an engine based on the plans and ideas of Prof. **Andreas von Baumgartner** (Austrian physicist; professor of physics and applied mathematics in Vienna from 1823). The apparatus was purchased by the University of Innsbruck in 1830 for 50 fl cm.

Around 1831, **Joseph Henry** (U.S. American) and **Michael Faraday** (English) discovered electromagnetic induction at about the same time and independently of each other: A changing magnetic field generates a voltage in a nearby wire. If the circuit is closed, this results in an electric current.

Henry also built a small electromagnetic rocker consisting of two batteries, two electromagnets, and a coil. As the coils leans to either the left or right side, contacts are closed and the current in the coil creates a Lorentz force that moves it to the other side.

In the summer of 1832, **William Ritchie** (English) built a rotating electromagnetic generator with four rotor coils, a commutator, and brushes. Ritchie is, therefore, generally regarded as the inventor of the commutator. At the end of his

[1] Source: Philosophical Magazine, 1822 (vol. 59).

Fig. 1.2 First DC generator by Pixii, 1832/33[2]

article, Ritchie describes how he made an electromagnet spin by exploiting the earth's magnetic field, lifting several ounces (between 50 and 100 g) of weight. The commutation (turning of current) was accomplished by the ends of the wires running in two semicircular channels of mercury.

In 1832, **Hippolyte Pixii** (French) was the first to succeed in developing a usable electrical generator. Originally, his device produced alternating current (AC). Later in the year, Pixii added a commutator and was then able to produce a pulsating direct current (DC) (Fig. 1.2).

[2] Source: F. Niethammer, *Ein- und Mehrphasen-Wechselstromerzeuger*, Verlag S. Hirzel, Leipzig, 1906.

Fig. 1.3 First engineered electric motor by Moritz Jacobi, 1834[3]

Today, it is generally known that all rotating electric machines can be operated as motors as well as generators. This insight dates to 1833 and was first published by **Heinrich Friedrich Emil Lenz** (German) as "Gesetz der Reziprozität der magnetoelektrischen und elektromagnetischen Erscheinungen" (Law of reciprocity of magnetoelectric and electromagnetic phenomena). Lenz operated a Pixii generator as a motor. His scientific text was presented at the end of 1833 at the Petersburg Academy of Sciences and published in 1834 in Poggendorff's "Annalen der Physik und Chemie". The law slowly becomes common knowledge among physicists, especially after several successful experiments were reported in 1838 (Fig. 1.3).

The first usable DC motor that delivered significant mechanical power was built in 1834 by the civil engineer **Moritz Hermann Jacobi** (German) in Königsberg (then Prussia, now Russia). Technologically, it was an axial flux motor. It delivered about 15 W of mechanical power.

The successor motor presented by Jacobi 4 years later in St. Petersburg (Russia) was already rated at 300 W. This motor was installed in an 8 m long boat which was widely used in the late summer of 1838 to ferry passengers across

[3]The small electric locomotive that Werner Siemens built for the Berlin Trade Exhibition (Berliner Gewerbeausstellung) in 1879 is often regarded as the starting point of the electric railroads. It pulled three carriages with six passengers each on a 300 m long circuit. This was most probably the first practical electric locomotive but by no means the first attempt in this direction.

the Neva River in St. Petersburg. The zinc battery of 320 pairs of plates weighed 200 kg. The plates were placed along the two side walls of the boat. The propulsion reached 1/5 to 1/4 horsepower (about 300 W). The boat traveled the 7.5 km distance at about 2.5 km/h and could carry a dozen passengers. Contemporary newspapers reported the zinc consumption after three months to be 24 pounds.

The first electric toy car was probably already built in 1827 by Istvan Jedlik but was not publicly demonstrated until many years later. In 1835, **Sibrandus Stratingh** and **Christopher Becker** (Dutch) built a small (30 × 25 cm) electrically powered three-wheeled vehicle weighing about 3 kg that was able to drive around on a table for about 15–20 min until the battery ran out. Stratingh and Becker published a report the same year. Stratingh was familiar with Jacobi's work and wanted to build a real electric car in 1840, but he died early and never succeeded.

The first patent for the construction of an electric motor was granted to **Thomas Davenport** (U.S. American) in 1837, although his motor was technically less advanced than other contemporary models.

Robert Davidson (Scotsman) also developed electric motors since 1837. He later manufactured several drives for a lathe and model vehicles. In 1839, he succeeded in building the first electrically driven railway carriage, which was able to carry two people. The rotors of his motors consist of wooden cylinders on which four iron rods are mounted. It can thus be said with some justification that Davidson built the first switched reluctance machine.

In September 1842, Davidson made test runs with a 5-ton, 4,8-m-long locomotive on the Edinburgh to Glasgow railroad. His engine produced about 1 hp (0,74 kW) and reached a speed of 4 mph (6,4 km/h).[4]

The typical design of DC machines we use today, the slotted drum armature rotor with stacked electrical sheets to suppress iron losses, was only introduced many years later in 1872 by the **Siemens** company. It originated from the development of DC generators. The designs of the early DC motor developers are hardly used today.

1.2 The AC System (Alternating Current)

1.2.1 Single-Phase AC

Early experiments with alternating current (AC) began as early as 1832, but at that time, there were no applications for it, and in subsequent years, the focus was on direct current (DC).

The basic principle of the transformer was invented and patented in 1875 by **Pavel Nikolayevich Yablochkov** (Russian). The first commercial products

[4] Replica, Institute of Electrical Engineering (ETI) of the KIT, Karlsruhe, Germany.

appeared in 1881 in France, making alternating current known in Europe and later also in the United States.

Galileo Ferraris (Italian) used these inventions in 1884 to build the first AC long-distance power supply line, covering a distance of 40 km from the town of Lanzo Torinese in Italy to the electro-technical exhibition in Torino.

Two years later, the industrial tycoon **George Westinghouse** (U.S.) built the first commercial single-phase AC transmission line in Great Barrington, Massachusetts. He used transformers from France which he had imported into the United States. The line was used for lighting purposes. By the end of 1887, his company had built 68 AC generator stations had been in theUnited States.

1.2.2 Poly-Phase AC

Friedrich August Haselwander (German from Offenburg in Baden) was the first to come up with the idea of using AC with three evenly shifted phases (rotary current) in 1887. He developed the first three-phase synchronous generator of the modern kind and put it into operation in October 1887. His generator was able to remove three of the original six conductors by internally connecting the star point. This made the three-phase system favorable to other electrical power transmissions, in particular the two-phase system which was later favored by Nicola Tesla and Westinghouse.

Haselwander envisioned to bridge long distances with higher voltages, lower currents, and consequently lower losses, which required three transformers at each end. Today, all electric power distribution lines are built according to Haselwander's ideas[5] and he must be considered the inventor of modern electrical power transmission (Fig. 1.4).

There are several important advantages of Haselwander's three-phase three-line transmission system compared to Tesla's two-phase four-line system: The transmissible power of three-phase systems increases by the square root of 3, i.e. by around 73 %, compared to a single phase while the additional conductor cross section only increases by 50 % (three lines instead of two). This results in a gain of more than 15 % in transmission capacity per weight of copper wire and therefore saves material and cost. As two-phase systems are only two single-phase systems, they do not have this advantage. Also, the symmetry of the three-phase system can be utilized to build transformers with only three legs. Two-phase systems, on the other hand, require two single-phase transformers with two legs each, which increases the cost by around a third. Finally, a three-phase electric motor has a better field distribution (lower harmonics) than a two-phase motor with the same number of slots and windings and hence it will run smoother.

[5] High-voltage direct current (HVDC) transmission is also used for certain applications. This type of transmission has become possible with the progressive development of power electronics in recent decades and is particularly suitable for long distances and high powers, as well as for coupling independent AC networks.

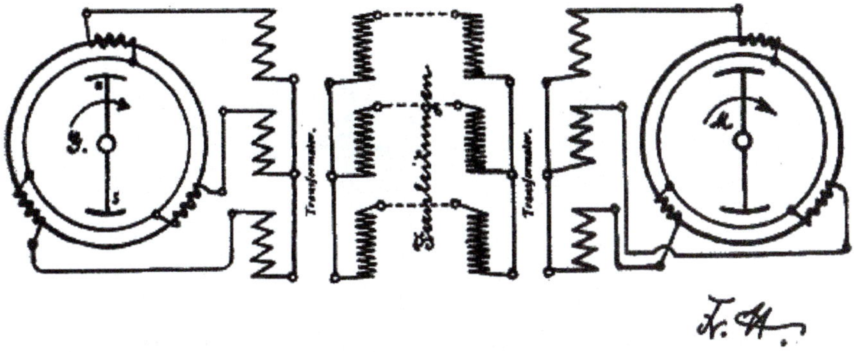

Bild 1 Drehstromübertragungssystem nach Haselwander

Fig. 1.4 Three-phase high voltage transmission with synchronous generator (left) and synchronous motor (right), six transformers, and three conductors (center)[6]

1.2.3 AC Motors

In the early years, all applications of AC were limited to lighting as there were no viable self-starting AC motors. The synchronous AC generator was already invented, and such machines were built by several companies. However, synchronous machines could not be operated as motors because they had no self-starting capabilities when fed by a constant voltage and frequency. Power electronics, which would be needed to gradually change voltage and frequency, was not yet available at that time.

The basic physical principle of the induction machine was presented to the Physical Society of London by **Walter Baily** (English) as early as June 1879 and published shortly thereafter.

However, it was not until 1885 that **Galileo Ferraris** developed a fully functional two-phase asynchronous induction motor. He presented his results to the Royal Academy of Sciences in Torino, Italy, at a highly acclaimed lecture in March 1888 and published a detailed paper shortly thereafter. George Westinghouse quickly became aware of Ferraris' invention and offered to acquire his patents but Ferraris, being an idealistic professor at the University of Torino, refused to protect or sell his inventions.

A few months after Ferrari's lecture, **Nicola Tesla** (Serbian, born and raised in Croatia) delivered his highly acclaimed speech at an event of the American Institute of Electrical Engineers, now IEEE, in New York. He presented two of his own two-phase motor prototypes that were quite like Ferraris' machines. Tesla,

[6] Sketch by Friedrich A. Haselwander, 1888.

Fig. 1.5 First modern-day induction machine by Dolivo-Dobrowolsky, 1889[7]

however, patented his inventions and was subsequently able to sell his patents to George Westinghouse.[8]

Westinghouse hired Tesla to further develop the prototypes into fully functional commercial electric motors at his Pittsburgh plant. In the end, however, both Ferraris in Europe and Tesla in the United States failed to develop commercially viable induction motors. After more than a year of intensive but ultimately unsuccessful work, Westinghouse discontinued all research activities on the induction motor. Tesla returned to his laboratory in New York in October 1889, frustrated. In particular, the problems with reliable self-starting and efficiency could not be solved satisfactorily (Fig. 1.5).

The development of the modern induction machine was achieved in early 1889 by **Michael Dolivo-Dobrowolsky** (Russian, later naturalized Swiss), who was a design engineer at AEG in Berlin, Germany, at that time. Dolivo-Dobrowolsky was born to Russian parents in St. Peterburg in 1861. He spent his first 11 years in St. Peterburg and then moved with his parents to Odessa. From 1878 until 1887 he studied chemistry and electrical engineering is Riga, Latvia and in Darmstadt, Germany. He was then discovered in Darmstadt by Emil Rathenau, the owner of the AEG. The "Allgemeine Elektrizitätsgesellschaft" (in English: *General Electricity Company*) was founded in 1883 after Rathenau acquired the rights to Thomas Alva Edison's inventions on incandescent lamps in Germany. AEG was one of Europe's leading electrotechnical companies for many decades. Years later, Dolivo-Dobrowolsky became the technical director of the AEG.

[7] Source: ETZ—Elektrotechnische Zeitschrift, 1917, Band 38.

[8] According to today's rules, Tesla might not have been able to maintain its patents because many ideas were already published by other inventors. Apparently, the timing of submission to the patent office mattered more back then than pre-publication.

Compared to the motor designs of Ferraris and Tesla, Dolivo-Dobrowolsky introduced three significant changes. Firstly, he used Haselwander's symmetrical three-wire three-phase system instead of the inferior two-phase system. Secondly, he used a cylindrical, distributed three-phase winding for the stator of his motors. Ferraris and Tesla had used Gramme ring windings that would need more copper, were more difficult to insulate, and produced less voltage (lower emf). Thirdly, Dolivo-Dobrowolsky invented the squirrel-cage rotor. Ferraris and Tesla had used pole windings in the rotor that were quite like the rotors of synchronous machines. The starting torque of these rotors was low and several of Tesla's motors did not even start on their own but had to be started by hand. The squirrel cage considerably improved the starting performance. Later, Dolivo-Dobrowosky also invented the three-phase wound-rotor with slip rings that was able to start with full-rated torque from standstill by adding series resistors.

To this day, many millions of induction machines are built each year according to Dolivo-Dobrowolsky's design principles. They are used in a wide range of applications in industry, as drives for pumps, fans, compressors, conveyor belts, elevators, robots, and more, and even in electric cars, for example in vehicles from Audi, Daimler, and Tesla Motors. Dolivo-Dobrowolsky can therefore be considered the true inventor of the modern-day induction motor.

1.3 Electric Power Transmission

Dolivo-Dobrowolsky was also instrumental in the creation of the world's first three-phase electrical transmission line, which was built in 1891 on the occasion of the Electrotechnical Exhibition in Frankfurt am Main. Oskar von Miller (German) initiated and organized the exhibition. He commissioned AEG and Maschinenfabrik Oerlikon (MFO) from Switzerland to build the three-phase transmission system.

The entire system consisted of a hydroelectric power station in Lauffen am Neckar, a 175 km long three-phase line along the railroad line to Frankfurt, the transformers, protection, and switching equipment required at both ends, and several demonstration motors on the exhibition grounds. AEG built the line, MFO under the leadership of Charles E.L. Brown (Englishman) was responsible for the generator in Lauffen. The line was put into operation for the first time on August 24, 1891. The generator achieved an output of 300 hp. Several induction motors from Dolivo-Dobrowolsky and a synchronous motor from Haselwander were exhibited in Frankfurt. One of the motors drove the pump for an artificial waterfall. In addition, 1000 light bulbs were supplied with electricity.

The whole exhibition was a great success and operated for several months. After the end of the exhibition, the power plant in Lauffen am Neckar and the first 15 km of the line continued to be used to supply the city of Heilbronn with three-phase current. Heilbronn was thus the first city in the world to use the new energy system. The power plant was operated for more than 50 years.

Dolivo-Dobrowolsky also invented the three-phase three-legged transformer that is widely used today. It is more compact and cheaper to manufacture than the three single-phase transformers that are otherwise required.

As early as 1891, Charles Brown set up his own business with his colleague Walter Boveri (Swiss). Together they founded the Swiss company Brown, Boveri & Cie, or BBC for short.

The world's first industrial use of the three-phase electric power system began at the end of 1893 in Sweden. The Swedish company ASEA built a waterpower electricity plant to supply energy to an iron ore mine. ASEA had been founded 10 years earlier by the Swede Ludvig Fredholm. The plant supplied several hundred lamps and six induction motors with electricity from a run-of-river power generator.

In 1988, BBC and ASEA merged to form today's ABB (abbreviation for Asea Brown Boveri).

Tesla also continued to develop AC motors and remote power transmission in the United States but concentrated on one and two phases. He never really recognized the advantages of three-phase rotary current. Looking back in 1917, Dolivo-Dobrowolsky commented: "It is difficult to understand why Tesla stubbornly clung to unlinked two-phase electricity, whereas here in Germany we used three phases right from the start, which don't need any more wires".

Together with Westinghouse, Tesla was instrumental in the first U.S. polyphase transmission line from Niagara Falls to the city of Buffalo, which was inaugurated in 1896. The synchronous generators were still two-phase, whereas the transmission line was already three-phase. The conversion was achieved by means of complicated interconnected transformers. It was not until the first power expansion of the plant, which happened 7 years later, that the two-phase generators were finally replaced by larger three-phase types.

Contrary to popular belief, Nicola Tesla is neither the inventor of the modern-day induction machine nor did he invent the modern-day AC electricity transmission system.

From 1900 onwards, the expansion of electrical AC power transmission began in all countries in the world. The European three-phase system quickly dominated the markets[9] and was later adapted also by Westinghouse, General Electric, and other U.S. American companies.

1.4 Batteries

The first functioning galvanic element and therefore the first battery was invented in 1800 by **Alessandro Volta** (Italian) (Volta's column). Three years later, **Johann Wilhelm Ritter** (German), the founder of electrochemistry, invented the Ritter

[9] 84% of the world's population use the European 50 Hz frequency.

column, the first accumulator, or battery for short. In subsequent years, the design was continuously improved, for example by William Cruickshank's trough battery, which avoided the disadvantage of the vertical design of the Voltaic column.

The development of lithium-ion batteries dates to the 1970s and the oil crisis. During this time, the British chemist **Stanley Whittingham**, who was teaching at Binghamton University in the US state of New York, developed a key component for lithium-ion battery by using a new type of material as a cathode.

In the 1980s, the American materials researcher **John Goodenough** at the University of Texas, Austin, further developed the materials of the cathode and anode, almost doubling the storage capacity of the batteries. On this basis, the Japanese engineer **Akira Yoshino** from the Asahi Kasei Corporation finally developed the first commercially available lithium-ion battery in 1985.

In 1985, Asahi Kasei Cooperation, Yoshino's employer, filed the Japanese patent application and began commercialization. The company licensed the patent to other companies, including Sony. The first commercial battery was developed by a Sony and Asahi Kasei team led by Yoshio Nishi in 1991. The first product to use the new battery type was a Sony handheld camcorder.

In 2006, Tesla Inc. introduced the Tesla Roadster. It was the first car to use lithium-ion batteries. The Roadster was available between 2008 and 2012. Around 2500 units were sold worldwide.

Nissan produced the first generation of the more mundane Leaf from 2010 to 2017. It was the first BEV to sell over 100,000 units. This target was reached in early 2014.

In 2019, John B. Goodenough, Stanley Whittingham, and Akira Yoshino were awarded the Nobel Prize in Physics for the development of the lithium-ion battery.

The development of the lithium-ion battery is far from complete. Since the first electric cars, the energy density of the cells has almost doubled and the cost per kWh has fallen by a factor of around 10. Today, cell prices of under 100 €/kWh and energy densities up to 360 Wh/kg are available. In the future, solid-state batteries will enable a further significant leap in energy density, charging speed, and lifetime.

1.5 Variable Speed Drives

Modern variable-speed drives require a variable voltage with a correspondingly adapted frequency. Power electronics and software control algorithms are used to convert the battery's direct currents and voltage (DC) into the required alternating currents and voltages (AC).

Power electronic technology is based on the invention of semiconductors, which began in 1947–1949 with the discovery of the transistor effect by the Americans **Walter Houser Brattain**, **John Bardeen,** and **William Shockley**.

In 1954, **Eberhard Spenke** and his team at Siemens & Halske AG in Germany succeeded in producing high-precision silicon using the zone melting process.

This, together with the availability of an insulating material (silicon dioxide) with favorable properties, brought about the breakthrough of silicon as a semiconductor material for the electronics industry in the mid-1950s.

The first commercial frequency converters of modern design were mass-produced by the Danish company **Danfoss** in 1968. They were cooled in oil, operated at 380 V three-phase voltage, and could initially supply up to 10 A phase current. They used thyristors as power semiconductors.

In 1974, the fundamentals of field-oriented control (FOC) of inverter-fed rotating field machines were developed by **Felix Blaschke** (German).

Today's frequency converters use IGBT semiconductor switches made of silicon (Si) or MOSFETs made of silicon carbide (SiC). The IGBT was invented by **B. Jayant Baliga** (Indian) at General Electric and developed ready-to-market by the end of the 1970s.

The wide availability of switchable power semiconductors (MOSFETs 1976, IGBTs 1982) led to the development of voltage link frequency converters in both two-quadrant and four-quadrant (braking energy recovery) operation.

In 1982, General Motors and **Masato Sagawa** (Japanese) of Sumitomo Special Metals independently developed neodymium-iron-boron (NdFeB) magnets that provided the highest specific energy values. These magnets were soon used for permanent magnet synchronous motors with particularly high specific torques and are the norm today in electric cars and Pedelecs (pedal electric bicycles—E-Bikes).

1.6 Electromobility

As early as 1835, there were experiments with electrically powered vehicles, initially small toy models and, as early as 1837, the first very simple rail vehicles.

In 1881, the first electric car was prototyped. The tricycle by **Gustave Trouvé** in Paris was powered by a converted Siemens motor with 70 W output. Trouvé had installed a 12 V lead accumulator as an energy storage device (Fig. 1.6).

A similar, somewhat more powerful vehicle was presented the following year in London by two English professors, **William Edward Ayrton** and **John Perry**.

The development of the electric automobile progressed rapidly. From 1895 onwards, electric cars were used in races (e.g. Paris–Bordeaux). In 1899, the 100 km/h mark was broken for the first time with a car, namely, with the electric car of the Belgian designer Camille Jenatzy (Fig. 1.7).

In 1900, there were many companies in America and Europe that had started industrial production of electric cars. A contemporary overview comes from more than 30 manufacturers.[10]

However, the design of most vehicles was still strongly reminiscent of horse-drawn carriages (Fig. 1.8).

[10] Another source names 565 different brands.

Fig. 1.6 First electric "car" by Gustave Trouvé in Paris, 1881

Competition arose to find the most favorable form of propulsion: combustion engines (powered by petrol, kerosene, alcohol, or acetylene), electric drives, steam drives, or drives with compressed gases.

At that time in Europe, the petrol engine already had a market share of 95%. In the United States, due to poor road conditions, the automobile first developed in the cities.[11] The steam drive was the most popular with a market share of 40%. The electric motor came in second place with a share of 38%. Cars with internal combustion engines were far behind with a market share of just 22%.

One reason for the initial success of electric vehicles in the United States was the possibility of a direct drive or at least the use of a simple gear transmission: "Since every petrol engine has the unfortunate obstinacy to run advantageously only at a certain number of revolutions per minute, it is indispensable to use intermediate gears [...] to achieve the different speeds and to overcome gradients. [...] in other words: transmissions, all of which are great power guzzlers [...] According to [...] measurements, this loss amounts to 40–60% of the engine power".

By 1900, the range of electric vehicles was already reaching 150 km. Batteries had capacities of 9–25 kWh[12] and weighed 450–600 kg. The disadvantages

[11] Source: Allgemeine Automobil-Zeitung, 14. Januar 1900.

[12] The Allgemeine Automobil-Zeitung states a typical capacity of 80–230 Ah at a voltage of 110 V DC.

Fig. 1.7 Electric racing car "La Jamais Contente" by Belgian engineer Camille Jenatzy, 1899[13]

Fig. 1.8 Advertisement in the Allgemeine Automobil-Zeitung, September 9, 1900

of the battery were well known: "For the time being, there is no talk of a radical improvement or reduction in the weight of the accumulators, the uncomfortably heavy carriers of electrical energy. For the time being, unfortunately, only lead can be used, and this, God be lamented, is still about eleven times heavier than water".

Usually, a central electric drive or two, rarely four, wheel hub motors were installed (Fig. 1.9).

---

[13] Source: http://depris.cephes.free.fr/presscom/060-02.htm.

Fig. 1.9 Hybrid drive with electric motor and horse: "The Haybrid", Columbia Electric Victoria, 1906

But there were also vehicles with range extenders (serial hybrids): "The small, rib-cooled petrol engine can also be used for transportable, electric charging stations, the first of which Bedovelli and Priestley demonstrated on their three-wheeled cab at the Paris exhibition last year (1899)".

Around the turn of the century, Siemens & Halske began to build electric omnibuses in Berlin with temporary power supply via overhead lines. In the following years, these "trackless streetcars" (today: trolleybuses) became fashionable. Corresponding routes were established in many cities in Germany (Fig. 1.10).

By 1920, the infrastructure in the United States had been developed and longer-distance journeys became commonplace. Electric vehicles could not compete in this discipline. Large oil deposits were found around the world, which greatly reduced the price of petrol. After the first electric starters were developed in 1912, which eliminated the tedious and dangerous cranking, another major disadvantage of the internal combustion engine was eliminated.

These developments also heralded the end of electric mobility in the United States. Virtually all manufacturers ceased production between 1910 and 1920. In Europe, battery-powered vehicles never played a major role.

In niche applications, however, the electric drive remained permanently in use, for example in forklifts, golf caddies, and milk delivery trucks, which were particularly popular in England.

Due to the oil crisis, the electric car experienced a certain renaissance in the 1970s. Many research and development projects were set up around the world to test modern electric cars. German manufacturers were also involved, including VW City-Stromer, Audi duo, BMW 1602e. However, the limitations of the lead-acid battery were still so significant that all projects were eventually abandoned.

Fig. 1.10 Carriages of the trackless Veischede Valley Railway between Grevenbrück and Kirchveischede in Westphalia, Germany, operated between 1904 and 1916 on 8 km of track with 600 V DC voltage[14]

Modern battery-electric passenger cars owe their triumph mainly to lithium-ion battery technology, which enables capacities of between 40 kWh and more than 100 kWh and thus ranges of 200–over 600 km.

Added to this is the powerful electric drive technology with three-phase motors and modern power electronics. Many disadvantages of the early DC motors have now been overcome.

[14] Source: Museum der Stadt Lennestadt, Grevenbrück, Germany.

Drives for Electric Vehicles

2

Abstract

This chapter presents the main operating characteristics of electric traction drives, in particular the speed/torque characteristics and the operating point-dependent losses. Similarities and differences to conventional internal combustion engines are shown. Emphasis is placed on the explanation of the constant torque and constant power range as well as the various operating limits in continuous and short-time operations.

The development of modern high-performance electric motors is a complex mechatronic task that requires a wide range of specialist skills.

The electromagnetic design determines the magnetic circuit, i.e. in particular the sheet metal geometry, the wires and windings, and the magnets. It also provides the data required for the inverter to control the motor. Due to the highly non-linear behavior of today's high-performance motors, this is a complex field of its own.

The mechanical design is primarily concerned with the strength of the rotor, which is critical due to the high speeds. Materials and manufacturing processes play a major role.

The cooling system of modern traction motors is complex (water jacket, internally cooled shaft, oil spray on winding overhangs, internal slot cooling) and requires complex thermodynamic models.

Finally, the production and component design must be considered: Windings with shaped coils or round wire, housing made of metal or plastics, possibly with functional integration, e.g. integrated electrical connections?

All four domains influence each other in many ways. They must be optimized together and iteratively (multi-physics optimization).

M. Doppelbauer, *Introduction to Electromobility*,
https://doi.org/10.1007/978-3-658-45482-1_2

This chapter starts with a detailed analysis of the most important characteristics (torque, speed, power) of combustion engines and electric motors. In later chapters, more details on the different types of electric motors and their design are presented.

2.1 Speed/Torque Characteristics of Internal Combustion Engines

Figure 2.1 uses the example of an older VW Golf 1.6 FSI with six-speed automatic transmission to show the wheel torque curve in the individual gear steps (blue). Smaller gear ratios i (higher gears) deliver a higher wheel speed, but also less torque.

The performance of a car depends on the power P (in kW) and the torque T (in Nm) available at the wheels. Gearboxes transform the torque and speed of rotation delivered by the engine, but they do not transform the power:

$$T_{wheels} = T_{engine} \cdot i \tag{2.1}$$

$$\Omega_{wheels} = \Omega_{engine}/i \tag{2.2}$$

$$P_{wheels} = T_{wheels} \cdot \Omega_{wheels} = T_{engine} \cdot \Omega_{engine} = P_{engine} \tag{2.3}$$

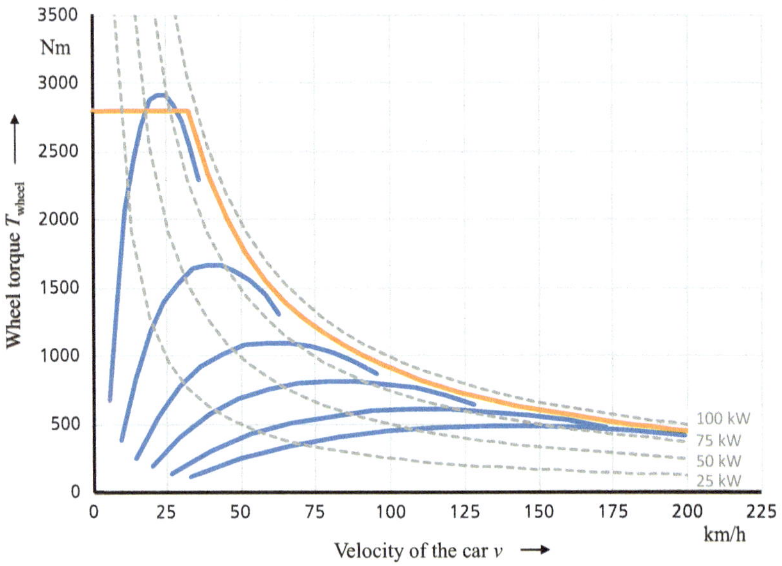

Fig. 2.1 Maximum wheel torque (blue) over speed for a combustion engine with multi-speed gearbox

i is the gearbox reduction ratio and $\Omega = 2\pi \cdot n$ is the mechanical angular velocity based on the speed of rotation n in 1/s. Typical values of i are in the range of 10 (low gear) to slightly below 1 (high gear).
Hence, engines with a higher torque but a low speed of rotation may deliver the same performance as engines with a lower torque and a high speed of rotation as long as the mechanical power $P = T \cdot 2\pi \cdot n$ remains the same and the gear ratio i is adapted according to the different speeds.

Note
There is a long-lasting debate weather torque or power is more important for the performance of a sports car.

Automobile magazines rave about the wonders of high-torque diesel engines. However, their performance is often disappointing when the car's acceleration is not nearly as high as expected based on the engines high torque.

The truth, however, is quite simple. It is the force between the wheel and the street that accelerates a car. When comparing different cars and assuming similar wheel sizes, this force is proportional to the wheel torque.

In case of direct drive electric motors, there is no gearbox and hence the motor torque is the same as the wheel torque. In this case, only the motor torque is decisive for the performance of the car, not the motor power.

However, nearly all cars have a gearbox between the wheel and the motor or engine. Depending on the reduction ratio, a gearbox increases the torque of the motor and reduces the speed at the same time. The power output of the motor, i.e. the product of speed and torque, is transferred to the wheels largely unchanged, apart from minor losses.

During the development phase of a new car, the developers must decide whether they want to build a motor with less torque and more speed, which requires a gearbox with a high reduction ratio, or a motor with more torque and less speed. This then requires a gearbox with a low reduction ratio.

By adapting the gearbox, the speed and torque of the motor can be played off against each other while still achieving the same torque at the wheels and hence a similar performance of the car.

To summarize, when the motor drives the wheels through a gearbox, it is both its speed and its torque that is decisive for the car's performance. Simply put, it is the motor's power that counts.

This also explains the disappointing performance of some cars with high torque diesel engines: The diesel engines typically run slower than petrol engines and so their gearbox ratio is lower. Therefore, the high torque of a diesel engine is not increased by the gearbox as much as the torque of a faster running petrol engine. Hence, high torque diesel engines have no advantage over similarly powered petrol engines in terms of acceleration, as long as their power is similar.

In the end, it is the wheel torque that wins races. Motor or engine power, not its torque, and a high gearbox ratio are needed to get as much wheel torque as possible.

These are all simplifications, of course. To precisely determine the performance of a car, the speed curve of the torque at the wheel for each individual gear is needed. In addition, the car's weight, the moment of inertia of all rotating masses, friction, and air resistance are required, as these counteract the acceleration. This information can then be used to create a simulation that precisely calculates the course of the acceleration process.

Lines of constant power become hyperbolas in the torque–velocity plane. The diagram shows four examples of such lines for outputs from 25 to 100 kW.

The available maximum torque (orange line) is roughly constant at low speeds and reduces at speeds above approximately 30 km/h in proportion to the speed, giving approximately constant power. This behavior, constant max torque at low speeds and constant max power at higher speeds, is typical for traction drives. It is generally accepted and even expected by customers.

This behavior is exploited in the design of electric motors by realizing a basic speed range with constant maximum torque and a field weakening range where torque decreases proportionally with the speed.

Due to this design, a wide speed range can be covered with nearly constant power. This has an advantageous effect on the dimensioning of the battery, cables, plugs, and power electronics.

If needed, an electric motor can be designed to provide constant torque up to the highest speed. In that case, however, the maximum power would only be available close to the maximum speed. The entire rest of the electric drive train would then have to be dimensioned for this peak, which would unnecessarily increase costs.

Note

For performance reasons, electric motors usually also use a gearbox to adapt the motor speed to the wheel speed. With a few exceptions, gearboxes for electric motors are single speed, i.e. not switchable.

The speed of electric motors is typically two to three times higher than the speed of combustion engines at the same vehicle speed. Therefore, the gear ratio is also two to three times higher.

Consequently, the motor torque of electric motors can be two to three times lower than the torque of combustion engines. When both produce the same engine power, the lower motor torque will still result in the same wheel torque and the same car performance.[1]

2.2 Speed/Torque Characteristics of Electric Drives

Figure 2.2 shows the torque curve over the speed of a typical electric drive. A drive usually consists of an electric motor, power electronics to convert from DC (battery) to three-phase AC, and control software to operate the motor according to the requirements of the vehicle control.

The vehicle controller will primarily attempt to satisfy the torque demand given by the position of the accelerator pedal. A secondary goal is the saving of energy.

The electric motor in the example was designed to match the combustion engine in Fig. 2.1. The electric motor requires a gearbox with a fixed transmission ratio of around $i = 10$ to achieve the same wheel torques and speeds as the combustion engine.

One key difference between electric motors and combustion engines is already apparent: while the combustion engine can deliver its maximum power continuously, the maximum power output of an electric motor is limited in time. Typically, peak power is available for around one minute, after which a cooling phase must take place. In practice, this has no negative consequences, as peak power is only rarely required for very fast acceleration processes (overtaking, starting at traffic lights). Racing cars are an exception. For these applications, the electric motor is designed for a high continuous output power.

Another difference is the dependence of the power of an electric motor on the available battery voltage. If the battery is discharged, the voltage level drops and the maximum power available to the drive is reduced accordingly. In practice, you will notice a slight reduction in peak power below a charge level of approximately 70%.

2.3 Limits of Operation

Traction motors are controlled via a so-called frequency inverter (also known as DC/AC converter).

[1] Similarly powered electric cars generally have a much better performance than combustion engine cars. This is because their maximum torque is already available at standstill and there are no torque interruptions during acceleration as no shifting between gears is required.

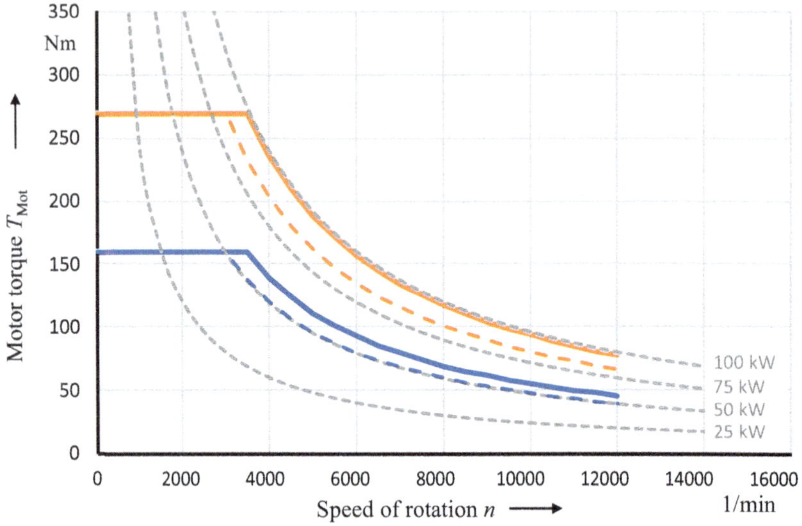

Fig. 2.2 Idealized torque/speed characteristics of an electric traction drive[2]

This is a power electronics module that converts the battery's direct current (DC) into a three-phase alternating current (AC) and in reverse. The frequency and effective value of the output voltage and current can be adjusted within wide limits by the motor control software.

There are several voltage and current limits that must be observed in an electric drive train.

2.3.1 Maximum Motor Voltage U_{max}

The level of the effective AC voltage available depends on the DC battery voltage.

In the best case, depending on the pulse pattern generation and voltage drop at the internal semiconductors and other components, the peak value of the AC voltage can be as high as the DC voltage. The effective value of the motor voltage is thus $U_{AC} = U_{DC}/\sqrt{2}$, see also Sect. A.2.

The maximum output voltage U_{max} provided by the inverter depends on the state of charge of the battery. When the battery is empty, around 10–20% less voltage is available than when the battery is full, depending on the battery type.

A typical configuration of current electric cars, for example, is $U_{max} = 400$ V AC_{eff} (full battery) to 320 V AC_{eff} (empty battery).

[2] Blue = continuous operation, orange = short-time operation, solid = maximum battery voltage, dashed = minimum battery voltage.

Note
U_{max} is the highest effective value of the motor voltage (line-to-line voltage) that the power electronics can supply when the battery is full.

2.3.2 Maximum Motor Current I_{max}

The current that the inverter can provide is also limited, namely by the heating and cooling of the power semiconductors.

As the thermal capacity of the silicon semiconductors is relatively small, there is hardly any possibility of briefly increasing the permissible continuous heating. The maximum permissible current is therefore only slightly greater than the continuous thermal current of the inverter.

It has therefore become common practice to specify the I_{max} current limit as the effective value of the continuous current permitted in the inverter.

2.3.3 Thermal Limits of Continuous Operation

The thermally permissible torque of an electric motor at a certain speed is determined by

- the total losses (efficiency),
- the capability of the cooling system (the thermally dissipatable losses),
- the thermal durability of the insulation materials (insulation papers, varnishes, resins etc.),
- the desired operating time,
- the desired service life of the motor.

Separate specifications are often made for the torque in continuous operation and short time operation (over minutes or even just some seconds).

Continuous operation requires continuous dissipation of the entire losses by the cooling system.

The heating in continuous operation should not exceed a temperature at which the minimum thermal service life of all insulation materials used is nominally 20,000 h. This theoretical value is typical for industrial motors. It is often reduced in drives for passenger cars, for example to 8000 h.

Common insulating materials suitable for electric motors correspond to thermal class 155 (F) or 180 (H). They may reach temperatures of up to 155 °C or 180 °C at the hottest point of the winding during continuous operation. The hottest point is usually located in the end areas of the winding, the so-called winding overhang. This is where the winding leaves the thermally conductive iron core and passes into the thermally insulating air.

According to the Montsinger rule, the service life halves or doubles for every 8–10 Kelvin change in temperature. Of course, this rule only applies to heating in continuous operation. Electric motors may occasionally and briefly become significantly hotter without significant effect on the lifetime. However, a sufficient safety margin to the softening limit of the insulation materials must be maintained, as otherwise the system would rapidly lose its service life.

The cooling system of an electric motor works largely independently of the speed and can therefore dissipate a constant thermal power loss. However, this only applies to motors with liquid cooling or external ventilation, which is standard for traction drives. Industrial motors usually have an internal fan on the shaft, the cooling effect of which is speed dependent.

Conventional oil or water cooling systems for traction drives in series-production cars operate with flow temperatures of around 65–75 °C, oil cooling in racing cars up to 110 °C. With a typical flow rate of 4–8 l/min, the temperature rise caused by the electric motor is up to 10 Kelvin.

Looking at the individual loss components of the electric motor, the I^2R-losses in the windings generally have a large impact. They are only dependent on the current, which is mostly proportional to the torque, and not on the speed.

Another cause for heating is iron losses. They are speed and load dependent.

Finally, there are friction losses, which are not dependent on the load but are proportional to the speed.

The total losses and hence the thermally permissible continuous current I_{th} are therefore mostly constant with a small dependency on speed.

Depending on the motor design, the continuous thermal current of the motor I_{th} is around 1.5–3 times lower than the maximum permissible current of the inverter I_{max}.

2.3.4 Thermal Limits of Short-Time Operation

Particularly when starting cold, electric motors can be heavily overloaded for a few seconds (short-time operation). The losses that cannot be dissipated by the cooling system during this time are stored in the heat capacity of the laminated core.

The temperature limits of the insulating materials may be exceeded briefly without harmful consequences if a sufficient safety margin to the softening limit is maintained. Temperatures of over 200 °C are possible for short periods of time.

2.4 Performance Characteristics of Electric Motors

2.4.1 Base Speed Range

As will be shown later, the torque of an electric drive results from the current in the stator coils and the total magnetic flux that is exchanged between stator and rotor.

In the base speed range, the magnetic flux is kept constant by changing the voltage in proportion to the speed and the frequency, see (A.70).

This ensures optimum magnetic utilization of the motor. As the flux is constant, the torque is proportional to the current. At very high currents the laminated sheets may become saturated and then the torque increases less than proportionally with the current.

The thermally permissible continuous current I_{th} therefore limits the continuous torque T_{th}, see the blue curve in Fig. 2.2. The maximum current I_{max} limits the maximum torque T_{max}, see the orange curve in Fig. 2.2. Both torques are approximately constant in the base speed range. This is why the base speed range is also referred to as the constant torque range.

Depending on the drive train design, the base speed range for typical battery electric vehicles ends between 40 and 60 km/h.

2.4.2 Field Weakening Range

The maximum base speed is reached when the inverter must set the maximum available voltage U_{max} to reach a constant flux.

A further increase in speed is still possible simply by increasing the frequency. However, the inverter can no longer set a voltage proportional to the frequency, as the highest voltage provided by the battery has already been utilized. A further increase in frequency then leads to a successive drop of the magnetic flux inversely proportional to the speed. This is called the field weakening range.

As will be shown later, the continuous torque T_{th} and the maximum torque T_{max} also fall in inverse proportion to the speed in this range.

When the available output voltage U_{max} is reduced at low states of charge, the limit speed is also reduced. The torque drop therefore already starts at lower speeds, see the dashed lines in Fig. 2.2. This effect also reduced the highest available power of the drive.

2.4.3 Breakdown Torque

Even if an arbitrarily high current was made available, electric motors can only develop a certain maximum torque due to the design of the magnetic circuit. This limit is referred to as the breakdown torque.

The breakdown torque of asynchronous motors is proportional to $(U_1/f)^2$, for synchronous machines it is proportional to $U_1 \cdot U_p/f^2$, i.e. terminal voltage U_1 and synchronous voltage U_p in relation to the square of the frequency. Both formulae result in approximately the same behavior of the machine.

Normally, the breakdown torque is far greater than the torque resulting from the current limit of the inverter. It therefore has no practical significance.

The following normally applies:

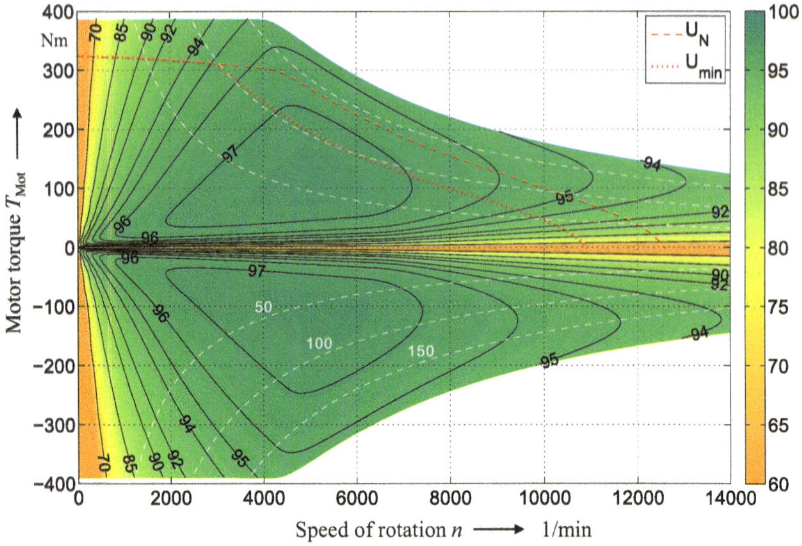

Fig. 2.3 Simulated torque–speed characteristic of a PMSM for an electric car

$$I_{th} < I_{max} < I_{breakdown}$$

and

$$T_{th} < T_{max} < T_{breakdown}$$

2.4.4 Practical Example of an Electric Drive

Figure 2.3 shows a calculated torque–speed characteristic of a permanent-magnet synchronous machine designed as a traction drive for an electric car. The upper half of the diagram applies to motor operation, the lower half to generator operation (braking energy recovery).

The drive achieves a maximum output power of 185 kW, see the power hyperbola at the top of the diagram. The continuous output with full battery voltage U_N is 140 kW, with minimum voltage U_{min} it is approximately 100 kW.

In the base speed range, the 30s torque is around 390 Nm and the continuous torque is around 320 Nm (red line). At full battery voltage, the base speed is around 4250/min. With an empty battery U_{min}, it drops to around 3000/min.

The colored areas indicate the efficiency of the motor. The calculated efficiency takes I^2R-losses and iron losses into account. Further losses in the drivetrain, for example in the battery, power electronics, gearbox, wheel bearings, and so on, are

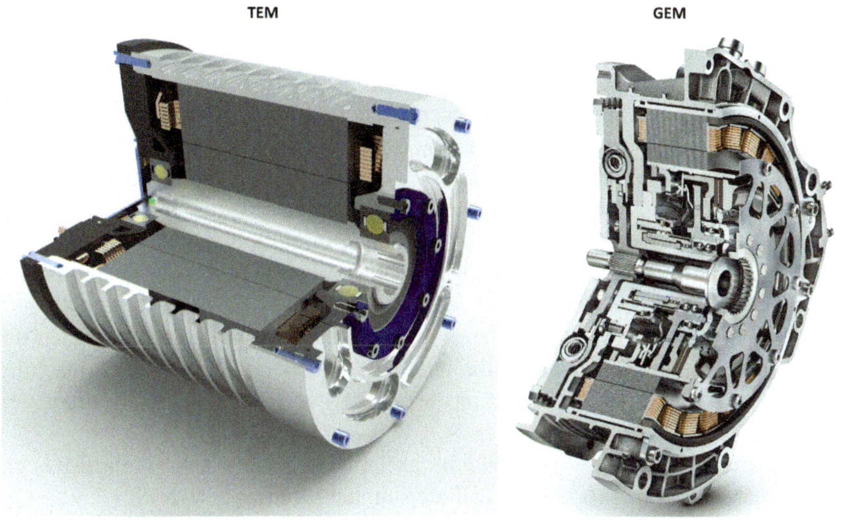

Fig. 2.4 Traction electric motor (TEM) and gearbox-integrated electric motor (GEM) in comparison[4]

not included. The motor achieves an efficiency of more than 95% in a wide range and only drops to 90% at very high torques.[3]

To analyze the performance of the motor in a driving cycle, the respective load points can be plotted into the diagram. Typically, the desired speed of a car in a driving cycle is given in intervals of 1 s. The required torque must now be calculated or measured for each speed. The efficiency of the drive is calculated by adding up the losses at each load point and relating them to the produced mechanical power over the entire driving cycle.

2.5 Electric Motors for All-Electric and Hybrid Drivetrains

It is useful to differentiate between traction electric machines (TEM) and gearbox-integrated electric machines (GEM).

TEMs are preferably used in fully electric vehicles, while GEMs are mostly used in hybrid vehicles. This differentiation is based purely on mechanical design constraints given by the drive train and not on the electromagnetic motor principle. GEMs and TEMs can be built equally as asynchronous or synchronous machines (Fig. 2.4).

---

[3] The efficiencies shown in the diagram are only the internal motor losses, i.e. I^2R and remagnetization losses. Friction losses, gearbox losses, and inverter losses must also be considered to obtain the total losses of the drive train.

[4] Source: KIT (left), Schäffler AG (right), with kind permission.

2.5.1 Gearbox-Integrated Electric Machines (GEM)

Today, gearbox-integrated motors are mostly designed as permanent magnet synchronous machines (PMSM) with concentrated windings (single-tooth windings).[5] This is due to the compact axial length, which is important for gearbox integration. The rotor is constructed as a hollow cylinder. Other technical elements (couplings, gear stages) are often integrated inside the rotor.

The disadvantages of concentrated windings, in particular the rather low rotational speed and the higher vibration and noise, are not significant when installed in a gearbox next to a noisy internal combustion engine (ICE).

GEMs can also be crankshaft starter generators, which are either integrated into the gearbox or flanged directly to the internal combustion engine. These machines are also called ISG (integrated starter generators) or IMG (integrated motor generators).

ISGs are often low-power units with an output of a few kW, while IMGs can have a peak output of over 50 kW and are suitable for plug-in hybrids.

2.5.2 Traction Electric Machines (TEM)

Traction machines are primarily used as the main drive of a fully electric vehicle but are also used in serial and power-split hybrids, for example in the Opel Ampera. Another name for TEM is SMG (separate motor generator). TEMs have a cylindrical design and rotate at high speeds in order to achieve a high power density (Table 2.1).[6]

TEMs are usually equipped with a distributed winding, which is ideal for high speeds with low noise and vibrations. This type of winding can be produced with round copper wires, flat wires or can be a hairpin winding (form wound winding).

The following illustration provides an overview of the power-to-weight ratios that can be achieved with TEMs today. The given power density refers to continuous operation, not short-time peak power. Depending on the design and operating time, the peak power is significantly higher, usually approximately 1.5–3 times higher (Fig. 2.5).

Current passenger car series motors achieve and exceed the power-to-weight ratio of ICE. Electric motors for electric racing cars are even better than the engines used in Formula 1.

[5] A well-known exception is the Toyota Prius, which contains two GEMs with distributed windings. A detailed discussion of the different types of windings is presented in Sect. 7.6.2.

[6] Current TEM series motors have a top speed of up to 18,000/min. Newer near-series developments go up to over 20,000/min. Prototype motors in the research and racing sector reach up to 30,000/min with outputs of 250 kW and more.

Table 2.1 Comparison of the main features of GEM and TEM

	GEM	TEM
Design	Short axial length, relatively large diameter	Length and diameter in a balanced ratio. Ideally, the active iron length is approximately equal to the rotor diameter
Cooling	Cooling via gear oil	Main cooling via outer jacket (water/glycol), secondary coolings in rotor shaft and/or spray cooling of winding overhangs
Speed	Speeds like ICE, typically 6000–7000/min	Speeds as high as possible to achieve good power density, typically up to 20,000/min
Torque	MCHEV and MHEV: High starting torque to start the ICE, then constant power over a wide speed range FHEV: Torque as constant as possible over the speed. Minimal use of the constant power range	Wide utilization of the constant power range, typically 3–4 times the base speed range
Stator winding	A short axial length is important, therefore almost exclusively concentrated single-tooth windings	Good electromagnetic properties are important (noise, vibration), therefore almost exclusively distributed windings, also hairpin
Machine type	Almost exclusively PMSM	Predominantly PMSM, but also IM and EESM
Power	Often only a few kW, maximum up to 30 kW continuous power (50 kW peak power)	Typically, 100 kW and more
Voltage level	12 and 48 V are common, for high power also up to 120 V	120 V up to 800 V

2.5.3 Belt Starter Generators (BSG)

The belt starter generator is a hybrid between GEM and TEM. As a separate unit, it has a cylindrical design like a TEM but often contains a concentrated single-tooth winding like a GEM.

Classic BSGs are connected to the 12 V vehicle electrical system and produce a few kW of short-time peak power to start the combustion engine.

New BSG developments use a separate 48 V on-board power supply and can deliver significantly higher power for longer periods of time. They are well suited for micro and mild hybrids.

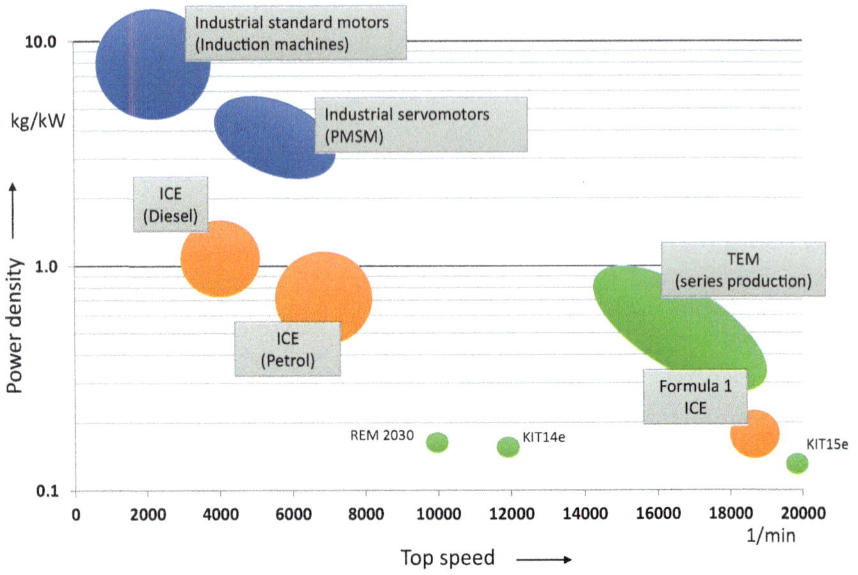

Fig. 2.5 Comparison of the continuous power-to-weight ratio of different motor technologies

2.6 Gearboxes

Gearboxes are torque and speed converters. Apart from small of losses, the mechanical power is transmitted from the input to the output shaft without change. Most gearboxes have a single input shaft and a single output shaft.[7]

If losses are neglected, the following applies:

$$P_{in} = T_{in} \cdot \Omega_{in} = T_{out} \cdot \Omega_{out} = P_{out} \tag{2.4}$$

The reduction of a gearbox i is defined as:

$$i = \frac{\Omega_{in}}{\Omega_{out}} = \frac{T_{out}}{T_{in}} \tag{2.5}$$

$\Omega = 2\pi \cdot n$ is the mechanical angular velocity in 1/s, n is the rotational speed in 1/s and T is the torque in Nm.

Due to the high speed of electric motors and the low speed of rotation of the wheels, the gearbox reduction i is typically larger than one in electric drivetrains.

In a multi-stage gearbox, different gear pairs are connected in series to increase the overall reduction ratio. Gear pairs of different technologies are often com-

[7] Planetary gearboxes with three shafts are used for power-split hybrid drives and torque converter automatic transmissions. This special type of gearbox is presented in detail later, see Sect. 3.5.7.1.

bined, for example a spur gear stage followed by a planetary stage, or two spur gear stages followed by a differential stage (bevel gear stage).

Multi-stage transmissions are often used in electric cars. A higher motor speed enables a higher power density and therefore less weight, less space requirement, and lower costs of the motor.

Single-stage spur gears with high NVH (noise–vibration–harshness) requirements for passenger cars can be realized with gearbox ratios up to $i=3$–4. In the industrial sector, ratios of up to $i=9$ per spur gear stage are common. Today's typical gearboxes for electric cars are usually designed with two spur gear stages or one spur gear stage and one additional bevel gear stage in the differential.

Due to their asymmetrical design and the resulting lateral forces, spur gearboxes are suitable for speeds up to a maximum of 20,000–25,000/min. Higher speeds can be achieved with planetary gearboxes, as they are symmetrical in design.

2.6.1 Multi-speed Gearboxes

Almost all gearboxes for electric cars are multi-stage single-speed. As the electric motor delivers full torque already at standstill and can cover a wide range of speeds, a shiftable multi-speed gearbox is normally not required but can be useful for high-performance cars.

Multi-speed transmissions, i.e. shiftable gearboxes, are equipped with at least one switching device to engage different gear pairs as required and thus provide different gear ratios i.

In a classic manual gearbox, the different ratios are realized by different pairs of gears that are constantly engaged and rotate at different speeds. In the picture, the countershaft rotates continuously at the speed of the first stage. The three lower gear wheels are also always engaged while the engine is running.

The output shaft can either be connected to one of the three gearwheels or directly to the input shaft by moving one of the two switching sleeves to the left or right side. The output shaft then takes over the speed of the coupled gear wheel.

To engage the coupling, the shaft must be load-free and already rotate at nearly the same speed of both gearwheels. A coupling between gearbox and engine will decouple the load momentarily. Synchronization is achieved by a friction element (not shown in the picture), ultimately a slipping clutch, which is actuated at the start of the shifting process when moving the shifting sleeve. An additional locking element ensures that shifting is only possible once the shafts are synchronized to a similar speed.

With co-axial gearboxes as in Fig. 2.6, two gear meshes are usually required for power transmission. However, a direct gear, in this case fourth gear, can be realized with a transmission ratio of 1 that needs no gear meshes.

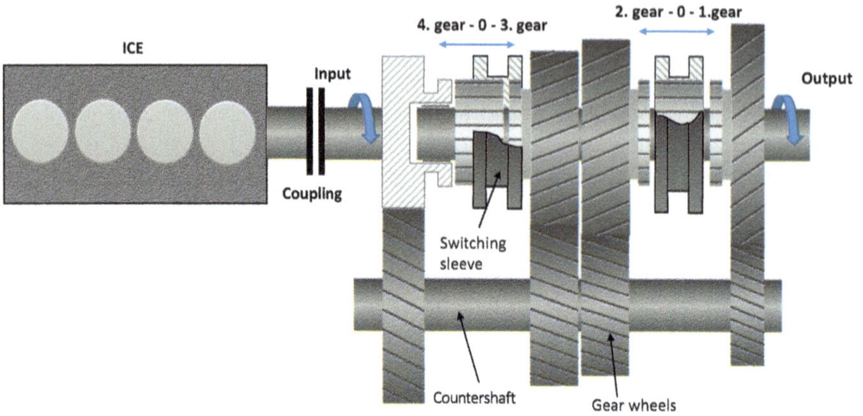

Fig. 2.6 Section through a parallel axle, four-speed gearbox with third gear engaged

If the axes do not have to be co-axial, the output of the gearbox is done via the countershaft. In that case, only one gear mesh is engaged and the efficiency of the gearbox is slightly improved. However, no direct gear can be realized.

Classic manual transmissions as shown in the picture are not used in electric cars. Most electric cars do not have switching gear in the first place and the few cars that do, have only two gears. Also, the constantly moving countershaft and the rotating gears result in high friction losses.

Shiftable transmissions for electric cars are designed either as double-clutch transmissions or as shiftable planetary transmissions. Both have the advantage of a faster-shifting process and therefore shorter interruption of power to the wheels. There is no need for a synchronizer, as the speed and angular position of the disengaged and therefore load-free electric motor can be adjusted in a fraction of a second to engage the next gear stage. Still, the driving experience of electric cars is compromised by the interruption in traction when changing gears.

2.6.2 Automatic Transmissions

Planetary gearsets are the heart of an automatic transmission. Unlike all other types of gearsets, they have three shafts: The ring gear, the planetary gear carrier, and the sun gear. Different ratios can be achieved by fixing one component or by connecting two components of a planetary gear set with clutches and freewheels.

A hydrodynamic converter is used in an automatic transmission for starting. Hydrodynamic converters were invented by Hermann Föttinger, an electrical engineer from Munich, around 1905 at the Vulcan AG shipyard in Szczecin. They operate wear-free.

The converter is a hydraulic component that enables power transmission between shafts of different speeds. It couples and decouples the engine from the gearbox. Unlike a conventional clutch, the converter achieves an increase in torque by a factor 2–3, making the hydraulic converter effectively a variable transmission.

The torque is transmitted hydrodynamically by a moving fluid only during the start-up phase. After starting, an integrated friction clutch is engaged to couple the two shafts directly thereby reducing hydraulic losses.

Today, torque converter automatic transmissions for passenger cars are manufactured with up to 9 gears as standard. They are mainly used in the P2 hybrids presented later.

Power-split two-mode hybrid drives work according to a similar functional principle as torque converter automatics and thus enable several shiftable gears with fixed ratios.

2.6.3 Continuously Variable Transmissions

Continuously variable transmissions (CVT) are a special type of multi-speed transmission, in which any desired transmission ratio can be freely set within certain limits.

The mechanical variation of the transmission ratio has been known for a long time. It is achieved by adjusting the position of two pairs of bevel gears. The two pairs are connected by a rubber or chain belt. As each pair moves closer, the belt runs on a larger diameter, when they are moved further apart, the belt runs on a smaller diameter. By changing the diameter of the belt, also the speed and therefore the gear ratio changes.

The disadvantages of stepless gearboxes are lower efficiency and higher wear compared to manual transmissions.

Power-split hybrid transmissions achieve a stepless adjustment of the transmission ratio by varying the speed of one of the three shafts of a planetary gearbox using an electric motor.

2.7 Electric Drivetrains

2.7.1 Gearboxes for Electric Traction Motors (TEM)

Electric traction drives (TEM) for fully electric vehicles (BEV, FCEV) are usually designed with a gearbox between the electric motor and the wheels. This allows the speed of the motor to be increased compared to the wheel speed and the required motor torque to be reduced at the same time.

In this way, the power density of the electric drive is significantly increased. This has a positive effect on space requirements, costs and weight.

The additional frictional losses of the gearbox are usually of secondary importance. Single-speed gearboxes with just two stages have a high efficiency of typically 95% and more. Otherwise, without gearbox a direct drive electric motor

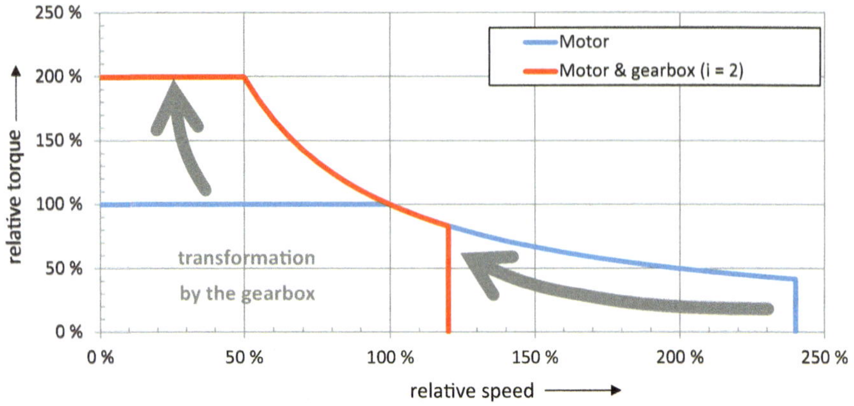

Fig. 2.7 Transformation of the torque/speed characteristic of a traction motor by a single-speed, single-stage gearbox

would have to be much larger to produce the higher torque. It would then develop more iron and bearing losses, so in the end the total losses are almost the same.

The following figure shows the change in the motor torque/speed curve due to the influence of a gearbox with the ratio $i=2$.

Obviously, a gearbox cannot add additional power into the drivetrain. Apart from the slight transmission losses, the motor characteristic curve shifts upwards and to the left along a power hyperbola.

Operating points above the motor's power hyperbola cannot be achieved by a gearbox. For example, the operating point of 100% relative speed and 150% relative torque, which lies above the motor power hyperbola in Fig. 2.7, is unattainable with this motor, regardless of the type of gearbox.

Note
The wheel speed of a car reaches around 1100–1500/min at top speed. For example, the speed of a 195/65 R15 tire at 130 km/h is 1090/min and the speed of a 245/45 R19 tire at 200 km/h is 1510/min.
As TEMs are typically built for top speeds of 16,000–20,000/min, gearboxes with reduction ratios of $i=10$–18 are required. This can be achieved with two spur gear stages, for example.

In some few cases, TEMs are also used with shiftable transmissions, i.e. multi-speed gearboxes, for example, dual-clutch transmissions or switchable planetary transmissions such as in the Porsche Taycan. This has two advantages:

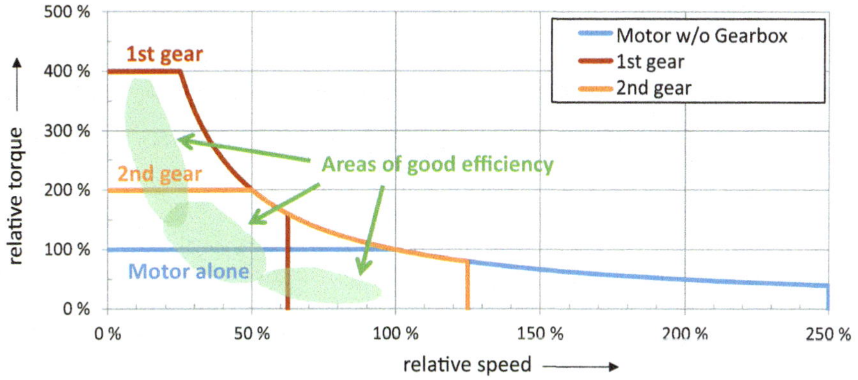

Fig. 2.8 Transformation of the torque/speed-characteristic of a traction motor by a multi-speed gearbox

- The range of high motor efficiencies can be spread, see Fig. 2.8.
- The electric motor can achieve both a high starting torque and a high top speed. This can be beneficial for sports cars.

The individual gears are normally shifted automatically by actuators.

However, the higher weight and the higher losses of the shiftable transmission offset the advantages of the different speeds in most applications. For this reason, shiftable transmissions are rarely used in fully electric production cars. Even the sporty vehicles of the Tesla brand and many Formula E racing cars drive with a single-speed, non-shiftable transmission. They achieve acceleration times of less than 3 s from 0 to 100 km/h and top speeds of more than 300 km/h.

Example 2.1—Electric cars with gearboxes

The Rimac Concept One sports car has a 250 kW electric motor on each of the four wheels and a 2-speed gearbox. The motors have a top speed of 12,000/min.

The Porsche Taycan has a single-speed planetary gearbox on the front axle with a reduction ratio of $i = 8$. A two-speed gearbox with three shafts is used on the rear axle. Two spur gear stages form second gear and a shiftable planetary gear set forms first gear. The first gear has a reduction ratio of $i = 15$, the second gear has a reduction ratio of $i = 8$. ◀

2.7.2 Powertrain Configuration

The following drivetrain configurations are used:

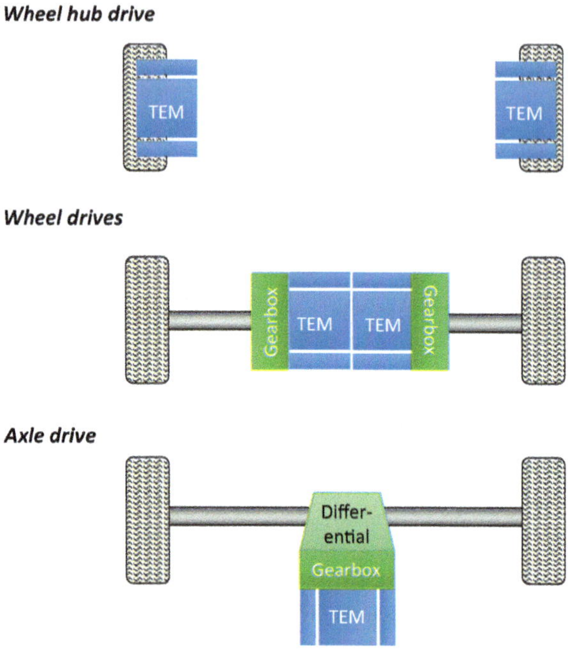

Fig. 2.9 Configurations of the electric drivetrain

- **Wheel hub drive**: The motors are integrated into the wheel hubs, either with or without a gearbox.
- **Wheel drives**: Two motors per axle are located inside the car. They act directly on the wheels via a drive shaft, usually through a gearbox.
- **Axle drive**: One central motor per axle is located inside the car. It acts on the drive axle via a gearbox and a differential (Fig. 2.9).

The distinction between the different configurations is made regarding the location of the motor(s) and regarding the application of the fore on the wheel hub, wheel, or axle. With wheel hub drives and wheel drives, all four or only two wheels can be driven. With axle drives, one or both axles can be driven.

These configurations can also be combined. For example, the Tesla Model S Plaid has wheel drives with two motors in the rear and a single axle drive in the front.

2.7.3 Wheel Hub Drives

Wheel hub drives are built both with gearboxes, e.g. *Michelin Active Wheel*, and as direct drives, e.g. *Schäffler eWheel Drive*, *Ziehl-Abegg ZAwheel*, *Fraunhofer e-concept*, *Continental DeepDrive*. They can also incorporate a mechanical brake, for example, *Michelin Active Wheel* and *Continental DeepDrive*.

Fig. 2.10 Wheel hub drive[8]

One of the first hybrid-electric vehicles, the Lohner Porsche from 1900, already had four-wheel hub motors, although each only had an output of 5.2 kW. Today's wheel hub motors achieve outputs of over 50 kW (Fig. 2.10).

Example 2.2—Michelin active wheel

In addition to the actual drive motor with 30 kW of power, the Michelin active wheel contains a gearbox, the mechanical disk brake and an active suspension in each wheel. A second electric motor is also integrated as an electric damper. ◀

Wheel hub drives take up no space in the vehicle so there should be more room for passengers and luggage.

However, a central electric traction motor with a fixed transmission is already much smaller than a conventional combustion engine with a complex multi-speed gearbox.

Furthermore, all auxiliary devices of a car, like power electronics, cooling systems, cables, hoses, etc., still must be installed in the engine compartment and take up space. All in all, the space benefit of wheel hub drives is quite small.

[8] Source: Schäffler AG, with kind permission.

The noise behavior is cited as another advantage. However, gearboxes and high-speed electric motors become louder at high speeds in particular, whereas they are whisper-quiet when driving slowly. When driving fast, wind and rolling noises mask the motor and gearbox noises, so that these do not play a significant role in practice.

Finally, better efficiency due to the lack of transmission losses is cited as an advantage for wheel hub drives. However, this consideration is largely based on a misunderstanding. Of course, removing the gearbox saves losses. However, the motor now must provide the high torque for the wheels directly. Due to the unfavorable speed and the high torque, the efficiency of the motor is reduced. The required increase in motor size and the fact that at least two and often even four motors are needed to achieve the desired overall performance lead to a significantly poorer efficiency in the motors. In total, two or even four direct drive wheel hub motors tend to have even more losses than a single, small, compact, and high-speed central drive with gears.

It should be noted that none of the purely conceptually developed wheel hub drives has gone into series production for passenger cars so far, for which there are many reasons:

- Splitting the drive power between two or four motors with a corresponding number of inverters results in higher costs for mechanical adaptation, for electrical cabling, for coolant connections and for the synchronization of the two or four motor controllers.
- The high unsprung mass results in poor car handling.
- There are high demands on the mechanical design of the motors: direct influence of the ambient temperature (low outside temperatures) on one side and high temperatures (brake disk) on the other side of the motor. High ingress of moisture, dirt, salt spray, and vibrations.
- Gearless direct drive electric motors generally have a low power density. They are therefore more expensive and heavier than a central, fast-running motor. The higher weight also means higher consumption in the driving cycle.
- A wheel hub drive is in an exposed position. It can be damaged by minor accidents and cause high repair costs.
- There is a risk of accidents due to yawing torques in the event of uncontrolled braking of only one motor. This can occur, for example, after a winding short-circuit or faults in the power electronics.
- Integrating the mechanical brake is difficult in terms of mechanical design and results in additional heat input and a further reduction in motor performance.

Today, wheel hub motors are only used in special applications, such as low-floor busses and trucks, prototype vehicles or in military applications. Another area is small three-wheeled cabin scooters, which usually have no space for a central drive. Wheel hub motors are not used in series-produced passenger cars.

2.7.4 Wheel Drives

With wheel drives, each wheel is driven by its own electric motor, which is located on the drive shaft inside the vehicle. A gearbox is usually connecting the motor and the shaft.

The advantages of wheel drives are:

- With two motors per axle, the braking and accelerating wheel torques can be freely controlled. In this way, the vehicle's cornering behavior can be improved. This is known as torque vectoring and is used in sports cars.
- High-speed motors with high power density can be used resulting in lower weight and cost.

Disadvantages are:

- Higher costs with two motors per axle, as two gearboxes and two inverters are required, as well as an additional synchronization of the two motor controllers.
- The failure of just one motor can result in yawing torques and thus dangerous driving conditions.
- Multiple motors result in more space being used in the vehicle.

Example 2.3—Electric cars with multiple motors

The Mercedes-Benz SLS AMG E-Cell is a battery-electric vehicle produced in small series with four-wheel drives. Each motor weighs around 80 kg with transmission and delivers 138 kW. The total output is 552 kW.

The Tesla Model S Plaid has wheel drives on the rear axle. ◄

2.7.5 Axle Drive

The axle drive is by far the most common type for the traction drive of fully electric vehicles.

There are basically three variants:

- Angular drive: Ideal arrangement from the perspective of the electric motor with maximum degrees of freedom in the electrical design.
- Parallel shaft drive: limitation of the motor diameter, preferably for high motor speeds.
- Coaxial drive: Restriction on the motor technology that can be used, as a large hollow shaft is required. Practically only possible as a PMSM. However, minimal space requirement (Fig. 2.11).

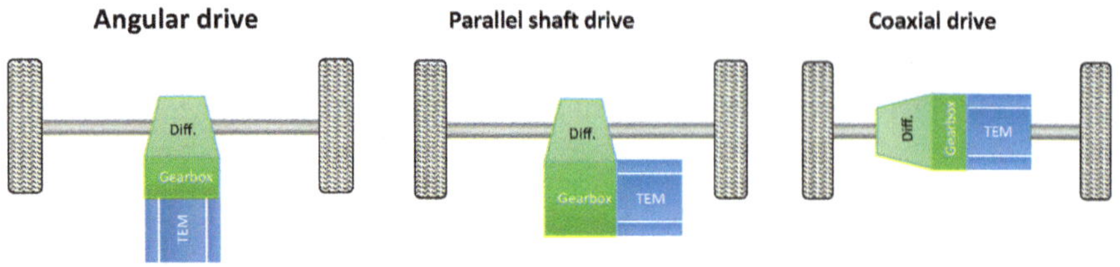

Fig. 2.11 Different variants of axle drives

Example 2.4—Axle drives in electric cars

An angular drive was prototypically implemented in the Porsche e-Ruf Greenster.

The parallel shaft drive can be found in most current vehicles, for example in all Teslas, Renaults, BMWs, VWs, and many more.

The coaxial drive is used, for example, in Hondas (Fit EV, FCX Clarity) and the Audi e-tron 55. ◀

In the coaxial drive, the drive shaft for one of the two wheels is fed back through the inside of the motor. The motor shaft is designed as a hollow shaft for this purpose. In this way, a particularly compact drive can be realized. However, the motor design must be suitable for this construction. Externally excited synchronous machines (EESM) require a large part of the rotor cross-section for the windings. They are not suitable for coaxial drives. Permanent magnet machines (PMSM) and induction machines (IM) are possible.

Further limitations of the coaxial design are the reduced ground clearance of the vehicle and the effort required for servicing, as every major repair to the motor and gearbox requires the complete axle to be removed.

Hybrid Vehicle Drives

3

Abstract

This chapter introduces the hybrid vehicle types, operating modes, and technologies. An overview is given of the common motor and transmission variants for hybrid and fully electric passenger cars. The hybrid vehicle classes from micro-hybrid to plug-in hybrid are presented in detail. The required motor power and battery capacities and the technical functionalities are presented. The shell diagram of the combustion engine is used to illustrate how fuel savings can be realized through hybrid drives and their potential. Finally, the hybrid powertrains (serial, parallel, and power-split hybrid drive) are discussed. The planetary gearbox, which is presented in detail in a separate section, is essential for power-split drives.

3.1 Fundamentals

Hybrid vehicles have at least two independent drive types. The degree of hybridization H is the ratio of the available electric traction power to the total available traction power, i.e.:

$$H = \frac{P_{el}}{P_{mech} + P_{el}} \tag{3.1}$$

The degree of hybridization H can range between 0 and 1. $H=0$ is a vehicle with an internal combustion engine only (ICE) and $H=1$ is a purely electric vehicle (EV = electric vehicle or BEV = battery electric vehicle or FCEV = fuel cell electric vehicle).

In hybridization, a distinction is made between hybrid **vehicle types**, hybrid **modes of operation**, and hybrid **drive structures (drive technologies)**.

M. Doppelbauer, *Introduction to Electromobility*,
https://doi.org/10.1007/978-3-658-45482-1_3

Hybrid vehicle types:

- Micro-hybrid
- Mild-hybrid
- Full-hybrid
- Plug-in-hybrid

Hybrid modes of operation:

- Automatic start–stop
- Load point shifting
- Sailing
- Braking energy recovery
- Boosting
- Electric driving/electric crawling

Hybrid drive structures:

- Serial hybrid
- Range extender
- Parallel hybrid
- Power split hybrid

Almost every hybrid operating mode can be achieved with different drive technologies. It is the task of the car manufacturer to design a suitable powertrain.

The following table shows a classification of the common hybrid vehicle types and the corresponding operating modes (Table 3.1).

3.2 Voltage Classes

Electrical energy is provided by the energy storage system (battery or fuel cell) as direct current (DC) and voltage.

A distinction is made between:

- Low-Voltage (LV) Systems: up to 30 V AC_{rms} or 60 V DC and
- High-Voltage (HV) Systems[1]: up to 1000 V AC_{rms} or 1500 V DC

HV cables and connectors are marked with orange color.

[1] The correct, industry-standard designations are extra-low voltage (ELV) and low voltage (LV). However, they are not established in the automotive sector and are therefore not used in this book.

Table 3.1 Technical properties of hybrid drive systems (typical values)

		Micro-hybrid MCHEV	Mild-hybrid MHEV	Full-hybrid FHEV	Plug-in-hybrid PHEV	Electric vehicle EV
Degree of hybridization *H*		0.03	0.1	>0.25	>0.5	1
Power of the electric motor	kW	<5	6 … 20	20 … 40	50 … 90	>100
Energy of the battery	kWh	<1	0,5 … 1,5	1,5 … 5	5 … 25	20 … 100
DC voltage	V	LV 12	LV 12/48	HV >100	HV >200	HV 400/800
Automatic start-stop		Yes	Yes	Yes	Yes	
Load point shifting			Yes	Yes	Yes	
Sailing			Yes	Yes	Yes	Yes
Braking energy recovery	kW		6 … 20	>20	>50	>50
Boosting	kW		<15	>15	>30	
Electric driving	km			1 … 5	20 … 60	>100

Basically, the higher the voltage, the lower the currents, and the less copper is required to transport the power. Losses are reduced in cables, connectors, and power electronic components, as the current heat losses decrease quadratically with the current. High charging power above 200 or 250 kW can practically only be transmitted with high voltages above 600 V.

The disadvantage of higher voltages is the higher requirements in electrical insulation. Therefore, a good compromise must be found when selecting the voltage level.

Nearly all cars including hybrids and battery electric vehicles have a 12 V battery (11–14,3 V normal operating range) and most larger trucks use 24 V (18–32 V normal operating range). This voltage level is required for many low-power consumers like lights, windshield lifters, seat adjustment, windshield wipers, and for all computerized control units.

LV systems up to 60 V DC are considered intrinsically safe. The required insulation and protection measures are therefore considerably lower than for higher voltages. This is exploited in 48 V DC hybrid drives. The 48 V system permits voltages between 36 and 52 V in normal operation and down to 24 V or up to 54 V with some functional limitations.

The 48 V voltage level is especially suited for mild-hybrids to supply loads between 3 and 12 kW (up to 24 kW peak) for start–stop function, boost and braking energy recovery, air conditioning compressors, electric heaters, pumps, power steering drives, and more.

For full-hybrids, voltage classes of 120 V DC up to 300 V DC have become established with continuous power of up to 75 kW.

Plug-in hybrids, fuel cell, and battery electric cars typically use voltages around 400 V DC or 800 V DC. These voltages can all still be controlled with industry-standard 1200 V semiconductor components, which are produced inexpensively in large series.

At 400 V system voltage, charging capacities of up to 200 kW are possible; Tesla allows 250 kW. An 800 V system is only required for even higher charging capacities up to 350 kW.[2]

It is considered to build battery electric cars with three voltage classes: 12 V, 48 V, and 400 (or 800) V. This allows for more flexibility in the optimization of current and insulation requirements depending on the consumer's power demand. Replacing the 12 V voltage level completely with 48 V is also being discussed. This would have advantages in terms of cabling, but there are hardly any suitable vehicle components available yet.[3]

Voltage classes above 800 V DC are not to be expected in cars. Suitable components are not available in sufficient quantities and at low prices. The motor insulation would have to be significantly reinforced for these high voltages, which would result in additional effort and therefore costs.

However, voltages of up to 1500 V DC are currently being used for fast charging of heavy commercial vehicles. With the new Megawatt Charging Standard (MCS), charging power of up to 3500 kW can be achieved, see Sect. 12.11.

Note

The advantage of a high system voltage is the lower currents for the same power. Lower currents have a positive effect on cables, plugs, and power electronics, as the required conductor size and semiconductor surface are proportional to the current. Very high charging powers over 250 kW can only be realized with 800 V systems.

However, contrary to popular belief, a high system voltage does not improve motor efficiency.

The motor coils must be adapted to the rated voltage during the design phase. At double the voltage, e.g. 800 V instead of 400 V, only half the

[2] It seems that the 800 V voltage level will be more often used in the future in upper middle class and luxury class vehicles for market differentiation, even though not technically required.

[3] The Tesla Cybertruck, launched at the end of 2023, is the first vehicle to remove the 12 V electrical system entirely and only use 48 V and 400 V.

current I flows. To achieve the same magnetic flux, which is proportional to $N \cdot I$, the number of turns N must be doubled. Therefore, the conductors must be twice as long and half as thick. So, the 800 V coils have four times the resistance R compared to the 400 V coils.

Heat losses, which result from $I^2 \cdot R$, are equal in both cases as the current is halved and the resistance is quadrupled.

Well-designed electric motors therefore do not benefit from a higher system voltage. They have the same losses, efficiency, power factor, continuous power, and peak power.

3.3 Hybrid Vehicle Types

Current forecasts assume that around half of all vehicles will have a 48 V system (MHEV) soon. 25% will be full and plug-in hybrids and the remaining 25% will be purely electric vehicles (BEV). In the long term, it can be assumed that full and plug-in hybrids will largely disappear from the market in favor of all-electric vehicles, while mild hybrids with 48 V technology will continue to be sold as a cost-effective alternative for some time to come.

3.3.1 Micro-Hybrid (MCHEV)

With the micro-hybrid (MCHEV), only an automatic start–stop system is added as a hybrid operating mode compared to a pure combustion engine vehicle. The system automatically stops the engine after a few seconds of standstill. As soon as the accelerator pedal is pressed, the combustion engine starts up again automatically. It can certainly be argued that micro-hybrids have no electric driving component and are therefore not hybrid vehicles in the strict sense.

Technically, the start-stop function is realized by a more powerful starter motor (BSG = belt starter generator or crankshaft starter generator) and corresponding software.

The electric motor's output is several kW for a short period of time. The system voltage level remains at the usual 12 V in passenger cars. When braking, the electric motor is initially used instead of the mechanical brake. This allows a small amount of electrical braking energy to be recovered to recharge the battery. Due to the frequent starting processes, the battery is subjected to a somewhat greater load than in a conventional vehicle.

The technical effort required to implement the micro-hybrid functions and the additional costs are low, which is why this design is used in many current vehicles. According to the manufacturers, fuel consumption can be reduced by up to 15% in the standardized driving cycle with a micro-hybrid. In real-world driving, the savings are usually lower, typically 5–7%. Many vehicles today are equipped with an automatic start–stop system as standard.

Example 3.1—Start/stop systems

An automatic start–stop system can be found in the "Blue Motion" vehicles from Volkswagen AG, in the "Blue Efficiency" vehicles from Daimler AG, the "ECOnetic" vehicles from Ford AG, and in many other vehicles from other manufacturers. ◀

3.3.2 Mild-Hybrid (MHEV)

With the mild hybrid (MHEV), load point shifting, boosting, and sailing functionality are added. Load point shifting includes two functions: load point boosting and load point lowering. These hybrid operating modes are presented in detail in a later chapter.

The electrical output of a mild hybrid is significantly higher than that of a micro-hybrid, around 6–20 kW, depending on the vehicle. The battery capacity is correspondingly higher. Mild hybrid drive systems are preferably realized with 48 V system voltage and therefore require a separate battery.

The electric motor is either designed as a separate starter generator (BSG), as in the micro-hybrid, or it is integrated into the transmission (ISG).

Manufacturers promise fuel savings of up to 25% in the standardized driving cycle for mild hybrids. Here too, the reality is somewhat different, typically around 10% fuel savings.

Example 3.2—Mild hybrids

Mild hybrids are available from Honda (Honda Insight), BMW (certain models of the 7 series), and Daimler (certain models of the S-Class), among others. ◀

3.3.3 Full-Hybrid (FHEV)

In contrast to the hybrid vehicles discussed so far, the full hybrid (FHEV) can also drive purely electrically. In most cases, only low speeds, typically up to a maximum of 80 km/h, and only short ranges of a few kilometers are achieved. However, this is sufficient for city traffic, driving in traffic jams, or slow-moving traffic (stop & go). Longer cross-country or highway journeys can only be carried out conventionally with the combustion engine.

Full hybrids are usually equipped with gearbox-integrated electric motors (GEM). Depending on the power requirements, they can be designed for a voltage of 48 V (up to approx. 30 kW output power), or 120 V with outputs of up to approx. 60 kW. Sometimes, even higher voltages up to 260 V are used in full hybrids.

48 V machines are hardly suitable for electric driving at higher speeds, as the power is generally not sufficient. Furthermore, a certain amount of power of the

electric motor must always be reserved in electric driving for starting the combustion engine. This reserve noticeably reduces the power available for electric driving. A starter generator can be installed in addition so that the 48 V machine does not have to maintain power reserves for the starting process.

It is quite possible to realize electric creeping with 48 V drives. This allows maneuvering electrically into or out of a parking space or drive-in a parking garage.

The battery capacity of full hybrids is higher than that of mild hybrids, typically 1.5–2.5 kWh. The battery is charged purely by braking energy recovery or by increasing the load point of the combustion engine.

Because the combustion engine is at a standstill during electric driving, all auxiliary units must be decoupled from the shaft and the belt of the combustion engine and must be driven separately by small electric motors. This includes the air conditioning compressor, the power steering pump, the oil lubrication pump, the brake booster, etc. In many modern vehicles in the upper mid-range and luxury class, however, the auxiliary units are electrified anyway to save fuel.

The technical realization of a full hybrid is relatively complex and cost-intensive. Many manufacturers therefore prefer to offer plug-in hybrids, which offer significantly enhanced functionality ("electric driving experience") at moderately more cost.

Example 3.3—Full hybrids

The Toyota Prius is the most successful full hybrid on the market. It was first introduced in 1997 and was the best-selling car in Japan for many years. Other examples are the Ford Fusion, the Ford Escape Hybrid, and certain models of the Mazda 6. ◀

3.3.4 Plug-In-Hybrid (PHEV)

Technically, it is only a small step from full hybrid to plug-in hybrid (PHEV). The electric drive power and battery capacity are increased, typically to around 50–90 kW and 5–25 kWh. The electric driving range is significantly higher, typically around 20 to over 60 km. These vehicles can also be driven purely electrically at higher speeds, typically up to 120 km/h.

The distinguishing feature compared to full hybrids is the ability to charge the battery electrically from a socket at an electric charging station or at home. Most plug-in hybrids are limited to AC charging but there are some that have DC fast charging capability as well.

Plug-in hybrids are usually equipped with a separate electric traction motor (TEM). Only very few plug-in hybrid vehicles have gearbox-integrated motors (GEM or IMG).

Example 3.4—Plug-in hybrids

One of the first popular and successful full hybrids was the Chevrolet Volt aka Opel Ampera. The Toyota Prius is also available as a plug-in hybrid in certain model variants.

All German car manufacturers now have many plug-in hybrids on offer.

The Swedish Polestar 1 was a plug-in hybrid with AC and DC-charging sockets. ◀

A special measurement method for fuel consumption applies to plug-in and full hybrids. This method shows exceptionally high savings but does not realistically reflect typical driving situations. As a result, manufacturers can give unrealistically low standard consumption figures, particularly for plug-in hybrids.

Example 3.5—Fuel consumption of large SUVs

The fuel consumption of the plug-in hybrid Porsche Cayenne Turbo S E-Hybrid is specified at 3.7 l/100 km. The vehicle has a 4 l eight-cylinder bi-turbo engine with an output of 400 kW (550 hp) and weighs almost 2.7 tons empty. According to an online consumption monitoring site, the actual consumption in real driving scenarios of this car is around 7.0 l/100 km.

An Audi Q7 3.0 TDI Quattro Hybrid with 373 hp engine power and an unladen weight of 2.5 tons is rated at 1.8 l/100 km. The real driving emissions are typically between 6.3 and 7.4 l/100 km.

Source: www.spritmonitor.de ◀

Detailed descriptions of the standardized driving cycles and consumption measurements are given in Sect. 4.5.

3.4 Hybrid Modes of Operation

3.4.1 Load Point Shifting

The aim of load point shifting is to operate the combustion engine at the most favorable load point even when the current power requirement of the car is lower or higher. When the load point is shifted, the load distribution between the engine and the electric motor is changed but the torque on the wheel, i.e. the traction force, remains constant. Ideally, the driver will not notice any change, but the fuel consumption is reduced.

3.4.2 Fuel Consumption Map of ICE

The following figure shows an example fuel consumption map (speed–torque diagram) of a typical internal combustion engine. The example shows an engine with around 110 kW peak power at 6000/min.

Curves of constant power appear as hyperbolas in the diagram. The full load curve starts at 1000/min with 140 Nm and extends up to the maximum speed of around 6000/min with around 175 Nm. Higher torques cannot be generated by the engine at the respective speeds.

The isolines of constant fuel consumption are drawn in the shape of a shell and are also colored. The figures correspond to the specific consumption in grams per kWh of mechanical energy generated.

The specific consumption increases in the partial load range due to the friction losses. Low combustion temperatures and a local lack of oxygen also lead to increased consumption and pollutant emissions at low partial loads. When idling at zero torque, the specific consumption is infinite, and the efficiency is zero.

The most favorable consumption of around 230 g/kWh is only achieved in a narrow speed and torque range: approximately between 2500 and 3500/min and 120–170 Nm.

The efficiency of an internal combustion engine is calculated from the reciprocal of the product of the calorific value of the fuel and the specific fuel consumption.

Example 3.6—Efficiency of combustion engines

At the best point ($b_e = 230$ g/kWh), the engine in Fig. 3.1 has an efficiency of $\eta = 1/(12.0 \text{ Wh/g} \cdot 0.230 \text{ g/Wh}) = 36.2\%$ when running on premium petrol (calorific value 12.0 Wh/g).

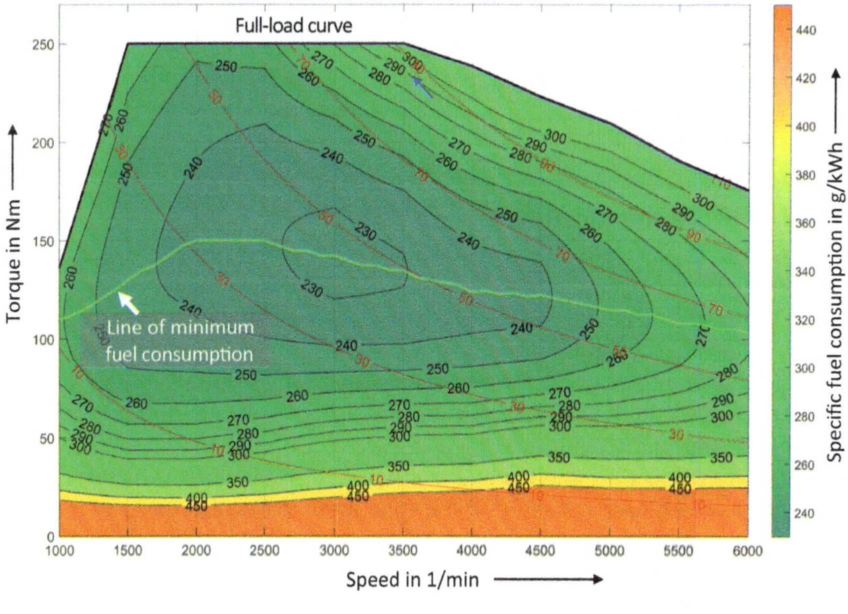

Fig. 3.1 Fuel consumption map of a 4-cylinder, 1.4 l internal combustion engine

Along the lowest line ($b_e = 450$ g/kWh), the efficiency is $\eta = 1/(12.0\text{ Wh/g} \cdot 0.450\text{ g/Wh}) = 18.5\%$. ◀

The specific consumption figures shown in the diagram only apply to a well-run-in engine at optimum operating temperature. During the first minutes after starting or during low-load driving in cold conditions, for example, city driving in winter, the consumption will be significantly higher. The efficiency is then only in the order of 20% even at the optimum load point.

Also, the efficiency of the multi-speed, multi-stage gearbox must be considered as well. It is typically between 80 and 90% and much lower than the efficiency of a small single-speed, two-stage gearbox used in electric cars.

Example 3.7—Fuel consumption of combustion engines

With an output of 30 kW and an engine speed of 4000/min, the engine in Fig. 3.1 has a specific fuel consumption of around 265 g/kWh.

Assuming a typical energy requirement for driving a small car of 16 kWh/100 km, this corresponds to a fuel consumption of 4.2 kg/100 km. Premium petrol has a density of 760 g/l, so the fuel consumption at this operating point is 5.6 l/100 km.

However, the efficiency of the gearbox, which is typically between 80 and 90% for automatic transmissions, must be considered as well. This brings the fuel consumption to around 6.6 l/100 km. ◀

Fuel consumption and CO_2 emissions are inextricably and directly linked. The combustion of regular petrol produces 0.276 kg of direct CO_2 emissions per liter, premium petrol 0.261 kg, and diesel 0.271 kg per liter, see Sect. 14.2. These values do not include the additional indirect emissions that occur during refining, which come on top.

3.4.3 Load Point Shifting Using a Gearbox

By changing the transmission ratio of the gearbox, whether shiftable or continuous, a relocation of the operating point can be achieved without changing output power. In the consumption diagram, the change takes place along a power hyperbola. The traction force on the wheel remains constant.

There are two options: load point increase (higher torque, lower speed) and load point decrease (less torque, higher speed). Load point increase generally leads to lower fuel consumption and load point decrease leads to better car performance as the torque reserve of the engine at the particular speed is increased (Fig. 3.2).

In the illustration, the load point P1 (40 kW, 5100/min, 75 Nm, 325 g/kWh) is changed to the new load point P2 (40 kW, 2500/min, 150 Nm, 250 g/kWh) by increasing the gear reduction (shifting to a higher gear).

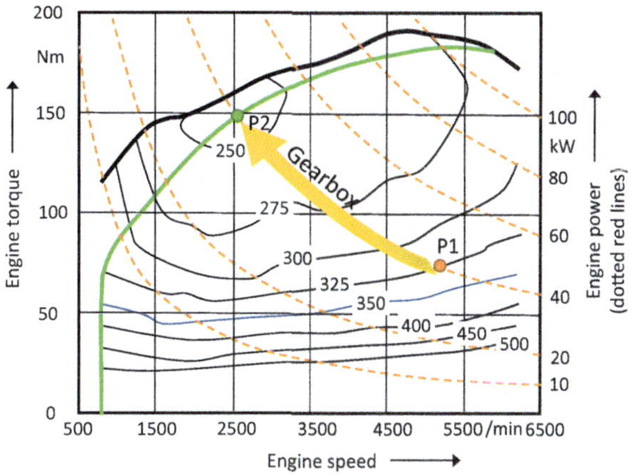

Fig. 3.2 Load point increase from P1 to P2 by changing the gear reduction ratio

This corresponds to everyday experience that low-speed engine operation saves fuel.

However, it should be noted that the new load point P2 offers hardly any torque and power reserve. Spontaneously pressing the accelerator pedal will be largely ineffective as only around 10 Nm is available as an additional acceleration reserve.

Acceleration reserve is the difference between the torque currently required for driving and the maximum torque available at the current engine speed.

The original load point P1 is very different: the power can be increased quickly from around 40 kW to around 100 kW and the torque from around 75 Nm to around 180 Nm if needed. The acceleration reserve is therefore more than 100 Nm, that is 10 times higher than at load point P2.

The more gear stages are available, the more effectively the transmission can shift the load point. Up to 9 gears are common in high-end automatic transmissions.

The transmission ratio can be adjusted even more sensitively with a continuously variable transmission (CVT = Continuous Velocity Transmission).

3.4.4 Load Point Increase by an Electric Motor

In hybrid vehicles, a load point increase is realized by operating the electric machine as a generator. The combustion engine increases its torque and transfers part of the generated power to the battery in the form of electrical energy. The speed of the engine and therefore the speed of the vehicle do not change.

In Fig. 3.3, the load point P1 (20 kW, 4500/min, 40 Nm, 400 g/kWh) is raised to P2 (80 kW, 4500/min, 170 Nm, 260 g/kWh) by using the electric machine in generator mode.

Engine and car speed and gear ratio do not change. The excess power of 60 kW is used to charge the battery.

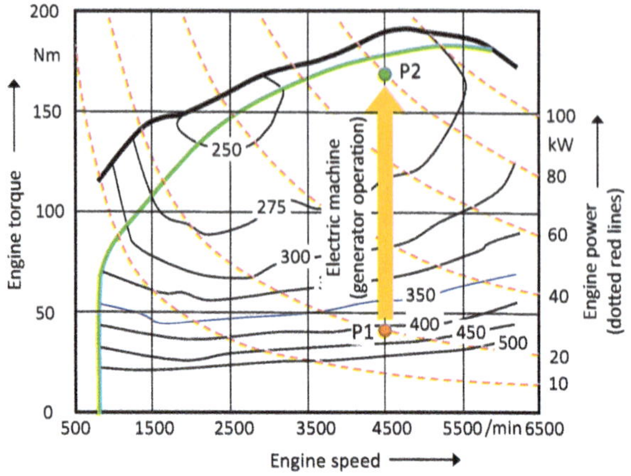

Fig. 3.3 Load point increase by means of an electric machine (parallel hybrid)

Note
A load point shift using an electric machine is only possible with parallel hybrid vehicles. In power-split hybrids, the speed and thus the velocity would change immediately if the torque of the electric motor were to be changed

For an overall energy efficiency analysis, the entire chain when storing and retrieving energy into and from the battery must be compared to the improvement in specific fuel consumption. Only if the losses in the complete electrical energy chain are lower than the gains due to the improvement in specific fuel consumption does the load point increase save energy.

Note
The efficiency chain in the electric drive train is utilized twice during the load point increase:

Firstly, to store the electrical energy in the battery. This results in losses in the electric generator, power electronics, and the charging efficiency of the battery.

Later, the stored energy must be reused for driving. This again results in losses in the power electronics and in the electric motor.

In terms of energy efficiency, load point increasing is only useful in the low partial load range where the specific fuel consumption is particularly high.

However, a load point increase must always be carried out to recharge the battery if the SOC (State of Charge) falls below a predefined limit value. Otherwise, some of the other hybrid operating modes, in particular boosting and electric driving, would no longer be possible.

3.4.5 Load Point Decrease by an Electric Motor

When the load point is lowered (load point decrease), the electric machine is operated as a motor. Its torque is therefore added to the torque of the combustion engine, which can then operate at a lower torque to achieve the same output torque as before.

The energy required to operate the electric motor is taken from the battery. Here too, the speed of the drive and the speed of the vehicle do not change.

A load point reduction is only useful in terms of energy consumption in very few operating points. It is therefore rarely utilized in practice.

However, a load point reduction must always be carried out if the battery charge level is too high. Otherwise, some of the other hybrid operating modes, in particular braking energy recovery, would no longer be possible.

3.4.6 Boosting

When boosting, the combustion engine operates at full torque and additional torque is generated by the electric motor. The total drive torque and the total drive power then exceed the maximum values possible with the combustion engine alone.

Boosting can be combined with a load point reduction depending on the operating point.

In Fig. 3.4, the combustion engine is already operating at the maximum torque at operating point P1. The operating point P2 is reached due to the additional torque of the electric motor. The output of the drive system increases from around 110 kW at 6000/min to 140 kW at the same speed.

In addition to briefly increasing the maximum drive torque, boosting can also be used to bridge dynamic start-up weaknesses of the combustion engine (turbo lag).

3.4.7 Sailing

When sailing, the drivetrain is operated without an active drive, i.e. neither the combustion engine nor the electric motor is activated. By eliminating the engine drag torque, energy can be saved when sailing and even with low load requirements.

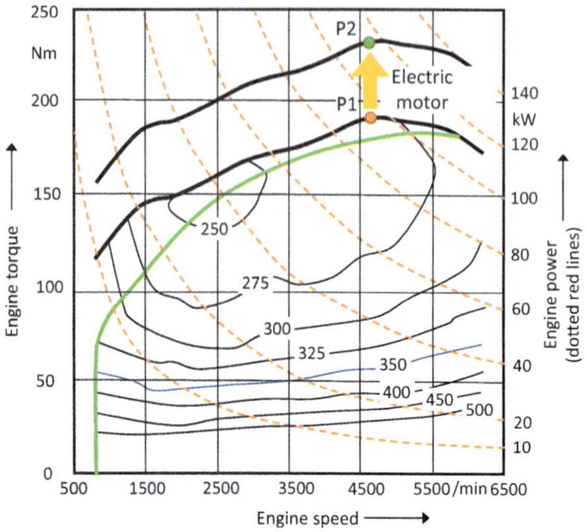

Fig. 3.4 Boosting by means of an electric motor (parallel hybrid)

The combustion engine must be decoupled from the drive train to avoid the high drag losses. Permanently excited synchronous motors can also have a noticeable drag torque and may have to be decoupled.

Important auxiliary consumers such as the brake booster and power steering must continue to operate without the combustion engine when sailing. They must therefore be electrically operated.

In practice, however, sailing often causes irritation for drivers when the combustion engine is suddenly switched off while driving. For this reason, sailing is no longer used in most vehicles today.

3.5 Hybrid Drive Structures

Drive structure refers to the interaction of electrical and mechanical power flows in the drivetrain. A distinction is made between three hybrid drive structures (drive technologies):

- **Serial hybrid**
  Mechanical torque and power are provided by an electric motor only. The combustion engine is used exclusively to generate electrical energy with the aid of a separate electric generator. No mechanical transmission from the combustion engine to the wheel.
- **Range extender**
  The range extender is a variation of the serial hybrid. These cars have a rather large battery and normally operate fully electric. Only when the battery is depleted, they become serial hybrids.

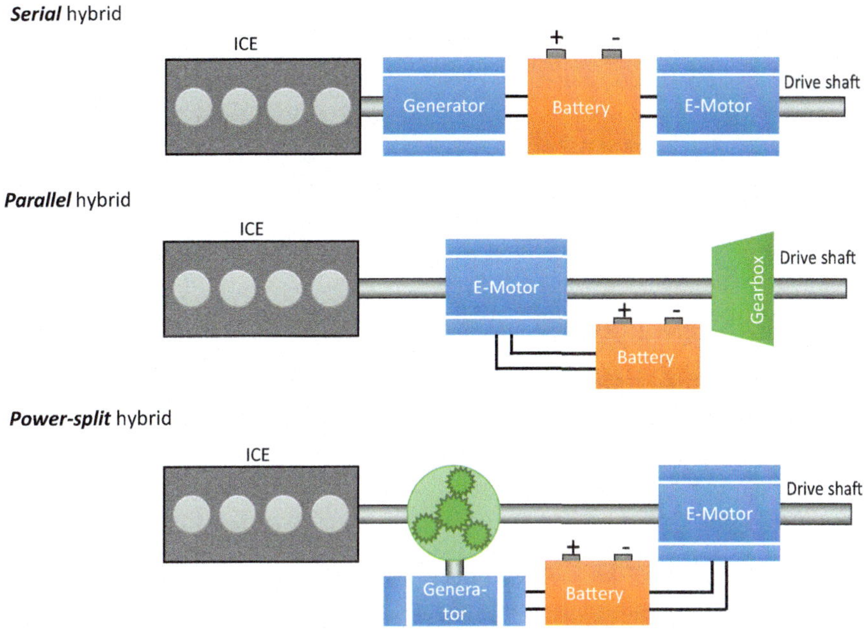

Fig. 3.5 Schematics of the three hybrid drive structures

- **Parallel hybrid**
  Direct mechanical transmission from the combustion engine to the wheel. The electric motor adds its power to the same transmission so that the total power is provided jointly by the ICE and the electric motor.
- **Power-split hybrid**
  The mechanical power of the engine is divided into a mechanical and an electrical power path. The wheel torque is provided jointly by the ICE and the electric motor (Fig. 3.5).

These three hybrid drive structures can also occur in mixed forms. In some vehicles, the drive structure is reconfigured depending on the operating point, e.g. in the Opel Ampera, which can operate as a power-split hybrid, serial hybrid, range extender, and parallel hybrid.

3.5.1 Serial Hybrid

In a serial hybrid, both energy converters are connected electrically in series. There is no mechanical connection between the combustion engine and the wheels. The combustion engine drives an electric generator that supplies the electric motor with energy.

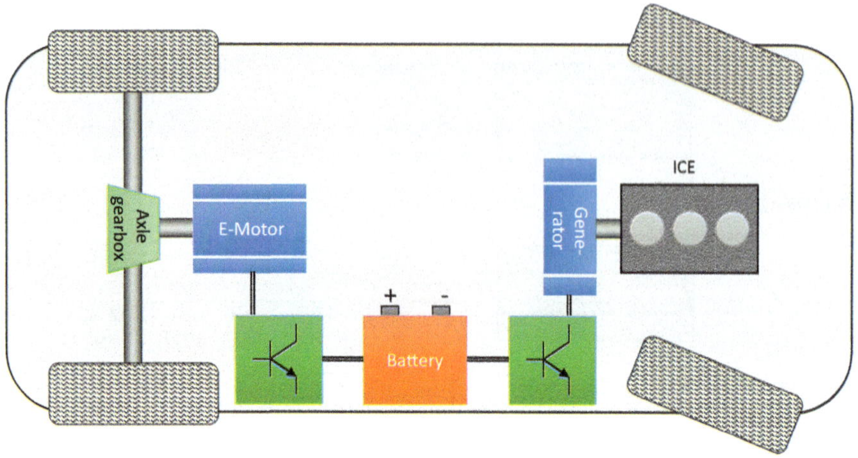

Fig. 3.6 Structure of a serial hybrid drive

This pure form of the serial hybrid is only found in diesel-electric ships and locomotives or in large commercial vehicles, for example, large mining trucks.

In passenger cars, an additional battery is always installed, which can temporarily store a significant amount of energy. This decouples the power of the ICE from the drive power, allowing it to be operated in an optimum operating range, i.e. a speed/torque range with favorable specific consumption.

In addition, due to the energy storage in the battery, the ICE only must provide the average drive power so it can be designed smaller and therefore more efficient (Fig. 3.6).

The electric motor can be arranged as a single central drive with gearbox and differential, as a dual or quadruple axle drive or as a wheel hub drive (dual or quadruple).

Advantages of the serial hybrid:

- Purely electric driving is possible.
- The ICE only works at energy-efficient operating points.
- The ICE only covers the average power, so it can be quite small.
- The peak power (acceleration) is produced from the battery and electric motor.
- Long driving ranges are possible.

The disadvantage of the serial hybrid is the long chain of conversions that are required before the mechanical power generated by the combustion engine reaches the wheels. This leads to relatively poor efficiency in constant operation with a high load at highway speeds, when neither braking energy recovery nor load point shifting is possible.

Serial hybrids are therefore particularly suitable for smaller city vehicles and low average driving speeds.

Example 3.8—Serial hybrid drives

The Nissan E-Power technology uses serial hybrid drives in various vehicle models.

For example, the Qashqai E-Power has a 1.5-L three-cylinder combustion engine that produces 116 kW (158 hp) and exclusively drives an electric generator. The generator charges a small 1.9 kWh battery (just about the size of four electric bicycle batteries) and directly supplies the energy for the electric drive motor. The maximum output power of the electric motor is 140 kW (190 hp).

Nissan's serial hybrid vehicles have low fuel consumption in city traffic but are at a disadvantage on interurban and highway routes compared to conventional ICE vehicles.

According to www.spritmonitor.de, the average fuel consumption of the Qashqai E-Power in Germany is between 5 and 9 l/100 km—not noticeably lower than any other typical middle-class SUV. ◀

3.5.2 Range Extender

A range extender is a serial hybrid with a large battery. The battery enables long-distance purely electrical driving, typically 100 km or more.

The combustion engine is normally switched off in a range extender car. It is only activated when the battery charge level falls below a certain threshold.

While driving with the ICE, the battery is usually not recharged because the efficiency of the drive chain would be too poor. Rather, the generated electricity is used solely for driving. The battery can only be recharged when the car is connected to an external AC- or DC-power source (Fig. 3.7).

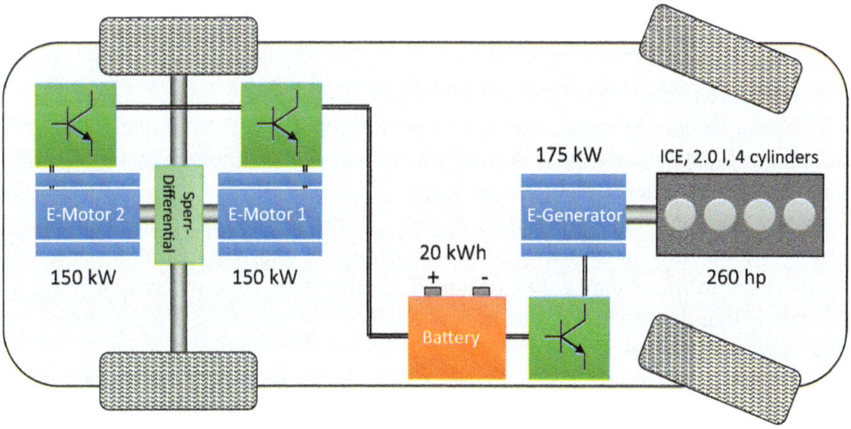

Fig. 3.7 Drive structure of the Fisker Karma (serial hybrid/range extender)

Example 3.9—Range extenders

The BMW i3 was available with an optional range extender.

The Chevrolet Volt/Opel Ampera also behaves like electric cars with a range extender in certain operating modes. ◀

Example 3.10—Fisker Karma with range extender

The range extender in the Fisker Karma shown in the picture switches on automatically during heavy acceleration to provide the required drive power of up to 300 kW and reduce the load on the battery.

In addition, the driver can activate the combustion engine at the touch of a button to charge the battery. ◀

Hardly any vehicles with range extenders are offered today. Overall, the cost of the drivetrain is too high compared to purely electric vehicles, which can also reach long driving distances.

3.5.3 Parallel Hybrid

In parallel hybrid drives, both the electric motor and the internal combustion engine have a direct mechanical transmission to the wheels.

The distribution of torque and power between the ICE and the electric motor can be adjusted so that load point shifting and boosting are possible to minimize consumption.

There are three different types of parallel hybrid drive: Single-shaft parallel hybrid, twin-shaft parallel hybrid and through-the-road (TTR) hybrid (Fig. 3.8).

3.5.4 Single-Shaft Parallel Hybrid/Torque Addition

In the single-shaft hybrid, the electric motor and ICE act on a common drive shaft. This structure is also known as torque addition (Fig. 3.9 and Table 3.2).

The single-shaft arrangement is particularly popular with mild and full hybrids because it is technically easy to implement. The electric motor can be integrated into the drivetrain in different positions.

Many manufacturers prefer full hybrids in the P2 arrangement. The electric motor is then fully integrated into the transmission and requires almost no additional space in the engine compartment. The advantage from the customer's point of view is that the separating clutch to the ICE enables electric driving, albeit usually only with short ranges.

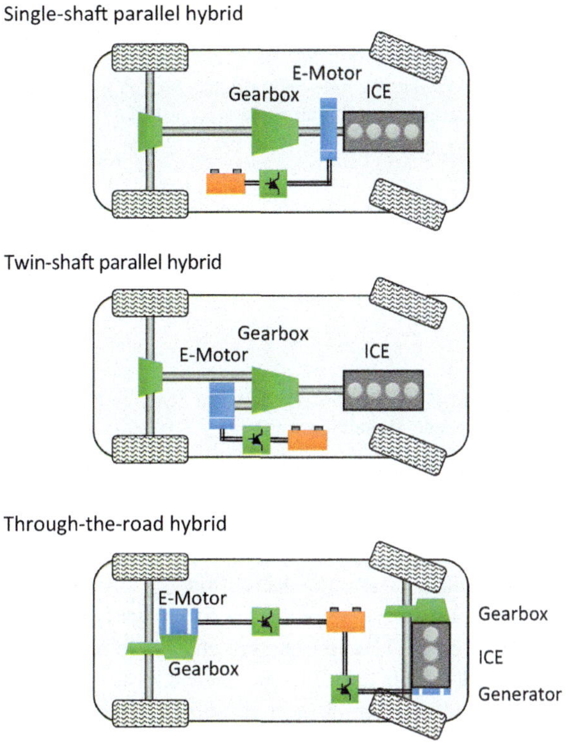

Fig. 3.8 Variants of parallel hybrid drives

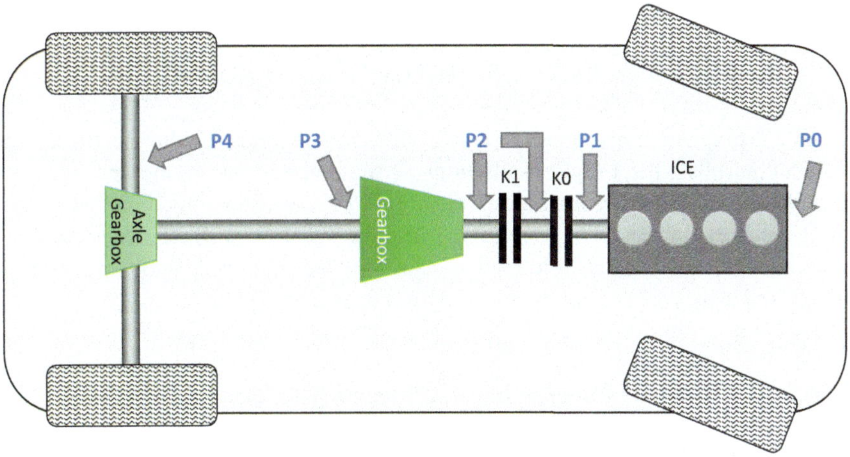

Fig. 3.9 Placement of the electric motor in the drivetrain of a single-shaft parallel hybrid

Table 3.2 Variants of the single-shaft parallel hybrid drive structure

P0	BSG (Belt Starter Generator)	Suitable for micro and mild hybrids. Mostly 12 V system voltage, sometimes 48 V
P1	GEM (Gearbox integrated Electric Motor), ISG (Integrated Starter Generator) or BSG	
P2	GEM, ISG or BSG	The ICE can be disconnected from the drivetrain via the K0 clutch during braking energy recovery phases or when coasting. Typical for mild hybrids. Electric driving is also possible (full hybrid) The K1 separating clutch also enables the battery to be charged by operating the ICE when stationary. Rarely used Typically, 48–120 V system voltage
P3	TEM (Traktion Electric Motor) or SMG (Separate Motor Generator)	Not normally used in series production vehicles
P4	TEM or SMG	Not normally used in series production vehicles

Example 3.11—Parallel full hybrid

The Audi A6 is available as a full hybrid with a P2 parallel arrangement.

Its 2.0 l TFSI combustion engine delivers 155 kW. A PMSM with an output of 40 kW is installed as an electric motor directly on the crankshaft between the ICE and in front of the hydrodynamic converter in the gearbox. The ICE can be decoupled via a separating clutch.

The nominal capacity of the lithium-ion battery is 1.3 kWh. The vehicle can travel electrically at speeds of up to 100 km/h and has a range of up to 3 km at a constant speed of 60 km/h.

The design of the BMW Active Hybrid 5 is very similar. ◀

The P2 arrangement also enables active vibration damping by the electric motor. For this purpose, the torque oscillations of the combustion engine are measured and then damped with maximum dynamics by corresponding counter-torques of the electric motor. This serves to improve comfort and efficiency. Vehicles with an internal combustion engine can also be equipped with two electric wheel hub drives. This is sometimes referred to as a P5 hybrid.

3.5.5 Twin-Shaft Parallel Hybrid/Speed Addition

In the twin-shaft hybrid (speed addition hybrid), the ICE and electric motor each drive the gearbox via their own shafts (summation gearbox). This arrangement requires a relatively large amount of space and is therefore not used in passenger cars.

However, the design is technically quite simple and well suited for expanding existing drivetrains, for example for prototypes or small series. Two-shaft arrangements are used in commercial vehicles, especially buses.

3.5.6 Trough-The-Road Hybrid (TTR)

In a through-the-road hybrid the ICE and electric motor act on different axles. The addition of the two drive powers is only achieved by the forces transmitted to the road.

The electric motor can be placed on the front axle, for example in the Porsche 918 Spyder, or on the rear axle, for example in the Peugeot 3008 Hybrid 4.

One advantage of this concept is the possibility of all-wheel drive.

However, the ability to shift the load point is limited because there is a risk of the wheels locking or spinning at high torques, as one of the two drives always acts on the rear axle, which can only be loaded with less torque.

If the electric motor engages the rear axle, the power limit for braking energy recovery is also reduced.

Example 3.12—Peugeot 3008

Figure 3.10 shows the drive structure of the Peugeot 3008 Hybrid 4.

The vehicle can be operated in hybrid all-wheel drive mode, in purely electric mode and purely with the ICE.

An 8.5 kW generator is connected to the V-belt of the ICE, which recharges the battery as required. Load point shifting can also be implemented with this machine.

There is also a conventional starter generator for charging the 12 V battery.

The electric traction motor is a permanent magnet synchronous machine (PMSM). It has an output of 27 kW and acts on the rear axle via a differential, a reduction gear, and a separating clutch. The separating clutch is required to avoid drag losses in pure ICE operation, which can be significant with the

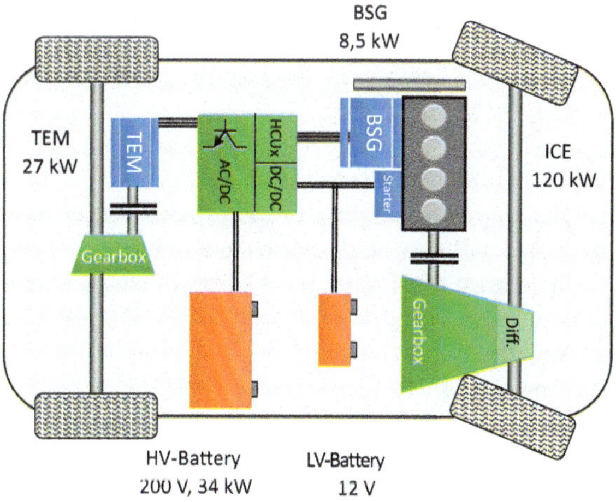

Fig. 3.10 Drive structure of the Peugeot 3008 Hybrid 4 (parallel TTR hybrid)

PMSM, especially at high speeds. In addition, the PMSM cannot technically cover the full speed range.

The clutch is designed as a claw clutch. Before engaging, the electric motor must perform an active synchronization. ◀

3.5.7 Power-Split Hybrid

The main feature of a power-split hybrid drive is the separation of the shaft power supplied by the combustion engine into an electrical and a mechanical path. This separation is carried out by a planetary gearbox.

The mechanical path leads to the wheels via further gears and a differential.

In the electrical path, the mechanical energy is first converted into electrical energy by an electrical generator. A second electrical machine then converts the electrical energy back into mechanical energy and feeds this back into the mechanical branch, similar to a parallel hybrid.

In this pure form, the first electric machine always works as a generator, and the second always as a motor, at least as long as the drivetrain delivers power to the wheels. When driving downhill, i.e. during braking energy recovery, the operating modes of the two electric machines are reversed.

At first glance, this design of the power-split hybrid drive may seem pointless. However, as will be explained in more detail later, it allows free speed selection of the sun gear of the planetary gearbox, whereby the transmission ratio of the planetary gearbox can be freely adjusted within certain limits. Ultimately, this results in a continuously variable transmission (CVT) between combustion engine and wheels.

In most cases, however, the power-split hybrid drive is combined with an electric battery. This enables additional operating modes like braking energy recovery, boosting, load point shifting, and pure electric driving (Fig. 3.11).

Depending on the driving status and hybrid operating mode, both electric motors can be operated as a motor or generator, even if there is no braking energy recovery. Nevertheless, it has become common practice to refer to the machine on the planetary gearbox as the electric generator and to the machine on the driven axle as the electric motor.

3.5.7.1 Planetary Gear Unit

As the mode of operation of the planetary gearbox is essential for the power-split hybrid, it will now be discussed in more detail.

In contrast to all other types of gearboxes, a planetary gearbox has three shafts:

1. Sun gear (S)
2. Carrier frame (C)
3. Ring gear (R)

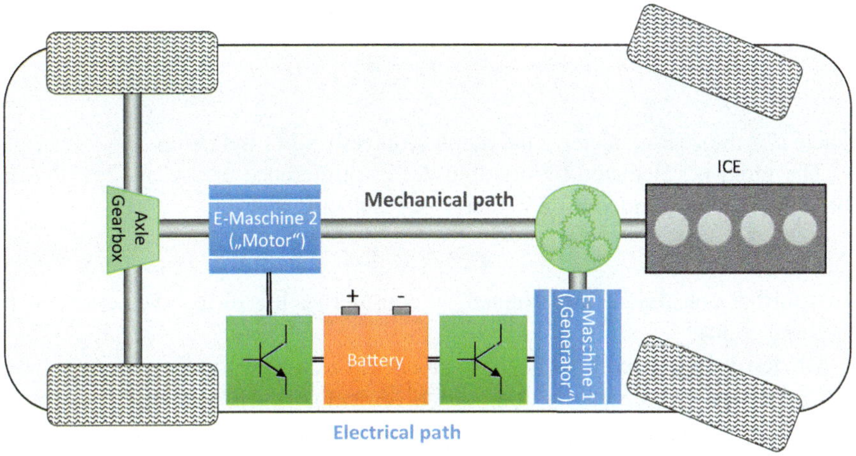

Fig. 3.11 Structure of a power-split hybrid drive (input split)

Fig. 3.12 Schematic diagram of a planetary gear unit

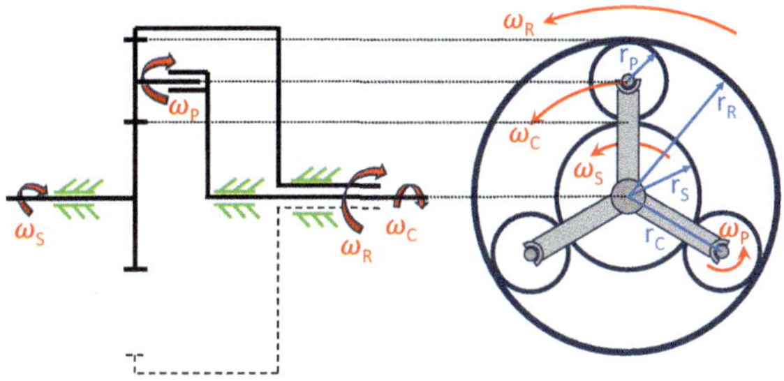

In addition, the planet gears (P) rotate with their own speed, but this rotation is not utilized.

The transmission ratio between the planet carrier and ring gear can be freely adjusted within certain limits by varying the speed of the sun gear (Fig. 3.12).

The planet gears revolve around the sun gear. The planets, usually three in number, are rotatably connected to each other by the bridge, also known as the planet wheel carrier, and are fixed to each other. The planets are also connected to the ring gear, exactly opposite the meshing with the sun gear.

This gives a planetary gearbox four different angular frequencies ω_S, ω_P, ω_{ST}, and ω_H and four different radii r_S, r_P, r_{ST}, and r_H.

However, these variables are not independent of each other. Obviously, the radii of the gears must satisfy the following geometric relationship:

$$r_R = r_S + 2r_P \tag{3.2}$$

The width of the teeth, i.e. the module, of all gears in the gearbox must also be the same so that they can all mesh with each other.

The module of a gear is defined as:

$$m = \frac{d}{z} = \frac{2r}{z} \tag{3.3}$$

where d is the gear diameter, z the number of teeth, and r the radius.

The tooth is calculated by dividing the circumference of a gear ($2\pi r$) by the number of teeth z. Equation (3.2) therefore results in:

$$z_R = z_S + 2z_P \tag{3.4}$$

Two further constraints are obtained by considering the surface velocities at the meshing points.

Obviously, the speed $v = r \cdot \omega$ of the teeth of the sun gear at the common meshing point must be the same as the speed of the teeth of the planet gear. This results in:

$$r_S\omega_S = (r_S + r_P)\omega_C - r_P\omega_P \tag{3.5}$$

The left-hand side of the equation describes the surface speed of the planetary gear. The surface speed of the planet gears results from the addition, with the correct sign, of the speed of the center of the planets $(r_S + r_P) \cdot \omega_C$ and the rotational speed of the planet gear itself, $r_P \cdot \omega_P$.

Similarly, the surface speeds in the tooth mesh between ring gear and planet gears can also be equated:

$$r_R\omega_R = (r_S + r_P)\omega_C + r_P\omega_P \tag{3.6}$$

The sign in the calculation of the planetary velocity is reversed. Inserting (3.5) into (3.6) results in:

$$-r_S\omega_S + 2(r_S + r_P)\omega_C - r_R\omega_R = 0 \tag{3.7}$$

Insertion of (3.2) results in:

$$-r_S\omega_S + (r_S + r_R)\omega_C - r_R\omega_R = 0 \tag{3.8}$$

Here, too, the radii can be exchanged with the numbers of teeth and, after a short transformation, the result is:

$$\omega_S - \frac{z_S + z_R}{z_S}\omega_C + \frac{z_R}{z_S}\omega_R = 0 \tag{3.9}$$

The standstill ratio (stand reduction, stall reduction) of a planetary gearbox is defined as:

$$i_{Stand} = -\frac{z_R}{z_S} \tag{3.10}$$

The standstill ratio is always negative due to the different direction of rotation of the ring gear (hollow gear) and sun gear.

Example 3.13—Gearbox of Toyota Prius

For the Toyota Prius, $z_{\mathrm{H}}=78$, $z_{\mathrm{S}}=30$ and $z_{\mathrm{ST}}=24$, resulting in a standstill ratio $i_{\mathrm{STAND}}=-2{,}6$. ◀

By substituting the standstill ratio in (3.9), we arrive at the Willis equation:

$$\omega_S - (1 - i_{Stand})\omega_C - i_{Stand}\omega_R = 0 \tag{3.11}$$

The Willis equation formulates a constraint relationship for the speeds of the three shafts ω_{S} (sun), ω_{C} (carrier) and ω_{R} (ring) of a planetary gearbox.

Note

In a planetary gearbox, any two shafts can be made to rotate at a freely adjustable speed. The speed of the third shaft is then determined by the Willis equation.

Now let's look at the torques of the three shafts of a planetary gearbox.

Similar to the equilibrium of forces on a lever arm, the torque equilibrium must apply to the entire gearbox:

$$T_S + T_C + T_R = 0 \tag{3.12}$$

The forces ($F = T/r$) at the two opposite meshing points of each planetary gear must also be equal:

$$\frac{T_R}{r_R} = \frac{T_S}{r_S} \tag{3.13}$$

From this follows:

$$T_R r_S = T_S r_R \tag{3.14}$$

Since all teeth must be the same size, i.e. have the same module, the same relationship also applies to the number of teeth instead of the radii:

$$T_R z_S = T_S z_R \tag{3.15}$$

If (3.15) is inserted into (3.12), the following two equations are obtained:

$$T_S = -\frac{z_S}{z_S + z_R} T_C \tag{3.16}$$

$$T_R = -\frac{z_R}{z_S + z_R} T_C \tag{3.17}$$

These equations represent two constraint relationships for the torques of the three shafts of the planetary gearbox.

Note
In a planetary gearbox, the torque of one shaft can be freely specified. The torques of the other two shafts are then determined by the ratio of the number of teeth.

The degrees of freedom of the planetary gearbox are utilized in the power-split hybrid.

The characteristics of a power-split hybrid drive with a planetary gearbox are presented below, in particular regarding power, speed, and efficiency in different operating conditions. We will first look at the simple case in which the battery is switched off, i.e. does not recharge or deliver any power.

3.5.8 Input-Split Hybrid

We first consider the widely used input-split variant, i.e. the planetary gear set is located directly behind the combustion engine and splits the power flow into a mechanical and an electrical branch, see Fig. 3.11. This variant is used in the Toyota Prius, for example.

The following assignment of the drivetrain components to the components of the planetary gearbox then applies:

- The sun gear S is connected to the electric generator (electric machine 1 (generator)),
- the carrier C is connected to the internal combustion engine and
- the ring R is connected to the electric axle drive (electric machine 2 (motor)) and to the axle itself.

By adjusting the speed of the electric machine 1, a gearbox with variable transmission is created between the ICE and the axle drive (differential):

The crankshaft speed of the ICE (planet carrier) can be variably adjusted by changing the speed of the electric machine 1 (sun gear) without changing the speed of the ring gear, that is, at constant driving speed.

On the other hand, different driving speeds are possible at constant crankshaft (ICE) speed by changing the speed of the electric machine 1.

3.5.8.1 Forward Power Flow

The power of the combustion engine P_{ICE} is split into an electrical and a mechanical branch by the planetary gearbox.

Only minor losses occur in the mechanical path, mainly due to bearing friction and in any additional gearboxes that may be present. We designate the efficiency of this branch as η_{mech}. It is typically 95% or higher.

In the electrical path, losses occur in the first and second electric drive, each of which consists of an electric machine and the associated power electronics (η_{E1} and η_{E2}). The intermediate storage of electrical energy in a battery, which results in further losses, is not considered here.

The following power flow results from the power splitting factor ε:

Depending on the operating point, electric machines achieve an efficiency of between 80 and 95%. This does not include low partial load. An inverter (AC/DC converter) is around 90–95%. This results in around 75–90% for both η_{E1} and η_{E2}. The overall efficiency in the electrical branch is then typically in the order of $\eta_{E1} \cdot \eta_{E2} = 55–80\%$.

ε_{EL} is the ratio of the power in the electrical path to the total power, i.e. the power of the ICE:

$$\varepsilon_{el} = \frac{P_{el}}{P_{ICE}} = \varepsilon \cdot \eta_{E1} \tag{3.18}$$

Then the overall efficiency, see Fig. 3.13, results in:

$$\eta_{total} = \frac{P_{wheel}}{P_{ICE}} = \left(1 - \frac{\varepsilon_{el}}{\eta_{E1}}\right) \cdot \eta_{mech} + \varepsilon_{el} \cdot \eta_{E2} \tag{3.19}$$

Due to the partial direct mechanical transmission, the efficiency is better than a serial hybrid but worse than a parallel hybrid.

Equation (3.19) shows that the overall efficiency is largely dependent on the power distribution ε_{el}. If no power flows via the electrical branch ($\varepsilon = 0 \rightarrow \varepsilon_{el} = 0$), then the overall efficiency is equal to η_{mech} and therefore relatively high.

If all the power flows via the electrical branch ($\varepsilon = 1 \rightarrow \varepsilon_{el} = \eta_{E1}$), then the overall efficiency is equal to $\eta_{E1} \cdot \eta_{E2}$ and therefore significantly lower than before.

The transmission ratio of the mechanical branch, i.e. the gear ratio, is calculated as follows:

$$i = \frac{\omega_{ICE}}{\omega_{wheel}} \tag{3.20}$$

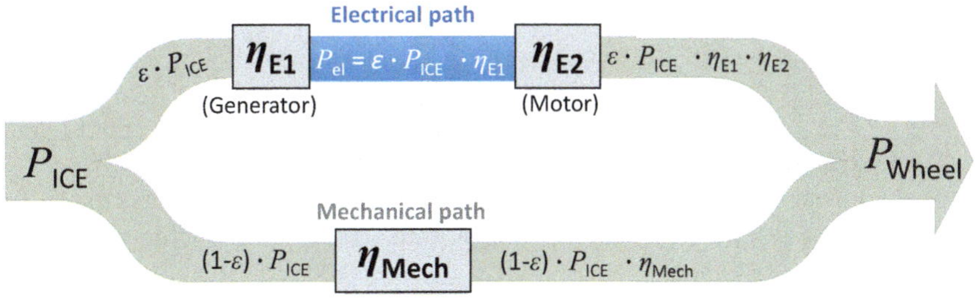

Fig. 3.13 Power flows in the power-split hybrid drive ($\varepsilon > 0$)

The Willis Eq. (3.11) gives the required speed of the first electric machine for any desired transmission ratio:

$$\omega_{E1} - (1 - i_{Stand})\omega_{ICE} - i_{Stand}\omega_{wheel} = 0 \tag{3.21}$$

This means that the gear ratio i can be varied freely within certain limits by varying the speed of the first electric machine (generator).

However, the distribution of the power flow between the electrical and mechanical paths is then also given by the desired gear ratio i.

Equation (3.16) results in:

$$T_{E1} = -\frac{z_S}{z_S + z_R} T_{ICE} \tag{3.22}$$

This means that the mechanical power fed into the first electrical machine by the planetary gearbox becomes:

$$P_{E1} = T_{E1} \cdot \omega_{E1} = -\frac{z_S}{z_S + z_R} T_{ICE} \cdot \left[(1 - i_{Stand}) + \frac{i_{Stand}}{i} \right] \cdot \omega_{ICE} \tag{3.23}$$

Simplified:

$$P_{E1} = -\frac{z_S}{z_S + z_R} \cdot P_{ICE} \cdot \left(1 - i_{Stand} \cdot \frac{i-1}{i} \right) \tag{3.24}$$

The power in the electrical branch is:

$$P_{el} = -P_{E1} \cdot \eta_{E1} \tag{3.25}$$

The negative sign results because the power P_{el} is counted as positive when it flows into the electrical branch. For electric motor 1, however, this means generator operation, i.e. negative power P_{E1}.

Therefore:

$$\varepsilon_{el} = -\frac{P_{E1}}{P_{ICE}} \cdot \eta_{E1} \tag{3.26}$$

Finally follows:

$$\varepsilon_{el} = \frac{z_S}{z_S + z_R} \cdot \left(1 - i_{Stand} \cdot \frac{i-1}{i} \right) \cdot \eta_{E1} \tag{3.27}$$

Example 3.14—Power split operation of the Toyota Prius

In the power-split Toyota Prius Hybrid, the speed of the electric machine 1 can be varied between −4000/min and +6000/min. The electric machine 2 reaches a top speed of 6000/min.

If the ICE rotates at 2000/min, this results in a transmission ratio range of $i = 0.60$ to $i = 4.33$ and thus a spread of 7.2.

At an ICE speed of 4000 min, the transmission ratio range only extends from $i=0.66$ to $i=1.24$, as the speed of EM1 must not be faster than 6000/min. The spread is now 1.9.

At the highest speed of the ICE ($n=6000$/min), the electric machine 1 must be operated constantly at maximum speed ($n=6000$/min). The only possible gear ratio is now $i=1.0$, which means that the output also rotates at the same speed.

In the Prius, further fixed gear stages are used to adapt the planetary ring speed to the wheel speed. ◀

3.5.8.2 Best Efficiency

At the point of best efficiency, no power flows via the electrical branch, i.e. $\varepsilon_{el}=0$. The corresponding gear ratio results from (3.27):

$$i_{Pel=0} = \frac{i_{Stand}}{i_{Stand} - 1} \tag{3.28}$$

For this operating point, the speed of the generator is zero, as the insertion of (3.28) in (3.20) and (3.21) immediately shows. This means that the electrical power ($P=T\cdot\omega$) is zero, as desired. However, this does not mean that the torque is also zero. Rather, the torque results from (3.22). The electric machine must therefore apply a holding torque to maintain this transmission ratio.

3.5.8.3 Negative Reactive Power

If the speed of the electrical machine 1 becomes negative, it works as a motor and draws its electrical power from machine 2, which thus becomes a generator.

The power of the ICE is thus fed backwards into the circuit and ε becomes negative. This operation is also referred to as "negative reactive power". Please note that this term refers purely to the mechanical power flow and is not related to electrical reactive power (Fig. 3.14).

The following now applies to the proportion of power in the electrical branch:

$$\varepsilon_{el} = \frac{P_{el}}{P_{ICE}} = \frac{\varepsilon}{\eta_{E1}} = \frac{z_S}{z_S + z_R} \cdot \left(1 - i_{Stand} \cdot \frac{i-1}{i}\right) / \eta_{E1} \tag{3.29}$$

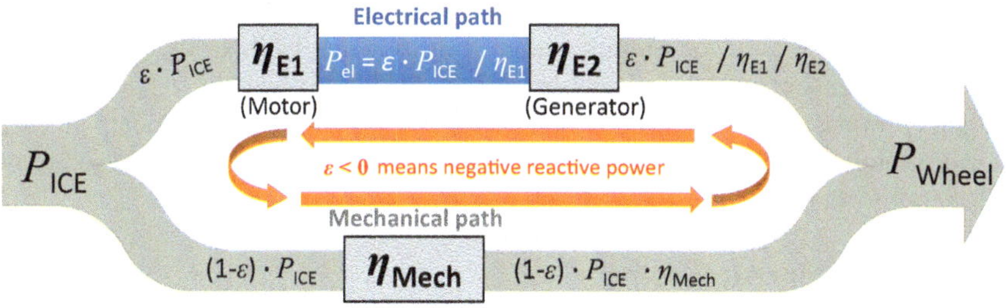

Fig. 3.14 Power flows in the power-split hybrid drive ($\varepsilon<0$)

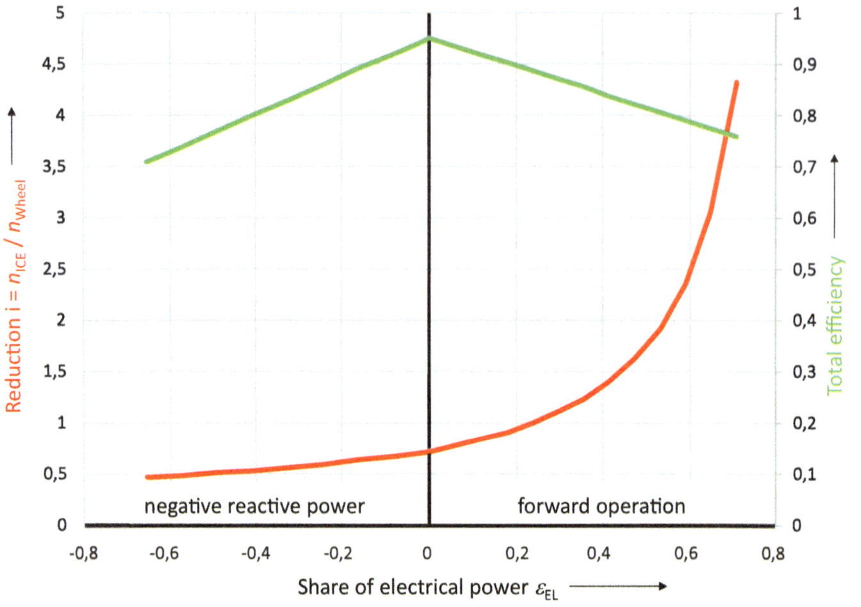

Fig. 3.15 Gear ratio i (red) and overall efficiency (green) over the proportion of electrical power

The overall efficiency in this operating mode is calculated from:

$$\eta_{total} = \frac{P_{wheel}}{P_{ICE}} = (1 - \varepsilon_{el} \cdot \eta_{E1}) \cdot \eta_{mech} - \varepsilon_{el}/\eta_{E2} \tag{3.30}$$

The efficiency decreases rapidly in the negative reactive power range with increasing power in the electrical path ε_{el}. This range can therefore only be utilized to a limited extent.

These relationships can be illustrated nicely in a diagram. It was calculated for the Toyota Prius at an ICE speed of 2000/min, $\eta_{E1} = \eta_{E1} = 0.85$ and $\eta_{mech} = 0.95$ (Fig. 3.15).

3.5.8.4 Starting

At low ICE speeds, the possible reduction range of the planetary gearbox extends to i = infinity, i.e. until the output shaft comes to a standstill. In the Toyota Prius, for example, this operation is possible for $n_{ICE} < 1700$/min.

The power-split transmission thus replaces both the starting clutch and the starter motor. The ICE is simply towed up by the electric machine 2, which draws its energy from the battery.

3.5.8.5 Pure Electric Driving

Electric driving is possible when the combustion engine is switched off ($n_{ICE} = 0$). The required energy is supplied by the battery.

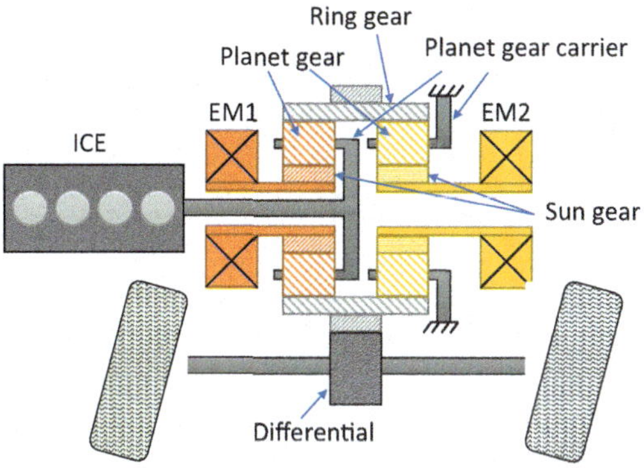

Fig. 3.16 Schematic layout of the drive structure of the Toyota Prius I–III

However, both electric machines must be running. Machine 1 is operated at negative speeds so that machine 2 generates positive speeds. The combustion engine and therefore the planetary gear carrier (C) are at a standstill.

The speed range of machine 2 and thus also the top speed is limited, because machine 1 would otherwise have to rotate too quickly. For example, the Toyota Prius can reach a maximum speed of 40 km/h with a purely electric drive when the ICE is stationary.

Higher purely electrically driven speeds are possible in principle if the switched-off ICE is being towed ($n_{ICE}>0$) by the electric machine 1. Obviously, a towed ICE will generate a strong friction torque and high losses. In practice, this operating mode is not used because the efficiency is poor.

Example: Toyota Prius Drive Structure
The Toyota Prius does not couple the second electric motor directly to the output drive train (ring gear), but via another planetary gearbox, which can be seen in the picture on the right (Fig. 3.16).

This gearbox has a fixed planetary gear carrier. It therefore does not work in a power-splitting manner, but like a conventional transmission gearbox with two shafts. The reason for this design is to increase the speed of the electric machine 2, which would otherwise have to run at the output speed. Due to the increased speed, the machine can be built more compactly and has a correspondingly higher power density.

3.5.9 Output-Split Hybrid

As an alternative to the input-split hybrid, in which the mechanical power of the ICE is split, the planetary gearbox can also be used to combine the power of the two drivetrains at the final stage. This results in the so-called output split hybrid (Fig. 3.17).

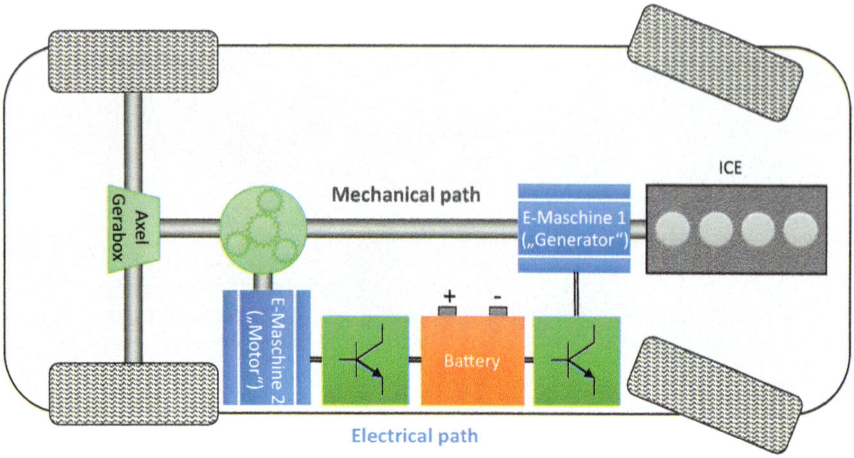

Fig. 3.17 Structure of a power-split hybrid drive (output split)

The following arrangement applies to this design:

- The sun gear S is connected to the electric machine 2,
- the planetary gear carrier (C) is connected to the axle, and
- the ring gear (R) is connected to the electric machine 1 and the combustion engine.

The output split is only used in a few vehicles. One example is the Chevrolet Volt/ Opel Ampera.

Example: Chevrolet Volt Drive Structure

The separation and classification into serial hybrids, parallel hybrids, and power-split hybrids are not always possible in practice. There are drive structures that can be operated either as serial, parallel, or power-split hybrid drive by the application of clutches and brakes.

A prominent example is the Chevrolet Volt, which is a mixture of a battery electric vehicle (BEV), serial and power-split hybrid (Fig. 3.18).

Depending on the position of the two clutches C2 and C3 and the brake C1, four different operating modes can be realized (Table 3.3):

Electric mode 1 is used at low speeds when the battery is sufficiently charged. The combustion engine and electric machine 1 are switched off. The vehicle runs on battery power with the electric machine 2.

At higher loads or higher speeds, electric mode 2 is selected. Brake C1 now opens and clutch C3 closes. This allows both electric machines to drive the vehicle.

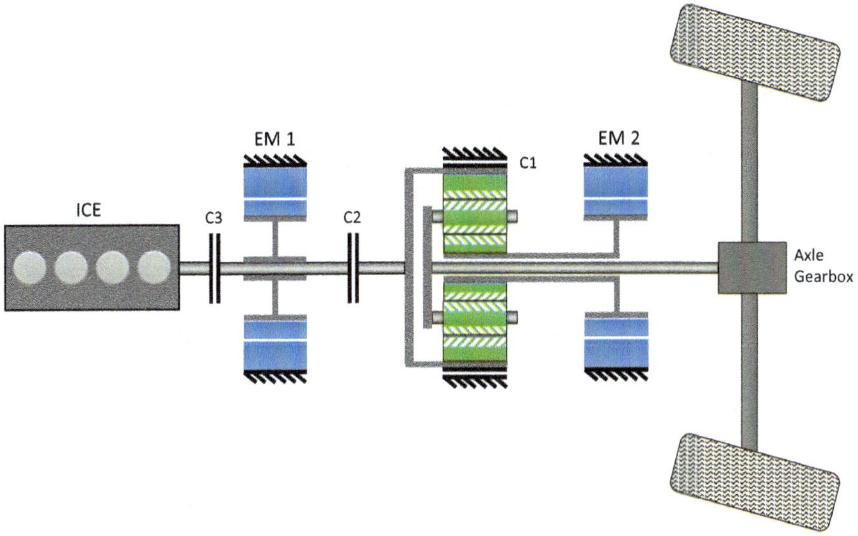

Fig. 3.18 Schematic layout of the drive structure of the Chevrolet Volt

Table 3.3 Hybrid and electric operating modes of the Chevrolet Volt

		Clutch C3	Clutch C2	Brake C1
EV-Mode 1	Single-motor BEV mode	Open	Open	Closed
EV-Mode 2	Twin-motor BEV mode	Open	Closed	Open
Range Extender Mode 1	Serial hybrid mode	Closed	Open	Closed
Range Extender Mode 2	Output-split hybrid mode	Closed	Closed	Open

The combustion engine switches on as soon as the SOC (state of charge of the battery) falls below a certain threshold value. The clutch C3 is closed for this purpose. At low speeds and low torques, the system operates in serial mode: the ICE generates electrical energy with the aid of the electric machine 1. The energy is used by the electric machine 2 to drive the vehicle. The clutch C2 is open and the brake C1 is applied.

Power-split mode is used for higher speed or torque requirements. Clutches C2 and C3 are closed so that the combustion engine can act directly on the output-side transmission together with the two electric machines (output-power split). Of course, brake C1 must then be open.

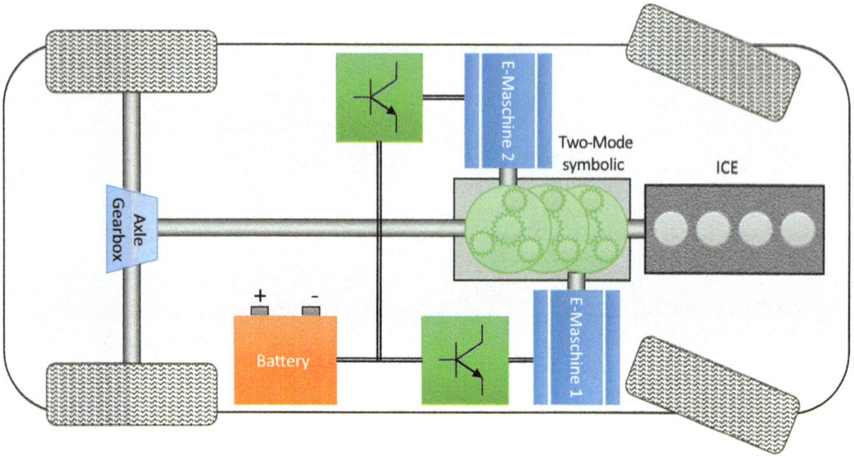

Fig. 3.19 Structure of a power-split hybrid drive (compound split)

3.5.10 Compound-Split Hybrid

The compound split hybrid, also known as a two-mode or dual-mode hybrid, uses multiple planetary gearboxes with multiple power splits (Fig. 3.19).

The usable transmission ratio range of the two-mode hybrid is larger than that of the one-mode hybrids discussed so far.

Both electric machines run through zero speed several times in the operating range, which means no power in the electrical path. This results in better efficiency over a wider transition range compared to one-mode hybrids. In addition, the power transfer in the electrical path is smaller, which means that the e-machines can be built smaller.

Additional clutches and brakes are often installed. They enable fixed mechanical ratios, like a manual gearbox. These ratios can be selected by the driver in certain operating modes ("sport mode").

The disadvantage of compound split hybrids is the complex design of the gearbox. In addition, there are various electromechanical actuators and a complex control system.

Two-mode hybrids have therefore not become established in the passenger car sector. However, this drive variant is still being pursued for commercial vehicles, especially buses.

Example: Daimler ML 450 Hybrid Drive Structure

A few years ago, there was a joint project between General Motors, Mercedes Benz, and BMW, from which the two-mode hybrid drives of several large SUV vehicles (BMW X6 Hybrid, Chevrolet Tahoe Hybrid, Daimler ML 450 Hybrid, and GMC Yukon Hybrid) emerged. However, the project has since been discontinued and the vehicles are no longer built with this drive train.

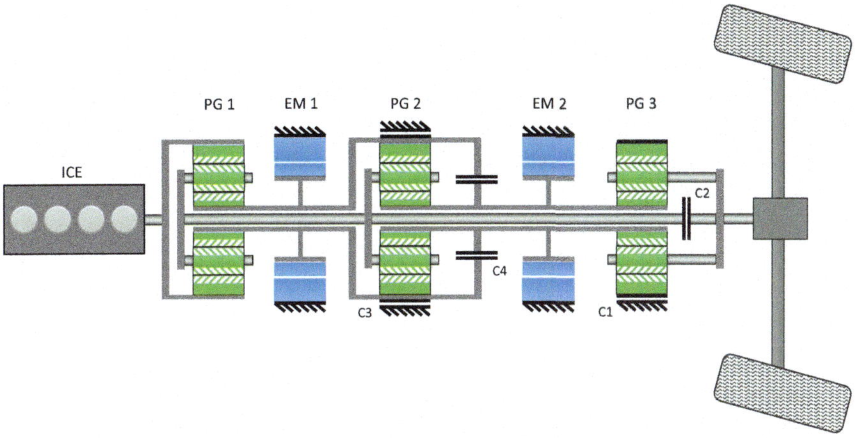

Fig. 3.20 Schematic layout of the drive structure of the Mercedes Benz ML 450 Hybrid

The Mercedes Benz ML 450 Hybrid has a power-split two-mode hybrid drive with three planetary gear sets and two electric motors. The output and torque of the two electric motors are 67 kW/250 Nm and 63 kW/280 Nm, respectively.

The two planetary gearboxes PG1 and PG2 are required for the two-mode compound power split. The third planetary gearbox is only used to combine the power of the ICE and the two electric motors. It also generates an additional transmission ratio.

The first clutch (brake) C1 activates the input split mode by holding the ring gear of the third planetary gear. The rotating clutch C2 activates the compound split mode. It connects the planetary gear carriers of the two first gearboxes with the output.

If both clutches are activated simultaneously, the transmission ratio of PG3 is no longer variably adjustable but is fixed by the number of teeth.

The two additional clutches/brakes C3 and C4 are required to enable three further fixed ratios by combining the gear wheels of PG2 and PG3 in different ways (Fig. 3.20).

This results in the following operating modes: two continuously variable modes (EVT 1 and 2) and four gears with a fixed transmission ratio (FG 1–4).

In addition, a third, smaller electric motor is coupled directly to the crankshaft of the combustion engine (not shown in the picture). It serves as a starter and 12 V generator but is not part of the power-split drive concept (Table 3.4).

Table 3.4 Operating modes of the Mercedes Benz ML 450 Hybrid[4]

		Brake C1	Coupling C2	Brake C3	Coupling C4
EVT Mode 1	EVT 1	On			
EVT Mode 2	EVT 2		On		
Fixed Gear 1	FG1	On			On
Fixed Gear 2	FG2	On	On		
Fixed Gear 3	FG3		On		On
Fixed Gear 4	FG4		On	On	

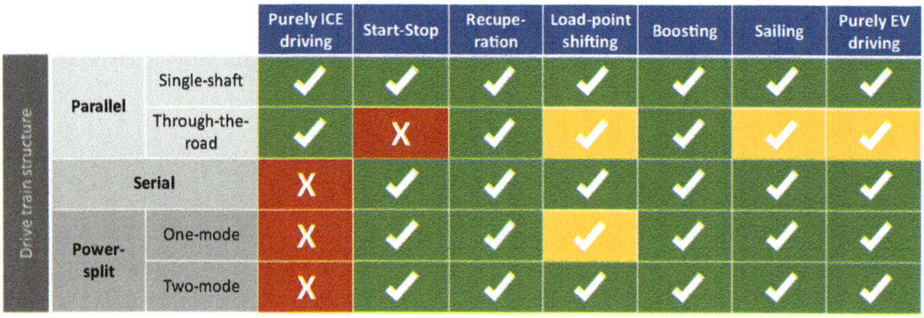

Fig. 3.21 Hybrid operating modes and typical drivetrain topologies

3.5.11 Summary Hybrid Drives

A distinction is made between serial, parallel, and power-split hybrid drive structures.

A well-known variant of the serial hybrid drive is the range extender, which, however, is not optimal in terms of efficiency and consumption.

With the parallel hybrid drive, there are a variety of options for positioning the electric motor. It can be in front of the internal combustion engine, between ICE and gearbox or close to the axle. Each placement has specific advantages and disadvantages.

The power-split hybrid drive is the most technically complex, but also the most flexible solution. The power-split planetary gearbox is essential for its function.

The degree of hybridization, i.e. hybrid functionality, and drive arrangement span a range of possible implementations. Theoretically, many combinations are conceivable, but only some of them are technically and economically viable, as the following tables show (Figs. 3.21 and 3.22).

[4] EVT = Electric continuously variable transmission.

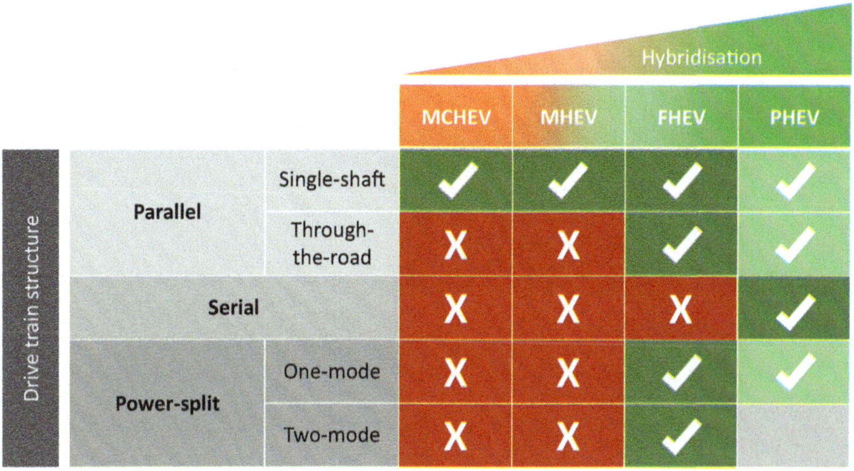

Fig. 3.22 Hybrid vehicle classes and typical powertrain topologies

Driving Resistances and Energy Consumption

4

Abstract

The driving resistance force is made up of various components. Together with the speed, this results in the required drive power. In electric cars, part of the energy used during braking can be recovered by braking energy recovery. The chapter introduces the most common driving cycles worldwide and describes how the energy consumption of hybrid vehicles is measured. Finally, the CO_2 efficiency classes of passenger cars are presented.

4.1 Driving Resistance Force

The traction power P_t required for a specific operating condition is the sum of the forces acting on the vehicle multiplied by the driving speed.

The traction force F_t of a car is equal to the driving resistance force. This is made up of four components:

- Aerodynamic resistance F_a
- Rolling resistance F_r
- Climbing resistance F_c
- Acceleration resistance F_{acc}

$$F_t = F_a + F_r + F_c + F_{acc} \tag{4.1}$$

These variables depend on the driving condition, i.e. the speed and acceleration, and are also partly dependent on external influences, for example, the road conditions, local wind conditions, etc.

The required traction power results from the traction force F_t and the driving speed v:

M. Doppelbauer, *Introduction to Electromobility*, https://doi.org/10.1007/978-3-658-45482-1_4

$$P_t = F_t \cdot v \tag{4.2}$$

The drive power P_d is calculated from the traction power P_t plus the losses in the drivetrain and plus the power required to operate the auxiliary units.

The wheel torque T_{wheel} can be calculated from the traction force:

$$T_{Wheel} = \frac{1}{2} \cdot F_t \cdot r_{dyn} \tag{4.3}$$

The pre-factor ½ applies to two-wheel drives. For all-wheel drives, ¼ must be used.

r_{dyn} is the dynamic wheel radius, which is slightly smaller than the theoretical wheel radius depending on the flattening, i.e. wheel pressure and load.

4.1.1 Aerodynamic Resistance

The force required to overcome the air resistance (drag force) can be calculated using the following formula:

$$F_a = \frac{1}{2} \rho_A \cdot A_f \cdot c_W \cdot v_{rel}^2 \tag{4.4}$$

with

ρ_A = air density in kg/m^3
A_f = front surface in m^2
c_W = drag coefficient, typical 0,25 up to 0,50
v_{rel} = relative speed of vehicle to wind speed in m/s

The air density depends on the temperature and altitude. Under standard conditions, that is 15 °C and an air pressure of 1.013 bar, the air density is 1.225 kg/m^3.

On a hot summer's day, the air density decreases by 6–7%. Especially when driving fast on the highway, when air resistance accounts for the largest part of the energy requirement, energy consumption decreases accordingly.

The front surface area, also known as the effective area, can be determined by placing the vehicle in front of a wall and then illuminating it with parallel light over a large area. The shadow cast on the wall corresponds to the desired area.

The drag coefficient is often stated in the technical data of vehicles. It should be noted that only the product of the drag coefficient and the frontal area is relevant for air resistance and not the drag coefficient alone. This combination is often called "cwA". For this reason, large vehicles (SUVs) always have a high aerodynamic resistance, even if their drag coefficient c_W is relatively low.

Table 4.1 Drag coefficients and front surfaces of various cars in comparison[1]

Car	Drag coefficient c_W	Front surface A_f	Effective air resistance surface $c_W A = c_W \cdot A_f$
Tesla Model S Plaid	0.208	2.34 m^2 (from old model)	0.49 m^2
Tesla Model 3 (Facelift 2023-)	0.219	2.27 m^2	0.50 m^2
Porsche Taycan	0.22	2.33 m^2	0.51 m^2
Tesla Model 3 (2017–2023)	0.225	2.27 m^2	0.51 m^2
Mercedes EQE	0.22	2.48 m^2	0.55 m^2
VW ID.7	0.23	2.46 m^2	0.57 m^2
VW ID.3	0.27	2.36 m^2	0.63 m^2
VW e-up	0.31	2.08 m^2	0.64 m^2
Polestar 2	0.278	2.34 m^2 (estimation)	0.65 m^2
Tesla Model X	0.24	2.82 m^2	0.68 m^2
BMW i3	0.29	2.38 m^2	0.69 m^2
BMW iX	0.25	2.81 m^2	0.70 m^2
Audi e-tron 55	0.28	2.65 m^2	0.74 m^2
Renault Zoe 2017	0.33	2.28 m^2	0.75 m^2
Mercedes EQC	0.29	2.67 m^2	0.77 m^2
Audi Q7	0.33	2.86 m^2	0.95 m^2
Porsche Cayenne	0.39	2.39 m^2	1.08 m^2
Mercedes G-class	0.54	2.97 m^2	1.60 m^2

Example 4.1—Historical development of drag coefficients

The drag coefficient of cars has become increasingly better in recent decades. Some 20 or 30 years ago, $c_w = 0.4$ or higher was quite normal.

Today, even monstrous vehicles like the Audi Q7 have a drag coefficient of 0.33. Due to the large size, however, its cwA value is quite poor at 0.95 m^2. Another negative example is the Porsche Cayenne with a cwA value of 1.08 m^2.

On the other end of the spectrum are cars like the Porsche Taycan (cwA = 0.50 m^2) and Tesla Model S (cwA = 0.49 m^2). At the same highway speed, these cars require less than half the driving power and consume less than half of the energy needed for a Porsche Cayenne (Table 4.1). ◄

[1] Source: Manufacturer's data.

4.1.2 Rolling Resistance

The force to overcome the rolling resistance is calculated from:

$$F_r = c_r \cdot m_v \cdot g \cdot \cos\alpha \tag{4.5}$$

with

c_r = rolling resistance coefficient (depending on tire and speed approx. 0.01–0.04)
m_v = vehicle mass in kg
g = gravitational acceleration = 9.81 m/s^2
α = slope of the road

Rolling resistance increases only moderately up to a speed of 100–120 km/h but increases significantly (roughly quadratically) at higher speeds. This is partly due to the increasing aerodynamic resistance, but also to the increase in tire vibrations.

Higher pressure in the tires improves the rolling resistance coefficient but reduces driving comfort.

The gradient of the road is included in the rolling friction by the sine function. On steeply inclined roads, only a part of the vehicle mass rests on the tires, which reduces friction.

4.1.3 Fuel Economy at Higher Speeds

At higher speeds, i.e. on the highway, the drag force predominates. The air resistance force increases quadratically with the speed, the power even cubically. At a speed of 200 km/h, a vehicle, therefore, requires around 8 times as much energy as at 100 km/h, if rolling resistance and acceleration are approximately neglected.

However, the vehicle also reaches its destination faster at higher speeds, so that the traveling time is reduced. The energy requirement therefore only increases quadratically with speed. In the example, the car would consume 4 times as much energy at 200 km/h compared to 100 km/h when traveling from one place to another.

However, these considerations only take into account the mechanical power and energy that must be supplied to the wheels. If the consumption of fuel or electrical energy is calculated, then the load-dependent efficiency of the engine or motor and battery or fuel cell system becomes important as well.

In general, the higher the load, the better the efficiency of a combustion engine. Only in the highest load range does the efficiency drop slightly again. For example, if a car is driving at 200 km/h, the combustion engine is at its most efficient, whereas at 100 km/h it is less efficient. As a result, fuel consumption at high speeds is not increased proportionally to the energy required at the wheels, but rather disproportionately. At twice the speed and 4 times the energy requirement, for example, only 3–3.5 times more fuel is consumed.

In a battery electric vehicle, on the other hand, the energy that can be drawn from the battery is reduced if the battery is operated at a high discharge rate. In addition, the efficiency of the entire drive is slightly reduced due to the high current. For example, if the wheels require 4 times the energy, the battery will have to supply 4.5 times the energy. The range of an electric car therefore decreases more at high speeds than in vehicles with a combustion engine. This is why most electric cars are limited in their top speed.

This effect is even more pronounced in vehicles with fuel cells than in battery electric vehicles because the fuel cell loses a great deal of efficiency under high loads. Fuel cell vehicles, therefore, lose a disproportionate amount of range at high speeds on the highway, even more so than battery electric cars.

Example 4.2—Driving power at higher speeds

A single car driving 160 km/h on a highway consumes about as much power as four cars driving 100 km/h. ◀

Example 4.3—Loss of range at higher speeds

The author has made a direct comparison of a BEV (Tesla Model 3) and a FCEV (Toyota Mirai, 1st gen). For this purpose, a circular route of around 50 km was covered with both vehicles in direct succession on a level highway in the Rhine Valley in southern Germany.

At constant 130 km/h, the Tesla consumes 21.5 kWh/100 km and the Toyota 1.3 kg_{H2}/100 km, which is equivalent to 43.3 kWh/100 km.

When the speed was increased to 160 km/h, the Tesla consumed 28.8 kWh/100 km (+34%) while the Toyota consumed 2.0 kg_{H2}/100 km (+54%). This is equivalent to 66.6 kWh/100 km—more than twice as much as the Tesla. Compared to their battery capacity and hydrogen storage, both cars now have a similar highway driving range of about 250 km.

Note that the increase in energy consumption for both vehicles was slightly less than quadratic with increased speed. This is due to a noticeable part of rolling friction in the total losses and rolling friction only increases linearly with speed. At even higher speeds, the proportion of rolling friction decreases, and the fuel economy develops a more and more quadratic behavior. ◀

Example 4.4—Fuel economy on the highway

Even cars with a higher effective air resistance surface, like the Polestar 2 (LRDM-MY23), can be driven quite efficiently at lower speeds. The author has made a 310 km long trip in cold and rainy winter weather, mainly on the highway with cruise control set to 100 km/h. The average energy consumption after 3.5 h was 17.5 kWh/100 km. The same route was driven in the summer with 15.1 kWh/100 km.

However, at 200 km/h, the fuel economy is 4 times worse compared to 100 km/h. For example, a Mercedes-Benz EQE had an average consumption of 62.2 kWh/100 km after around half an hour of highway driving at an average speed of 192 km/h. ◀

4.1.4 Climbing Resistance

The following formula applies to the force required to overcome a slope, the so-called downhill force:

$$F_c = m_v \cdot g \cdot \sin\alpha \tag{4.6}$$

with

m_v = vehicle mass in kg
g = gravitational acceleration = 9.81 m/s²
α = slope of the road

The gradient resistance goes with the sine function of the gradient angle. The steeper the road, the higher the gradient resistance. The gradient of the road is given as a percentage. A 100% gradient corresponds to a gradient angle of 45° or π/4.

The gradient resistance is not considered in all normative driving cycles, i.e. the determination of the energy requirement relates purely to driving on a flat surface.

4.1.5 Acceleration Resistance

The acceleration force is determined using the following formula. It also applies equally to deceleration. The sign of the force is then naturally reversed.

$$F_{acc} = (m_v + m_{acc}) \cdot \frac{dv}{dt} \tag{4.7}$$

with

m_v = vehicle mass in kg
m_{acc} = equivalent acceleration mass in kg
v = velocity in m/s

In addition to the linear acceleration of the vehicle's mass m_v, all rotating masses must also be brought to a higher rotational speed when the vehicle accelerates. The motor must generate the torque required for both effects.

This torque can be computed in a single formula (4.7) by introducing an equivalent acceleration mass m_{acc}, in which all inertia masses of the rotating parts in the drivetrain are included. In practice, the influences of gears and shafts are small, so that only the inertia of the wheels and the rotor of the electric motor must be considered.

The calculation of the equivalent acceleration mass of the wheels and the motor is based on the principle of conservation of energy. The result is:

$$m_{acc,W} = 4 \cdot \frac{J_W}{r_{dyn}^2} \tag{4.8}$$

$$m_{acc,Mot} = J_{Mot} \cdot \frac{i^2}{r_{dyn}^2} \tag{4.9}$$

with

$m_{\text{acc,W}}$ = equivalent acceleration mass of all four wheels in kg.
J_{W} = moment of inertia of the wheel including tire and rim in kg·m^2
$m_{\text{acc,Mot}}$ = equivalent acceleration mass of the motor in kg.
i = gear reduction from the motor shaft to the wheel shaft including axle reduction
r_{dyn} = dynamic wheel radius in m

The dynamic wheel radius is the radius of the unloaded wheel reduced by the value of the flattening due to the influence of the vehicle weight.

The moment of inertia J of a solid cylinder with the outer radius R, for example, the rim of a wheel or the rotor of an electric motor, is calculated from:

$$J = \frac{1}{2} m \cdot R^2 \tag{4.10}$$

The moment of inertia J of a hollow cylinder with inner radius r and outer radius R, for example a tire, is calculated from:

$$J = \frac{1}{2} m \cdot (r^2 + R^2) \tag{4.11}$$

4.1.6 Summary

Depending on the vehicle speed, air resistance, rolling resistance, and acceleration resistance account for a different proportion of the total drive power or drive force required (Table 4.2).

4.2 Energy and Power Flow

The following diagrams show approximate values for the typical energy flow of a car with an internal combustion engine. All percentages and the widths of the bars are related to the energy of a full fuel tank (Fig. 4.1).

Figure 4.2 shows a typical battery for electric vehicles.

Table 4.2 Typical breakdown of the components of a car's driving resistance

	Air resistance	Rolling resistance	Acceleration resistance
Constant 50 km/h	40%	60%	–
Constant 130 km/h	82%	18%	–
Inner-city (ECE)	12%	35%	53%
Country road (ECE)	48%	26%	26%

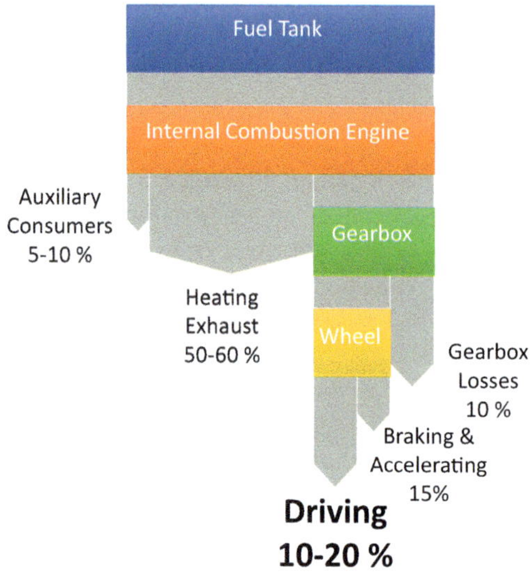

Fig. 4.1 Energy flow in an ICE car

All percentages in this figure are related to 100% energy content of the battery, the width of the bars corresponds to the previous figure. The higher percentages given for the auxiliary consumers result from the reference to the smaller amount of energy in the battery. The total energy consumption of auxiliary consumers is not different between an ICE and BEV vehicle.

The amount of energy from the battery that can be used for propulsion is around 2–4 times higher compared to ICE due to the better efficiency of the electric drive system and braking energy recovery.

Roughly speaking, a battery must carry 2–3 times less energy to cover the same distance as an ICE vehicle.

4.3 Auxiliary Consumers

Auxiliary consumers are divided into **continuous consumers**, e.g. fans for cooling, engine management, **long-term consumers**, e.g. air conditioning compressor, windshield wipers, headlights, and **short-term consumers**, e.g. starter-generator, rear window heater.

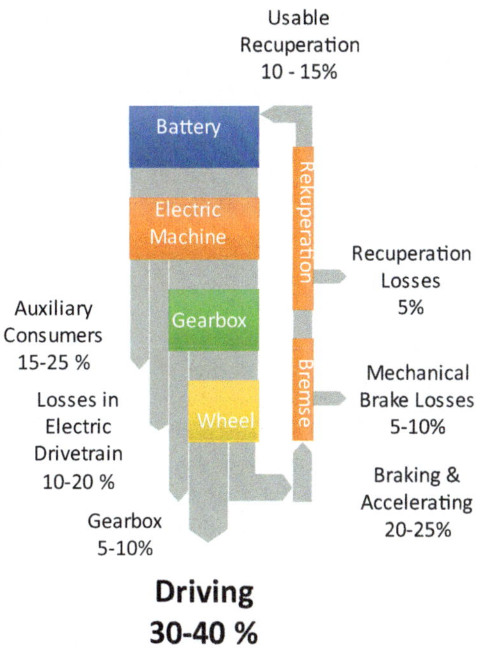

Fig. 4.2 Energy flow in a BEV car

4.3.1 Heating

Due to the low losses and therefor low waste heat of electrically powered vehicles, the heating of the cabin is a significant additional consumer compared to a vehicle with a combustion engine at low outside temperatures.

The required heating power can be several kW. Depending on the outside temperature, up to 6 kW and more are possible even for a passenger car. At temperatures below freezing, the total energy consumption of the car in kWh/100 km typically increases by 20–30% due to heating, even with a heat pump. Most of the energy is needed at the beginning to heat up the cold vehicle. This energy loss is therefore particularly noticeable on short journeys of less than 20 km. If possible, it is advisable to preheat the stationary vehicle while plugged in for charging.

PTC heating elements are the simplest and most cost-effective solution for heating. They have a low ohmic resistance when cold. As a result, a high electric current flow and the desired heat output is achieved quickly. However, the more PTCs heat up, the greater their resistance becomes, which reduces the current and therefore the heating output. Thanks to this self-regulation, additional control units or temperature sensors are often not required. The disadvantage of PTC elements is the high energy requirement and the associated lower range of the output.

Heat pumps are often used in modern electric cars. In heating mode, heat pumps extract heat from the environment and transfer it into the vehicle. In cooling mode, they release heat from inside the vehicle to the environment.

Instead of heating the entire air inside the car with a high energy input, heating can be limited to the contact surfaces of the body, i.e. seat and steering wheel heating. The comfort level is similar, especially for short journeys, but the required energy is much lower.

In many electrically powered buses, the heating requirement is so high due to the frequent opening of the doors that a gas or liquid fuel-powered heating unit is used on particularly cold days.

4.3.2 Heat Pumps

A heat pump works in the same way as a refrigerator: At low pressure, the working medium is vaporized and extracts heat from the environment. The gaseous working medium is then compressed by an electrically driven piston compressor. This increases the pressure and temperature. The working medium condenses and releases heat at a higher temperature level than before. Finally, the working medium is expanded in a throttle. In many cars, the compressor can be heard when running. The process is carried out continuously.

The heat output of a heat pump, typically some 2.5–3 kW, is significantly greater than the electrical energy consumption of the pump motor, typically around 1 kW, as thermal energy is extracted from the environment. However, heat pumps usually only work up to a certain ambient temperature, typically down to around −18 °C. At lower temperatures, a PTC high-voltage heater is still required.

Figure 4.3 demonstrates the improvement of driving range in cold weather that can be achieved with a heat pump compared to a PTC heater. Volkswagen claims a 20% increase in range at −7 °C ambient.

However, it should be noted that these values are very optimistic. In reality, the energy savings of heat pumps are not too noticeable, especially if one is prepared to rely primarily on seat and steering wheel heating instead of air heating.

R134a, R1234yf or R744 are used as refrigerants in heat pumps in cars, with the latter being practically pure CO_2. R744 is technically much more difficult to process in the refrigeration system but has a GWP value (Global Warming Potential) of only 1 when released. Also, it can be used in very low ambient temperatures, it is non-flammable and does not create any harmful vapors. The GWP value of the chemical refrigerant R134a is 1430 and that of R1234yf is 4. R1234yf can only be used at ambient temperatures of −7 °C or more.

4.3.3 Other Auxiliary Units

In electric vehicles, many auxiliary units that would otherwise be driven directly by the combustion engine via a belt have to be converted to electric drives, as the main electric motor does not turn when the vehicle is stationary.

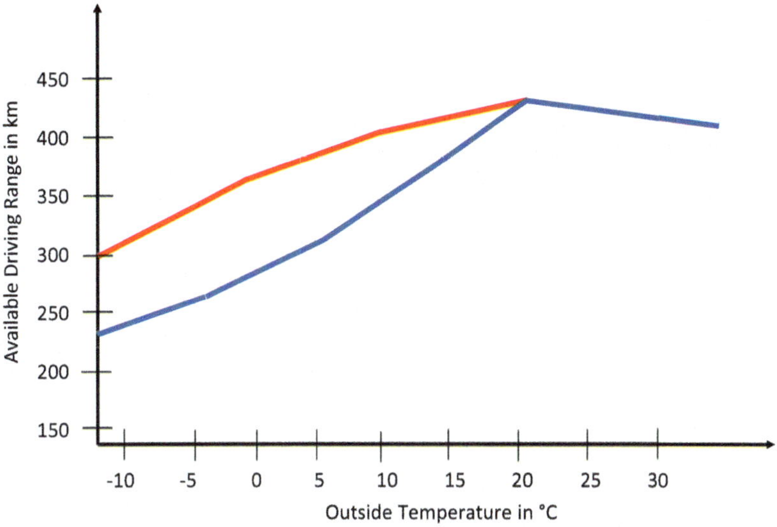

Fig. 4.3 Driving range of a VW ID.3 with (red) and without (blue) heat pump depending on ambient temperature[2]

This applies to

- Power steering → Electric/electrohydraulic steering
- Brake booster → Electric vacuum pump
- Air conditioning → Electric refrigerant compressor
- Oil support → Electric oil pump
- Coolant support → Electric coolant pump

The transition to electric auxiliary units is also required for full and plug-in hybrid vehicles that can drive short distances purely electrically.

Even in ICE vehicles, the electrification of auxiliary units can help save fuel, firstly because the efficiency of electric drives is better than that of belt drives and secondly, because electrically driven auxiliary units can be speed controlled. Auxiliary units that are connected to the combustion engine via belts, on the other hand, run continuously and proportionally to the speed of the engine, regardless of the current demand. One example is the air conditioning compressor, which can be adjusted precisely to the ambient temperature by an electric drive.

There is, therefore, a tendency to electrify more and more auxiliary units in modern ICE vehicles.

The user experience of electric auxiliary drives is also better. For example, with power steering, you want as much assistance as possible at low speed, but only a little assistance at high speed. This is only possible with electric power steering pumps. If the pump is connected to the combustion engine by a belt, it behaves in exactly the wrong way: the effect is weak at low speed and strong at high speed.

[2] Data source: Volkswagen AG.

The same applies to the heating system in cold weather. A car with an internal combustion engine takes a long time to heat up. An electric car has a PTC heater and heats up immediately. Almost all electric cars can be controlled remotely via a cell phone app and can thus be preheated in an environmentally friendly way without exhaust fumes before the start of the drive.

4.4 Braking Energy Recovery

Through braking energy recovery, a part of the energy used during acceleration can be recovered during braking. The recuperated energy can then be used later to drive the vehicle. In conventional vehicles with ICE, this energy is converted into heat at the wheel brakes and is therefore lost.

The maximum recuperable power is limited by the charging currents of the battery. Lithium-ion batteries allow significantly lower charging currents than discharging currents, especially at cold temperatures. Under such conditions, the braking energy recovery performance may be limited. In winter, this is noticeable in a change in driving behavior. The battery warms up while driving, so that the reduction in braking energy recovery performance decreases over the course of the journey.

A typical battery can accept a maximum charging current of 2–3 C. For example, a 60-kWh battery would be able to charge between 120 and 180 kW, depending on the state-of-charge and the temperature. If the required braking power is higher than the available braking energy recovery power of the battery, mechanical braking must be carried out.

It should be noted that the recuperated energy must pass twice through the efficiency chain from the wheel/road contact via the gearbox, motor/generator, power electronics to the battery, which increases the losses (Fig. 4.4).

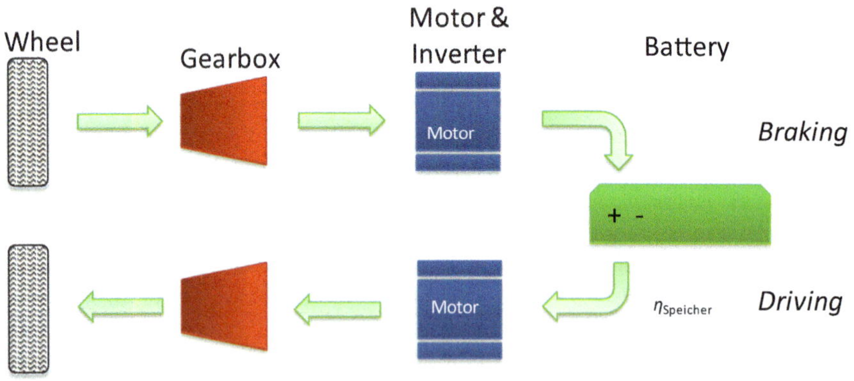

Fig. 4.4 Chain efficiency during braking energy recovery and subsequent use of braking energy

The whole chain efficiency of the recuperated energy, starting at the braking wheel, feeding into the battery, taken from the batteries, going through the electric drivetrain, and finally ending up at the driven wheel is:

$$\eta_{recup} = \eta_{wheel} \cdot \eta_{gear} \cdot \eta_{generator} \cdot \eta_{inverter} \cdot \eta_{battery} \cdot \eta_{inverter} \cdot \eta_{motor} \cdot \eta_{gear} \cdot \eta_{wheel} \tag{4.12}$$

The round-trip efficiency is typically only some 25%. In mixed operation of city, interurban, and highway driving, battery electric cars can be expected to increase their range by 10–15% through regeneration. This estimate is of course heavily dependent on the height structure of the terrain.

It should be noted that hydrogen cars cannot benefit from braking energy recovery unless they carry a large battery in addition to the fuel cell, as the conversion from hydrogen to electricity in the fuel cell is irreversible.

Example 4.5—Braking energy recovery in high mountains

When driving in the high mountains, braking energy recovery can have a much greater impact on range. The author drove an electric car from Pas-de-la-Casa in Andorra in the Pyrenees to Vernet-les-Bains, close to the sea in South France. At the end of the 90 km journey, the SOC of the battery was even slightly higher than at the beginning. However, the starting point is at an altitude of 2100 m, and the finishing point is at only 600 m. ◀

An electric machine can maintain its maximum torque for braking and even for holding the vehicle at a standstill. However, the recovered mechanical power P_{Rekup} decreases with decreasing speed v, even when a constant braking force F is applied, because $P = F \cdot v$.

Below a speed of 5–10 km/h, the recuperable mechanical power is no longer sufficient to cover the electrical losses of the electric machine and the inverter. To continue braking electrically at even lower speeds, electric energy must be supplied from the battery to cover the losses. This would be technically possible but is useless in terms of recovering braking energy.

This is why modern electric vehicles switch from electric to mechanical braking below a certain speed limit.

The brake pedal in many electric cars is designed as a complex electromechanical system. Its operation is automatically adjusted by an electric actuator depending on the driving situation. With constant pressure of the foot on the brake pedal, a control system regulates how much braking force is applied by the electric machine and how much by the mechanical brake, depending on the speed and desired braking torque. Ideally, this happens without the driver noticing, i.e. constant pressure on the brake pedal results in uniform deceleration.

In any case, there is an emergency braking function: if the brake pedal is pressed hard, the mechanical brakes are activated hydraulically, so that mechanical braking is always possible even if the on-board electrical system fails.

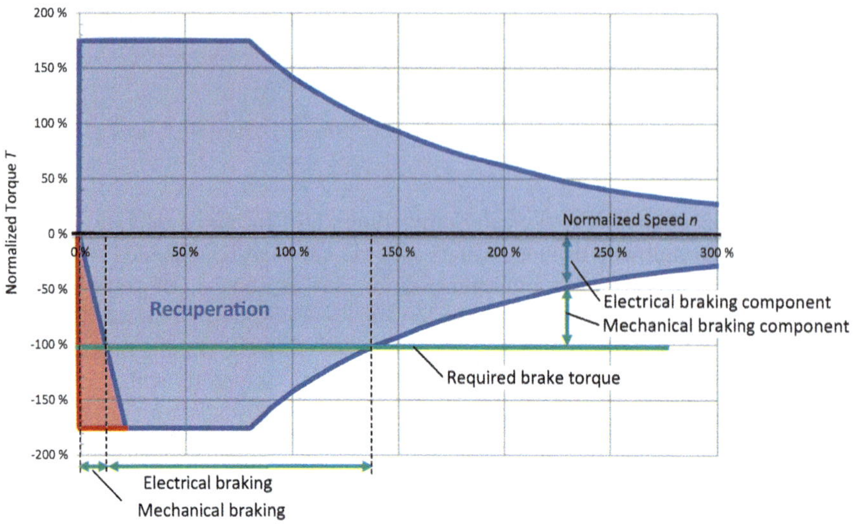

Fig. 4.5 Distribution of mechanical and electrical braking depending on the desired braking torque and speed

There are also some electric cars, which rely solely on one-pedal driving (see next chapter) and do not have an electromechanical braking system, for example, all cars of the Tesla company. The brake pedal of these vehicles works purely on a hydraulic basis, just like in a conventional combustion engine car.

> **Note**
> In terms of an electric car control system, the brake and power (gas) pedals can be thought of purely as setpoint generators. Depending on how far the pedals are depressed, a setpoint for the braking or motor torque is generated, which is then processed by the vehicle control system. The battery, inverter, and electric machine are controlled on this basis and the brake shoes are also activated hydraulically if required (Fig. 4.5).

4.4.1 One-Pedal Driving

Many electric cars offer so-called one-pedal driving: when the foot is lifted of the brake pedal, the car is not coasting but rather braking electrically (recuperating). Depending on the rate of deceleration if required by law, the brake lights are switched on automatically.[3] Some cars even come to a full stop when both pedals

[3]Typically, the brake lights are switched on automatically at decelerations above 1.3 m/s^2 and switched off below 0.7 m/s^2.

are not engaged, others start costing at very low speeds. In this way, the speed of the vehicle can be controlled in city traffic largely without having to switch the foot from the power to the brake pedal. After a short familiarization phase, most drivers find this a pleasant experience. Also, efficiency and driving range are improved thanks to the extensive use of braking energy recovery.

The rate of deceleration can be adjusted in many vehicles. Some cars have shifting paddles behind the steering wheel (for example Hyundai), others have two positions of the drive lever (typically marked "D" for coasting and "B" for braking/one-pedal driving, for example, Volkswagen) and yet other cars can be controlled in the setup menu (for example Polestar or Tesla).

Electric braking and one-pedal driving also significantly reduce brake wear and abrasion, i.e. particle pollution.

However, the greatly reduced usage of the mechanical brake can lead to new problems: As the friction discs and brake pads are rarely ever engaged, rust can accumulate resulting in reduced braking forces and ultimately in costly repairs. It is, therefore, recommended to actively use the mechanical brakes every couple of weeks to rub away any rust.

4.5 Driving Cycles

Precisely defined driving cycles are required for a comparable and reproducible determination of the energy consumption (fuel economy) of different vehicles and drive concepts. Driving cycles are used to determine the average drive power P_a, from which the specific consumption is then derived.

A standardized driving cycle specifies the conditions under which the vehicle must be operated to determine its energy consumption.

A speed-time profile specifies at what time and at what speed the vehicle should drive. In most cases, the required speeds are specified at 1-second intervals.

In addition, the operational boundary conditions of the vehicle are specified, for example, the operation of the air conditioning and heating, the tire pressure, attachments, payload, ambient temperature, etc.

Consumption is determined on a chassis dynamometer test bench so that comparable conditions can be created. Of course, the vehicle does not experience any air resistance on the chassis dynamometer and the rolling friction is also different from normal driving conditions. For this reason, the speed-dependent rolling and air resistance of the vehicle is determined in advance on the road. These data are then programmed into the brake roller control of the test bench.

The energy consumption of electric vehicles can be determined directly by measuring the current and voltage or the battery charge level.

In vehicles with internal combustion engines, fuel consumption is not normally measured directly in liters as this would be too imprecise. Instead, the CO_2 exhaust emissions are captured by a bag attached to the exhaust. The consumption is then calculated back from the amount of CO_2 gas in the bag. Diesel produces 2.62 kg_{CO2} per liter, petrol 2.32 kg_{CO2} per liter. Note that the weight of the carbon

Table 4.3 Excerpt from the WLTP Class 3 driving profile (version 5)

Phase	Elapsed time (Total) s	Elapsed time (Phase) s	Vehicle speed km/h	Acceleration m/s²
...				
Low	25	25	30.8	0.39
Low	26	26	31.9	0.46
Low	27	27	34.1	0.65
Low	28	28	36.6	0.69
Low	29	29	39.1	0.65
Low	30	30	41.3	0.47
Low	31	31	42.5	0.28
Low	32	32	43.3	0.19
Low	33	33	43.9	0.15
Low	34	34	44.4	0.08
Low	35	35	44.5	−0.03
Low	36	36	44.2	−0.25
Low	37	37	45.7	−0.60
Low	38	38	39.9	−0.79
Low	39	39	37.0	−0.74
Low	40	40	34.6	−0.65
Low	41	41	32.3	−0.78
...				

dioxide produced is much higher than the weight of the fuel burned. The difference is the oxygen, which is taken from the air (Table 4.3).

4.5.1 NEDC for ICE-Cars

The NEDC (New European Driving Cycle) was introduced in Europe in the 1990s.

It comprises four identical urban cycles (ECE), each lasting 195 s at constant speeds of 15, 32, 40, and 50 km/h, and a 400-s cross-country drive (EUDC) at speeds of up to 120 km/h. The total driving time is just under 20 min, during which around 11 km are covered.

As consumption measurements according to the NEDC lead to unrealistically low results, the testing cycle has now been largely abandoned.

Some of the reasons are[4]:

[4] Source: Jos Dings, *Mind the Gap! Why official car fuel economy figures don't match up to reality*, Study, March 2013, Transport and Environment.

- The driving cycle contains a large proportion of unusually low speeds and only slow accelerations, e.g. from 0 to 50 km/h in 26 s,
- the alternator is disconnected during the measurement,
- all non-essential auxiliary units, e.g. heating, air conditioning, lights, and radio, are switched off,
- joints on the outer shell of the vehicle are taped to reduce air resistance,
- the toe and camber settings of the wheels are manipulated to reduce rolling resistance,
- the air pressure in the wheels is increased up to 4.5 bar,
- vehicles with minimal weight are used, without any special equipment,
- all measured values are reduced by 4% tolerance,
- the brake pads are pushed back manually before the test drive to avoid any grinding,
- and the engine control unit is adapted, sometimes illegally, to the test bench operation to reduce emission optimization in favor of efficiency optimization.

Figure 4.6 shows an example of the NEDC operating points and their frequency in the speed-torque diagram for a lower mid-range electric vehicle.

Only around half of the maximum speed possible with this vehicle is utilized and only around a third of the maximum torque. The focus of the converted energy is below 40 Nm, i.e. only a fifth of the maximum torque of 200 Nm.

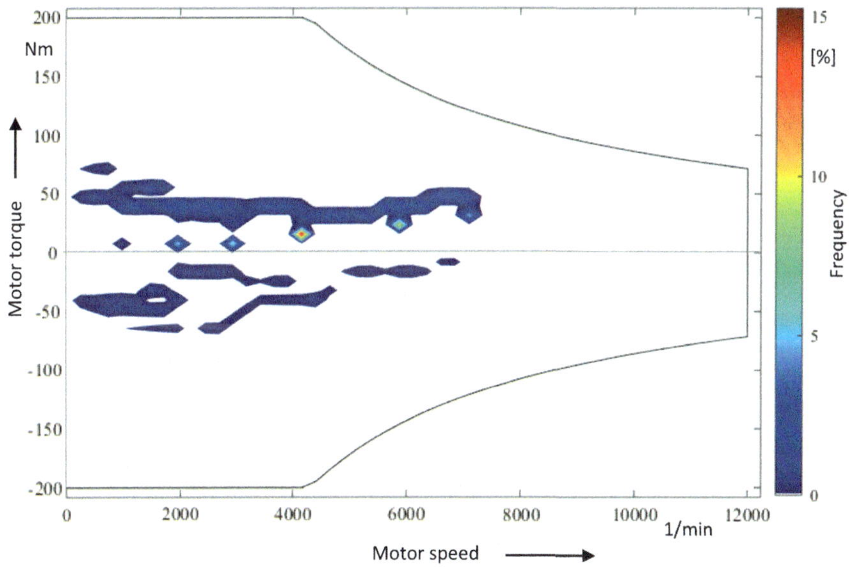

Fig. 4.6 Load points in the NEDC driving cycle for a compact-class electric vehicle (BEV)

4.5.2 NEDC for Hybrid Cars

There is an adapted version of the NEDC measurement for plug-in hybrids. The standard cycle is run twice: Once with a full battery, i.e. purely electric (Condition A), and once with an empty battery, i.e. with the internal combustion engine (Condition B).

The theoretical electric range is determined from the electric drive by dividing the battery capacity by the energy consumption and multiplying by 11 km of driving distance.

The specific consumption is determined from the distance with the combustion engine and extrapolated to 25 km, which is the assumed distance to the next charging point.

The specific total consumption V is then calculated from the weighted average value:

$$V = \frac{D_e \cdot M_1 + D_{av} \cdot M_2}{D_e + D_{av}} \tag{4.13}$$

with

D_e = Electric range in km, calculated from average consumption in the test cycle and nominal battery capacity
M_1 = Specific consumption in g/km when driving electrically
D_{av} = Hypothetical distance to the next charging point = 25 km
M_2 = Specific consumption in g/km when driving with a combustion engine

M_1 is generally set to zero, as electricity does not create any pollution according to the inventors of this driving cycle.

This results in the final calculation formula:

$$V = \frac{25\ \text{km}}{25\ \text{km} + D_e} \cdot M_2 \tag{4.14}$$

These complicated calculations have rather simple consequences: The measured consumption of the ICE is halved for an electric range of 25 km, tripled for a range of 50 km, and so on.

This procedure thus assumes the following user behavior:

The vehicle is always started with a full battery and the driven electrically until the battery is depleted, typically 15–30 km for plug-in hybrids. Afterwards, another 25 km is driven with the combustion engine. Finally, the vehicle is parked, and the battery is fully recharged. If the desired driving distance is not covered, the cycle is continued.

Furthermore, the CO_2 emissions from electricity production are not considered. In other words, the electricity cannot be taken from the public power grid, but rather from an individual PV or wind power plant. The equivalent GHG emissions from the construction of the energy system are also not considered.

It must, therefore, be concluded that the NEDC consumption measurement for hybrid vehicles is completely unrealistic and produces downright ridiculous consumption values.

Example 4.6—NEDC data for ICE and hybrid vehicles in comparison

The Mercedes-Benz E 350 e plug-in hybrid (2 l cylinder capacity, 155 kW (ICE) + 65 kW (EM) = 240 kW combined power) is specified with an average fuel consumption of 2.1–2.5 l/100 km. The vehicle has an unladen weight of just under 2 tons, a top speed of 250 km/h, and acceleration from 0 to 100 in 6.2 s.

The battery capacity is 6.4 kWh. According to measurements by *auto, motor and sport magazine*, this is sufficient for a range of 17 km. The manufacturer states an electric range of 33 km and can therefore reduce the measured fuel consumption by almost 60% (25/(25 + 33)).

The largely comparable E 300 (2 l cylinder capacity, 180 kW)—but without the small battery and electric motor—is quoted at 6.3–6.9 l/100 km, although the total weight of this vehicle is around 300 kg lower than that of the hybrid variant. ◀

4.5.3 WLTP for ICE Cars

In autumn 2017, the WLTP (Worldwide Harmonized Light-Duty Vehicle Test Procedure) was introduced in the European Union, replacing the non-realistic NEDC.

The WLTP was developed to consider all driving patterns occurring worldwide. In addition to the actual driving cycle (WLTC = driving speed profile over time), a new measurement procedure (WLTP = definition of the framework conditions of the test, shift patterns, etc.) is also being introduced.

The new global emissions and consumption standard will not only apply in Europe, but also in Australia, China, India, Japan, South Korea, Russia, South Africa, Turkey, and some other countries.

The WLTP defines measurement procedures for vans, buses (class 2), cars (class 3), and low-power vehicles (class 1). The differentiation is based on the engine power in relation to the total weight of the vehicle. The WLTP also describes more realistic measurement procedures for plug-in hybrid vehicles and fuel cell vehicles.

The class 3 cycle consists of four phases: low (589 s), medium (433 s), high (455 s), and extra high (323 s) with a total of 30 min. The top speed varies between 56 km/h and 131 km/h (Fig. 4.7 and Table 4.4).

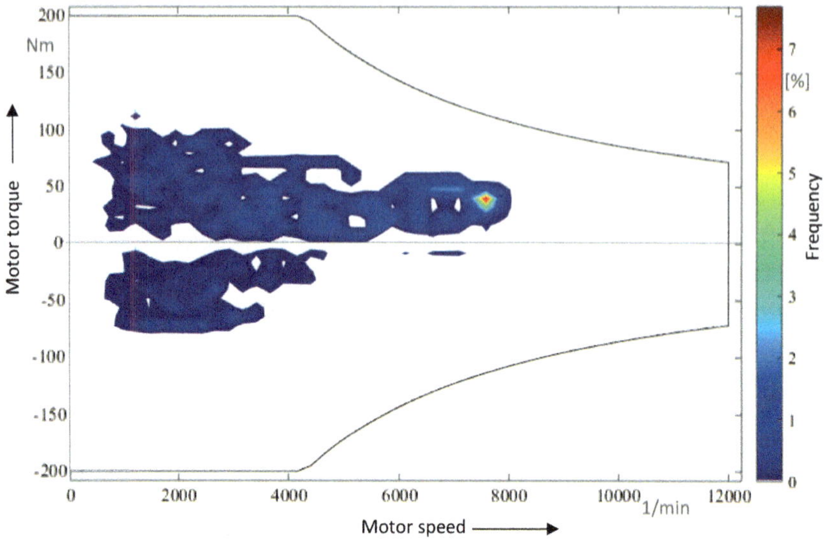

Fig. 4.7 Load points in the WLTP driving cycle for a compact-class electric vehicle (BEV)

Table 4.4 Comparison of WLTP and NEDC features

	WLTP Class 3	NEDC
Starting temperature	23 and 14 °C, cold start	20 … 30 °C, cold start
Cycle time	30 min	20 min
Standstill	13%	25%
Cycle length	23.25 km	11 km
Speed	Avg: 46.5 km/h max: 131 km/h	Avg: 33.6 km/h max: 120 km/h
Power	Avg: 7 kW max: 47 kW	Avg: 4 kW max: 34 kW
Special equipment and air conditioning	… considered for weight, aerodynamics, on-board power supply requirements. Air conditioning system switched off	… everything switched off or dismantled
Plug-In hybrids	More realistic measurement	Entirely unrealistic measurement

4.5.4 Real Driving Emissions

Together with the introduction of the WLTP, measurements of emissions in real driving conditions, i.e. on the road, and not under artificial conditions on the test bench are becoming mandatory.

For many years, it was common for vehicles to be equipped with special software to detect test bench operation, which then switched the engine to a special low-emission operating mode. In real-world driving, this mode would have significantly worsened engine performance and, in the case of diesel vehicles, would also have consumed an excessive amount of AdBlue, so that it was not normally activated. In 2015, an investigation by the ICCT (International Council of Clean Transportation) found that the emission of nitrogen oxides from diesel vehicles in real-world driving was up to 7 times higher than measured on the test bench, see also Sect. 4.6.

The aim of RDE is to prove the effectiveness of the exhaust gas cleaning system (especially regarding NOx) in real operation to detect defeat devices and cheating software. A Portable Emission Measuring System (PEMS) is installed in the vehicle and continuously measures pollutant emissions. The cost of these systems is typically in the six-digit range. It also takes some time to install the equipment in a car. Therefore, the measurement costs for RDE are quite high.

The drives are 90–120 min long and are carried out in the city, overland, and on the highway, one third each. The highway speed is 90–110 km/h minimum and 145 km/h maximum. The average city speed must be 15–30 km/h. Several traffic lights stop of at least 10 s are planned. The route must not be above 700 m above sea level and the maximum difference in altitude must not exceed 100 m. A range of 0–30 °C is permitted for the ambient temperature. The air conditioning must be switched on (normal operation).

The previous NOx limit value for NEDC test bench measurements was 80 mg/km. It is known that 4–5 times this value is measured during real-world driving, even without cheating software.

The EU Commission has defined limit values for the RDE measurements by applying correction coefficients, so-called conformity factors (CF), to the limit values for nitrogen oxides (NO_x) and particle numbers (PN). These tolerance allowances are intended to compensate for the statistical and technical inaccuracies that can occur during RDE measurements.

The CF factor for NO_x is 1.43 and PN 1.5 for the Euro 6d-ISC-FCM, which is still valid for new registrations until 31.8.2024, and 1.1 (NO_x) and 1.34 (PN) for the upcoming Euro 6e (type approval from 1.9.2023, new registration from 1.9.2024).

These changes will make it more credible to comply with the limit values and in fact significantly tighten them.

4.5.5 WLTP for Hybrid Cars

The procedure for measuring the consumption of plug-in hybrids has been changed in the WLTP compared to the NEDC. The aim is to reward vehicles with a larger battery by assuming an increasingly larger proportion of electrically driven distances in real driving conditions. A benefit factor is introduced for this.

For the consumption test, the WLTP driving cycle is repeated several times starting from a full battery (cycle 1) until the battery is completely discharged after a certain number of cycles n (CD test = charge-depleting test). In plug-in and full hybrids, the cycles are initially operated entirely by the electric motor. However, before the battery depletes entirely, the combustion engine is switched on to complete the last cycle. The fuel consumption measured in the CD test is therefore not zero, but very low.

A final run of the WLTP cycle is then performed with an empty battery exclusively by the combustion engine (CS test, confirmation cycle, cycle $n+1$). The battery is recharged slightly by using braking energy recovery. Overall, the net power drawn from the battery from the start to the end of this cycle may only account for 4% of the total energy required at the wheels.

The vehicle is then switched off and the battery is fully recharged. The energy drawn from the grid is measured so that charging losses are considered.

The electric range and total range as well as fuel consumption and CO_2 emissions can now be calculated.

The CO_2 value is calculated considering the so-called utility factor (UF). This factor is fictitious and represents the proportion of drives that are made purely electrically in practice.

The old NEDC contains an implicit weighting factor based on a fictitious range of 25 km. In the new WLTP, the weighting factor varies depending on the electric range: for ranges of less than around 40 km, the electric component is weighted lower than in the NEDC; for longer electric ranges, it is weighted higher.

The UF is determined using the following formula:

$$UF_i(d_i) = 1 - \exp\left(-\left(\sum_{j=1}^{k} C_j \cdot \left(\frac{d_i}{d_n}\right)^j\right)\right) - \sum_{l=1}^{i-1} UF_l \qquad (4.15)$$

with

UF_i Utility factor in phase i

d_i Covered distance until the end of phase i in km

C_j j coefficient: C1 = 26.25; C2 = 38.91; C3 = 631.05; C4 = 5964.83; C5 = 25,094.60; C6 = 60,380.21; C7 = 87,517.16; C8 = 75,513.77; C9 = 35,748.77; C10 = 7154.94

d_n Normalized distance: $d_n = 800$ km

k Number of terms and coefficients in the exponent: $k = 10$

i Number of the phase under consideration

j Number of the term/coefficient under consideration

$\sum_{l=1}^{i-1} UF_l$ Sum of the calculated utility factors up to the phase (i-1)

In practice, this formula for the Utility Factor results in a value between zero and 1, depending on the range of the battery. The diagram below shows, for example, that the value of UF is 0.9 for a range of 100 km (Fig. 4.8).

The weighted fuel consumption C is determined as follows:

$$C = UF \cdot C_1 + (1 - UF) \cdot C_2 \qquad (4.16)$$

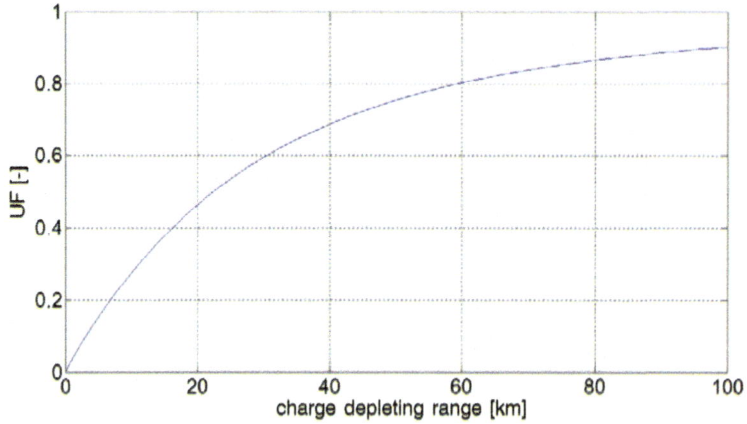

Fig. 4.8 Utility factor as a function of the purely electric range in the WLTP cycle[5]

with

C	Weighted fuel consumption in l/100 km
UF	Utility factor according to the previous chart or formula
C_1	Fuel consumption in CD test in l/100 km
C_2	Fuel consumption in the CS test in l/100 km

The equivalent electric range (EAER = Equivalent All Electric Range) is calculated from:

$$EAER = \left(\frac{M_{CO_2,CS} - M_{CO_2,CD,AVG}}{M_{CO_2,CS}} \right) \cdot R_{CDC} \tag{4.17}$$

with

EAER	Electric range in km
R_{CDC}	Measured electric range in the driving cycles up to the first switching on the combustion engine for the first time in km
$M_{\text{CO}_2,\text{CS}}$	CO_2 emissions in the CS test in g/km
$M_{\text{CO}_2,\text{CD,AVG}}$	Average CO_2 emissions in the CD test in g/km

The electrical energy consumption results from the ratio of the electrical energy required to charge the battery and the equivalent electrical range EAER.

The WLTP, therefore, causes a noticeable increase in the reported fuel consumption compared to the NEDC, especially for plug-in hybrids with a low electric range: on the one hand, the cycle itself is more demanding with faster sections and higher acceleration, and on the other hand, the weighting factor of the electric

[5] Source: EU regulation 2017/1151.

driving components decreases. Plug-in hybrids with a long electric range, on the other hand, are likely to benefit more from the WLTP, which also seems to make ecological sense.

4.5.6 FTP-72 (UDDS) and FTP-75 (Federal Test Procedure)

The current EPA (US Environmental Protection Agency) approval procedure for passenger cars includes measurements according to FTP-75 and three additional measurements (supplemental tests = HWFET, US06, SC03) to record the real consumption more realistically than before.

These U.S. American cycles were first standardized in 1972 and 1975. The FTP-72 (also known as UDDS = Urban Dynamometer Driving Schedule) consists partly of real test drives up to 90 km/h and partly of synthetic components. The driving distance is around 12 km. Speeds of up to approx. 92 km/h are covered.

The further developed FTP-75 contains the FTP-72 cycle plus a further 505 s of an additional test.

In contrast to the NEDC and WLTP, one of the EPA tests, the SC03, is specifically designed to determine the consumption of the air conditioning system when driving in urban areas. For this purpose, the ambient temperature is raised to 35 °C. The driving cycle covers a 5.8 km circuit at an average speed of 35 km/h. A maximum of 88.2 km/h is achieved.

4.5.7 CADC (Common Artemis Driving Cycle)

The CADC was developed years ago based on a project involving driving patterns in Europe. It consists of various driving scenarios (urban, interurban, and highway driving) with widely varying speeds. The cycle lasts around 55 min at speeds of up to 130 km/h. The CADC is no longer in use today.

4.5.8 10–15 Mode/JC08

The 10–15 mode cycle is used in Japan. It is purely synthetic and consists of three identical city modes (10-mode) and a cross-country drive (15-mode). The cycle lasts around 11 min and covers speeds of up to 70 km/h. It was replaced by the JC08 cycle in 2011.

Table 4.5 Discrepancy between type-approval and real-world CO_2 emission levels for new passenger cars[6]

Testing procedure	Year	Average underreporting of fuel economy by manufacturers
NEDC	2002	−10%
	2010	−24%
	2014	−35%
	2020 (estimation)	−49%
WLTP	2020	−23%
	2025 (estimation)	−31%

4.6 Deviations Between Tested and Real Driving Fuel Consumption

Over the years, manufacturers have become better at finding loopholes and inaccuracies in the legally required test procedures. In particular, the NEDC offered many opportunities for fraud, which were shamelessly exploited by almost all manufacturers.

On the other hand, there are independent organizations such as the ICCT (International Council on Clean Transportation), which expose and publish these manipulations. A variety of data sources are used to determine the real-world consumption, for example, fleet consumption of company cars, car club measurements, measurements from trade journals, etc.

It turns out that the deviations between real consumption and manufacturer's figures according to the NEDC have increased significantly in recent years (Table 4.5).

The deviation between reality and the manufacturer's specification is dependent on the driving cycle used and the approval procedure. The main points of attack for manipulation are road load determination (friction and air resistance), chassis dynamometer testing, and technology deployment.

In the November 2017 update of the analysis by the ICCT, it was found that the Japanese 10–15 and JC08 measurement regulations in particular deliver completely unrealistic results. By 2011, the difference between the manufacturer's specifications and reality had risen to almost 60% with the 10–15 procedure. The procedure was then replaced by the JC08, but this measurement method had also allowed deviations of around 45% in 2014.

[6] Source: *Real-world consumption and CO_2 Emissions of new passenger cars in Europe*, ICCT fact sheet, November 2016 and update November 2017 and: *From laboratory to road—A 2018 update of the official and "real-world" fuel consumption and co_2 values for passenger cars in Europe*, ICCT white paper, January 2019.

The only measurement method that gives realistic consumption values is the US EPA procedure. With this method, careful drivers even manage to undercut the consumption values.

The situation for plug-in hybrids is even worse. The ICCT together with Fraunhofer ISI in Karlsruhe, Germany conducted a study including more than 100,000 vehicles in China, Germany, Norway, United States & Canada, and the Netherlands including private cars and company cars.[7]

For private vehicles, the real fuel consumption and CO_2 emissions as tested with the NEDC procedure are 2–4 times higher than in the test cycle, while the deviation for company cars differs by a factor of three to four. Unfortunately, the new WLTP testing procedure has not improved the situation but rather shows a similar deviation.

The drivers of company cars often receive a fuel card from the company to fill up gas free of charge. However, if they want to charge the battery, they must pay for the electricity out of their own pocket. This is why many plug-in hybrids are bought for tax reasons only and never charged during their operation. When the leased vehicles are returned after a few years, the charging cable is often in its original packaging and placed unused in the trunk.

4.7 CO_2 Efficiency Classes

Since a long time, the European Union requires labeling of CO_2 efficiency classes using letters and colors (A+, A, B, C, …, G) to display the energy efficiency to the consumers. Cars have also been affected by this labeling requirement since 2011 but each country was allowed to come up with their own procedure.

In Germany, the automotive industry was able to ensure that the actual CO_2 emissions of their vehicles are not used for labeling. Instead, a reference value R (in g_{CO_2}/km) is determined based on the vehicle mass m (in kg), to which the CO_2 efficiency class is then related:

$$R = 36.59079 + 0.08987 \cdot m \tag{4.18}$$

In the second step, the deviation of the actual, vehicle-specific consumption NEDC from the reference value is determined:

$$\Delta_V = \frac{V_{NEFZ} - R}{R}. \tag{4.19}$$

Table 4.6 can then be used to determine the CO_2 efficiency class:

This approach leads to the absurd situation that the heavier a vehicle is, the easier it is to achieve a higher efficiency class, regardless of the number of seats or the possible payload (Table 4.7):

[7] Source: *Reale Nutzung von Plug-In-Elektrofahrzeugen*, ICCT, fact sheet Europa, September 2020 and: *Reale Nutzung von Plug-In-Hybrid Elektrofahrzeugen*, Fraunhofer ISI, policy brief, September 2020.

Table 4.6 Determination of the CO_2 efficiency class from the deviation from the reference value

CO_2 efficiency class	Percentage deviation Δ_V
A+	≤ −37%
A	−36.99% to −28%
B	−27.99% to −19%
C	−18.99% to −10%
D	−9.99% to −1%
E	−0.99% to +8%
F	+8.01% to +17%
G	≥ +17.01%

Table 4.7 CO_2 efficiency classes of some vehicles

Vehicle	Weight in kg	NEDC in g_{CO2}/km	Efficiency class
Porsche Cayenne Turbo S Hybrid (680 PS)	2565	88.1	A+
Audi Q7 3.0 TDI Quattro Hybrid (373 PS)	2520	47.7	A+
Toyota Aygo (69 PS)	915	100	C
Battle Tank Leopard 2	App. 62,000	App. 1500	E
VW Golf R 2.0 TSI	1541	195	F
Mercedes Benz A 200 Turbo	1315	178	G

It would have been sound to use the permitted payload or the number of seats as a reference value rather than the total weight. But then the poor environmental performance of heavy vehicles, especially SUVs, would have become all too obvious.

In February 2024, a new regulation finally came into force in Germany in which the vehicle weight no longer plays a role in the labeling of energy consumption.[8] In future, fuel consumption figures will be based on the more realistic WLTP cycle. For plug-in hybrid vehicles, the energy efficiency class for pure combustion mode, which is predominantly and often even exclusively used in practice, will be shown in addition to the mixed mode labeling.

[8] Source: Pkw-Energieverbrauchskennzeichnungs-Verordnung (Pkw-EnVKV) des Bundesministeriums für Wirtschaft und Klimaschutz (BMWK) as of 23.02.2024.

4.8 Typical Energy Consumption of Vehicles

4.8.1 Bicycles

A well-trained bicycle rider can continuously produce between 150 and 250 W, a Tour-de-France rider up to 350 W. At a speed of 30 km/h, this results in a production of 0.5–0.8 kWh/100 km mechanical power.

An electric PEDELEC (Pedal Electric Cycle) typically has a battery of 500 Wh capacity. The average power supply is 250 W (peak power up to 500–700 W for a few minutes). The speed of a PEDELEC is 25 km/h and so the average electric energy consumption is about 1 kWh/100 km.

4.8.2 ICE Passenger Cars

The average fuel consumption of passenger cars is very much depending on the country. In Germany, with its high driving speeds on highways, the consumption is quite high. The average consumption of combustion engine cars is 7.4 L per 100 km in 2021, see Sect. 13.4. This is equivalent to about 70 kWh/100 km.

For comparison: the average consumption in Austria is 6.7 L per 100 km, in Switzerland only 6.1 L per 100 km (both as of 2021).

It can be concluded that a highway speed limit has a significant positive effect on fuel consumption.

The annual fuel economy in the United States was 25.4 miles per gallon in 2020 and 2021 alike, as published by the EPA. This translates into 9.3 L per 100 km—even more than in Germany. However, the average vehicle in the United States is much larger than in Europe.

4.8.3 BEV Passenger Cars

The electricity consumption of battery electric vehicles is not yet as well recorded statistically as the energy consumption of conventional vehicles. In the report on the results of the electric mobility model region,[9] an average consumption of just under 17 kWh/100 km for compact- and mid-range electric vehicles was determined from the summarized evaluation of various projects. Other studies[10] arrive at 14.5 kWh/100 km for compact-class vehicles.

---

[9] Source: Tenkhoff, Braune, Wilhelm, *Ergebnisbericht 2011 der Modellregion Elektromobilität*, Bundesministerium für Verkehr, Bau und Stadtentwicklung, Hg. 5. Dezember 2011.

[10] Source: Pfriem, *Analyse der Realnutzung von Elektrofahrzeugen*, Dissertationsschrift, KIT, 2016.

A study in Denmark[11] with 200 vehicles (small cars) and 741 drivers showed an average consumption of 18.3 kWh/100 km over 2 years.

According to the German car club ADAC Ecotest[12] measurements, luxury class vehicles consume between 20.4 and 26.3 kWh/100 km. Mid-range vehicles use between 18.5 and 22.8 kWh/100 km and compact-class vehicles between 16.7 and 19.3 kWh/100 km. These values already include AC charging losses.

Taking all the findings together, a realistic average fleet consumption seems to be around 20.0 kWh/100 km not considering charging losses. When charging losses of typically 10% (the range is 5–15% depending on the charging system) are included, the average BEV consumption from the electricity grid is around 22 kWh/100 km.

Example 4.7—Average consumption of Tesla Model 3

The author drove a Tesla Model 3 Performance privately for around 4 years. After around 76,000 km, the average consumption was exactly 20.0 kWh/100 km not accounting for charging losses. ◀

4.8.4 FCEV Passenger Cars

Due to the small number of production vehicles—only the Toyota Mirai and the Hyundai Nexo are or were available on the open market in small series—there is currently hardly any reliable information on the real energy consumption of hydrogen cars.

The manufacturers' figures range from 0.9 to 1.1 kg_{H2}/100 km. However, these figures should be viewed with caution, as fuel cells have a relatively good efficiency in artificial driving cycles with a low partial load. If the load increases in real-world driving, the efficiency drops rapidly. This is why the real consumption of FCEVs is significantly higher than the factory specifications.

Table 4.8 shows the ranges measured according to the EPA method. However, these figures do not consider air conditioning and bad weather conditions, which means that the average real-world consumption is higher.

Example 4.8—H_2 consumption of the Toyota Mirai (1st gen)

Toyota states a combined NEDC consumption of 0.76 kg_{H2}/100 km for the Mirai. In the United States, a range of 312 miles is stated when filled up (2 tanks of 2.5 kg H_2 each) according to EPA certification, which corresponds to the equivalent of 1.0 kg_{H2}/100 km. The same value was also found in the ADAC Ecotest.

[11] Source: Fetene, Kaplan, Mabit, Jensen, Prato, Dänemark, 2012–2014.

[12] Source: https://www.adac.de/rund-ums-fahrzeug/autokatalog/ecotest/.

Table 4.8 Hydrogen consumption of all publicly available FCEVs, measured according to the US EPA procedure

Car	Class	Tank volume	EPA range	Consumption
Toyota Mirai	Mid-range	5.0 kg_{H2}	312 miles	1.00 kg_{H2}/100 km
Hyundai Nexo	Crossover SUV	6.3 kg_{H2}	354 miles	1.11 kg_{H2}/100 km
Hyundai Tucson	Compact crossover SUV	5.6 kg_{H2}	265 miles	1.31 kg_{H2}/100 km
Honda FCX Clarity	Mid-range	4.1 kg_{H2}	240 miles	1.07 kg_{H2}/100 km

According to the on-board computer, a rental Toyota Mirai from Stadtmobil Karlsruhe, Germany, has an average consumption of 1.34 kg_{H2}/100 km after thousands of kilometers of driving as a car-sharing vehicle at the beginning of 2021. During a highway test drive, the Mirai consumed around 1.3 kg_{H2}/100 km at a constant 130 km/h and around 2.0 kg_{H2}/100 km at a constant 160 km/h. ◄

For mid-range FCEVs, a yearly average consumption of 1.2 kW_{H2}/100 km, therefore, seems realistic.

As with BEVs, charging losses must be added. For the high pressure (700 bar) gas filling, the losses are 12% resulting in a total consumption of 1.36 kg_{H2}/100 km. This is equivalent to 45 kWh/100 km—about twice as much energy as needed for BEVs.

4.8.5 Trucks and Busses

Light and heavy-duty trucks of all classes are already available as fully battery electric vehicles.

Example 4.9—Heavy electric trucks

A heavy truck by the Swiss company Designwerk (based on a Volvo FH truck) comes with a 1017 kWh LFP Battery (864 kWh usable capacity) at a total weight of 5620 kg. The 42t truck has a motor power of 500 kW (680 PS) and a reported range of 576 km. The charging time from 10 to 80% is about 1 hour with regular CCS charging. The consumption of the truck is about 150 kWh/100 km without consideration of charging losses. ◄

Heavy trucks have a consumption of around 120–150 kWh/100 km. A comparable fuel cell truck consumes around 8.0 kg_{H2}/100 km, that is 270 kWh/100 km. In both cases, charging losses must be added.

The consumption of intercity buses (coaches) is similar to heavy-duty trucks. City buses have a higher consumption due to the many start–stop cycles, up to 200 kWh/100 km are possible.

4.8.6 Ships

As of 2024, there are a fast-growing number of battery electric ships in service around the world. Their energy consumption is highly dependent on the size of the ship and the application and can therefore only be given as an example.

Example 4.10—Electric ships

In Norway, there is the tourist ship "Rygerelektra" with space for 300 passengers. With its 2200 kWh battery, it has a range of around 40 nautical miles (74 km). The ship reaches a top speed of 43 km/h and is powered by 2 motors with an output of 375 kW each. At an average speed of 33 km/h, the consumption is 2400 kWh/100 km.

COSCO Shipping Heavy Industry, China, is currently (as of 2024) building two identical 10,000 grt (gross register tons) container ships with a loading capacity of 700 standard containers. The ships have a length of 120 m and a width of 23.6 m. They are powered by two 900 kW electric motors which get their energy out of a 57,600 kWh LFP battery, which is placed in 36 swappable containers. The ships have a consumption of about 5800 kWh/100 km and a range of 1000 km, enough to travel all the way down the Yangtze River to the sea. ◀

5 Operating Strategy

Abstract

The selection of the best operating strategy by the hybrid controller depending on the situation is essential for the success in minimizing the consumption of an electric or hybrid vehicle. Many concepts and algorithms are known for this purpose, some of which are presented in this section.

The operating strategy is the targeted control of the hybrid operating modes by the hybrid controller. The main task of the operating strategy is to minimize fuel consumption. The operating strategy is particularly important for hybrid vehicles because there are usually many options for intervention and optimization.

The hybrid controller records a series of input variables and generates output variables, see the following tables.

Several secondary conditions must be observed, such as

- SOC (state of charge = state of charge in percent) of the battery
- Maximum available power of the battery
- Current limit of the inverter
- Voltage limit of the battery and the inverter
- Thermal state of the battery and electric motor(s) (Table 5.1)

A prerequisite for the optimal use of the various hybrid operating modes is a suitable state of charge of the battery. If the battery is discharged, i.e. low SOC, boosting and load point reduction can no longer be carried out. If, on the other hand, the battery is fully charged, load point boosting and braking energy recovery can no longer be activated.

M. Doppelbauer, *Introduction to Electromobility*,
https://doi.org/10.1007/978-3-658-45482-1_5

Table 5.1 Input variables of the operating strategy

Driver request	Accelerator pedal position
	Brake pedal position
	Desired direction of travel
	Operating mode (gear)
	Heating/air conditioning request
	Navigation destination data
Internal data	Speeds of all drives
	Torques of all drives
	Vehicle speed
	Operating temperatures
	Battery charge status
	Status of internal simulations
Learned data	History/driving behavior
	Navigation route guidance
	Traffic information systems
	Car2Car communication

The other essential task of optimizing the operating strategy of hybrid vehicles is therefore to keep the state of charge of the battery in the medium range as far as possible (Table 5.2).

5.1 Optimization of the Operating Strategy

5.1.1 Rule-Based Optimization

Rule-based optimization uses characteristic maps or heuristics, i.e. laws of if-then rules, to select the most favorable operating strategy. Boolean variables or fuzzy logic can be used (Fig. 5.1).

The individual operating modes are activated according to the desired torque (pedal position) and speed depending on the characteristic map, if this is possible due to the boundary conditions (battery charge status, temperatures, etc.).

The advantages of rule-based optimization are good reproducibility and predictability. A major advantage is the reliable vehicle response. A disadvantage is the non-optimal control, especially with variable usage profiles.

5.1.2 Online Optimization

During online optimization, the actual and alternative driving states are continuously calculated and evaluated by using a cost function.

Table 5.2 Output variables of the operating strategy

Combustion engine	On/off
	Target operating point (torque specification)
Electric machine(s)	On/off
	Setpoint speed or setpoint torque (speed or torque control)
Hybrid gearbox	Clutch position(s)
	Brake position(s)
	Gear selection
Other components	Clutch position(s)
	Braking force
	Steering aid
	Accelerator pedal pressure Braking pedal pressure

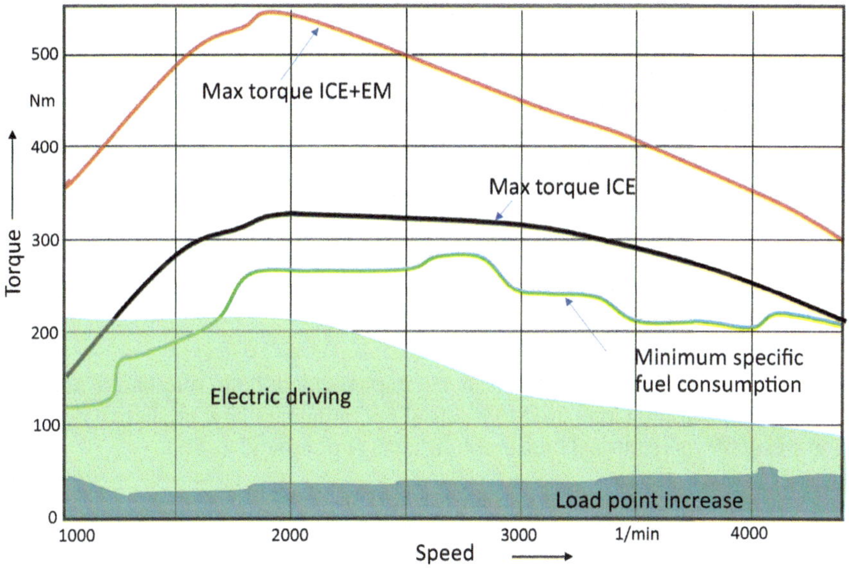

Fig. 5.1 Operating areas of a rule-based operating strategy

The cost function includes, for example, terms for the energy used in the combustion engine and in the electric motor. In addition, penalty terms are built in, which increase with increasing deviation from the average battery charge level.

The cost function is calculated at certain time intervals for many possible operating states and the "most cost-effective" state is then selected. Online optimization can be carried out statically, numerically, or analytically.

5.2 Predictive Operating Strategy

If the operating strategy works purely based on currently available vehicle data, it is referred to as causal, i.e. not predictive.

A predictive operating strategy is based on the forecast or foresight of the future driving profile. The operating strategy is then non-causal.

Possible data sources for prediction are

- the vehicle condition (yaw rate, speed, driver preference),
- the driving behavior of the driver (preferred acceleration, maximum speed),
- monitoring the vehicle in front (distance, speed, Car2Car communication),
- the road situation (city, interurban, highway),
- traffic conditions (traffic jam, stop-and-go, free flow), and
- route information (altitude profile from the navigation system, route to destination, traffic jam messages from the RDS radio data system, speed limits, curve radii).

A predictive operating strategy can also be used to optimize the state of charge of the battery.

Example 5.1—Navigation data in predictive operating strategies

If, for example, the control system knows from the route planning that a long downhill section is coming up soon, the battery can be drained shortly beforehand without any disadvantage.

If the traffic information indicates that a traffic jam is to be expected, the battery should be as full as possible in order to be able to drive purely electrically in slow traffic. ◀

A predictive operating strategy can be implemented both rule-based and with online optimization. Compared to a causal operating strategy, predictive optimization can achieve additional consumption savings of around 2–5%.

5.3 Driver Interaction

5.3.1 Visualization of Vehicle Status

One way of interacting with the driver is to influence the mechanical counterpressure of the accelerator and brake pedals.

When the pedals are pressed, an electric actuator builds up counter-pressure, which provides the driver with additional feedback depending on the operating point. In this way, for example, excessive pushing of the pedals (excessive accelerating) can be penalized.

Example 5.2—Application of the accelerator pedal in a Renault Zoe

In the Renault Zoe, the back pressure of the accelerator pedal is noticeably increased above a defined threshold value (torque limit) in "Eco" mode. Higher torque can still be called up by the driver but requires the pedal to be pressed uncomfortably hard.

If the "Eco" mode is deactivated, the accelerator pedal can be depressed with evenly low force all the way. ◀

A similar behavior can be implemented for the brake pedal. The initial actuation path exclusively controls the braking energy recovery of the electric drive. It can be overcome with little effort. The mechanical brakes come into play if the pedal is depressed further. This is signaled to the driver by a significantly increased actuation force.

Motivation for energy-saving driving can also be achieved by appropriately designing the status display in the vehicle (HMI = human–machine interface).

Example 5.3—Driver display in a Renault Zoe

The Renault Zoe displays the average consumption in kWh/100 km after each journey. In addition, two bars assess driver efficiency during braking and the average speed. A schematized plant leaf shows the overall assessment of the journey by its size and a code number. In Fig. 5.2, for example, this is 91 out of 100 points. ◀

Example 5.4—Driver display in a Chevrolet Volt

With the Chevrolet Volt, the summarized driving performance data can be automatically uploaded to an Internet portal after each journey if desired. This has led to competitions in internet forums regarding the most efficient driving style possible. ◀

Figure 5.3 shows an example of the consumption display of a Polestar 2. The average consumption over the last 20, 40, or 100 km can be displayed as a bar graph. Orange bars show the energy consumption for the drive, and light blue bars for the air conditioning.

5.3.2 Consideration of Driver Input

Some hybrid vehicles allow the operating strategy to be adapted by user input.

For example, purely electric operation may be desired for short journeys.

Or the battery charge level (SOC) can be kept constant over a longer highway trip to be able to drive into a low emission zone in a city purely electrically.

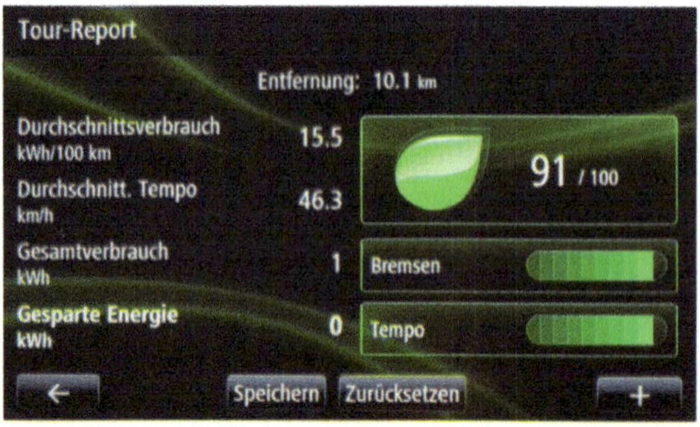

Fig. 5.2 HMI of the Renault Zoe

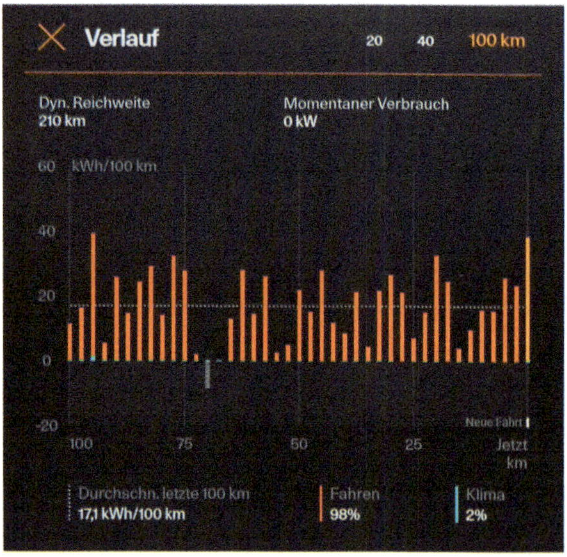

Fig. 5.3 HMI of a Polestar 2, average consumption

Many electric vehicles offer a "normal mode" and an "economy mode". In the latter, for example, the available acceleration torque, maximum speed, and climate control are reduced (Fig. 5.4).

Example 5.5—Driving modes in a Chevrolet Volt

With the Chevrolet Volt, you can select "Hold" mode to save the energy stored in the battery for a later city trip or a long traffic jam.

Alternatively, a "Mountain" mode is available for actively charging the battery by increasing the load point so that sufficient energy is available for boosting when driving uphill later.

Finally, the "Sport" setting prefers small gear ratios and fully enables the boost function even when the battery charge level is low. ◄

Fig. 5.4 HMI of the Chevrolet Volt (Opel Ampera) with selection of the drive mode

Example 5.6—Driving modes in a Mercedes Benz

Mercedes Benz hybrid cars offer four selectable driving modes:

Hybrid: Electric operation or driving with the combustion engine is possible. Automatic selection of operating mode. Optimum use of combustion engine and electric motor.

E-mode: Pure electric operation. Electric power control via the haptic accelerator pedal (variable pressure point). Maximum availability of the electric drive.

E-Save: The current charge status is retained for future electric driving. Electric operation is possible to a limited extent.

Charge: The HV battery is charged via the combustion engine for future electric driving. Electric operation is not possible. ◀

5.3.3 Accelerator Pedal

BEV manufacturers follow different strategies for how much coasting or braking energy recovery is performed when the accelerator pedal is not pressed.

- No noticeable deceleration.
  This can be irritating for unfamiliar drivers, as there is no drag torque as with ICE vehicles.
- Simulation of combustion engine drag torque.
  Braking energy recovery is carried out so that the resulting deceleration corresponds to a conventional ICE car (gearbox stage D).

- Maximum braking energy recovery = one-pedal driving, see Sect. 4.4.1 L mode in Chevrolet Volt, B mode in Volkswagen ID.3, ID.4, etc.).

After a brief adjustment period, most drivers prefer strong regenerative braking. With a forward-thinking driving style, the car can be driven in city traffic without needing to switch from the accelerator to the brake pedal. Some vehicles even automatically engage the parking brake when coming to a stop and release it automatically upon starting, reducing the need to use the brake pedal except in rare extreme situations.

Energy Storage Systems 6

Abstract

This chapter covers the basics of electrochemical energy storage systems. The most important variants—lead-acid batteries, nickel–metal hydride batteries, and lithium-ion batteries—are presented in detail. The lithium-ion battery in particular has a variety of different materials for the anode, cathode, and electrolyte, resulting in some very different operating characteristics. Separate sections are devoted to the various charging methods, charge balancing between the various parallel-connected cells, cell designs, and the structure of battery storage systems (stack, battery management). Modeling the operating behavior of lithium-ion batteries as accurately as possible is necessary to control fast charging in such a way that no irreversible damage occurs. Suitable models are presented. In the final sections of the chapter, the hydrogen storage is discussed.

Table 6.1 shows some of the most prominent options to store energy in technical systems. Many of these storage systems are used in cars today, but some are only available in prototypes or in special applications, for example in motor racing.

Thermal latent heat accumulators store energy through the phase transformation of a material, i.e. by melting or freezing. Thermal sensible storage accumulates energy in the heat capacity of a material, for example in concrete.

Thermochemical storage systems use the energy of reversible chemical reactions that absorb heat in one direction and release it in the other, for example, the hydration of calcium chloride (CaCl).

Electrochemical storage systems are particularly suitable for use in hybrid and electric vehicles. A distinction is made between primary, secondary, and tertiary elements:

Primary elements are electrochemical energy storage devices with irreversible cell reactions, i.e. non-rechargeable batteries.

M. Doppelbauer, *Introduction to Electromobility*,
https://doi.org/10.1007/978-3-658-45482-1_6

Table 6.1 Classification of technical energy storage systems

Physical principle	Storage medium	Storage time
Thermal	Latent Sensitive Thermochemical	Long-term
Chemical	Diesel Petrol (gasoline) Methane gas Hydrogen	Long-term
Electrochemical	Lead-acid NiMH Lithium-ion Redox flow Fuel cell	Long-term
Electrostatic	SuperCaps	Short-term
Inductive	Superconducting coils	Short-term
Kinetic	Flywheel	Short-term
Static	Spring pressure	Short-term
Pneumatic/hydraulic	Pressure	Short-term

Secondary elements are electrochemical energy storage devices with reversible cell reaction. They are often simply called batteries, but the correct term is accumulator.

Tertiary elements are electrochemical energy storage devices in which the cell reaction can take place continuously, for example fuel cells and redox flow batteries.

Electrostatic and inductive storage devices use the energy of electric or magnetic fields for storage. They are used in special applications but rarely in hybrid vehicles and are discussed briefly later in the chapter.

Kinetic storage systems use the energy stored in a flywheel. There are also applications for these devices in the vehicle sector. Some examples are given below.

Static, pneumatic, and hydraulic accumulators use the compression of a medium, such as mechanical springs, air, or liquids, to store energy. It should be noted that water and oil is not compressible and therefore cannot be used to store energy.

6.1 Terms and Definitions

Gravimetric Energy Density in Wh/kg
The energy storage capacity in relation to its weight. Particularly important for the electric range. Please note the difference between gravimetric system energy density and gravimetric cell energy density. The first includes the whole battery

system that is the cells, housing, cooling, cabling, safety devices, battery management, etc. The latter is related only to the cells.

Volumetric Energy Density in Wh/l
Volume required to accommodate the energy storage unit in the vehicle. Together with the gravimetric energy density, the volumetric energy density is also decisive for the autonomy and possible range of purely electric vehicles. Here again, a distinction must be made regarding system and cell level.

Gravimetric Power Density in W/kg
Important for start/stop, acceleration, and braking energy recovery processes.

Volumetric Power Density in W/l
Particularly important for hybrid vehicles as there is little space for the battery and high peak power is required.

Energy Throughput/Working Capacity in kWh
Total energy throughput of the cell over the entire service life. Sometimes the charge throughput or the so-called number of full cycles is given as an alternative.

The lifetime of a battery before its capacity falls under a defined threshold, typically 70% of its initial capacity. A decrease in the available maximum power can also mean the end of the service life.

Capacity in Ah or in C-rate
The amount of stored charges (ions). Usually specified in ampere-hours, sometimes in coulombs (1 C = 1 A · s). A distinction is made between theoretical (gross) and utilized capacity (net). The latter is typically between 50 and 90% of the gross capacity.

The capacity of a battery is not equivalent to the stored energy. To calculate the energy, the voltage must be considered, see (6.2).

Energy Content (Gross or Net) in kWh
The amount of energy released by the battery when discharging from 100 to 0% SOC. The gross energy refers to the theoretical energy content of all cells installed in the battery pack. The net energy is the amount of energy released by the battery management system. It is usually somewhat lower.

The energy stored in a battery is the result of its capacity multiplied by the voltage:

$$W = Q \cdot U \tag{6.1}$$

In most batteries, the voltage is not constant, but decreases with increasing discharge, so that the integral must be calculated to determine the available energy exactly:

$$W = \int_{Q_0}^{Q_1} U \, dQ. \tag{6.2}$$

In addition, the available energy of a battery cell also depends on the temperature and the discharge rate, the so-called C-rate, because both influence the voltage.

Calendar Life in Years
Lifetime of the energy storage device without operation. The lifetime limit is defined by the loss of capacity (typically more than 30%) or loss of output power due to an increase in the internal resistances.

Cycle Life in Number of Full Cycles
Number of equivalent full cycles or total energy throughput until the energy storage system is unusable.

Cost in €/kWh
Battery costs are usually related to the battery capacity to compare different batteries. The typical cell cost for automotive applications is currently below 100 €/kWh.

C-Rate in 1/h
The C-rate describes the charging/discharging current of the battery in relation to the nominal capacity. 1 C corresponds to charging with a charging current that fully charges or discharges the battery with a charging time of one hour.

Example: Charging a 20 Ah battery with 0.1 C corresponds to 2 A charging current, 2 C corresponds to 40 A and 0.5 C would be 10 A.

The C-rate can also be interpreted as charging time. Charging with 2 C means a full charge in around 1/2 h, charging with 0.5 C corresponds to 2 h.

Finally, the C-rate also determines the charging power. For example, a 40 kWh battery with a charging rate of 2 C requires an average charging power of 80 kW. It should be noted that high-power charging of a battery must follow a complex time curve, especially during fast charging, in order to avoid damaging the battery. The specification of an average charging capacity is therefore often not very useful.

Open-Circuit Voltage (OCV) in V
Stationary voltage at the terminals in no-load operation, i.e. without current flow, after completion of all internal equalization processes, which can take several hours.

Nominal Voltage in V
Typical cell voltage of the energy storage device according to the manufacturer. The nominal voltage lies between the end-of-discharge voltage and the end-of-charge voltage.

Terminal Voltage in V
Actual, momentary cell voltage. Under dynamic conditions, i.e. under load with corresponding current flow, it will always be lower than the open-circuit voltage.

End of Discharge Voltage in V
The discharge process must be stopped when this voltage is reached. The final discharge voltage must not be undercut, otherwise the battery will age quickly.

End of Charge Voltage in V
The end-of-charge voltage must not be exceeded during charging, otherwise the battery will age rapidly. If it is significantly exceeded, this can lead to outgassing of the electrolyte, bursting of the cell and ultimately even fire. 3.2–4.2 V per cell is typical for most lithium-ion batteries depending on the cell chemistry.

State of Charge in % (SOC)
Charge remaining in the battery in relation to the capacity.

State of Health in % or kWh (SOH)
Available full charge capacity in percent of new or in kWh.

State of Power in kW (SOP)
Maximum available battery power (depending on SOC, SOH, temperature).

Depth of Discharge in % (DOD)
Depth of charge in relation to the capacity: DOD = 100% − SOC.

Internal Resistance in Ω
The complex internal resistance of the energy storage system changes during operation depending primarily on temperature, SOC, and aging condition.

Efficiency in %
The cycle efficiency (round-trip efficiency) is the ratio of the energy that can be retrieved from a battery compared to the energy that was previously supplied while charging. It depends on the cell chemistry and on external conditions (charge/discharge current, temperature, SOC).

6.2 Electrochemical Storage Systems in Comparison

In the Ragone diagram, the gravimetric power density, i.e. the power that is available per kg of cell weight, is compared with the gravimetric energy density, i.e. the energy stored per kg of cell weight, see Fig. 6.1.

The respective C-rates are also shown in the diagram. These are lines of constant charging and discharging times. The bending of the lines results from the semi-logarithmic representation of the diagram.

On closer inspection, the lithium-ion battery type consists of many variants, which differ from each other in terms of anode and cathode material as well as the electrolyte and separator being used. The technical properties of these variants vary greatly. In the diagram, only a rough distinction is made between "high-

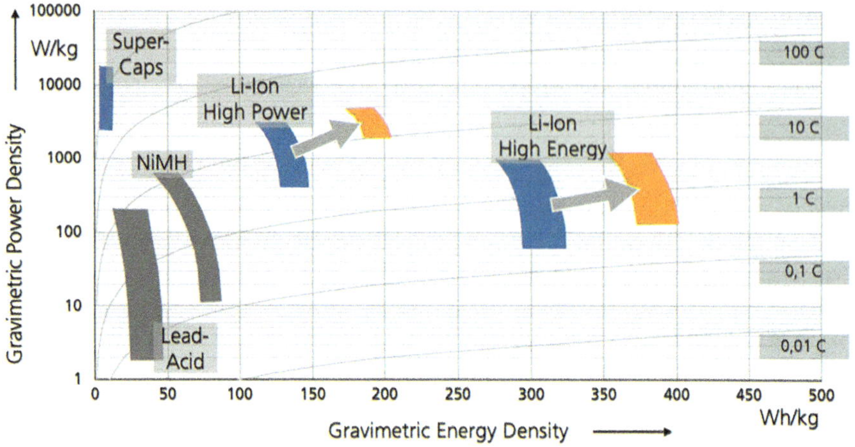

Fig. 6.1 Ragone diagram of typical electrochemical cells

power" and "high-energy". The orange areas are an outlook on the next 5 years of expected development.

The currently most powerful vehicle-compatible lithium-ion battery cells achieve around 360 Wh/kg gravimetric energy density and 700 Wh/l volumetric energy density. Future developments are leading towards 500 Wh/kg and 1000 Wh/l.

An asymmetry can be observed in lithium-ion batteries, which is not shown in the diagram: The maximum permissible C rate during charging is usually around 2–3 C. This means that charging takes between 20 and 30 min. However, the C rate during discharging, for example when starting at full power at a traffic light, can be much higher for a short time, typically up to 10 C or more. Certain variants (LiPo batteries) can even reach 50 C.

The lead-acid accumulator is still used as a 12 V battery in many vehicles today. However, lithium-ion 12 V batteries are becoming more and more common.

The nickel–metal hydride (NiMH) battery is reliable and cost-effective in series production. However, it is now obsolete and is only used in older hybrid vehicles.

Note
When comparing literature data on energy and power density, it should always be noted whether the figures refer to the cell alone or to a complete battery system, i.e. cells plus housing with impact protection, cooling system, battery management system, safety devices, and more.
The difference is about a factor of 1.5.

Example 6.1—Battery of a Tesla Model 3

The battery in the Tesla Model 3 with a gross capacity of approximately 78 kWh, of which approximately 75 kWh are effectively available net, weighs 478 kg. This corresponds to a gravimetric energy density at system level of around 165 Wh/kg. ◀

Example 6.2—Battery of a Renault Zoe

Renault specifies a battery weight of 305 kg with a nominal capacity of 41 kWh for the Zoe in model year 2017. This corresponds to a gravimetric energy density at system level of 134 Wh/kg. ◀

Example 6.3—State-of-the-art batteries

The Farasis P73 pouch cell (size $294 \times 103 \times 14.5$ mm) has a weight of 0.95 kg, a capacity of 73 Ah and can store 280 Wh of energy. Its energy density is therefore 294 Wh/kg or 715 Wh/l.

The SKI E556 pouch cell (size $355 \times 100 \times 9.5$ mm) has a weight of 0.745 kg, a capacity of 55.6 Ah and can store 212.2 Wh of energy. Its energy density is therefore 285 Wh/kg or 712 Wh/l. This cell is being used in the Hyundai Ioniq 5.

Source: https://www.batemo.com/products/batemo-cell-explorer/

In the first half of 2024, the Chinese battery manufacturer WeLion New Energy Technology started the production of the first semi-solid-state battery for the NIO ET7. According to the manufacturer, a 150 kWh battery pack weighs 575 kg. The energy density of the cell is reportedly 360 Wh/kg. NIO will use the batteries in a cell-to-pack (CTP) design that removes the need for modules, thereby increasing the overall energy density even more. ◀

For comparison: Petrol and diesel achieve a gravimetric energy density of 10,000 Wh/kg. This is 30 times more than the best batteries currently in series production.

However, due to the significantly better efficiency of 80–90% on average, electric drives only require around a third of the energy of combustion engines, whose efficiency is typically limited to around 25–35% by the Carnot process. In addition, electric vehicles are built with better drag coefficients to save power. Also, batteries of electric cars are heavier than typical fuel tanks.

These measures can largely compensate for the different energy density of the two systems. Today, electric cars achieve real-world ranges of around 300–500 km.

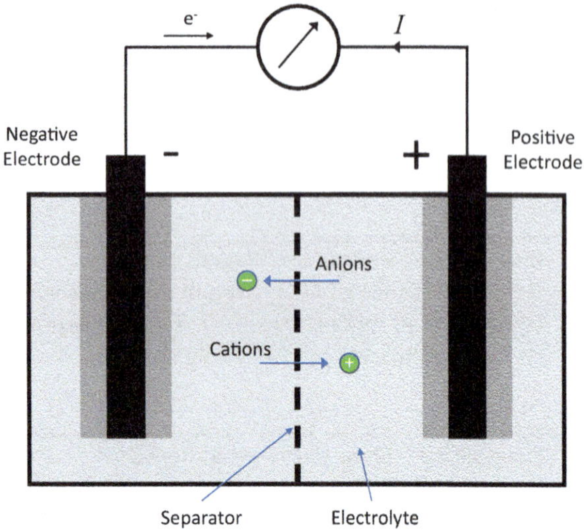

Fig. 6.2 Schematic structure of a battery cell

6.3 Basics of Battery Cells

6.3.1 Design and Functional Principle

The basic unit of the secondary element is the galvanic or electrochemical cell. This is where the conversion of chemically bound energy into electrical energy and back takes place.

Cells consist of two different metals and an electrolyte solution. The voltage results from the difference between the two metals in the electrochemical voltage series. The two electrodes are both energy storage devices and energy converters (Fig. 6.2).

The electrolyte must only conduct ions, but not electrons. Electrons flow via the consumer. The electrolyte can be liquid, solid, or gelled and must not interact with the electrodes.

Note
The positive counting direction of the electric current is traditionally opposite to the direction of movement of the electrons, as these carry negative charges.

The separator prevents direct contact between the positive and negative electrodes, which would lead to a short circuit. The separator should be as thin as possible and have good ionic conductivity.

When discharging, the electrons flow from the less noble metal (negative electrode/anode) to the more noble metal (positive electrode/cathode). The electrons are replenished by dissolution, i.e. oxidation, of the base metal.

At the positive electrode, the electrons are attracted by the positive ions in the electrolyte solution. The electrons reduce the ions and deposit on the electrode. Ion exchange takes place via the electrolyte so that electric current can flow on the outside.

Note
Oxidation = release of electrons
Reduction = absorption of electrons

There is always a negative and a positive electrode in an accumulator. This designation is fixed and does not change when charging or discharging. The negative electrode always has lower potential compared to the positive electrode.

Note
During discharging: anode = negative electrode, cathode = positive electrode
During charging: anode = positive electrode, cathode = negative electrode

In many non-scientific publications and in everyday language, the negative electrode is generally referred to as the anode and the positive electrode as the cathode, i.e. only the discharge process is considered. This book also follows this convention.

In lithium-ion batteries, the anode material, usually graphite or graphite–silicon, is applied to a current collector foil made of copper. The cathode material, usually some variant of lithium-metal, is applied to a current collector foil made of aluminum.

The capacity of a battery cell is determined by the absorption capacity of ions in the anode and in the cathode material. To achieve the highest possible energy density, all components not involved in the storage process, particularly the housing, the separator and any current collector foils, must be as thin as possible.

A battery storage system usually consists of many cells connected in parallel and/or in series, a housing, a cooling system, electrical connections, a thermal safety system and a battery management system (BMS). This regulates temperature control, balancing, protection against overcharging and deep discharge, temperature monitoring, determination of the state of charge, determination of the current permissible charging and discharging capacity and more.

6.3.2 Battery Losses

Essentially, three different loss mechanisms occur in a battery:

- Resistive heating: heating of the battery due to the internal ohmic resistance
- Charging losses: for example, due to gassing of the electrolyte
- Self-discharging: electrochemical process (temperature-dependent)

Cycle efficiencies of around 95 up to more than 99% are possible with lithium-ion batteries, but only around 70% with NiMH batteries. The cycle efficiency of lead batteries is around 80%.

6.3.3 Battery Voltage

The nominal voltage of a battery results from the position of the anode and cathode elements in the electrochemical voltage series. Some selected elements are listed in Table 6.2.

By definition, the electrochemical voltage series has its zero point at hydrogen (H). Lithium is the element with the lowest voltage (−3.04 V) and fluorine (F) is the one with the highest voltage (+2.87 V). Theoretically, an optimal battery would be lithium–fluorine with a voltage difference of almost 6 V. Unfortunately, the reaction of these elements results in a salt (lithium fluoride = LiF) and is irreversible.

Table 6.2 Electrochemical voltage series (extract)

Element	Standard potential
Fluorine (F)	+2.87 V
Oxygen (O)	+2.00 V
Gold (Au)	+1.69 V
Chlorine (Cl)	+1.36 V
Chrome (Cr)	+1.33 V
Lithiated graphite (LiC_6)	+0.75 V
Hydrogen (H_2)	0 V per definition
Lead (Pb)	−0.13 V
Nickel (Ni)	−0.23 V
Cobalt (Co)	−0.28 V
Vanadium (V)	−1.17 V
Manganese (Mn)	−2.362 V
Lanthanum (La)	−2.522 V
Sodium (Na)	−2.71 V
Lithium (Li)	−3.04 V

There is also no electrolyte that can withstand such a high voltage without decomposing. Lithium fluoride is therefore an unsuitable material combination for rechargeable batteries.

Oxygen also has a high electrochemical voltage (+1.23 V). Research on the lithium–air battery is ongoing, but the efficiency is relatively poor, and the reversibility of such cells is unresolved for the foreseeable future. So far, only three to four charging cycles are possible. This battery variant is therefore still a long way from being ready for series production.

In lithium-ion batteries, lithium (−3.04 V) is used against graphite (+0.75 V). This results in a cell voltage of 3.8 V. However, instead of pure lithium, a lithium alloy with iron, chromium, phosphate, nickel, or manganese is used on the cathode. This increases the full-load cell voltage to around 4.2 V.

The following figure shows the typical curve of the battery voltage of a lithium-ion cell without load (OCV), i.e. no current flow, depending on the state of charge (SOC).

The course of the open-circuit voltage is characteristic of the cell chemistry used in the battery and depends on the temperature and the state of aging (Fig. 6.3).

The measurement is carried out by means of constant current measurement with low current (<C/20) or by current interruption measurement with relaxation phases.

The area under the OCV discharge curve corresponds to the energy stored in the battery. The area between the charging and discharging curves corresponds to the energy loss (cycle efficiency), see Eq. (6.2).

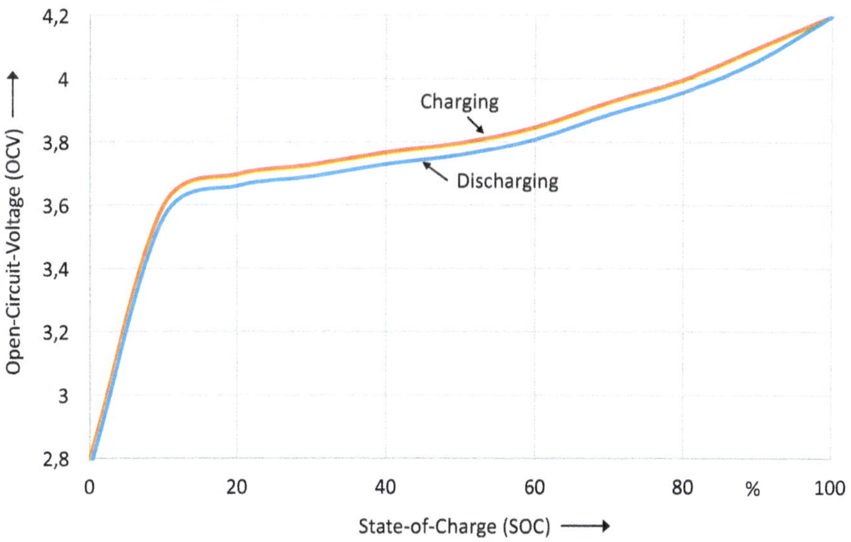

Fig. 6.3 Typical charge/discharge curve of a lithium-ion battery cell

As soon as a current is applied to the battery, the terminal voltage decreases slightly and thus deviates from the open-circuit voltage. This voltage difference, which leads to losses in the battery, can be divided into two groups:

- Current-proportional voltage drop due to the internal resistance of the battery

$$\Delta U_{Ri} = R_i \cdot I \tag{6.3}$$

- Further, non-linear voltage drop, η_k, in electrochemistry also called overvoltage.

In electrochemistry, the term "overvoltage" refers to the additional charging voltage required compared to the open-circuit voltage.[1] When discharging, an "undervoltage" occurs because the output voltage is lower than the open-circuit voltage. In both cases, the term "non-linear voltage drop" would be more appropriate.

The overvoltage depends on many factors, including the state of charge, the current, the temperature, or the state-of-health (aging). Causes of overvoltage are diffusion, charge transfer through the SEI layer, and others. For the modeling of overvoltage effects, complex battery models are required, see also Sect. 6.7.3.

The terminal voltage of a battery can be described by the following equation:

$$U_{terminals} = U_{OCV} - \Delta U_{Ri} - \sum_k \eta_k \tag{6.4}$$

There are various measurement methods for the recording and separation of the various loss processes:

- Constant current measurement: charging and discharging the energy storage device with different currents.
- Impedance spectroscopy: Excitation of the cell with small, sinusoidal current signals of variable frequency and simultaneous measurement of the voltage response.
- Step response: Measurement of the current response to a voltage step. A steep step excites all frequencies simultaneously. Therefore, the frequency-dependent cell impedance can be determined with suitable filtering and Fourier analysis of the current signal.

Due to the overvoltages and the electrochemical processes in the cell, the battery capacity depends on the current curve.

[1] In electrical engineering, the term "overvoltage" or surge is used differently, namely for a voltage that is significantly higher than the normally prevailing measured voltage, for example in the event of a lightning strike.

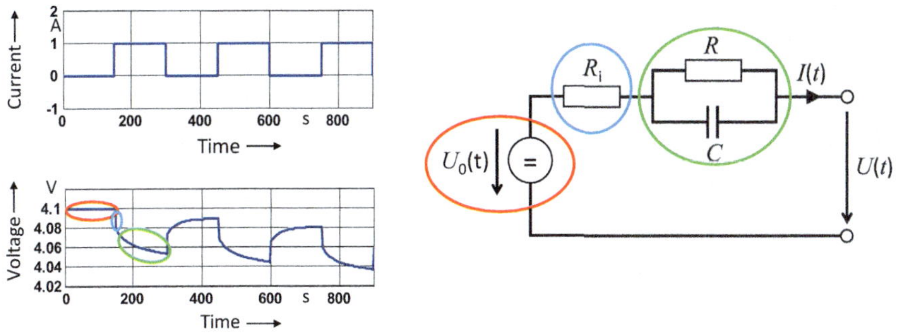

Fig. 6.4 Extended equivalent circuit diagram of a lithium-ion battery

6.3.4 Simple Behavior-Based Battery Models

An electrical equivalent circuit diagram can be used to model the static and dynamic effects of the voltage at the terminals of a battery. In the simplest case, the equivalent circuit diagram for lithium-ion batteries only consists of an ideal voltage source and an internal resistance. However, this does not adequately reflect the actual, measurable operating behavior.

Figure 6.4 shows a slightly extended equivalent circuit diagram:

The two images on the left show measured values. The equivalent circuit diagram elements on the right model the various sections of the measured curves:

The red circle represents the stationary part, i.e. the open-circuit voltage. It is not constant but must be adjusted depending on the state of charge, temperature, and aging.

The blue circle models the voltage drop across the ohmic resistance. If the current is not kept constant, the cell voltage drops further to a final value. The green circle models this non-linear voltage drop.

Realistic models used in the automotive industry use several RC elements, typically 2–3, to map the voltage curves even more accurately. Models with up to 100 elements are examined for more precise investigations and for estimating the state of charge.

More precise statements on voltages and losses are possible with physical/electrochemical battery models.

6.3.5 Battery Lifetime

The lifetime of a battery storage system is limited by time (calendar life) and by the number of charging/discharging cycles (cycle life). The separation of calendar and cycle life is difficult in practice.

The cycle life is usually more limiting and is either given as the number of full-load (100% DOD) cycles or as the total energy throughput. For example, a 60 kWh battery with a lifetime of 1000 cycles has an equivalent of 60,000 kWh total energy throughput, which corresponds to 300,000 km of an average electric car. In practice, the lifetime will be achieved typically by more and smaller cycles, for example 1250 cycles between 90% and 10% SOC.

The main cause of cycle aging is mechanical damage to the active masses of the anode and cathode and secondary reactions. The typical guaranteed cycle life of automotive-grade cells today is more than 3000 full cycles.[2]

The main cause of calendar aging is the temperature-dependent, continuous decomposition of the electrolyte components. With lithium-ion batteries, a calendar life of 10–15 years is now realistically achievable.

Causes of aging are, for example:

- Loss of contact of the current collector foils (copper, aluminum) due to decomposition of the binder.
- Damage to the current collector foils due to corrosion and cracking
- Dissolution of copper atoms from the current collector foil and transport into the anode.
- Formation of metallic deposits (lumps) on the surface of the graphite anode (lithium plating), which locally prevent the passage of lithium ions.
- Formation of growths (dendrites) on the anode, which can originate from both dissolved copper and lithium.
- Crack formation in the graphite particles of the anode.
- Decomposition and partial reformation of the SEI (solid electrolyte interphase).
- Local short circuits within the cathode.
- Cracking in the particles of the cathode.
- Formation of cathodic surface films.

As a rule of thumb (Montsinger rule): If the end-of-charge voltage is regularly increased by 0.1 V above the nominal value, the lifetime of the battery is halved.

A battery is usually considered to be faulty if its capacity has fallen below 70 or 80% of the nominal capacity. The exact value is a matter of definition. It differs slightly depending on the manufacturer. The lifetime is also limited by the available output power of the battery. Due to aging, the internal ohmic resistances increase and consequently the available current at a given voltage reduces which also reduces the available output power.

The successively decreasing battery capacity is sometimes concealed from the user by stating a lower net capacity than the installed gross capacity.

Example 6.4—Gross vs. Net Capacity in a Chevrolet Volt

The Chevrolet Volt reports an "empty battery" after consuming 10 kWh. Then, the combustion engine is turned on. However, the nominal capacity of the battery when new is 16 kWh.

[2]The Chinese manufacturer BYD even quotes a typical lifetime of 5000 full cycles for its LFP car batteries.

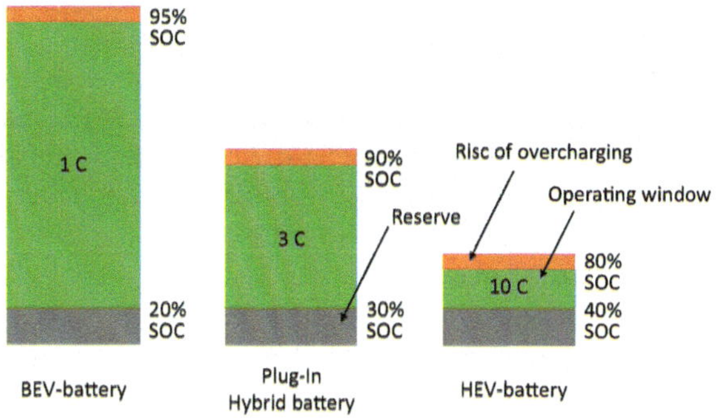

Fig. 6.5 Typical use of batteries in various hybrid and electric cars

The remaining capacity is used to operate the hybrid modes (braking energy recovery, load point shifting, boosting). It is also available as a reserve for capacity loss due to aging. ◀

The lifetime of battery cells depends very much on the charge per each cycle (DOD). Also, temperature is important. In general, the warmer a battery is, the faster it ages (Fig. 6.5).

Typical batteries for hybrid vehicles (MHEV and FHEV) only have a low capacity. They are flat cycled, i.e. the net capacity used (DOD) is only around 30–40% of the gross capacity. This enables a high number of cycles for hundreds of thousands of traffic light starts and other driving maneuvers. Ideally, the battery is kept in a medium state of charge by the control system to ensure maximum flexibility for the various hybrid operating modes, such as load point increase, load point reduction, boosting, or braking energy recovery. Batteries for HEVs are therefore primarily optimized for a high number of cycles.

Batteries for fully electric vehicles (BEV) and plug-in hybrids (PHEV), on the other hand, have a much larger capacity and are usually operated much closer to their gross capacity. The DOD is typically around 60% to over 90%. Accordingly, the required number of cycles is significantly lower.

Of course, a BEV battery is also subject to the micro cycles of a hybrid battery, for example during traffic light starts. However, due to the low charge turnover in relation to the total capacity, these do not have a noticeable life-limiting effect.

Note
At 1–3 C, the discharge rate of typical BEV and PHEV batteries is significantly lower than that of typical HEV batteries (around 10 C). However, due to the higher capacity, this corresponds to a similar power output:

1 C of 60 kWh (typical BEV battery) is 60 kW peak power.
20 C of 3 kWh (typical HEV battery) is also 60 kW.

Example 6.5—Battery lifetime of Renault Zoe

The car manufacturer Renault announced in mid-2022 that 99% of all batteries installed in the Renault Zoe since its market launch in 2013 are still fully functional. They still have at least 70% of their original capacity. This applies to all battery types of the Zoe, from the 22 kWh battery of the first years of production to the current 52 kWh battery.

Renault grants a warranty of eight years or 160,000 km on batteries purchased for the Zoe—after the specified period or mileage, the batteries should have a minimum capacity of 70% of the new value. However, less than one percent of all Zoe batteries ever put into circulation in Germany have fallen below this mark and had to be replaced. This also applies to vehicles whose battery warranty has already expired. ◀

Example 6.6—Battery lifetime study by Varta

The company Varta/Johnson Control carried out a detailed investigation into the lifetime of NiMH cells. At a temperature of 20 °C and a depth of discharge (DOD) of 100%, the cell was able to deliver the energy of around 2500 full charge cycles until the end of its service life.

If the depth of discharge was reduced to 80% DOD, the lifetime increased to 5000 full charge cycles.

If the depth of discharge was reduced even further to just 12% DOD, i.e. the SOC range was just 44–56%, the lifetime was increased to 10,000 full charge cycles. However, this operating window is far too small for most applications.

At colder temperatures around 10 °C instead of 20 °C, the lifetime increased by around 50%. At temperatures around 40 °C, it decreased to around 50%. Temperatures around 50 °C even led to a reduction in lifetime to a quarter. ◀

As a rule of thumb, the lifetime of the battery decreases by 50% for every 10 K above 25 °C if the battery is operated continuously at these temperatures. Therefore, batteries should not be regularly operated at over 80 or 90% SOC and not regularly below 20 or 10% SOC.[3] For longer periods of storage without use, temperatures under 20 °C and charge levels between 40 and 60% are recommended.

Batteries do not suffer damage at very low temperatures even far below 0 °C but can then only deliver low power. Also, the available energy is significantly reduced, see the following chapter. Batteries can be stored at temperatures of around −20 °C or less without negative impact on lifetime. However, the battery must be heated prior to usage.

[3] The exception to this rule is LFP-batteries. LFP-cathodes are very cycle stable and can be charged regularly to 100% SOC without a significant reduction in lifetime.

Example 6.7—Battery cycles over their lifetime

The average lifetime of a car in Europe is nearly 12 years. The oldest fleet operates in Lithuania and Romania, where the average age is 17 years.

The average annual mileage of a car is around 13,000 km. This results in a typical total milcagc of around 160,000 km.

Assuming a typical BEV car with a 60 kWh battery and a consumption of 20 kWh per 100 km, this corresponds to around 530 full cycles over the entire lifetime. ◀

In general, the lifetime of today's vehicle lithium-ion batteries is significantly longer than that of cell phone batteries, for example. This is partly due to the changed chemical composition of the internal components, see Sect. 6.7.4. In addition, the battery management system (BMS) of automotive batteries is more sophisticated, see Sect. 6.13. The state of charge of each individual cell is constantly monitored and compensated for any deviations (balancing). During charging and discharging, the maximum currents are precisely controlled depending on the state of charge, temperature, and aging.

Vehicle batteries also always have an internal buffer so that they are practically never charged to 100% or discharged to 0%.

Example 6.8—Battery warranty

Audi, Mercedes, Tesla, and VW provide a warranty of 8 years or up to 160,000 km on their batteries. At BMW, the 8 years even apply without a mileage limit. ◀

Example 6.9—High-mileage electric cars

Sometimes electric cars are driven with quite astonishing mileage. In Karlsruhe, Germany, a Tesla Model S was driven an average of 150,000 km a year and passed the 1 million km mark at the end of 2019.

The first battery had to be replaced after 290,000 km. The manufacturer's loaner battery that was then installed was replaced after a further 250,000 km. The third battery has now covered 460,000 km and still has a remaining capacity of 86%.[4]

The remaining battery capacity of all Tesla S and X cars after 320,000 km (200,000 miles) is still 88% on a statistical average according to Tesla Inc.[5] ◀

6.3.6 Available Power and Energy

Depending on the C-rate during discharge, a battery provides a different amount of energy. With OCV measurement, the energy, i.e. the area under the curve of

[4] *Elektro-Auto spult eine Million Kilometer ab*, Badische Neueste Nachrichten BNN, Wolfgang Voigt, issue 277, 29.11.2019, page 11.

[5] *Impact Report 2022*, Tesla Inc., https://www.tesla.com/ns_videos/2022-tesla-impact-report.pdf.

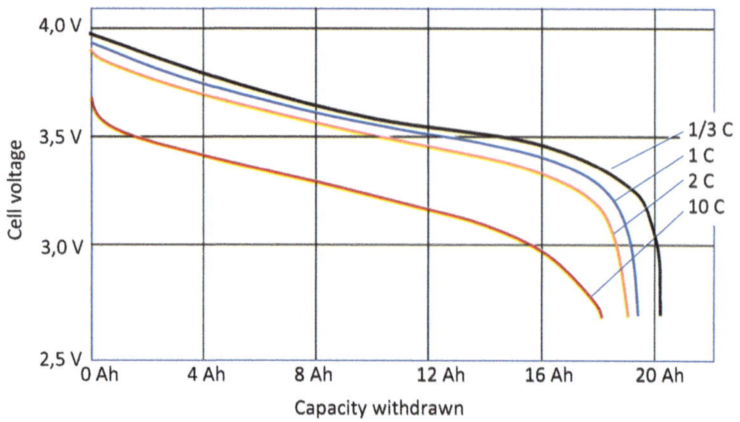

Fig. 6.6 Cell voltage curve over the capacity withdrawn depending on the discharge rate[6]

the cell voltage above the SOC, is the highest. If the discharge current is increased and thus the C-rate, less energy is available because the final discharge voltage is reached earlier. The effect is reversible: After a few hours of idling, the OCV voltage returns to a level in line with the previously drawn energy. The cause of this effect is the time required for the uniform diffusion of the lithium ions in the graphite layers of the anode. However, the reduction in the available energy only becomes noticeable at relatively high C-ratios of around 10.

Because all electrochemical processes in the battery are thermally activated, they run faster at higher temperatures.

When charging at high charging rates (2–3 C or even more) is desired, the battery must be brought up to a suitable temperature, typically around 30–50 °C. Nearly all electric cars today have built-in battery thermal management system. In many cars, the system automatically starts a heating process around 20–30 min before reaching a fast charging station, provided the destination has been programmed into the navigation computer. Because these temperatures are only reached for a short time, they do not have too negative an effect on the lifetime. However, frequent fast charging reduces the lifetime of a battery to some extent.

The temperature also has a significant influence on the available energy. At high temperatures of around 60 °C, the available energy also increases by around 10%, but aging also increases significantly.

At low temperatures, the amount of energy available decreases because the electrochemical processes slow down. At temperatures around freezing point and a discharge rate of 1 C, for example, around 10–20% less energy is available. At temperatures around −20 °C and below, the amount of energy available can be halved (Figs. 6.6 and 6.7).

[6] Data source: Information from the company Saft.

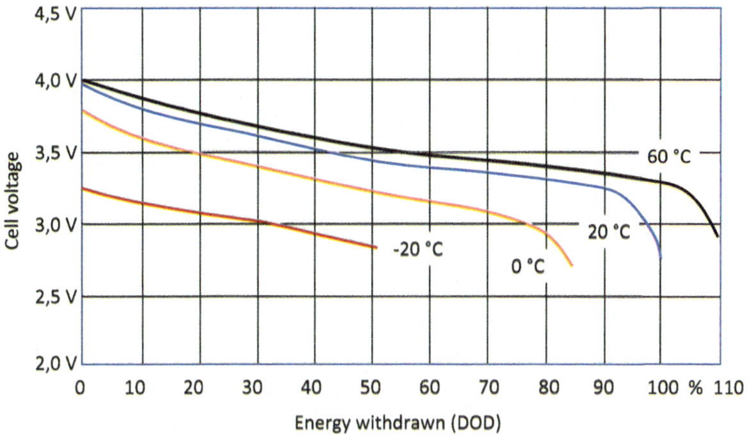

Fig. 6.7 Cell voltage curve over the energy withdrawn depending on the ambient temperature[43]

It is therefore quite useful to heat batteries during driving in cold weather. Although energy is required for heating, this is usually far less than the loss of capacity when the battery remains cold. The battery management system must find a good compromise to optimize energy.

Most mid-range and luxury electric cars today have battery temperature control. However, some smaller and cheaper battery electric vehicles do not for cost reasons.

6.3.7 Battery Packages

There are three distinctly different designs of battery cells: cylindrical, pouch, and prismatic.

6.3.7.1 Cylindrical Cells

In classic round cells (cylindrical cells), long strips of the individual materials are rolled up in a cylindrical metal housing. The strips consist of the negative electrode, which is a copper foil coated on both sides with graphite, the separator (plastic foil) and the positive electrode, which is an aluminum foil coated on both sides with lithium cobalt oxide. During winding, all electrodes are separated from each other by a separator layer.

The advantage of cylindrical cells is their automated production, which is based on extensive production experience with batteries for consumer electronics, as well as their long lifetime. Disadvantages are the complex cooling process (the interior of the cells cannot be cooled at all) and the loss of installation space due to the resulting gaps.

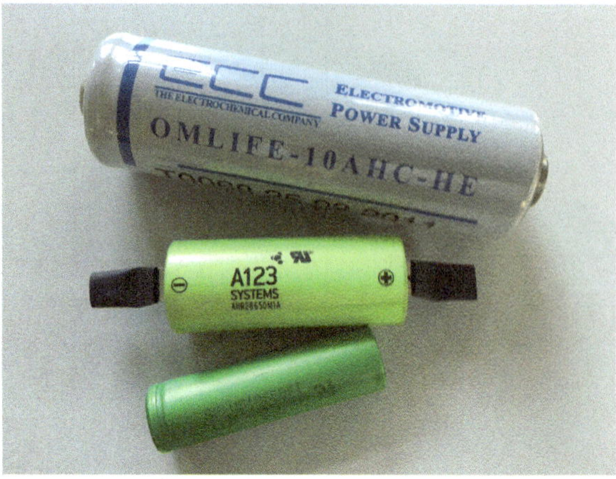

Fig. 6.8 Cylindrical lithium-ion cells of various sizes (contacts taped for safety reasons)

Cylindrical cells are named according to their diameter and length, e.g. 18650 = 18 mm diameter with 65 mm length or 21700 = 21 mm diameter with 70 mm length. The final 0 indicates that these are round cells. Other, less common designs are 14500, 26650 and 32650. Newer designs tend to increase the diameter, for example 4695(0) and 46120(0), which will be used in Gen6 of BMW cars from 2025.

The manufacturer Tesla Motors uses round cells in its vehicles. In the first model, the Tesla Roadster from 2008, these were 18650 cells from the battery production for notebooks. Since the collaboration between Tesla and Panasonic with the introduction of the Model S, however, the cell chemistry has been optimized specifically for automotive applications. In the newer vehicles, i.e. the Model 3 and the Model S Raven, 21700 cells are used, which enable a slightly higher packing density. The latest generation from Tesla is 4680 cells (Fig. 6.8).

6.3.7.2 Pouch Cells

In pouch cells, several rectangular layers of negative and positive electrodes are stacked with corresponding layers of separators like in a sandwich. This increases the capacitance and energy content compared to cylindrical cells. Pouch cells have a packaging utilization rate up to 90–95%, which is the highest utilization rate of all battery packaging methods.

The electrodes are usually connected to one side. The cells are housed in sealed and welded metal foils, usually aluminum. This is also known as coffee bag design.

Pouch cells can be easily cooled and stacked. They have a high volumetric energy density. However, achieving tightness through the foil is somewhat more difficult. Today, many manufacturers of electric cars use pouch cells.

The most common design is 162 mm (length) × 330 mm (width) × 7 mm (thickness) (Fig. 6.9).

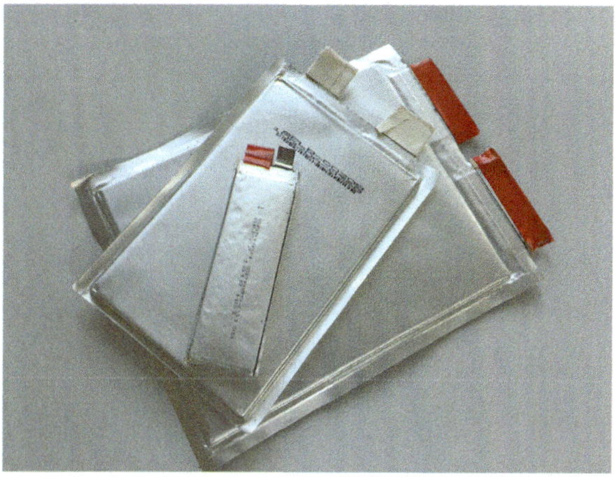

Fig. 6.9 Lithium-ion pouch cells of different sizes (contacts taped for safety reasons)

6.3.7.3 Prismatic Cells

Prismatic cells are fundamentally constructed like pouch cells. A jelly pack typically consists of four jelly rolls, that is, a large-size battery is folded into four layers. The housing is made of a solid material, usually aluminum or steel. This simplifies the design of battery packs, as the mechanical stability is quite high, but the cooling properties are poorer. Safety features are usually integrated into the housing.

There are two common designs for prismatic cells: 91 mm × 148 mm × 26.5 mm and 115 mm × 172 mm × 45 mm (length × width × diameter) (Table 6.3).

To further increase the energy density of the battery cells, prismatic cells are being designed with ever larger dimensions (cell-to-pack design). So-called blade cells have standard widths of around 90–120 mm and standard thicknesses of around 13 mm. However, at around one meter, their length is considerably greater than that of conventional pouch cells. Such cells are often integrated into complete battery packs and cell modules are not required. The production of such cells places high demands on the accuracy of all process steps. Blade cells can reach a gravimetric cell-to-pack ratio of more than 85%.

6.3.7.4 Integration of Batteries into the Car Body

In modern cars, the battery pack is mounted directly into the chassis as a compact sandwich between the axles. This design is also known as a skateboard design. The seating position is slightly raised, but otherwise the battery does not take up any usable space.

In addition, the skateboard design increases the stability of the car body, and the low center of gravity ensures better cornering stability and increased safety against overturning of the vehicle after an accident. The skateboard design is a key reason why electric cars regularly perform so well in crash tests.

Table 6.3 Technical data of the battery systems of selected BEVs

Car	Market entry	Gross capacity kWh	Net capacity kWh	System weight kg	Gross energy density Wh/kg	Cell design	Battery type
Tesla Roadster	2006	56	53	410	137	18650	
BMW i3	2013	22	18.8	230	96	Pouch	NCM111
Renault Zoe	2013	25.9	23.3	290	89	Pouch	
BMW i3	2016	33.2	27.2	256	130	Pouch	NCM111
Renault Zoe	2017	45.6	41	305	150	Pouch	
Tesla Model S	2016	102	98	590	173	18650	NCA
Tesla Model 3	2017	80.5	75	478	168	21700	NCA811
Mercedes EQC	2019	80	76	650	123	Pouch	NCM622
Audi e-tron 55	2019	95	83.6	700	136	Pouch	
Jaguar i-Pace	2019	90	84.7	605	149	Pouch	
Polestar 2	2020	78	75	500	156	Pouch	
VW ID.3	2020	62	58	364	170	Pouch	NCM712
BMW iX	2021	111.5	105.2	600	186	Prismatic	
BMW	2025	tba	tba	tba	tba	4695(0) 46120(0)	NCM811
NIO ET7	2024	150	Unknown	575	260	Unknown	Semi-solid state

The electric drive requires less space than a conventional combustion engine with an automatic transmission, so the overhang for the hood can be shorter and the cab can extend further forward towards the wheels. This is known as cab-forward design. The interior can typically be made a whole car class larger than in a vehicle with a combustion engine. Many electric cars even have an additional front trunk (frunk) under the hood.

In the event of a head-on collision, there is no longer any risk of the heavy and stiff engine/transmission block being pushed into the interior of the vehicle. This also contributes to the improved occupant safety in electric cars.

Skateboard and cab-forward design has become the standard in all pure electric car platforms from companies like Tesla, Jaguar, Volkswagen, Mercedes

Benz, Volvo, Nio, and many more. There are only a few manufacturers left who still offer "multi-purpose" bodies that accommodate several drive systems (ICE, PHEV, BEV, and sometimes even FCEV) as poor compromises.

The next generation in electric car body construction is battery housings that become part of the vehicle's support structure (cell-to-body design). The first steps in the direction of integrated battery structures have already been taken, for example since mid-2022 in the Tesla Model Y from the production in the Grünheide factory near Berlin, Germany. The latest BYD cars, for example the BYD Seal, have LFP blade-cells with a cell-to-body design.

The final step in car body construction may one day be multifunctional energy storage composite (MESC) structures with embedded battery layers. The battery stack (multiple layers of anode, separator, and cathode) is inserted in fiberglass or carbon-fiber composites and becomes an integral part of car body structures like doors, hoods, roofs, fenders, etc. Initial attempts in this direction are already being made, but series implementation is still a long way off at present.

6.3.8 Battery Safety

Lithium in its liquid state and as metal dust spontaneously ignites when it comes into contact with oxygen (air) even at normal room temperature. However, this effect does not occur in lithium batteries because lithium is only used as a metal oxide.

Nevertheless, severe mechanical damage, for example after an accident, mechanical production faults, excessive charging (dendrite building), and excessive currents can lead to short circuits inside the battery that cause even higher currents, high temperatures, and ultimately a fire.

The fire risk of the various materials used in batteries varies greatly. Batteries with NCA or LCO cathodes are particularly critical, LMO and LFP cathodes are much more save, see Sect. 6.7.4.

Modern batteries are equipped with several safety measures to reduce the risk of fires as will be shown below. For a general discussion of car fires, see Sect. 6.17.

Note

The electrolyte begins to vaporize at 80 °C. This leads to a build-up of pressure, which can cause the cell to burst. At 130 °C, the separator (PE, PP, or PE/PP) melts, additional heating occurs due to the short circuit. An autocatalytic (self-increasing) rise in temperature occurs.

At 230 or 300 °C (NCA, LCO) to 450 °C (LFP, LMO), the cathode material reacts exothermically with the electrolyte leading to decomposition. Pressure increases in the cell due to vaporization and decomposition gases.

Next, swelling of the cell housing occurs that leads to a possible opening (escaping decomposition gases are ignitable). Some cathode materials decompose spontaneously at temperatures below 200 °C and release heat and oxygen in an exothermic reaction, which can lead to a thermal runaway.

At 600 °C, the cathode materials decompose and change their crystal structure. Oxygen is released. A cell fire will start within a short time and a thermal runaway is inevitable.

At 660 °C, the aluminum current collector foils melt. Subsequently, graphite is released, leading to a risk of dust explosion.

Finally, a further increase in temperature results in burning aluminum foils (metal fire).[7]

6.3.8.1 Charging and Discharging

When charging a battery, both the maximum current limit and the voltage limit of each individual cell must be observed precisely to avoid permanent damage.

If the end-of-charge voltage is exceeded significantly and for several minutes, the electrolyte will decompose, leading to the formation of gases. If overcharging is continued for several minutes, more and more gases are produced and the cell inflates. Ultimately, the cell can burst open and the toxic electrolyte is sprayed out. At this moment, contact with the airborne oxygen can lead to a fire or even to an explosion.

When a lithium-ion battery cell is deeply discharged below its end-of-discharge voltage, initially, nothing happens. The cell is intrinsically safe. However, the current collector foils can dissolve during deep discharge and copper (anode) becomes Cu^{2+}. When the cell is subsequently charged again, the Cu^{2+} can deposit on the cathode and then lead to dendrite formation. These dendrites can short-circuit the cell and lead to fires and even explosions later. Therefore, deeply discharged cells must never be recharged but rather must be recycled.

6.3.8.2 Ceramic Coating of Separators

Modern separators are composite materials of polypropylene (PP) and polyethylene (PE). Some of them have additional aluminum oxide ceramic coatings (Al_2O_3), which prevent the penetration of dendrites and therefore short circuits.

If a breakdown still occurs, the heat from the short-circuit current immediately closes the gap by melting the separator locally (shut-down separator). PP melts from about 130 °C, PE from about 160 °C.

6.3.8.3 Safety Valves

Most cylindrical cells have a safety vent that interrupts the current (CID = current interruption device) when excessive gas pressure is building up inside the cell. If

[7] Source and further reading: https://www.brand-feuer.de/index.php/Lithium_Batterien.

too much current flows, heat is generated and the electrolyte is partially vaporized. Like a crack-frog, a metal disk bulges outwards and opens an electrical contact thereby disconnecting the plus and minus pole. The release depends on the internal gas pressure of the cell and is not reversible.

In typical automotive applications, many cylindrical cells are connected in parallel. If one of them triggers the safety valve due to an internal fault, the external contacts are electrically disconnected. However, the remaining cells continue to function and the vehicle can continue to be operated with a slightly reduced battery capacity.

Prismatic cells are typically connected in series and only a few strands of many serial cells are then connected in parallel. Therefore, in case of excessive gas pressure, a thermal valve disconnects the internal cell just like in cylindrical cells. However, the outer contacts of the prismatic housing are then short-circuited. The short circuit is produced by a snap-action contact, like the operating principle of the safety vent in cylindrical cells. In a typical car application, the affected strand will then have a lower voltage but can still be used.

Most cells have an emergency burst valve, usually a built-in weak point in the housing, which opens at very high internal pressure and releases the gas to the outside, preventing an explosion. However, this process usually leads to a cell fire but prevents an explosion.

6.3.8.4 Thermal Fuses in Cylindrical Cells

Thermal fuses (PTC = positive temperature coefficient) consist of a material that suddenly increases its electrical resistance when heated. In cylindrical cells, this material is inserted on the top and/or bottom of the cylinder between the outer electric connections and the internal current collection foils. When temperature builds up due to an excessive current, the increasing resistance reduces the current and consequently the temperature.

This security feature is activated before the CID. The process is reversible.

6.3.8.5 Internal Short Circuit Tolerance

Car batteries must be tested for security against penetration by external objects. This can be caused by external sharp metal objects penetrating the battery in a traffic accident. The nail penetration test is a safety test that tests the internal short circuit tolerance of lithium-ion batteries. It has been widely used across the battery industry and battery-user community to assess lithium-ion battery safety and is regarded as one of the most rigorous ways to test the thermal runaway of batteries.

The SAE Standard J2464 (Electric and Hybrid Electric Vehicle Rechargeable Energy Storage System (RESS) Safety and Abuse Testing) defines abuse tests for car batteries including the nail penetration test. This test requires a nail of 3 mm diameter penetrating the full thickness of the fully charged battery perpendicular to the electrodes at a speed of 80 mm/s. Any released gases will be ignited by a nearly spark source. The cell/chemistry is deemed to have passed if there is no smoke or flame following the nail penetration.

6.3.8.6 Fire Containment by the Battery Pack

In addition to the requirements for individual cells, the battery pack must also be designed in such a way that a possible fire is contained for at least 5 min and does not lead to smoke gases or flames inside the vehicle. This is achieved by using phase change materials (PCM) between the cells and modules to delay the propagation of a fire. PCMs are substances that absorb large amounts of latent heat when they go through a change in their physical state, i.e. from solid to liquid. For example, paraffin wax or hydrated salt can be used. If a cell starts to burn, the heat generated is initially absorbed by the surrounding PCM. Only when this has completely melted does the heat reach the neighboring cells and can then also set them on fire in the worst case.

6.4 Lead-Acid Batteries

Lead-acid batteries have low material costs and are the cheapest secondary elements available. Technical modifications, such as lower electrode width and improved grid design, can improve performance and lifetime. Nevertheless, the energy density and cycle stability are too low for hybrid and BEV vehicles.

The positive electrode consists of lead oxide (PbO_2), the negative of a fine-pole, sponge-like lead metal. The electrolyte is 35% sulfuric acid. A highly porous polyether film or a glass fiber fleece is used as a separator.

As the battery discharges, the PbO_2 on the positive electrode and the metallic lead on the negative electrode are converted into lead sulfate. In contrast to other battery systems, the electrolyte actively participates in the discharge reaction. This leads to a reduction in the concentration of sulfuric acid as the discharge continues.

The typical open-circuit voltage of the full battery is 12.6 V. It drops to 11.8 V when fully discharged.

During discharge, the reaction proceeds as follows:

Lead and lead oxide and sulfuric acid, dissolved in the electrolyte, are converted into lead sulfate and water. The lead changes its oxidation number: from Pb^{4+} (oxygen O is 2−) and Pb0 to Pb^{2+}.

At the negative electrode, metallic lead (Pb is 0 +neutral) changes to Pb^{2+} ($1PbSO_4$). Two electrons are released in the process.

At the positive electrode, lead dioxide (PbO_2) (Pb is 4+) transforms into lead sulfate Pb 2+ ($1PbSO_4$). Two electrons are taken up in the process. An ion current of H^+ takes place in the electrolyte.

Lead gel batteries are a closed system and completely maintenance-free. They are used in luxury vehicles and motorcycles. In conventional batteries (lead–liquid acid), water from electrolysis is decomposed to protect against overcharging. Occasionally, water must be refilled.

6.5 Nickel–Metal Hydride Batteries (NiMH)

The nickel–metal hydride battery can be seen as the successor to the nickel–cadmium battery, which is no longer used for environmental reasons.

One characteristic is the strong, electrochemically induced voltage hysteresis. This theoretically limits the energy efficiency to around 92%. Practically implemented systems are significantly lower and typically have a cycle efficiency of around 70%. The advantage of this battery is the wide operating temperature range down to −20 °C.

The positive electrode consists of nickel(III) oxide hydroxide, the negative of a hydrogen-storing alloy (metal hydride). The electrolyte is a highly concentrated alkaline solution with potassium hydroxide (KOH) as the main component. It mainly serves as a transport medium for the charge carriers (protons) and is not involved in the basic reaction. Polyethylene (PE)/polypropylene (PP)-based polymer fiber fleece is used as a separator.

The chemical cell reaction is as follows: $Ni(OH)_2 + M \longleftrightarrow NiOOH + MH$

The open-circuit voltage is 1.25–1.35 V. C ratios of 1–5 are possible.

NiMH batteries are technically mature and significant improvements are not expected in the future. Developments are concentrating on optimizing the overall system in terms of costs, packaging, and battery management.

Today, NiMH batteries are still used in mild and full hybrids from Toyota, which has its own production facilities. In the long term, NiMH batteries in passenger cars will be completely replaced by lithium-ion batteries.

6.6 Vanadium Redox Flow Batteries

The vanadium redox flow battery (VRFB) is a not actually a battery but rather a tertiary element like a fuel cell. The energy is stored in two aqueous electrolytes that circulate in separate circuits and separate tanks. The size of the tanks determines the amount of energy stored.

The energy conversion takes place in a stack whose output is determined by the size of the electrodes. Ion exchange takes place via a membrane.

In a redox flow battery, power and energy are therefore independent of each other.

The negative electrode is vanadium 2+ (V^{2+}), which transforms into vanadium 3+ ($V^{3+} + e^-$) by releasing an electron. The positive electrode is vanadium 5+ ($VO_2^+ + 2H^+$), which changes to vanadium 4+ ($VO^{2+} + H_2O$) by absorbing an electron. A highly porous polyethylene film (Nafion), which is permeable to hydrogen ions H^+, is used as a separator.

The high cost of the aqueous electrolytes and the low energy density, which is roughly equivalent to that of a lead–acid battery, are problematic. The redox flow battery is therefore hardly suitable for use in vehicles. However, it can be used as a static energy storage system to stabilize power grids, as the high amount of energy required can then be stored quite easily in correspondingly large tanks. The electrolyte can be reprocessed many times and has a long service life. At the end of its lifetime, vanadium, the most expensive element of the system, can be recycled rather easily. In this way, redox-flow storage systems can be operated economically on a large scale.

Example 6.10—Redox flow battery at KIT, Karlsruhe, Germany

A redox flow battery with an electrolyte volume of 53 m^3 and a discharge capacity of 860 kWh was installed at the grid connection point at the Karlsruhe Institute of Technology (KIT), Germany, for testing purposes in 2019.

The battery has a DC voltage of $700\,V \pm 50\,V$ and a DC-rated power of ± 200 kW. The AC cycle efficiency at the grid connection point (3×400 V AC) is 68%. ◀

Recent developments in vanadium–air redox flow batteries are promising. One of the tanks can be omitted, which doubles the energy density. When used in cars, the "charging time" is short, as the charged liquid is simply filled into the tank. However, these batteries are still at the research stage and are not yet commercially available.

6.7 Lithium-Ion Batteries

6.7.1 Introduction

Lithium-ion batteries (LIB) are in several ways different from other battery types.

The lithium is always ionic and is not oxidized or reduced. Instead, the electrodes are oxidized and reduced, the lithium ions are merely stored (intercalated) in the anode and cathode. As a result, the reaction is faster, and more often reversible and the storage capacity is higher.

Secondly, at up to 4.2 V, the voltage of lithium-ion batteries is significantly higher than that of all other electrochemical cells. In fact, it is so high that all known organic electrolytes decompose. A chemical passivation layer (SEI = solid electrolyte interphase) is therefore deposited on the anode. The SEI is practically an electrical insulator, but it allows lithium ions to pass through unhindered.

These two properties give the lithium-ion battery superiority over other battery types. Due to their high energy density, lithium-ion batteries are ideal for plug-in hybrids (PHEV) and battery electric vehicles (BEV) and in almost all other areas (hand tools, smartphones, flashlights, etc.). They are now replacing all other battery types. Even the typical 12 V lead car batteries are already being replaced by lithium-ion batteries in many vehicles.

Unlike all other common battery types, whose chemistry and key technical properties are largely the same regardless of the manufacturer, lithium-ion batteries use a variety of different anode and cathode materials, which are also combined with different electrolytes and additives, and production technologies.

Depending on the manufacturer and the battery variant, different properties in terms of power density, energy density, fire risk, fast charging capability, service life, costs, etc. can be found. The generic term "lithium-ion battery" therefore refers to a class of quite different battery variants.

6.7.2 Components and Operating Principles

In modern lithium-ion batteries, metallic lithium (metal oxide) in various alloys with iron, chromium, cobalt, manganese, or phosphate is used as the positive electrode, for example, $LiMO_2$ or LiM_2O_4, or a metal phosphate ($LiMPO_4$). The lithium ions are intercalated in and out without changing the metal oxide structure.

The negative electrode is usually a graphite intercalation compound (LiC_6), i.e. a carbon material that can store lithium ions. In modern cells, the graphite is enriched with silicon to improve storage capacity even more.

For electrical contacting, an aluminum current collector foil is used on the positive electrode, and a copper foil is used on the negative electrode. During the manufacturing process, the anode and cathode materials are spread onto these foils as a paste-like mass and then dried (Fig. 6.10).

The thickness of the individual layers can vary considerably depending on the manufacturer. The values given in the figure are therefore only approximate.

The electrolyte used is a mixture of organic solvents, e.g. EC/DMC (ethylene carbonate/dimethyl carbonate), with a conducting salt, e.g. $LiPF_6$ (lithium hexafluorophosphate).

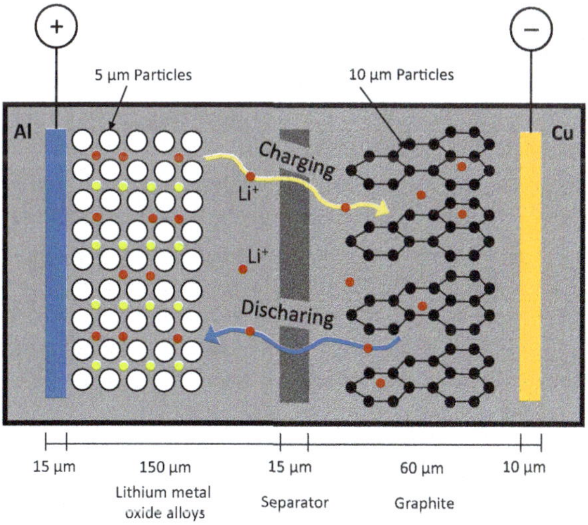

Fig. 6.10 Schematic structure of a typical lithium-ion cell

The separator is an electrically insulating, lithium-ion permeable membrane.
The cell reaction is: $LiMO_2 + C_6 \longleftrightarrow MO_2 + LiC_6$
The individual steps of the discharging process are as follows:

1. First, lithium ions (Li^+) in the anode material (graphite intercalation compound) are transported to the boundary layer with the electrolyte. In the process, an electron is separated from the negative electrode.
2. The lithium-ion is then released from the grid structure of the anode and enters the electrolyte.
3. The electron travels backward through the anode to the current collector foil (copper) behind it.
4. The electron continues to flow via the external electrical connection and finally to the positive cathode thus generating the external electrical current.
5. At the same time, electrons are transported from the current collector foil (aluminum) on the positive electrode into the metal oxide.
6. Meanwhile, the lithium ions move through the electrolyte and the separator to the boundary layer on the cathode surface.
7. The lithium ions are deposited in the metal grid of the cathode. This results in a charge reaction and the electrons are absorbed by the metal compound of the positive electrode.

During the process, the lithium is always ionic (Li^+). It is merely stored in the layers of the anode and cathode (intercalation). The host electrodes, i.e. the graphite and the metal oxides, are oxidized and reduced. This is different from all other types of batteries. For example, in a lead–acid battery, the lead is oxidized (electron release) and reduced (electron absorption). The same happens in the NiMH battery with Nickel.

To improve electrical conductivity, lithium-ion cells must be made thin. This leads to a more complex internal structure of the cells, a comparatively high proportion of passive elements (current collector foils), and therefore higher costs.

Lithium-ion batteries do not behave symmetrically when charging and discharging. They can generally be discharged faster than they can be charged. The reason for this is the relatively slow rate at which the lithium ions are stored in the graphite. Typical charging rates are 1–2 C, typical discharging rates up to 10 C.

Small traces of water and oxygen are permanently damaging to the cell, which is why there are high demands on the tightness of the housing.

The comparatively high voltage of the individual cells has the advantage that a smaller number of cells is already sufficient to generate a high system voltage. However, the voltage of each individual cell must be precisely maintained. This causes additional effort in the battery management system when connected in series. Ideally, the voltage of each cell connected in series should be monitored individually (balancing, see Sect. 6.12).

Figure 6.11 gives an overview of common anode and cathode materials and the approximate electrochemical voltage.

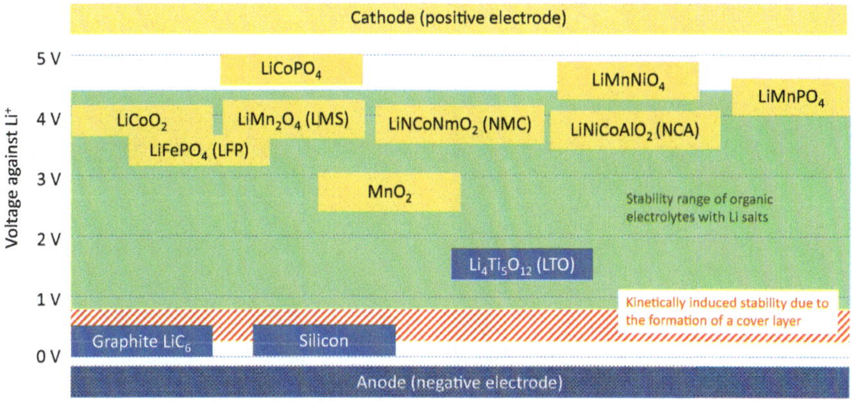

Fig. 6.11 Overview of anodes and cathodes for lithium-ion batteries[8]

6.7.3 Equivalent Circuit Diagram and Performance

The simple behavior model shown in Fig. 6.4 can be used as a first approximation to describe the terminal behavior of lithium-ion batteries. The terminal voltage in lithium-ion batteries is made up of an open-circuit voltage and a dynamic component:

- If the terminals of the battery are open, the open-circuit voltage (OCV) U_0(t) is present at the terminals. The open-circuit voltage of lithium-ion batteries changes with the state of charge.
- If the battery is discharged, the terminal voltage U(t) drops; if it is charged, the terminal voltage U(t) rises. Part of this voltage change occurs immediately when current flows, and part of it decays over time. This behavior is modeled in the equivalent circuit diagram by an ohmic resistor R_i and an RC element.
- The final discharge voltage is typically in the range of 2.0 (LFP) up to 3.3 (NMC) V, the final charging voltage is typically in the range of 3.2 (LFP) up to 4.2 (NMC) V, depending on the materials of the anode and cathode.

Such a model of a lithium-ion battery imitates the terminal behavior, but it does not describe the physical processes within the battery.

In physically motivated battery models, the physical, chemical, and thermodynamic processes within a battery are simulated and linked together. With the help of physical models, the behavior of a battery can be depicted much more accurately than with simple equivalent circuit diagrams. They are also necessary for predicting and preventing lithium plating during fast charging.

[8] Source: Holzapfel, M.; Tübke, J.; *Stand der Batterieentwicklung*; REM2030, Karlsruhe, 24.10.2013.

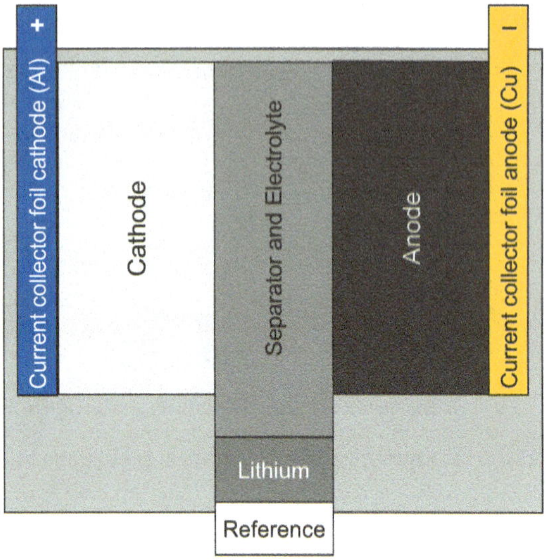

Fig. 6.12 Separation of anode and cathode voltage by introducing a reference electrode[9]

In the following, the behavior of lithium-ion batteries is derived directly from the physical conditions.

For this purpose, it is first necessary to improve the modeling of the open-circuit voltage. There are two electrodes in a lithium-ion battery, and both form a voltage with respect to a fictitious metallic lithium reference electrode. This structure is shown schematically in Fig. 6.12.

In addition to the anode and cathode, a fictitious third reference electrode made of metallic lithium is inserted in the electrolyte. This allows the open-circuit voltage $U_{0,C}$ on the cathode side to be measured between the positive terminal and the reference and the open-circuit voltage $U_{0,A}$ on the anode side to be measured between the negative terminal and the reference. The open-circuit voltage of the battery is the difference between the open-circuit voltage on the anode side and the open-circuit voltage on the cathode side.

Both open-circuit voltages change with the state of charge of the cell as shown in Fig. 6.13.

A lithium-ion battery is always operated between a lower voltage limit U_{min} and an upper voltage limit U_{max}. The resulting operating range is marked in gray in Fig. 6.13. Both the cathode and the anode are not fully utilized in the cell shown, as only part of the characteristic curves is used.

If the cell is discharged, the terminal voltage reduces as the anode voltage rises dynamically and the cathode voltage drops. The variation of the voltages of both electrodes is indicated by the blue arrows. When charging, the terminal voltage increases accordingly, as indicated by the orange arrows, the cathode voltage increases and the anode voltage decreases.

[9] Source: Batemo GmbH.

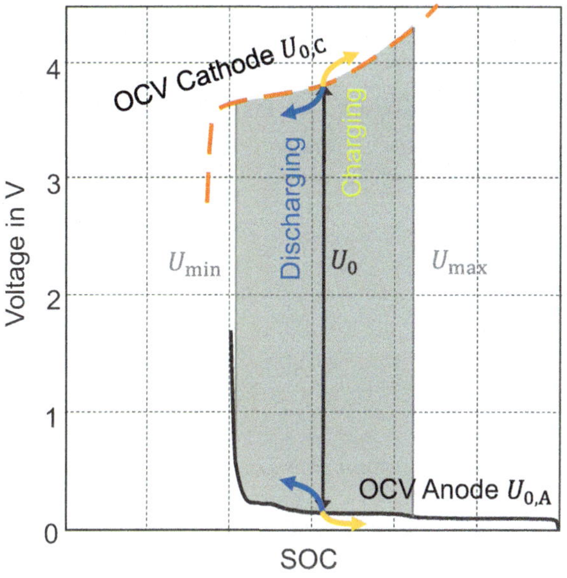

Fig. 6.13 Behavior of anode and cathode voltage in a new cell[41]

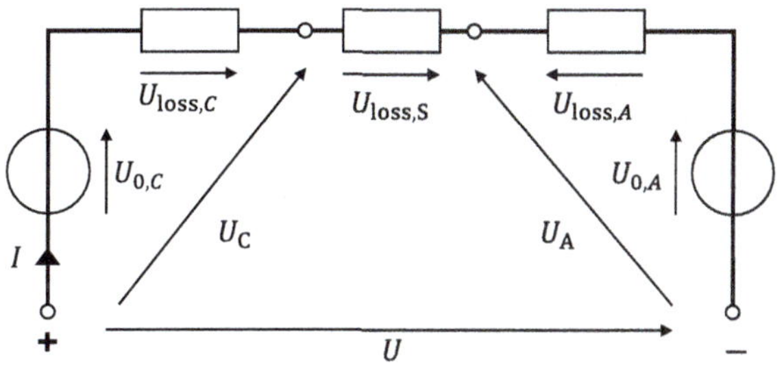

Fig. 6.14 Extended equivalent circuit diagram of the lithium-ion cell[41]

Physical battery models are based on a system of non-linear, coupled algebraic and partial differential equations. In particular, the anode voltage is of high importance. If it falls below the reference potential of zero volts, there is a risk of lithium plating, see Sect. 6.7.6. The description of the exact dynamic processes is beyond the scope of this book.

The equivalent circuit diagram is extended as shown in Fig. 6.14. It includes the two voltage sources of both electrodes and the dynamic voltage drops of the anode $U_{loss,A}$, the separator $U_{loss,S}$ and the cathode $U_{loss,C}$ as ohmic resistors for simplification.

with

U	Measurable cell voltage at the terminals
I	Measurable current into the cell
$U_{0,A}$	Open-circuit voltage of the anode
$U_{0,C}$	Open-circuit voltage of the cathode

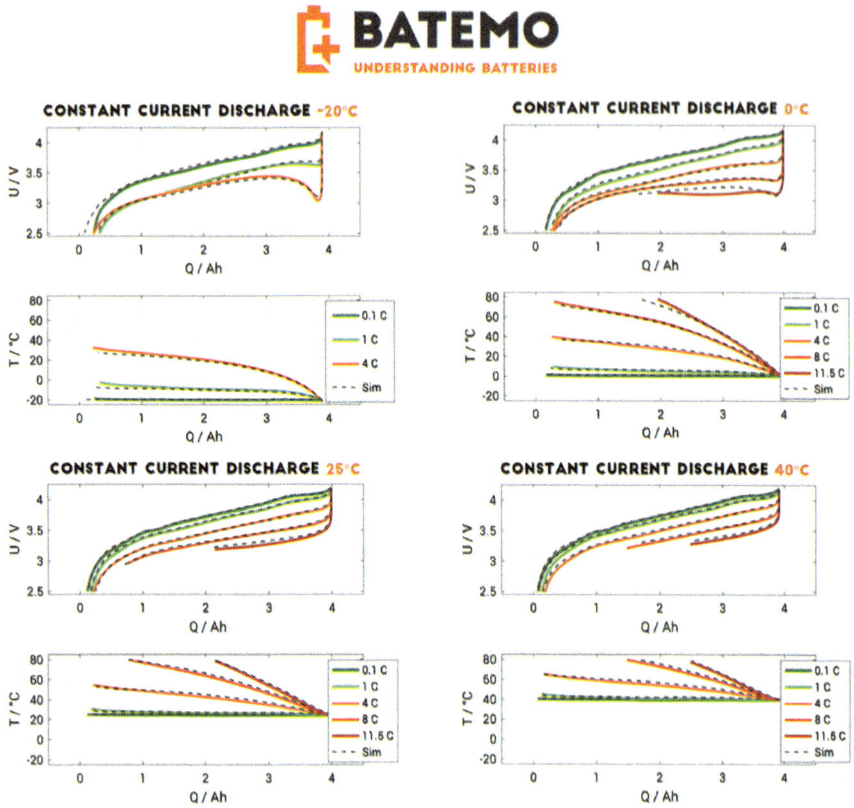

Fig. 6.15 Measured discharge behavior of a commercial 21700 lithium-ion cell and model prediction (black dashed line)[41]

$U_{\text{loss,A}}$ Voltage drop of the anode
$U_{\text{loss,S}}$ Voltage drop of separator and electrolyte
$U_{\text{loss,C}}$ Voltage drop of the cathode
U_{A} Voltage of the anode to the reference electrode
U_{C} Voltage of the cathode to the reference electrode

The overall behavior of a lithium-ion battery when discharging at a constant current in the entire current, temperature, and state of charge range is shown as an example in Fig. 6.15.

The complex behavior of the battery, which changes considerably with the ambient temperature and the discharge current, can be clearly seen. The graphs show both the terminal voltage and the surface temperature of the cells.

At an ambient temperature of −20 °C and a discharge current of 4 C, an undershoot occurs at the start of discharge. This is due to the low temperature of the battery at the start of discharge and therefore a high total voltage drop

$U_{loss,A} + U_{loss,S} + U_{loss,C}$. Consequently, the voltage drops sharply. The resulting thermal losses cause the battery to heat up, which reduces the loss of voltage and increases the terminal voltage. The complex interaction of electrochemical and thermal processes is particularly visible during these processes.

At an ambient temperature of 0 °C, the shape of the terminal voltage changes significantly. The visible steps at a low current of 0.1 C initially disappear before the characteristic curves tilt differently and finally run almost horizontally when discharging at 11.5 C.

The development of the surface temperature varies significantly, as the thermal losses occurring at the respective operating points change greatly with the physical processes in the battery.

Due to the complexity of these processes, it is not possible to reliably predict the behavior of a battery in all operating states using simple characteristic-based behavior models like in Fig. 6.4. Instead, models that correctly depict all relevant physical processes of the batteries are required. The simulation results of such a model are shown in dashed lines in Fig. 6.15. The correct description of all occurring effects and the global validity of the underlying model is clearly recognizable.

Such models form the basis of simulation-based battery system development and optimization in industry and research, for example for pack design, temperature control, fast charging, balancing, operating strategies, and service life optimization. Today, many car manufacturers implement physical battery models in the battery management systems, allowing them to safely increase the fast charging speed of their vehicles.

6.7.4 Cathode Materials (Positive Electrode)

$LiCoO_2$ (Lithium Cobalt Oxid = **LCO**) was the first commercially available cathode material. It was introduced in 1991 by Sony for video cameras and still represents a good compromise between capacity, cyclability, and safety. However, the cost is relatively high due to the cobalt required. LCO is still the most used cathode material in the consumer sector, e.g. in many round cells of 18,650 notebook batteries. The specific capacity of the cell is 160 mAh/g.

$LiNiMnCoO_2$ (Nickel Manganese Cobalt = NMC) is a replacement for LCO with lower costs and improved safety, with a high energy density of 180 mAh/g. NMC is a good compromise between energy density, cost, and safety. It is currently the standard material for batteries in many electric cars, power tools, and smartphones. NCM622, a blend of six parts nickel and two parts each of cobalt and manganese, is commonly used today. The trend is towards NCM712 and NCM811. Blends of NMC and NCA (NCMA) have a cobalt content of less than 10%.

$LiNiCoAlO_2$ (Nickel Cobalt Aluminum = NCA) with a specific capacity of 186 mAh/g is used for high-energy batteries. It was developed from LCO by substituting nickel for cobalt. However, is less safe. For example, cells of this type are produced by Panasonic and used in Tesla brand vehicles.

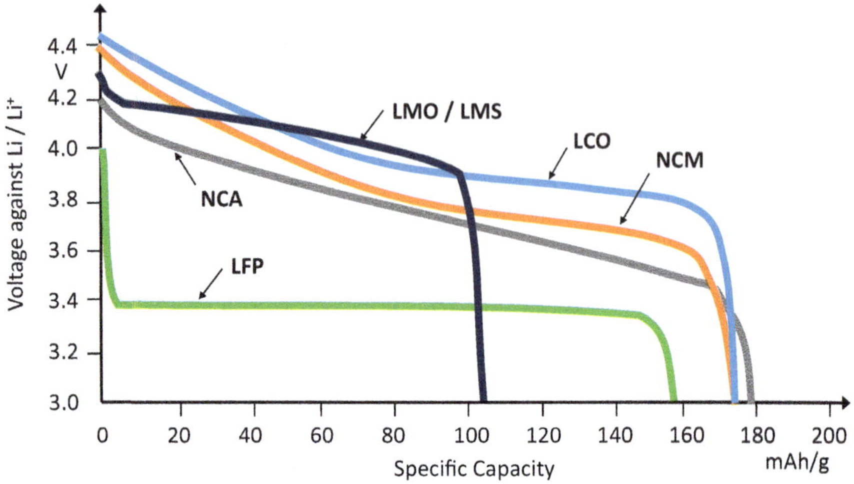

Fig. 6.16 Cathode voltage against specific capacity[10]

$LiFePO_4$ (Lithium Iron Phosphate = LFP) batteries can provide high discharge currents (C rates of up to 100). Cycle stability and costs are also favorable. As the material does not decompose thermally, it is very safe and difficult to ignite. The capacity of LFP has been steadily increased in recent years and reaches up to 170 mAh/g. The nominal cell voltage is lower compared to the other cathode materials and so the energy density is also lower. LFP cells are becoming more and more popular in cars and are also used in other applications like busses, trucks, ships, and in stationary energy packs for electrical home storage in photovoltaic (PV) systems.

$LiMn_2O_4$ (Lithium Manganese Oxide = LMO/LMS) has a lower energy density (130 mAh/g) but is comparatively inexpensive and safe. Due to its lower cycle stability, LMO is often used in blends with other materials, e.g. with NMC.

Figure 6.16 compares the different cathode voltages over their SOC. In principle, cells with a higher nominal voltage have a higher energy density, because energy and power are the product of voltage and current.

> **Note**
> Determining the state of charge (SOC) of LFP batteries is a particular challenge. While measuring the open-circuit voltage is a good indicator of the state of charge for most battery variants, the LFP battery has a constant voltage of around 3.4 V over a wide SOC range.

[10] Image based on *K. Reif, Kraftfahrzeug-Hybridantriebe, Springer Vieweg, 2012.*

Table 6.4 Proportion of cobalt in the total active material (cathode and anode) of a lithium-ion cell[11]

Cathode material	Cobalt content
LCO	39%
NMC111	13%
NMC433	11%
NMC532	8%
NMC622	7%
NCA	6%
NMC811	5%
LFP	–
LMO	–

Therefore, the state of charge can only be estimated by an integrator running in the battery controller, which adds up all the currents flowing into or out of the battery (current counting). Naturally, the actual state of charge will deviate from the integrator over time. LFP batteries must therefore be fully charged from time to time to reset the integrator and correct the SOC estimation.

A current research topic is the development of more sophisticated methods for the determination of the SOC of LFP batteries that do not require regular resets.

Some of the cathode variants contain cobalt. Cobalt is being criticized because of the mining conditions, especially in the DR Congo. It is also an expensive component that has a significant impact on the price of the battery (Table 6.4).

Incidentally, lithium-ion batteries do not contain any rare earth materials, even if this is occasionally claimed in the press.

Intensive research is being carried out into the development of completely new cathode materials. $LiCoPO_4$ or $LiMnNiO_4$, for example, are promising candidates because they can achieve voltages of around 5 V and potential energy densities of over 300 Wh/kg per cell. However, both materials are not yet in commercial use as their electrochemical voltage is above the stability limit of known electrolytes.

In summary, it can be said that there are many cathode materials, some of which have very different properties (energy density, raw materials, intrinsic safety, fast charging capability, cost, calendar life, cycle life and more). It always depends on the application as to which material is most suitable.

[11] Data source: Bloomberg New Energy Finance.

6.7.5 Anode Materials (Negative Electrode)

The most used material in commercial cells is graphite (LiC_6). During the electrochemical intercalation of lithium atoms in graphite, the ions move into the gaps between the carbon atoms and widen the grid structure of the carbon layers and thus also the band structure. This is noticeable in a change in the color of the graphite electrode. If there is only a small amount of lithium, the graphite has the typical grey–black color. If the proportion of lithium is increased, the graphite first turns reddish, then copper-colored and finally gold-colored.

A maximum of one lithium ion can be intercalated onto 6 C atoms in highly ordered graphite. This increases the volume of the anode by around 10%.

Charging rates of up to 2–3 C are possible without damaging the material, which results in a charging time of around 20 min. At higher charging rates, there is a risk of lithium plating, see the next section. The discharge rates can be much higher at up to 10 C or more.

A variant of the graphite anode is hard carbon (HC = hard carbon). Here, the lithium ions are deposited in nanopores. Due to the large surface area, the material has a greater specific capacity and faster kinetics than graphite. However, the lifetime, i.e. the number of full cycles, is shorter.

Graphene is created by modifying the grid structure of carbon into a honeycomb pattern. These developments focus on improving electron transport and the kinetics of diffusion of the lithium ions. Graphene can also be combined with hard carbon to form a sandwich-like hybrid material (HC/G).

At present, anodes made of hard carbon and graphene are the subject of research and development. They are not yet used in production-ready batteries for cars.

Silicon can store considerably more lithium than graphite: 4200 mAh/g compared to 372 mAh/g. The permissible charging rate of silicon is around 5 C and therefore significantly higher than that of graphite anodes. However, pure silicon is not suitable as an anode material for various reasons. Instead, a combination of silicon and graphite is used (SiGr composite).

The disadvantage of this material is the considerable volume change during charging and discharging, which can lead to cracking in the anode material and thus to a destruction of the cell.

There are research approaches to mechanically stabilize the material, for example using silicon-based nanoparticles. In recent years, it has been possible to add an increasingly higher proportion of silicon to the graphite without material decomposing. This was achieved through microstructuring. The size change primarily takes place within the grains and not at a macroscopic level.

Today, combined graphite–silicon anodes (Si–C) are being used in several series of production cells, for example by Panasonic/Tesla Motors. In the future, the proportion of silicon will continue to increase.

Another standard anode material is $Li_4Ti_5O_{12}$. Lithium is stored in ionic form in the crystal structure of a nanostructured lithium titanate (LTO). Titanate is a titanium oxide with a large reactive surface. The potential of charged LTO is 1 V compared to metallic lithium. The capacity is lower than that of graphite (only 175 mAh/g), but cycle stability, temperature stability (−40 °C to +55 °C), and safety are better. Higher charging and discharging currents (C rates) can also be permitted. Lithium titanate currently only has a small market share of around 1% compared to graphite.

6.7.6 Lithium Plating and Dendrite Formation

Lithium plating refers to the conversion of free Li^+ ions into metallic lithium, which is deposited on the graphite anode. It occurs when the anode potential $U_{0,A}$ falls below 0 V. Lithium plating is particularly critical for cycle aging of batteries.

Lithium plating mainly occurs during fast charging with high charging currents in the range of higher states of charge above around 50%. The transport rate of the Li^+ ions into the graphite of the negative electrode can then exceed the maximum intercalation rate. There is a particular danger at low temperatures when the intercalation rate is reduced for chemical reasons.

The process is largely non-reversible. It results in a loss of free lithium ions and a loss of accessible anode material and thus a loss of capacity in the cell.

The continued deposition of metallic lithium on the negative electrode can also lead to dendrite formation, that is finger-like growths. This process also reduces the capacity, the remaining lifetime but also the reliability of the battery. In the worst case, dendrites can short-circuit the cell and provoke an explosion-like reaction.

Lithium plating and dendrite formation can be prevented by a suitable battery management. The charging currents must be precisely adapted to the battery state based on precise dynamic simulation models. The state of charge and the battery history also have a significant influence. Thin electrodes and homogeneous metal conductors, which distribute the currents more evenly, are generally advantageous. Local areas of mechanical pressure are problematic.

6.7.7 Electrolyte and Conducting Salt

Today, the electrolyte is the key component in the development of high-energy cells. Its chemistry is decisive for the lifetime, safety and performance of lithium-ion batteries.

The electrolyte consists of a mixture of organic solvent and a conducting salt. It also includes various additives. Their exact composition is a well-kept secret of the manufacturers.

Common solvents are ethylene carbonate (EC), dimethyl carbonate (DMC), diethyl carbamazine (DEC), ethyl methyl carbonate (EMC), ethyl acetate (EA) and propylene carbonate (PC).

To improve ionic conductivity, a salt is added to the electrolyte, for example lithium hexafluorophosphate ($LiPF_6$), $LiCF_3SO_3$ or $LiN(CF_3SO_2)_2$.

Additives to increase cycle stability and service life are vinylene carbonate (VC), vinyl acetate (VA) and fluoro-ethylene carbonate (F-EC). The temperature resistance can be improved by adding propane sultone (PS). Biphenyl (BP) is used as overload protection.

6.7.8 Solid Electrolyte Interphase (SEI)

Covering layers of decomposition residues of the electrolyte form on the surface of the anode and, to a much lesser extent, on the cathode. Without further measures, these layers are not cycle-stable and would destroy the battery in a short time. However, by adding additives to the electrolyte, the formation of the covering layers can be controlled and ultimately stabilized. The composition of these additives is a well-kept secret of the battery manufacturers.

The organic carbonates of the electrolyte are thermodynamically unstable to reduction from a potential below 0.7 V against Li/Li^+. The continuous decomposition of the electrolyte can be stopped by building an insulating layer (passivating cover layer) at the surface of the graphite anode.

In the electrolyte, the lithium ions are not bare cations but are surrounded by solvent molecules (solvate shell). During the very first charging, the solvated lithium ions penetrate the outer structures of the graphite anode and the solvents decompose to form organic lithium salts. These salts are poorly soluble in the electrolyte and are therefore deposited on the electrode and the outer structures of the graphite. The so-called SEI layer = Solid Electrolyte Interphase is formed. This process is a special feature of lithium-ion batteries and is not found in other batteries.

The SEI layer is permeable to lithium ions but electrically insulating. It prevents direct contact between the electrode and the electrolyte and thus further decomposition of the electrolyte. The lithium-ion strips off the solvate shell as it passes through the SEI layer and migrates into the electrode as a bare ion. Without this effect, the battery would be destroyed after a few charging cycles.

A disadvantage of the SEI layer is the increase in the internal resistance of the battery.

SEI formation is primarily a phenomenon on the anode. The structure and formation conditions are an ongoing subject of research. The chemical reactions are very complex still not fully understood. The effectiveness of additives is therefore investigated in extensive test series. A large part of the battery manufacturers' know-how lies in the formation of the SEI and the additives required for an optimization of the SEI.

The formation of the SEI layer takes place primarily in the first charging cycle. It leads to an irreversible and unavoidable capacity loss of approximately 10%. After the initial formation, the cycle efficiency is high, typically over 99.9%.

Every lithium-ion battery is cycled several times before delivery at the manufacturing plant to ensure the SEI is built up properly. This process is time-consuming and contributes to the high cost of the battery.

During subsequent operation, the SEI layer may be damaged in some places. It then forms again as long as the corresponding additive is still present in the electrolyte in sufficient quantities.

In research, attempts are being made to increase the voltage stability of the electrolyte in the direction of 5 V. This requires new materials and additives.

6.7.9 Separators

The separator is an electrically insulating, porous layer between the anode and cathode. It is one of the most expensive individual components in lithium-ion batteries.

Separators must meet the following requirements: As thin as possible, high porosity, good wettability with the electrolyte solution, high mechanical and thermal stability, and low shrinkage.

Polyolefin films from dry or wet processes and polymer nonwovens with a ceramic coating are mainly used. Research is being conducted into direct ceramic coatings for the cathode or anode material (solid-state batteries).

The thinner the separator is made, the more volume is available for the anode and cathode and the higher the energy density of the cell. However, this can lead to the foil becoming too thin in series production and therefore permeable, which can result in short circuits and even spontaneous fires.

Example 6.11—Smartphone recall

In 2016, a large Asian smartphone manufacturer had to recall 2.5 million devices due to several incidents where a fire was suddenly ignited. The root cause of the problems was never fully revealed but it is likely that an overly thin separator led to manufacturing problems. Samsung wanted to increase their battery capacity to compete with Apple smartphones. ◀

6.7.10 Lithium Polymer Batteries (LiPo)

LiPo batteries are not to be confused with lithium iron phosphate (LFP) batteries.

The electrolyte of Lithium polymer batteries (LiPo or LiPoly) batteries is a gel-like polymer-based film. These batteries do not require a solid housing and can be built particularly flat. For cost reasons, the cathode material is usually LCO. Graphite is used as the anode material. In commercial LiPo batteries, the battery management system is built into the housing.

The polymer foil is sensitive to low (below 0 °C) and high (above 60 °C) temperatures. It has a comparatively high internal resistance, which is why charging currents above 1–3 C lead to excessive heating and should be avoided. However, the possible discharge rates are very high, typically 10–25 C continuous current, 25–50 C for short periods.

LiPo batteries are not used in cars due to their limited durability and temperature stability but are very popular in model-making due to the high discharge currents.

6.7.11 Solid-State Batteries

Solid-state batteries have been touted as the "next big thing" for some time now. However, implementation is apparently more difficult than expected. At the end of 2023, the Chinese car manufacturer NIO was the first to present production-ready semisolid-state battery (SSB). Mass production will start in the first half of 2024.

Example 6.12—NIO semi-solid-state battery

The 150 kWh semi-solid-state battery available in the NIO ET7 was developed and produced by Chinese battery manufacturer WeLion New Energy Technology.

The battery pack weighs 575 kg, giving it an impressive energy density of 360 Wh/kg, around a third more than the best previous battery packs.

With dimensions of 2.062 × 1.539 × 0.186 m, the volumetric energy density of the pack is 0.25 kWh per liter, also a new record.

However, the battery is currently still expensive. NIO quotes manufacturing costs of around US$ 42,000 for the whole pack. This corresponds to a cell price of around 180 US$ per kWh—about twice as much as today's conventional cells. ◀

A battery with a solid-state electrolyte is not a fundamentally new type of battery, but rather a variant of the lithium-ion battery.

Solid-state electrolytes provide ionic conductivity in a non-liquid solid. The advantages are robustness, a long service life even with many charge/discharge cycles, and high calendar life. A separator foil is no longer required as the solid electrolyte insulates anode and cathode electrically so energy density can be improved. Safety is also much better because, unlike liquid electrolytes, solids are hardly flammable and dendrites cannot form and cause short circuits.

The disadvantages of solid-state electrolytes are their lower ionic conductivity and the currently still high manufacturing costs. In particular, the ion transition between the solid-solid interface with the anode and cathode causes problems due to the poor transfer of high currents. The ionic conductivity of most glass-ceramic electrolytes is also still too low for series production. One solution could be to apply the electrolyte as a paste-like mass that only hardens after the cell has been manufactured.

The first available solid-state batteries are achieving energy densities of around 400 Wh/kg. In the medium term, 500 Wh/kg and more seem to be realistically achievable.[12]

Because of its promising properties, many manufacturers and research groups are looking into the subject. Some car manufacturers have announced the introduction of solid-state batteries over the next few years. Due to the higher manufacturing costs, however, they will initially be sold in the luxury vehicle segment.

6.8 Lithium Sulfur Batteries

One promising research is the lithium–sulfur battery (Li–S)[13], in which the cathode consists of sulfur and the anode of lithium metal.

Lithium and sulfur react differently to lithium and carbon in conventional lithium-ion batteries. In carbon, up to six lithium ions are incorporated into the grid structure. The reaction of lithium and sulfur takes place via a multiple electron transfer mechanism, which should result in a higher energy density. However, the theoretical values of around 2500 Wh/kg and 2800 Wh/l have by no means been achieved in the batteries built to date. The poor electrical conductivity of sulfur is particularly critical, which is why this material cannot normally be used in its pure form for the cathode, but rather combined with lithium metal.

The cycle stability of lithium-sulfur batteries is also critical. When discharging, lithium and sulfur interact and the ring-shaped sulfur molecules transform into chain-like structures, so-called polysulfides. As the battery goes through several charging and discharging cycles, parts of the polysulfide can enter the electrolyte, causing the battery to gradually lose active material over time. Great progress has been made in this area in recent years through various modifications.

Another difficulty is the mechanical stability of the cathode, which expands by around 78% during the absorption and release of lithium and therefore tends to crack. But progress is being made here too, as the report by an Australian research group shows.[14]

One advantage of lithium–sulfur batteries is the absence of cobalt. This should make it possible to achieve lower manufacturing costs in the long term.

There is cautious optimism regarding the future commercialization of lithium–sulfur batteries. An initial application could be in aviation because the weight advantage is particularly important and the significantly higher costs during the start of series production are more justifiable.

[12] IBM Battery 500 Project.

[13] Peng, H.; Huang, J.; Cheng, X.; Zhang, Q.; *Review on High-Loading and High-Energy Lithium-Sulfur Batteries; Advanced Energy Materials*, Volume 7, Issue 24, December 2017.

[14] Shaibani, M.; Mirshekarloo, M. S.; Singh; R. Easton, Ch.; Dilusha Cooray, M.C.; *Expansion-tolerant architectures for stable cycling of ultrahigh-loading sulfur cathodes in lithium-sulfur batteries*; Sciences Advances Vol. 6, no. 1, 03 Jan 2020.

6.9 Sodium-Ion Batteries

Sodium-ion batteries (SIB) are on the way to becoming an alternative to lithium-ion batteries (LIB) in some areas. These batteries contain neither lithium nor cobalt and are therefore particularly environmentally friendly and also cheaper. Sodium is found on earth in large quantities in the form of common salt (sodium chloride) and is easy to extract without harming the environment.

Production is largely compatible with the existing systems and processes for manufacturing lithium batteries. Initial series production has already started in China in 2023 and the first cars with this type of battery will be available from 2024.

Three different variants of the cathode are currently being investigated:

- Layered oxides, comparable to the cathode materials of LIB. Iron and manganese are favored and critical raw materials such as cobalt and nickel are avoided.
- Prussian blue analogs are the most commonly used variant ($Na_xFe_y[Fe(CN)_6]_z$. The raw materials are inexpensive, but some are critical due to the cyano groups. The volumetric energy density is low.
- Polyanionic cathodes, the most common representative is $Na_3V_2(PO_4)F_3$, have good cycle stability, but are critical due to the use of vanadium and fluorine.

The electrolyte is an organic or aqueous inorganic substance, for example, carbonate-based $NaPF_6$.

The anode is either graphite or a hard carbon with a specific capacity between 350 and 530 mAh/g, quite like LIB batteries.

The disadvantage of sodium-ion batteries is the energy density, which can be only around 2/3 of that of lithium cells. The reason lies in the sodium ions, which are around 3 times heavier and 33% larger than lithium ions. This requires thicker electrodes.

Currently, Chinese manufacturers achieve an energy density of 145–160 Wh/kg with sodium-ion batteries. This is comparable to lithium–iron–phosphate batteries (LFP) but at around 30% lower costs.

Energy densities of up to 200 Wh/kg are expected for the next generation of SIB.

6.10 Metal Air Batteries

In metal–air batteries, electrical energy is released from the chemical reaction of metals with oxygen. They occupy a special position among batteries, as one of the two reaction partners (in this case oxygen) is obtained from the ambient air via a special electrode and does not have to be stored in the battery. Therefore, at least theoretically, significantly higher energy densities can be achieved with these systems than with conventional battery types.

Some metal–air batteries are already in use, such as zinc–air batteries for hearing aids. However, these are not rechargeable. Attempts are currently being made worldwide to develop rechargeable metal–air batteries.

One of the biggest challenges is to increase the very short lifetimes of these battery types. The complex chemical charging and discharging processes are not yet sufficiently understood and mastered.

Intensive research is being carried out into the realization of lithium–air batteries in particular. As lithium has the highest electrochemical potential of all metals, these batteries offer by far the highest energy density of all metal–air systems that can theoretically be achieved. In the long run, energy densities well over 1000 Wh/kg should be within reach. The theoretical limit is even higher, around 14,000 Wh/kg.

Whether it will ever be possible to commercialize these batteries is still undetermined. As of today, there are simply too many unsolved problems.

6.11 Battery Charging

6.11.1 Constant Current, Constant Voltage Charging (CCCV)

In the widely used CCCV procedure, the battery is first charged at a constant current (CC) corresponding to the desired C-rate until the final charging voltage, for example, 4.2 V per cell for many types of lithium-ion batteries, is reached.

The charging voltage must always be higher than the battery voltage. It is controlled so the desired current is driven depending on the battery's internal resistance. Typical batteries charge to around 80% before the voltage limit is reached. This first part of the charging process takes place quite quickly as the maximum C-rate can be used continuously.

In the second step, the voltage is kept constant (CV) at the maximum permissible level. The charging current and thus also the C-rate now gradually decrease. The end of the charging is reached when the current drops below a specified minimum. This second part of the charging process is noticeably slower than the first part because the C-rate reduces more and more (Fig. 6.17).

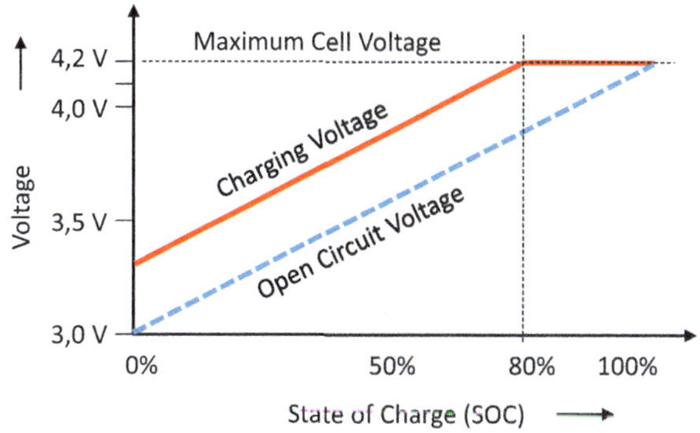

Fig. 6.17 Theoretical voltage curve for constant voltage, constant current charging (CCCV)

The charging power is the product of the DC current and the voltage difference between the open-circuit voltage and the charging voltage. It is constant during the first charging phase when charging with a constant current. In the second phase, the current continues to decrease and at the same time, the voltage difference between the open-circuit voltage and the charging voltage also decreases. As a result, the charging power now drops rapidly.

6.11.2 Improved CCCV with Adaptive Current

A further reduction of the charging time is possible by increasing the charging current during low SOC when a sufficient voltage margin is available.

However, higher charging currents can lead to lithium plating and permanent damages to the cell. As shown in Fig. 6.13, the anode voltage U_A drops during charging. If it falls below 0 V, lithium plating occurs, resulting in a partially irreversible loss of battery capacity.

It is therefore essential for the development of fast charging processes that the anode potential never reduces below 0 V. Globally valid, physical battery models are used that correctly predict the anode voltage U_A under all conditions. With these models, the charging current can be controlled so that a certain safety margin of the anode potential is always maintained.

Figure 6.18 gives examples of rapid charging of a lithium-ion battery.

The standard CCCV charging process is shown in the black curve. The battery is charged from 0 to 80% SOC in 32 min. During the first 30 min, there is ample distance between the anode voltage and 0 V potential, so faster charging is permissible. However, at about 35 min the anode voltage drops slightly below 0 V before the charging switches to the CV phase. In this situation, lithium plating can occur.

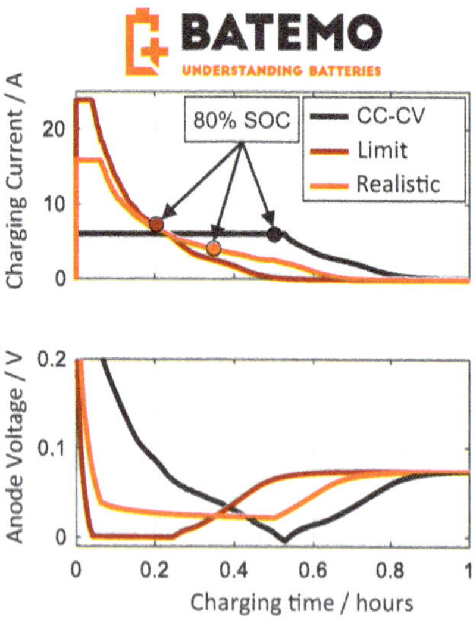

Fig. 6.18 Accelerated CCCV charging[46]

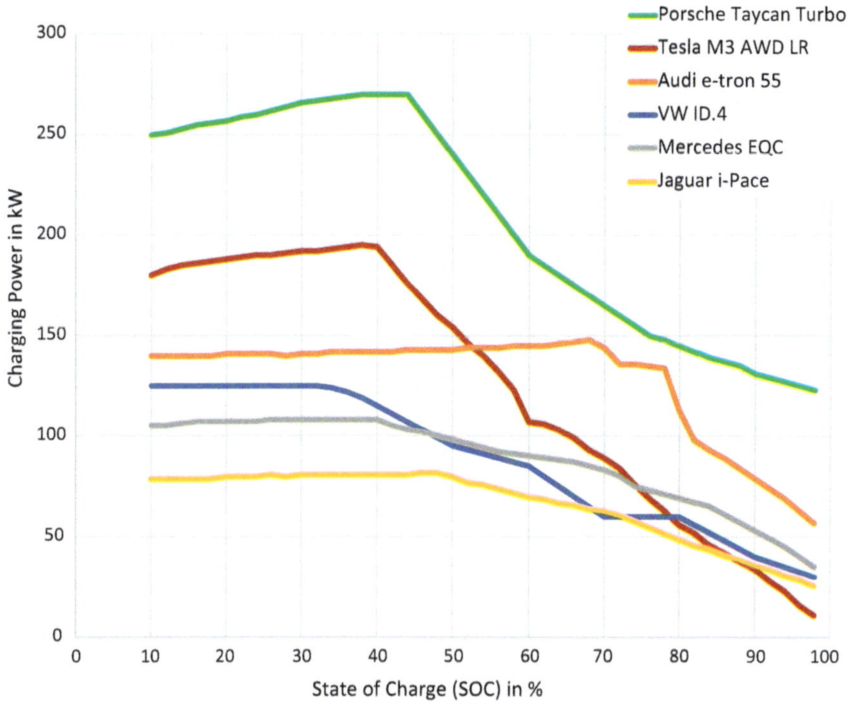

Fig. 6.19 Measured charging power curves of selected vehicles over the charge level (SOC)

By controlling the anode voltage and keeping it always just above 0 V, a faster charge can be achieved that follows the physical limits of the battery. The dark red curve demonstrates how the 0–80% charging time can be reduced from 32 to 13 min. Note that the anode potential is very close to the 0 V limit over a long period of time.

In practice, a sufficient safety distance to the 0 V potential of the anode voltage should always be maintained. This has been considered in the orange-colored charging profile. The 0–80% charging time is then 21 min. Compared to the original black curve (CCCV), the charging time is reduced by about one third and yet the safety regarding lithium plating is improved considerably.

The curves in Fig. 6.19 show the examples of the real charging behavior of different vehicles measured at ionity DC fast charging stations. Charging always started at 10% SOC.

The charging current can reach a maximum of 500 A. The cables between the car and the charging station are liquid-cooled.

At 800 V system voltage, the charger provides up to 350 kW of charging power. This voltage level is currently used in the Porsche Taycan and cars from Kia/Hyundai. At 400 V system voltage, the available charging power is 200 kW.

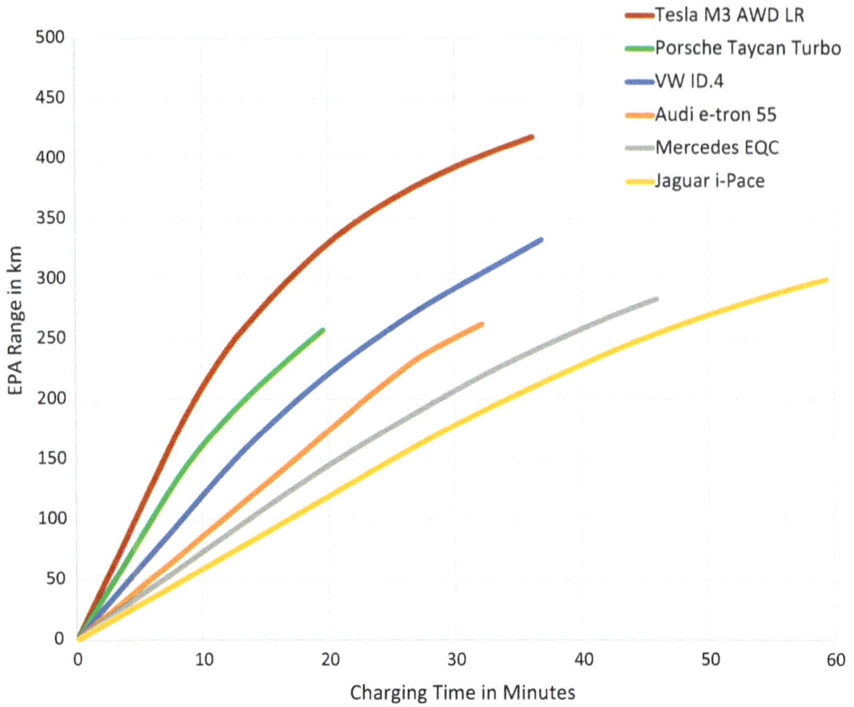

Fig. 6.20 Charged EPA range in km depending on the charging time (charging from 10 to 90% SOC)

The curves indicate that Porsche and Tesla use the enhanced CCCV charging process, as power increases during the lower SOC stages. The other manufacturers use the classic CCCV method and exhibit a constant power over a wide SOC range.[15]

However, for the driver, the distance covered by charging is far more important than charging power. The energy consumption of the vehicle plays an important role here. The charging speed can then be converted to kilometers of range per hour of charging time. Some vehicles display this value in km/h during charging. With suitable fast charging stations, the theoretical range gain per hour of charging time can briefly exceed 1000 km.

The curves in Fig. 6.20 are based on the same measured values as those in the previous graph. However, the energy charged has been converted into a charged range. The energy consumption per km was based on the realistic EPA measurements, see Sect. 4.5.6. All charging processes shown start at 10% and end at 90% SOC.

[15] In recent years, various manufacturers have accelerated the charging speed of their vehicles via software updates even after delivery. By implementing better battery models, an improved CCCV procedure could be used.

6.11.3 Pulse and Step Charging

In addition to the CCCV charging procedure, other, more complex charging processes are available. These are intended to charge the battery more gently in terms of cycle life and dendrite formation and/or even faster. Even more detailed knowledge of the internal chemical processes in the battery is required for the targeted use of these processes. They are, however, not generally used in electric cars for safety reasons.

In pulse charging, the charging current is pulsed during the constant voltage phase. The charging pulse has a fixed length of a few fractions of a second. Each pulse is followed by a charging pause until the control voltage is undershot or the charging pause has reached a certain maximum pause duration. The control voltage is usually the same as the end-of-charge voltage.

During step-charging, the cell is charged with a voltage value above its voltage limit. Lithium NMC cells, for example, are charged with 4.3 V, i.e. 0.1 V above the actual voltage limit. There is practically no current limit during this charging phase. As soon as the current has fallen below a certain value, the second constant voltage charging phase begins. In this phase, charging then takes place at the actual limit voltage. As with CCCV charging, charging is terminated when the current falls below the end-of-charge currents.

Like pulse charging, step charging is not used for cars. However, it is occasionally used for example in model making.

6.12 Balancing

The internal resistance of battery cells manufactured in large-scale production will vary slightly due to manufacturing tolerances, in particular slight variations in the thickness of the individual layers. This also results in slight fluctuations in the cell capacity.

If only the total voltage is regulated when charging many cells connected in series, individual cells would be charged at higher voltages than others, resulting in selective overcharging, see the following figure.

This would lead to a greatly reduced service life of the affected cells because the electrolyte decomposes quickly at higher voltages. If one single cell fails, the entire battery module is unusable.

Example 6.13—Balancing

When charging 16 cells connected in series with the same internal resistance and a total voltage of 67.2 V, each cell receives a voltage of $U_C = 67.2\ \text{V}/16 = 4.2\ \text{V}$. This corresponds to the end-of-charge voltage of typical lithium-ion cells.

However, if the internal resistance of a single cell were only 10% greater than that of the other cells due to manufacturing tolerances, this cell would have a voltage of $U_C = 67.2\,V \cdot 1.1/(16.1) = 4.59\,V$. The voltage of the remaining 15 cells would be reduced accordingly to 4.17 V.

This 4.59 V is well above the permissible end-of-charge voltage of 4.2 V and would significantly reduce the lifetime of the affected cell after just a few charging cycles. When this cell fails, the entire string becomes unusable. ◀

The charging voltage of each cell must therefore be controlled individually and an equalization of the charges between all cells must be ensured. This task is performed by the balancing circuit in the module controller.

Balancing is the forced charge equalization between cells connected in series until all cells are in the same electrochemical state.

Balancing is not required for parallel connections, as all parallel cells receive the same voltage. Balancing is only required for the serial connected cells within a string. Strings connected in parallel are not balanced.

Note

When connecting cells in parallel, it is advantageous to use the Z circuit: The two contact plates that connect all parallel cells at the top and bottom are each connected to the load or charger from the opposite direction on the plus and minus sides. This ensures that the line resistance to all cells is the same on average so that all cells have the same voltage.

If the two metal strips were connected from the same direction, the rear cells would have an increasingly lower voltage than the front cells due to the voltage drop across the lead resistance. This would have a negative effect on the lifetime and overall capacity of the battery pack.

There are various circuit variants for carrying out balancing in practice. The technically most simple solution is to run wires from each set of parallel connected cells to the balancing controller. The controller measures the voltages and when one serial cell reaches the maximum permissible voltage, the controller switches a discharge resistor across the cell to reduce the charge, see the following figure (discharge balancing).

In this way, the faster-charging cells are slightly discharged again and again until the slower-charging cells have reached the same voltage level. This method is used in many vehicles, for example in Tesla cars. However, the efficiency of this method at high SOC is low as excess charges are just wasted in the resistor (Fig. 6.21).

Another option is to connect a capacitor in rapid succession in parallel to each individual cell or set of parallel cells (shuttle cell balancing circuit). If the cell voltage is higher than the capacitor voltage, the capacitor is automatically recharged by drawing charge from the cell(s). However, if the cell voltage is lower than the capacitor voltage, the capacitor releases its excess charge to the cell(s). In

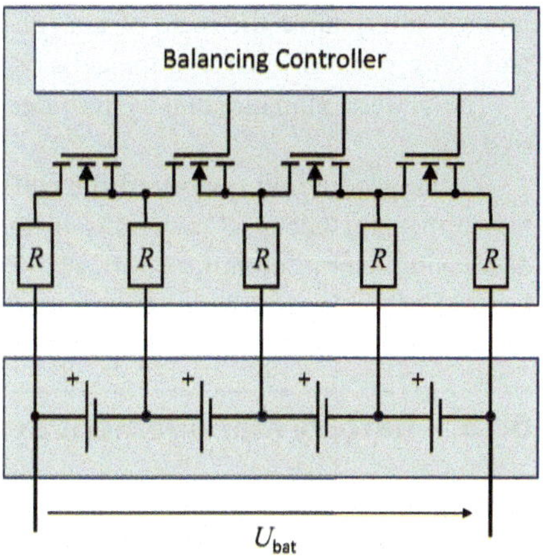

Fig. 6.21 Cell balancing using resistors and by-pass FETs[16]

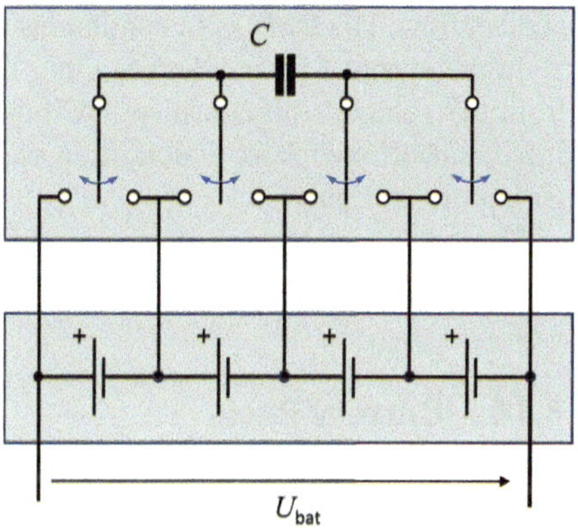

Fig. 6.22 Capacitor-based shuttle cell balancing circuit[54]

this way, all cells are gradually brought to the same state-of-charge. The capacitor must be switched from one cell to the next quickly so that the voltage jumps are not too large. In practice, several thousand switching operations per second can be achieved. This method does not require any voltage measurements or control (Fig. 6.22).

Alternative to switching the voltage with a capacitor, a coil can also be used as an intermediate storage device.

Balancing can take place continuously during the whole charging process. But it is only required when the SOC and consequently the voltage is relatively high, as there is an increased risk of the individual cell voltage being above the limit.

In practice, the charging process of a car is often stopped before full charge is reached, for example at 80 or 90% SOC, to protect the battery or to save charging time. In this case, it is useful to continue balancing after the end of the charging

[16] Source: *Battery Cell Balancing: What to balance and How*, Yevgen Barsukov, Texas Instruments.

process to equalize the state of charge of the cells. It is also possible to carry out balancing when discharging the cells or during idle phases without drawing current.

The battery management system decides when and how much balancing is carried out.

As the production quality of the cells improves, balancing is becoming less and less important because variations in the internal resistance between the cells are decreasing. Some manufacturers are no longer performing individual cell balancing and only carry out balancing at module level.

6.13 Battery Management System

A battery management system (BMS) continuously determines the state of charge (SOC), state of health (SOH), and temperature of the battery system. It determines the permissible charging and discharging currents (SOP) and regulates the required coolant flow. The BMS is in continuous communication with all module controllers.

In the event of an accident or other serious incident, the battery is disconnected from the vehicle's high-voltage DC bus in a fraction of a second. This is done by safety cut-off switches, which often work internally pyrotechnically, like the triggering of an airbag.

The battery management system also includes the battery's data interface to the outside, i.e. to the vehicle's controllers, which is normally carried out via a CAN bus.

6.14 Battery Pack

An entire battery Pack consists of many battery modules, each of which contains multiple cells. Depending on the desired capacity and voltage, the cells inside the modules and the modules themselves are connected in serial strings and in parallel groups.

The number of cells connected in series (s) is called a string. The total number of serial cells across all modules determines the system voltage. The number of strings connected in parallel (p) determines the capacity or, more precisely, the energy content of the storage system.

For example, the specification 50s4p refers to a battery system consisting of 4 parallel strings, each with 50 individual cells connected in series. The system voltage is $50 \cdot 4.2\,\mathrm{V} = 210\,\mathrm{V}$. The total energy results from the energy content of the total of 200 cells.

Example 6.14—Battery structure of Renault Zoe

One module in the Renault Zoe consists of 16 pouch cells. The entire drive battery series has 12 modules, divided into 2 strings.

This results in a total of 192 cells, each with a capacity of 36 Ah at 3.75 V open circuit voltage. The total capacity of the battery is approximately: 192 · 36 Ah · 3.75 V = 25.9 kWh.

In each case 96 cells (= one string) are connected in series, resulting in a voltage of 96 · 3.75 V = 360 V. The two strings are connected in parallel.

The following Table 6.5 and Figs. 6.23, 6.24, 6.25) illustrate battery cell configurations and the components of a battery pack. ◀

Table 6.5 Battery cell configuration of various cars

Vehicle	Battery gross capacity	Topology
Mitsubishi iMiEV	16 kWh	88s1p
Mercedes Smart ED	17.6 kWh	98s1p
BMW i3	21.6 kWh	96s1p
Nissan Leaf	24 kWh	96s2p
VW e-Golf	24.2 kWh	88s3p
Renault Zoe	25.9 kWh	96s2p
Tesla Model 3 LR	60 kWh	106s1p (LFP)
Hyundai Ioniq 5	77.4 kWh	192s2p (800 V)
Tesla Model S	85 kWh	96s74p
Porsche Taycan	93.4 kWh	198s2p (800 V)

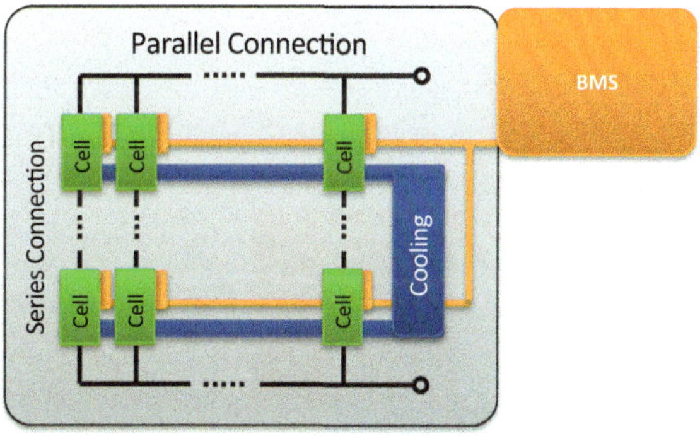

Fig. 6.23 Construction of a battery module from parallel strings of cells

The battery housing creates the mechanical and electrical connection between the cells, known as packaging. It also contains the battery management system, the safety system, and the cooling system.

A cell controller is assigned to each cell for monitoring purposes. It continuously checks the voltage and temperature and thus prevents the cell from being overloaded.

Each module is equipped with module electronics, which are responsible for balancing.

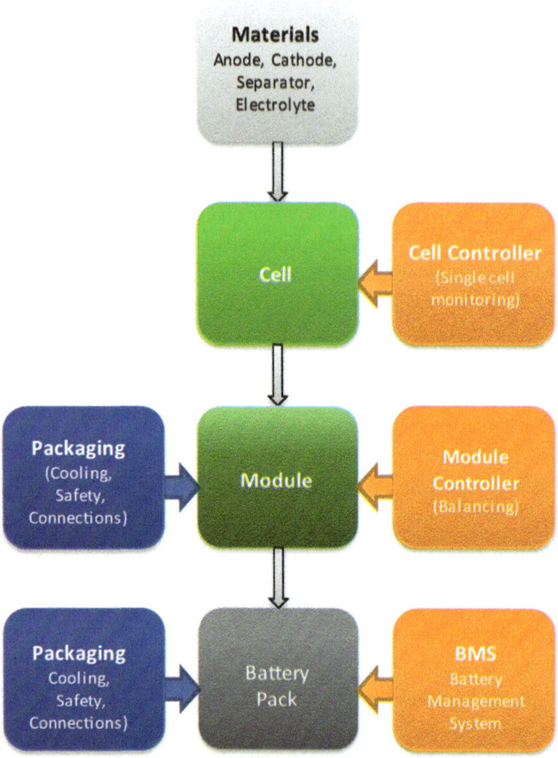

Fig. 6.24 From the materials to the cell and module to the battery pack

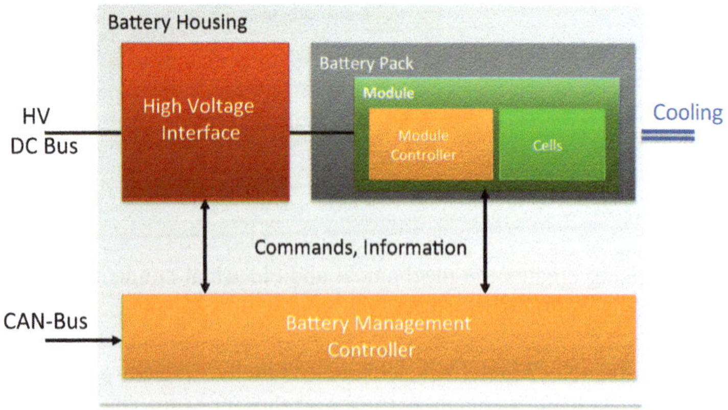

Fig. 6.25 Construction of a complete battery pack for use in BEVs

6.15 Hydrogen

For six decades now, hydrogen has been actively promoted as an energy carrier for road vehicles and investigated by many companies and research groups. The modern hydrogen fuel cell has been under development since the 1950s.

As early as 1966, General Motors presented the Electrovan, the first car with an electric drive that received its energy from a hydrogen fuel cell. The technology is much older than the lithium-ion battery cell, which was invented in the 1980s and slowly began to be used in commercial products from the 1990s.

From around 2007, Honda, Kia, and Toyota produced fuel cell cars in small series, but all of these have since been for the most part discontinued. German manufacturers, especially Mercedes-Benz and BMW, have also built multiple prototypes. The European Union, the United States, and various countries in Asia have pumped billions of tax dollars into the development of fuel cells and hydrogen cars. Nevertheless, they have all failed to make an impact on the market. There are many good reasons for this.

6.15.1 Chemical Properties of Hydrogen

The poor chemical properties of hydrogen are a particular challenge for fuel cell cars. Table 6.6 provides an overview in comparison with methane. Around 80% of natural gas (CNG, LNG) consists of methane.

Hydrogen is the element with the lowest density. It is therefore complex to reliably seal a large hydrogen system with its many pipes, pumps, and valves. Hydrogen systems require many sensors to detect possible leaks.

Together with atmospheric oxygen, hydrogen can form an explosive mixture (oxyhydrogen) just like natural gas. However, while natural gas is only explosive in a narrow mixing ratio (around 4–17 vol%), the ignitable mixing ratio for hydrogen is very broad (4–78 vol%, most dangerous at 30 vol%). Unlike a leak in a nat-

Table 6.6 Comparison of the chemical properties of hydrogen and methane

			CH_4 (Methane)	H_2 (Hydrogen)
Reactivity	Molar mass	g/mol	16	2
Explosion hazard	Ignitable mixture with air	Vol%	5 … 15	4 … 78
	Ignition temperature	°C	595	560
Gaseous transportation and use	Heating value (by volume)	kWh/m^3	10.0	3.0
	Density at 200 bar and 20 °C	kg/m^3	162	15
Liquid transportation	Boiling point	°C	−162	−252.8
	Heating value (by mass)	kWh/kg	13.9	33.3

Table 6.7 General technical properties of hydrogen

Hydrogen	
Density (liquid LH_2)	70.8 kg/m^3
Density (gaseous CH_2 700 bar) (gaseous CH_2 250 bar)	40 kg/m^3 18 kg/m^3
Density (gaseous 1 bar, 15 °C)	0.0841 kg/Nm^3
Heating value (at 25 °C)	33.3 kWh/kg 120 MJ/kg 3.00 kWh/m^3

ural gas pipeline, which is usually harmless, a leak in a hydrogen pipeline almost inevitably results in an explosive mixture.

When the hydrogen leak occurs inside a building or another container, e.g. in a car, the risk of explosion is very high. The accumulation of hydrogen under a canopy in calm conditions is sufficient for the formation of an explosive mixture. Only in the open air does hydrogen escape quickly into the atmosphere at around 50 km/h due to its low density.

If hydrogen is released into the atmosphere, its long-term impact on the climate is considerable and significantly more harmful than that of carbon dioxide. Although hydrogen does not have an immediate effect, it changes the composition of the atmosphere and therefore has an indirect effect on the climate, about half as much as methane. See also Sect. 15.1.

Another difference between hydrogen and natural gas is the effect of hydrogen embrittlement: Hydrogen can penetrate the lattice of metals, become embedded and significantly reduce the strength. This mainly affects steel. Hydrogen also reacts with oxygen-containing types of copper (hydrogen sickness). For this reason, steel pipelines carrying hydrogen must be lined with plastic and copper pipelines may only be made of expensive, oxygen-free pure copper or copper alloys.

The energy content of hydrogen is only less than a third of that of natural gas with the same volume. Therefore, hydrogen must either be compressed to extremely high pressures at great expense or liquefied at even greater expense for storage and transportation. The low weight of hydrogen doesn't help because the high pressures or low temperatures require complex storage systems that significantly increase the system weight and cost compared to natural gas storage systems. Including the tank, hydrogen only achieves around half the energy density of conventional liquid fuels.

For vehicle applications, hydrogen is either stored and refueled in gaseous form at high pressure (CH_2—Compressed Hydrogen at 200, 350 or even 700 bar pressure) or liquefied at extremely low temperatures (LH_2—Liquid Hydrogen with −253 °C = 20.4 K) (Table 6.7).

Depending on the pressure, the density of gaseous hydrogen increases or decreases, and the mass (kg) stored per cubic meter (m^3) changes. An ambient pressure of 1.013 bar and a temperature of 15 °C is referred to as the standard volume (Nm^3), not to be confused with Newton meters (Table 6.8).

Table 6.8 Compression factors of hydrogen gas at 0 °C ambient temperature

Hydrogen gas (CH_2)	
Pressure	Compression
1.013 bar	1
50 bar	1.032
350 bar	1.236
700 bar	1.489

Note
Compared to conventional liquid fuels (petrol, diesel, kerosene), the gravimetric energy density of hydrogen is around 3 times better: 33.3 kWh/kg versus 11–12 kWh/kg. However, due to the extreme storage conditions of hydrogen, expensive and very heavy tanks are required, which is why the total storage system weight is much worse.

The volumetric energy density of liquid hydrogen (around 2360 kWh/m^3) is more than 4 times lower than that of diesel fuel (around 9800 kWh/m^3). For pressurized hydrogen CH_2 at 700 bar the increase in tank volume compared to diesel fuel is almost sevenfold for the same amount of stored energy.

6.15.2 Fuel Cells

The fuel cell is a tertiary element, i.e. it converts chemically bound energy into electrical energy. It is therefore not an energy storage device in the true sense of the word but is nevertheless discussed in this section because, together with the hydrogen tank, it serves as an energy source for fuel cell electric vehicles (FCEV).

The fuel (anode gas, usually molecular hydrogen H_2) is oxidized with an oxidizing agent (cathode gas, usually atmospheric oxygen O_2) with direct generation of electrical energy.

The negative pole (anode) is in contact with the fuel, while the oxidizing agent is supplied to the positive pole (cathode). The hydrogen molecules release electrons, which flow through the consumer to the cathode.

If a reformer is connected upstream of the fuel cell, complex hydrocarbons can also be used as energy sources: Natural gas, liquid gas, methane gas, methanol, ethanol, diesel reference mixture, modified petrol. However, this reduces the efficiency even further.

The fuel cell is not subject to the Carnot cycle of a heat-power machine like the diesel engine, so its efficiency can be better. However, a complete fuel cell system requires a variety of supporting components, like filters, heaters, coolers, pumps, which reduce the achievable efficiency.

Today, only the polymer electrolyte fuel cell (proton exchange fuel cell/polymer electrolyte membrane = PEMFC), or PEM cell for short, is used for vehicles.

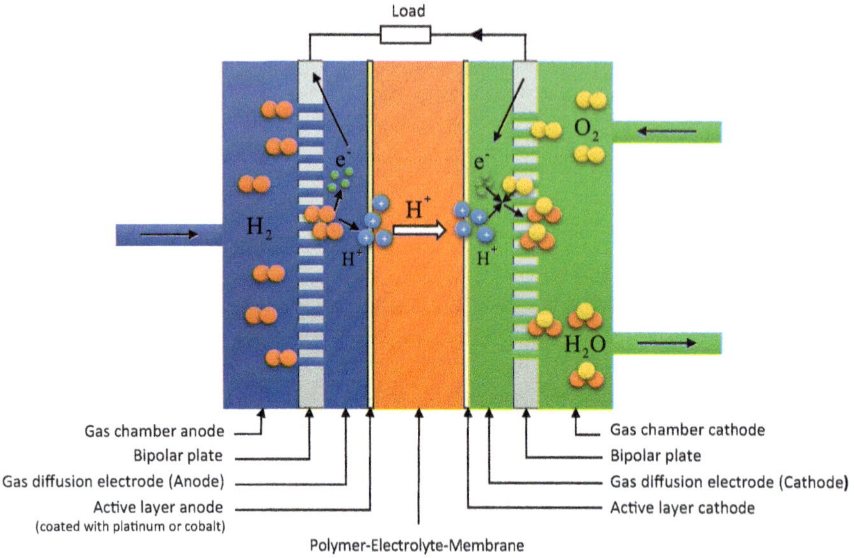

Fig. 6.26 Schematic structure of a PEM fuel cell

The electrolyte is a solid polymer, the mobile ion is hydrogen H^+. The optimum operating temperature is around 80 °C and the cell efficiency can reach more than 80%, depending on the power output. At full load, however, the efficiency drops noticeably.

There are also alkaline fuel cells (AFC) and phosphoric acid fuel cells (PAFC), which also work with hydrogen and oxygen. However, they have a lower efficiency and/or require much higher operating temperatures so they are not used in cars. Such cells are used as stationary electrolyzers.

The process of the PEM cell is basically reversible. However, actual fuel cell system in cars can only convert chemical energy (H_2) into electrical energy and not the other way around for a variety of practical reasons. A fuel cell vehicle therefore always needs a sufficiently large battery to store the recovered braking energy.

The battery is also required for cold starting the stack and for rapid increase of power output (quick start). Around 10–40 s after the cold start, the fuel cell is at a sufficient temperature and switches on. It takes a few seconds to start up an idling fuel cell system to produce full power.

Figure 6.26 shows the schematic structure of a single cell. It has an open-circuit voltage of around 1 V. Therefore, 300–400 such cells are connected in series in a complete fuel cell stack.

The anode and cathode are each a network of highly porous carbons coated with a catalyst for hydrogen: H_2 converts to H^+ and O_2+4H^+ converts to $2H_2O$. Platinum was and is frequently used as a catalyst material. However, platinum is expensive and its environmental impact is critical. Some manufacturers are therefore switching to using cobalt as a catalyst material.

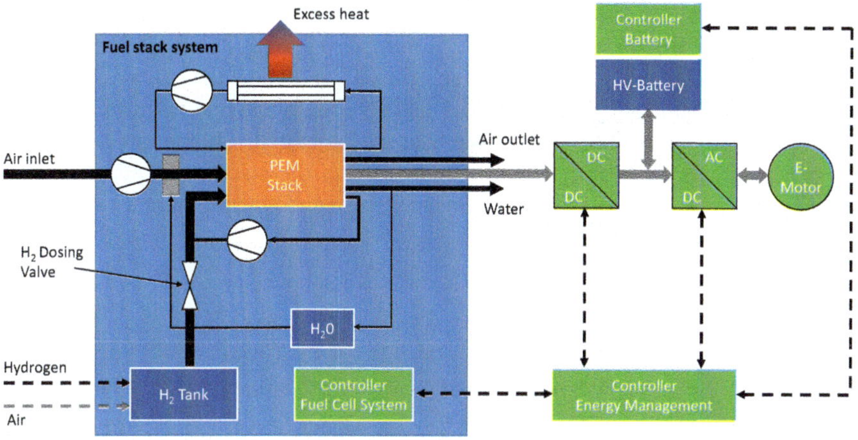

Fig. 6.27 Schematic structure of a fuel cell system

The two bipolar plates are electrically conductive and serve as current collectors. The PE membrane insulates electrons but conducts hydrogen ions.

At the end of the process, pure distilled water is produced. It is usually collected in a small tank in the vehicle and can be drained at the touch of a button if required.

6.15.3 Fuel Cell System

In a car, hydrogen is usually stored in gaseous form at a high pressure of up to 700 bar when the tank is full. When hydrogen is withdrawn, the tank pressure is gradually reduced to approximately 20 bar (Fig. 6.27).

The hydrogen is fed into the PEM stack via a metering valve (hydrogen gas injector = HGI) with electrical control for continuous mass flows. Due to the expansion, hydrogen gas becomes hot and must be cooled.

Note

The Joule–Thomson effect describes the change in temperature of a gas when the pressure is reduced. Under normal conditions, the temperature of most gases, including air for example, falls when the pressure is released.

For hydrogen, helium, and neon, however, the temperature rises when pressure is reduced, for example during refilling.

Ambient air is also fed into the stack. The air must be cleaned, humidified, tempered, and brought to a suitable pressure by means of a cathode air compressor. Depending on the system design, the pressure is several bars. The air mass flow depends on the required stack capacity and is typically around 400 kg/h at 85 kW power output of the stack. The required pump power is several kW up to more than 10 kW.

Due to the moderate efficiency of a fuel cell system and due to the air compression, relatively large amounts of waste heat are generated, which must be removed by a separate cooling circuit. The temperature in the stack must not rise above 95 °C, otherwise there is a risk of permanent damage to the membranes.

Note

A fuel cell is very sensitive to impurities in the hydrogen and the supply air. Sulfur, ammonia, salt, and all halogens are particularly dangerous and cause irreversible damage to the membrane.

Hydrocarbons, carbon dioxide, and carbon monoxide also cause damage to the membrane, but the damage is reversible and will heal, if pure hydrogen and air are used later over a longer period.

Water in the hydrogen causes irreversible damage in the H_2 storage and supply system.[17]

It is therefore essential that the hydrogen has a high degree of purity, otherwise, the available power and the lifetime of the cells will be reduced.

As natural gas contains between 10 and 20 mg/m^3 of sulfur and sometimes also ammonia, hydrogen that is transported via the natural gas network must be separated from the methane and purified at great expense before being used in fuel cells.

Due to the high costs, it is not possible to clean the hydrogen in the vehicle or to check its purity. One must trust the filling station and hope that the hydrogen is sufficiently pure. The lifetime of fuel cells is still a challenge today.

The cathode air cooler either heats or cools depending on the operating point. An electric coolant pump delivers a coolant flow rates of 10,000–12,000 l/h. The coolant must be electrically non-conductive and non-corrosive, as it is in direct contact with the electrodes of the stack. A water–glycol mixture is usually used. Because the coolant releases ions from the materials, permanent cleaning by an ion exchanger is required. For this reason, the cooling circuits of the hydrogen fuel cell system, the other drive units (battery, power electronics, electric motor) and the passenger compartment are usually designed as separate units.

Some of the water produced in the process is used to humidify the air because the cell quickly becomes defective in case the electrolyte dries out. The residues of unused hydrogen are separated from the water and fed back into the cell by a pump.

[17] Source: https://emcel.com/de/wasserstoffverunreinigung-in-brennstoffzellen/

As the voltage of the fuel cell system is highly load-dependent, a stabilizing DC/DC converter is always required, unlike in a battery storage system. This is followed by the usual inverter (DC/AC converter) and the electric motor.

To bridge the cold start phase of the fuel cell stack, which can last for minutes, and to absorb the recuperated energy, a high-voltage battery with the associated power electronics is also required. The battery system must be designed for full drive power and must have enough capacity to hold the recovered braking energy. In 2019, Daimler AG presented a plug-in hybrid vehicle with a fuel cell and 13.5 kWh HV battery capacity. The Toyota Mirai (2nd gen) uses a small 1.2 kWh battery, but with the disadvantage that the vehicle can only recover a small amount of braking energy.

Example 6.15—Integration of fuel cell systems in passenger cars

A fuel cell system in current passenger cars, for example in the Toyota Mirai or the Hyundai Nexo, requires an installation space comparable to a large 6-cylinder diesel engine with automatic transmission and auxiliary units. It delivers around 120 kW of electric output power.

The passenger compartment and trunk are correspondingly restricted. Cab-forward or skateboard chassis designs are impossible. The front luggage compartment popular with many electric cars cannot be realized. The Toyota Mirai does not even have folding rear seats because one of the hydrogen tanks is installed right behind the seats. ◀

6.15.4 Operating Characteristics

The theoretical open circuit voltage per cell is 1.23 V at 25 °C. Practically, it is closer to 0.5–1 V due to voltage losses caused by internal resistance, reaction inhibition, and gas diffusion.

The stack efficiency at low partial load is around 65% or more. It drops to around 55% at full load. A fuel cell has its highest specific power output at around 0.8 A/cm^2.

Figure 6.28 shows the typical efficiency of the stack and of the whole fuel cell system. The system efficiency includes the losses of the many auxiliary devices, particularly the cathode air compressor and cooling system.

Typical fuel cell systems in passenger cars achieve close to 60% efficiency in the partial load range. The efficiency gradually decreases with increasing load and reaches around 45% at full load. This characteristic is advantageous for use in urban traffic or overland, as little power is required on average and the maximum power is only required for a short time. On highways with higher speeds and correspondingly higher power requirements, FCEVs significantly lose efficiency and range.

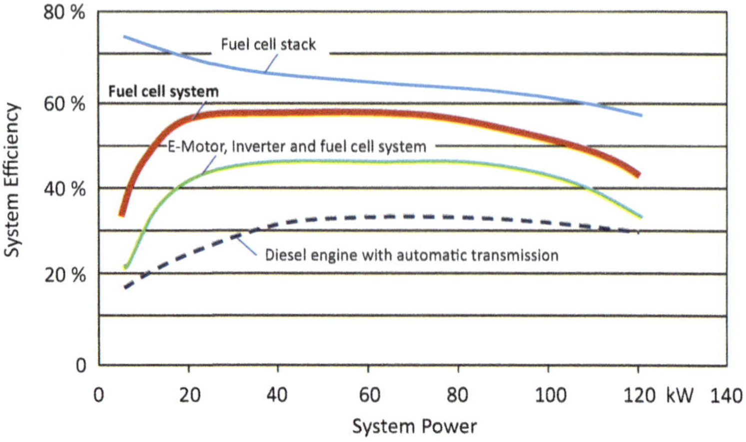

Fig. 6.28 Typical efficiencies of fuel cells, fuel cell systems, and, for comparison, a passenger car diesel engine including automatic transmission

6.15.5 Hydrogen Storage Systems

6.15.5.1 Compressed Hydrogen (CH_2)

All current fuel cell cars are equipped with compressed gas storage systems (CH_2 = Compressed Hydrogen). A pressure level of 700 bar when fully charged has become the standard. In commercial vehicles, especially buses, compressed gas storage tanks are often operated at 350 bar. This halves the volumetric energy density but accounts for a simpler, safer, and cheaper tank system.

At 700 bar, hydrogen has a density of 40 kg_{H2}/m^3, under atmospheric pressure it is only 0.090 kg_{H2}/m^3. A 60 l tank can hold 2.4 kg of hydrogen, which is good for a range of around 200 km. The volumetric energy density of 700 bar CH_2 is around 1.3 kWh/l.

The total weight of a CH_2 tank is largely determined by the weight of the cylindrical pressure vessel, not by the stored hydrogen. In current cars, around 5.7 wt% is achieved. This means around 5.7% of the total weight is the weight of hydrogen, the remaining 94.3% of the weight is contributed by the tank. The gravimetric energy density of the tank system is around 1.8 kWh/kg.

Example 6.16—Hydrogen tanks in the Toyota Mirai

The two cylindrical tanks in the Toyota Mirai (1st gen) have a capacity of 60 and 62.4 L. Together they hold 5 kg of hydrogen. The EPA range is 500 km.

A diesel car of a similar size would need around 7 l/diesel per 100 km, i.e. 35 L of tank volume. CH_2 tanks therefore require approximately four times the volume of conventional fuel tanks. In addition, the tank must be largely cylindrical due to the extremely high pressures. This makes spatial integration into the vehicle much more complicated.

The two tanks of the Mirai weigh just under 90 kg in total. Of this, 5 kg is hydrogen (5.7 wt%) and the remaining 85 kg is the weight of the tanks themselves. A comparable diesel tank would weigh around 30 kg fully loaded. CH_2 tanks therefore have around three times the mass of conventional fuel tanks. ◀

Compressed gas tanks are made of carbon fiber and are becoming mature products. A safety inspection is required every 3 years and after every accident. The typical lifetime is 15 years.

6.15.5.2 Cryp-Compressed Hydrogen (CcH_2)

Another variant is transcritical hydrogen storage (cryo-compressed = CcH_2). Hydrogen is compressed at a high pressure of around 300 bar and simultaneously cooled down to about −220 °C. This method provides the best energy density (around 72 kg_{H2}/m^3 and 2.4 kWh/l) but requires both high-pressure and low-temperature tanks. CcH_2 is currently being discussed primarily for the refueling of heavy commercial vehicles. It is not used in cars.

6.15.5.3 Liquid Hydrogen (LH_2)

Hydrogen can also be stored in liquid form with low pressure, which significantly increases the volumetric and gravimetric energy densities of the tank system compared to gaseous storage (cryogenic tanks = liquid hydrogen = LH_2). The storage temperature is 20 K, which is around −250 °C, at a pressure of 8–16 bar.

At around 65 kg_{H2}/m^3 and 2.2 kWh/l, the storage density of LH_2 is 60% more than CH_2 at 700 bar and nearly three times greater than that of CH_2 at 200 bar. If the heavy weight of the high-pressure CH_2 tanks is considered, the weight advantage of LH_2 increases even further.

However, the energy required for the liquefaction of hydrogen is considerably higher than that required for compressed gas storage. It amounts to around one third of the total energy contained in the hydrogen in the first place. The best H_2-liquefiers today achieve an efficiency of around 64%, for example the Linde AG plant in Leuna, west of Leipzig in Germany.

In addition, boiling, i.e. vaporization, leads to an increase in pressure due to sub-optimal cooling, so that gaseous hydrogen must be constantly released into the surrounding air. The daily loss of a typical liquid hydrogen car tank corresponds to around 3–4% of the stored volume. The tank of a vehicle that is parked for 14 days will lose around one third to one half of its contents through evaporation. Especially when the vehicle is parked in closed rooms, an explosive H_2 gas cloud can form.

For this reason, after several unsuccessful experiments with liquid hydrogen in recent decades, only compressed hydrogen is used as a tank medium in cars and buses today.

6.15.5.4 Liquid Organic Hydrogen Carrier (LOHC)

LOHC (Liquid Organic Hydrogen Carrier) is a process for storing hydrogen chemically in an aqueous carrier liquid at ambient temperature. The charged liquid can be filled up like conventional fuel and the bound hydrogen can be released later in the car. As LOHC is only discharged but not consumed, the discharged liquid needs to be stored in a second tank in the car. It is later pumped back into the filling station and recharged either directly on-site or at central collection points.

The origins of LOHC go back to a patent from 1968. The first test plants were built in the 1980s. Today, 60 years later, the process is still at the research stage. No commercial applications are known.

The absorption of hydrogen by the carrier liquid is referred to as hydrogenation and the release as dehydrogenation.

Hydrogenation, i.e. the charging of the liquid, is an exothermic reaction, i.e. energy is released in form of heat, and is carried out at pressures of around 50 bar and temperatures of around 150–200 °C. The heat must be dissipated via a cooling system.

Dehydrogenation, i.e. the removal of hydrogen from the carrier, is an endothermic reaction, i.e. it requires heating to temperatures of around 250–320 °C. Due to the high temperatures, suitable insulation from the vehicle interior is required. The discharged liquid must then be cooled down again before being returned to the tanks. These processes require a significant amount of energy and reduce the efficiency of the hydrogen extraction.

The technology required for the hydrogen extraction is of a considerable size and completely fills the trunk of a large car. Added to this is the complex fuel cell system, which also requires a lot of space.

In addition, the volume of the LOHC tanks is very large. In the last decades, the density of hydrogen has been increased by around 20% from 52 g_{H2} to 63 g_{H2} per kg of LOHC. Considering a typical FCEV car consumption of 1.2 kg_{H2}/100 km, around 100 kg of liquid is required for a range of 500 km. This corresponds to a LOHC throughput of around 19 l/100 km. The required tank volume is therefore around three times larger compared to conventional vehicles with a combustion engine.

Further challenges of LOHC are the production, disposal, and, of course, the costs of the carrier liquid, the exact chemical composition of which is still a subject of research. In the past, N-ethylcarbazole was used, but this is harmful to the environment and must not be allowed to enter groundwater. In addition, the material decomposes when hydrogen is extracted and must be replaced regularly. It is highly flammable and has a relatively low flash point below 200 °C. Today, other carrier fluids with better properties are being investigated.

The question arises as to whether LOHC has the potential to compensate for the considerable infrastructural disadvantages of hydrogen as an energy storage medium for cars compared to batteries.

If green hydrogen is used for LOHC, this requires around 3.5 times the amount of electricity compared to BEVs. Added to this are the losses when refueling the liquid and later further losses when extracting the hydrogen in the car. The author is not aware of any realistic data on the efficiency of these two processes. It is to be expected that the energy requirement of LOHC will be between that of CH_2 and LH_2. LOHC therefore does not solve the hydrogen production problem.

The second challenge of fuel cell cars is the complex infrastructure for distribution and refueling, which would have to be completely rebuilt for both gaseous and liquid distribution. LOHC promises a simple distribution, as in principle the existing infrastructure for gasoline and diesel fuels can be used. However, at up to 20 L per 100 km, the tank capacity of LOHC is around three times higher. The existing tank infrastructure for diesel and petrol is therefore not sufficient to cover the needs of millions of vehicles and would have to be significantly expanded. LOHC also requires a circulation system to return the discharged fuel. This is not yet available. The facilities for charging LOHC with hydrogen are also lacking. All in all, we can hope to reduce infrastructure costs with LOHC compared to CH_2 or LH_2. However, it is currently difficult to estimate the extent of this reduction.

The third problem area of hydrogen cars is the space required and the cost of the fuel cell system in the vehicles. This aspect will worsen with LOHC because, firstly, the tanks must hold three times as much liquid as in conventional cars, secondly, new types of tanks are needed to separate fresh and discharged liquid and, thirdly, there is the technology for extracting the hydrogen, which is operated at high temperatures and requires correspondingly complex heating and cooling systems. It is not yet clear whether it will ever be possible to integrate the necessary technology to a size that is acceptable for passenger cars at low cost and high efficiency.

All in all, LOHC is hardly suitable for road traffic and certainly not for passenger cars. It remains an interesting research topic that could perhaps one day become interesting for other applications (ships, planes, …).

6.15.5.5 Metal Hydride Storage

Hydrogen can be stored in a highly porous metal compound and released again later. Work on metal hydride storage systems for hydrogen has been going on for several decades.

The metal is finely ground and has a consistency like flour. The challenge of this storage concept is to find a suitable metal with a high storage capacity for the hydrogen. Promising candidates include lithium hydride, lithium aluminum hydride, magnesium hydride, and sodium borohydride.

Storage takes place at a relatively low pressure below 100 bar and at medium temperatures. This requires a catalyst, usually titanium. The low pressure and thus the high level of safety are advantageous.

The hydrogen is released through pressure reduction and heating.

Current developments achieve a storage volume of around 12.6 g_{H2} per liter of metal powder. A 100 l tank can therefore hold 1.26 kg of hydrogen, which is 100 km range for a car.

In addition to the large volume, the high mass of the storage tank is also a disadvantage. Storage systems based on magnesium or magnesium alloys can theoretically achieve a hydrogen content of up to 8% per weight, but currently only reach around 3.7% in practice. Another critical factor is the long time required to refuel the storage system.

Metal hydride storage systems are still a research topic and are not in commercial production. However, they have already been used in submarines. The volumetric energy density and even more the poor gravimetric energy density practically rule out their use in road vehicles.

6.16 Other Short-Time Energy Storage Systems

In addition to batteries and hydrogen, energy can also be stored in electric and magnetic fields and as kinetic energy. Such devices cannot hold nearly as much energy but have other advantages, like a long calendar or cycle life or a high-power output.

Electrostatic energy storage (capacitor):

$$W_e = \frac{1}{2} \cdot C \cdot \left(U_{max}^2 - U_{min}^2 \right) \tag{6.5}$$

with C = capacity in F and U = voltage in 1/s.

Magnetostatic energy storage (coil):

$$W_m = \frac{1}{2} \cdot L \cdot \left(I_{max}^2 - I_{min}^2 \right) \tag{6.6}$$

with L = inductivity in H and I = current in A.

Kinetic energy storage (flywheel):

$$W_k = \frac{1}{2} \cdot J \cdot \left(\Omega_{max}^2 - \Omega_{min}^2 \right) \tag{6.7}$$

with J = moment of inertia in kg·m^2 and Ω = angular velocity in 1/s.

Electrostatic storage devices are fundamentally capacitors. To increase the energy density, the distance between the electrodes (gap size) must be reduced and the surface area increased. The so-called double-layer capacitors use a Helmholtz layer to achieve effective gaps of just a few nanometers. Graphite, which has a very porous surface, is used as the electrode material. This results in an effective surface area of up to 3000 m^2 per gram (!) of graphite. Finally, an electrolyte with a very high dielectric constant is used, around 10,000 times higher than vacuum. If these measures are combined, the result is a supercap or ultracap. Such storage devices are manufactured by Maxwell Technologies, for example, and achieve capacities of around 500 F per g and energy densities of up to 5 Wh/kg. These capacitors are still a long way from lithium-ion batteries, which have energy densities of up to 300 Wh/kg. The advantages of double-layer capacitors are their high power density (up to 17 kW/kg) and high cycle stability (up to over 1 million cycles).

However, double-layer capacitors are only used in a few applications today. For example, to store recovered braking energy in hybrid buses or electric locomotives.

Magnetostatic storage devices store energy in a coil with the help of an electric current. In practice, they only work if the current can circulate in the coil without producing ohmic losses. Low-temperature superconductors are used for this purpose. This technology is still in its infancy and larger areas of application have not yet emerged. Magnetostatic storage systems are not used in vehicles.

Flywheel storage systems are often used in stationary applications. For example, in data centers and hospitals flywheels temporarily bridge a power outage until the diesel generators have started up. Modern flywheels reach surface velocities far above the speed of sound, i.e. over 330 m/s and speeds up to 250,000/min. Their energy density is around 5–55 Wh/kg, which is 5 times lower than lithium-ion batteries. However, their specific power is quite high with up to 4 kW/kg. They also have a long calendar and cycle lifetime.

Small flywheel storage systems have been used in racing cars to store recuperated braking energy. Today, in-car applications are almost exclusively served by lithium-ion batteries.

Example 6.17—Flywheel storage systems for cars

The Porsche 911 GT3 R Hybrid racing car had a flywheel storage for recuperating braking energy. The storage unit had a capacity of 0.25 kWh and weighed 40 kg. However, it was able to deliver a power output of 120 kW for up to 8 s, which is quite remarkable.

The company Compact Dynamics from Munich, Germany, supplies a flywheel storage with 0.14 kWh capacity and 60 kW output power at a total weight of 26 kg. The unit is designed for over 5 million full load cycles and operates in a temperature range from −40 to +150 °C. ◀

6.17 Fire Hazard of Cars

In principle, there is a risk of fire for any transportation of electrochemically bound energy. This applies equally to petrol, diesel, batteries and hydrogen.

The question arises as to whether battery electric vehicles are more dangerous than cars powered by fossil fuels. In western countries with a high standard of living, around 1 per thousand cars burn every year. In Germany, this amounts to around 40,000 car fires per year, 15,000 of which are serious fires.

A study by the Chair of "Production Engineering of E-Mobility Components" at RWTH Aachen University has shown[18] that the fire risk of electric cars is

[18] Source: https://edison.media/verkehr/brandrisiko-beim-verbrenner-viel-groesser-als-beim-e-auto/25239585/.

only around 0.9–1.2 per 10,000 cars. For conventional vehicles with combustion engines, there are statistically 7.3 fire accidents per 10,000 cars, that is 7 times more.

A U.S. American study has shown that around 25 out of every 100,000 electric cars sold in 2023 were involved in fires. For cars with internal combustion engines, the figure was 1530 out of 100,000 cars, i.e. around 61 times more.[19]

Batteries for cars must meet high safety requirements and pass extensive crash tests before approval.

Spontaneous battery fires of electric cars following serious accidents have not been reported to date. Before each battery fire, there was a time window of several minutes for rescuing the occupants. Petrol or diesel fuel that leaks after serious accidents often ignites on hot surfaces which can lead to spontaneous and fierce explosions.

It can therefore be concluded that occupant safety in electric cars is significantly higher than in cars with combustion engines.

On the other hand, battery fires can start even days after a severe accident. Damaged batteries should be dismantled as quickly as possible and intact cells discharged. Battery fires cannot be extinguished as the oxygen required for the reaction is supplied by the burning materials themselves. Battery fires continue to burn even under water. However, water cooling is still needed to prevent the fire from spreading to neighboring cells.

Small amounts of hydrofluoric acid (hydrogen fluoride HF) can be produced inside a battery fire, which is extremely dangerous. However, it is only hazardous in the white smoke that may be produced. Smoke gas must never be inhaled in any case and firefighters are trained to always wear breathing apparatus in such cases.

The firefighting procedure and the risks for accident responders are different for electric cars than for petrol-powered vehicles. Appropriate training is advisable. But ultimately, the fire risks of electric cars are not fundamentally greater than those of gasoline cars.

In hydrogen cars, there are two critical paths after an accident: Firstly, there is the possibility of a mechanical explosion of the tanks, which are under an enormous pressure of up to 700 bar. The second risk is the explosion of escaping hydrogen.

Hydrogen is highly flammable in air and causes catastrophic explosions. In parking garages or tunnels, there is a great danger that also affects bystanders (unlike burning batteries, which normally do not spread further).

Many hydrogen sensors are installed in hydrogen vehicles for safety reasons. Due to its low density, hydrogen gas escapes into the atmosphere at a high speed of around 50 km/h in the open air. It is then harmless. However, in enclosed

---

[19] Source: *Extinguishing the EV Battery Fire Hype*, Willie Jones, IEEE Spectrum, 04 Dec 2023 mit Daten aus https://www.autoinsuranceez.com/gas-vs-electric-car-fires/.

spaces, for example in parking garages, tunnels, but also under canopies or eaves, a dangerous concentration of hydrogen gas and atmospheric oxygen can quickly form as ignition is possible in a wide mixing ratio of 4–78 vol%. In addition, only a low ignition energy is required.

A commercial hydrogen filling station for cars exploded near Oslo, Norway, in the early summer of 2019.

All in all, fire safety is best for electric cars, followed at a clear distance by cars with combustion engines. There are no statistics for hydrogen cars due to the very small number of cars on the roads. However, based on the chemical properties of hydrogen and the extremely high tank pressures, it can be assumed that the risk of fire and explosion is the highest.

6.18 Summary of Energy Storage Systems for Vehicles

PEM fuel cells have been developed since the early 1960s. So far, they have not been able to secure a significant market share in any application due to their high costs and low reliability. This may change in the future thanks to the intense development work that continues to be invested in this technology.

Lithium-ion batteries are much younger, only about half as old as fuel cells, and have already been able to establish themselves on a broad front. Today's lithium-ion cells are reliable and cost-effective. They also show considerably more potential for future improvements. Therefore, the technological gap between batteries and fuel cells will even widen in the coming decades.

Batteries are well suited for applications that require a relatively small amount of energy and a relatively high output power. With fuel cells, the opposite is true, as the amount of energy is determined by the not-so-expensive tanks, while the expensive and less reliable fuel cell system determines the available output power.

Energy applications can be rated according to their C-rate. High C-rates (short storage times, high power-to-energy ratios) are perfect for batteries. As of today, that includes all light road vehicles (electric bicycles, passenger cars, light commercial trucks) and probably also heavy-duty trucks.

For low C-rates (long storage times and/or high continuous energy demand, relatively low power-to-energy ratios) fuel cells may be the better choice in the future. These may be applications in overseas vessels or large long-haul aircrafts. Also, large-scale seasonal electricity storage systems could become an interesting field for fuel cells[20] as well as the supply of energy-intensive industrial processes (chemical industry, cement and steel production).

[20] Short-time stationary electricity storage systems are already being equipped with batteries and used in more than one million home storage photovoltaic installations, see Sect. 14.4.6.

Table 6.9 Comparison of energy systems for cars

	Diesel fuel	Lithium-ion batteries 2025	700 bar hydrogen CH_2 and fuel cell system
Energy density fuel/battery cells	12 kWh/kg 9.8 kWh/l	0.40 kWh/kg 1.0 kWh/l	33.3 kWh/kg 1.33 kWh/l
Energy density tank/battery pack	10 kWh/kg 8 kWh/l	0.26 kWh/kg 0.25 kWh/l	1.85 kWh/kg 1.1 kWh/l
Average car consumption	7.6 l/100 km 75 kWh/100 km	20 kWh/100 km	1.2 kg_{H2}/100 km 40 kWh/100 km
Energy system weight for 500 km	20 kg tank + 30 kg fuel = **50 kg**[b]	100 kWh battery pack = **380 kg**	100 kg tank + 300 kg stack[a] = **400 kg**
Energy system volume for 500 km	**44 L**[b]	**400 L**	180 l tank + 500 l stack[a] = **680 L**

[a] "stack" refers to the fuel cell and all auxiliary components (pipes, pumps, filters, valves, DC/DC converter, heating and cooling systems, controllers etc.) excluding the H_2 tanks. However, a car fuel cell system can deliver significantly less power (typically 100 to 150 kW) than a battery or diesel fuel. In this respect, the comparison is somewhat unfair.

[b] When comparing ICEs and BEVs, the additional weight and size of the combustion engine, the automatic transmission, and the exhaust aftertreatment must also be considered. In the end, battery electric cars are still heavier, but the gap to ICE cars is narrowing.

Table 6.9 provides comparative estimates of the different energy technologies available in passenger cars. The technical properties of the energy carriers will be discussed in later chapters.

The table presents the best values for battery vehicles in 2024/25 based on semi-solid-state batteries (SSB). As battery technology is currently developing at a rapid pace, the numbers can be expected to improve significantly over the next years.

This direct comparison clearly shows that the often-cited weight advantages of fuel cell cars over battery cars disappear with the latest generation of battery cells. The trend is even reversing: Soon, BEVs will be lighter than FCEVs with the same range and require less raw material. In terms of cost, available passenger and trunk space, durability, safety, public and private charging infrastructure, battery cars have long been far ahead of fuel cell cars.

In the field of battery electric vehicles, three market segments are served with different technologies: entry-level cars will have LFP cells or sodium batteries with energy densities of up to 180 Wh/kg. Mid-range cars will use NMC or NCA cells, which will be available up to 350 Wh/kg. Luxury vehicles will be equipped with solid-state cells with energy densities of more than 400 Wh/kg.

Note
The number of passenger cars in use worldwide (by the end of 2022) breaks down as follows:

Petrol or diesel combustion engine cars (ICE): 1400 Mio. (including hybrids).

Battery electric cars (BEV&PHEV): 28 Mio. (Estimation for 2025: 76 Mio).

Natural gas combustion engine cars (NG-ICE): 23 Mio.

Hydrogen cars (FCEV): 0.058 Mio.

While battery electric cars are becoming more and more common place in many countries in the world there is still a long way to go before a significant proportion of the global car fleet will be converted to climate-friendly drives.

Fundamentals of Electric Machines

7

Abstract

After a brief introduction to the basic functional principle of electric machines and the three-phase system, the three main basic variants of electric machines, the induction machine, the externally excited, and the permanent magnet synchronous machine, are presented. This is followed by an overview of the designs of the stator and rotor of these three motor variants. The physical background of braking energy recovery (voltage generation) and motor operation (magnetic forces) is explained in more detail. The base speed range and the field weakening range, which were presented purely phenomenologically in the second chapter, are now introduced in detail on the basis of the magnetic properties of the motor. The chapter concludes with an overview of the properties of the materials used and the losses that occur in electric machines.

7.1 Introduction

Since the 2000s, there has been an astonishing development in the field of electric drives. Through the intensive efforts of the automotive industry, various new concepts have been developed, resulting in an increase in the power density of electric motors by nearly two orders of magnitude— despite the technology being almost 200 years old. The reasons are in a nutshell:

- Improved numerical calculation methods and the use of significantly more computer power in design, resulting in better system understanding and closer approaches to physical limits, affecting both electromagnetism (magnetic circuit) and mechanical strength (speed) as well as thermal utilization.

M. Doppelbauer, *Introduction to Electromobility*,
https://doi.org/10.1007/978-3-658-45482-1_7

- Better and novel cooling methods (water jacket, liquid-cooled shaft, direct internal laminated core cooling, internal slot cooling, winding overhang cooling with oil spray mist) enabling higher current densities.
- Improved materials: New hard and soft magnetic materials allow for higher electromagnetic utilization, and new construction materials (glass and carbon fiber as well as speciality steels, e.g. Inconel) enable higher speeds.
- New winding techniques, particularly rectangular conductors with higher filling factors, also allow for higher current densities.
- New and improved control methods enable safe operation of the magnetic circuit deep into saturation with a high degree of non-linearity.

The series drives of modern electric vehicles have power densities that match and sometimes exceed those of Formula 1 combustion engines. This results in lower costs and reduced space requirements, leading to immediate customer benefits.

To illustrate these developments with numbers: a 250 kW industrial motor (4-pole 50 Hz asynchronous motor) currently weighs around 1300 kg, while a 250 kW motor for a modern electric car weighs about 30–50 kg.

7.2 Operating Principle

The basic operating principle of electric machines is based on the attraction of magnetic poles of different polarity, see Fig. 7.1a.

The rotating part (rotor) consists of a magnet with north and south poles. Magnetic poles can never exist individually. A pair of poles ($p=1$) is equal to two poles ($2p=2$). Both designations are equivalent and common. Real machines usually have more than two poles, typically 6 or 8, sometimes many more. The functional principle does not change.

In the stator in Fig. 7.1a, a magnet is attached above and below the rotor. The gray yoke made of iron is used to conduct the magnetic flux. The magnetic flux

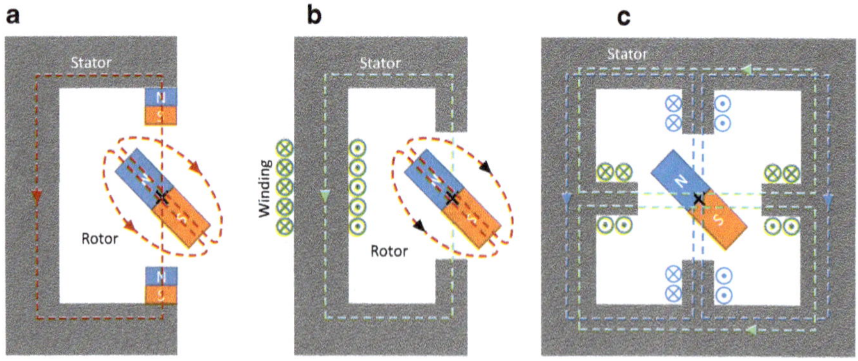

Fig. 7.1 Fundamental operating principle of an electric motor

of the two stator magnets and the rotor magnet is shown as red dashed lines. For better understanding, the two fluxes from stator and rotor are drawn separately. In reality, they overlap to form a resulting total flux.

The rotor in Fig. 7.1a will rotate clockwise due to the attractive force of the different poles. However, the movement ends as soon as it is completely aligned vertically.

An improvement of the arrangement can be seen in Fig. 7.1b. The two magnets of the stator have been replaced by an electric coil. This produces the magnetic flux shown in green. It is not different in its function and effect from that of the stator magnets in Fig. 7.1a.

However, the rotor only performs a short movement and then remains in a vertical position. This problem is solved by introducing a second pair of coils in the stator, see Fig. 7.1c. This is also referred to as two phases.

Initially, only the two blue coils, which are connected in series, are supplied with current and the rotor turns clockwise to the vertical position.

Next, the two blue coils are switched off and the two green coils are switched on. The rotor continues to turn horizontally.

In the third step, the blue coils are switched on again, but this time with the current flowing in the opposite direction. The rotor continues to turn clockwise until it is vertical again, but now upside down.

Finally, the cycle is completed by switching on the green coils with reversed current flow. The rotor continues clockwise and turns to the horizontal position. The cycle then starts again from the beginning, resulting in a continuous rotary movement.

7.3 Three-Phase Current (Rotary Current)

The motor in Fig. 7.1c has two electrical phases, each of which feeds two coils. This rotor generates torque but has various disadvantages. A motor with three electrical phases, each of which has a sinusoidal current flowing through it that is 120° out of phase, works much better. Such a current system is also referred to as three-phase current or rotary current.

Note
Motors with more than three phases are also technically possible. However, this only minimally improves motor performance. Three phases represent the economic optimum.

Exception: In the traction sector, $2 \cdot 3 = 6$-phase machines are currently being investigated, because if one of the two three-phase systems fails, the other system still provides an emergency driving function.

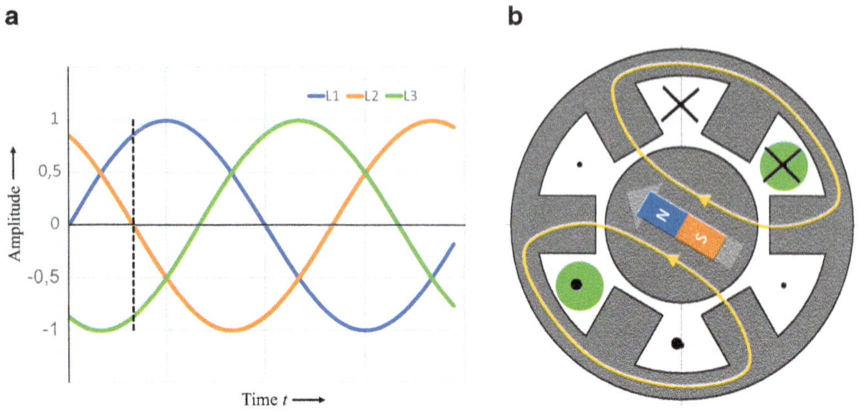

Fig. 7.2 Three-phase current (left) and basic configuration of a three-phase motor (right) at $\varphi = 60°$

Figure 7.2a shows the time curves of the currents in the three phases (L1 = blue, L2 = orange, L3 = green). Figure 7.2b shows a three-phase motor in its simplest form. This winding is also referred to as the fundamental winding. There is only a single coil in the stator for each of the three phases. The forward and return conductors are arranged opposite to each other.

The rotor of this three-phase motor is not shown in further detail. It could be constructed similarly to Fig. 7.1) and include a magnet, for example. Other rotor variants will be discussed later.

The black dashed line in Fig. 7.2a indicates a randomly selected moment in time. At this moment, the current in phase L2 (orange) is practically zero and the current in the other two phases is the opposite, equal, and almost maximum. The figure on the right indicates the current strength by the size of the circles.

The magnetic fluxes of the blue and green coils are superimposed to form the total field shown in yellow.

Figure 7.3 shows the situation at a later point in time when the currents have been rotated by 60°. Assuming a normal 50 Hz network, this would correspond to a time difference of 3.3 ms.

In this position, the current of phase L3 is practically zero. The magnetic fields of coils L1 and L2 are superimposed to form the field shown in yellow, which, like the currents, has rotated 60° counterclockwise. The rotor follows this rotation.

Note

The functional principle of a three-phase winding is based on two properties:

- The coils of the phases are spatially offset by 120° along the circumference and

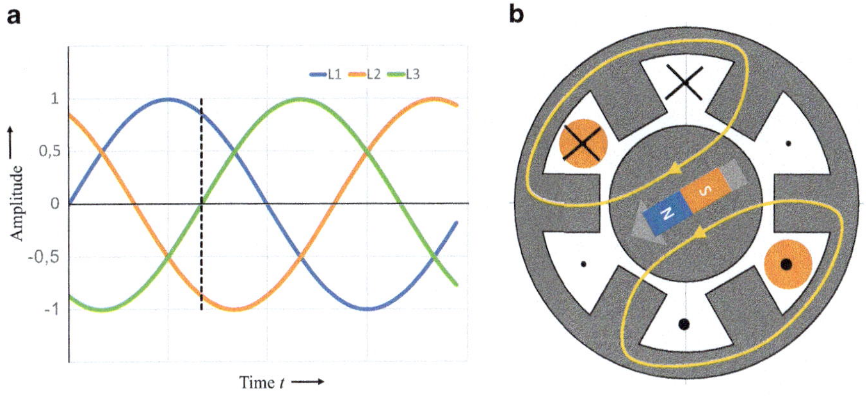

Fig. 7.3 Three-phase current (left) and basic configuration of a three-phase motor (right) at $\varphi = 120°$

- the alternating currents in the three phases have a 120° time delay against each other.

Under these conditions, a uniform rotating magnetic field with a constant amplitude is created.

The width of a magnetic pole, i.e. the distance along the rotor from one magnetic pole to the next, is also referred to as the pole pitch τ_p. It is simply calculated from the rotor diameter d and the number of poles $2p$ to give:

$$\tau_p = \frac{\pi \cdot d}{2p} \tag{7.1}$$

The speed of rotation of the stator field n_1 is:

$$n_1 = f_1 / p \tag{7.2}$$

where f_1 is the electrical frequency of the stator current and p is the number of pole pairs.

Although it may seem confusing at first, a three-phase system requires three coils to generate two rotating magnetic poles. Correspondingly, six spatially offset coils are required for four magnetic poles, and so on. Electric machines in cars often have 6 or 8 poles for various reasons. Accordingly, 9 or 12 coils are required.

However, it turns out that a motor built with this minimum number of coils has poor operating characteristics. We will see later that its harmonic fields are quite large, which leads to increased vibrations, noise, and torque oscillations. It is therefore advantageous to divide the individual coils into several neighboring subcoils. The operation of all adjacent subcoils of a phase can be regarded as a single coil.

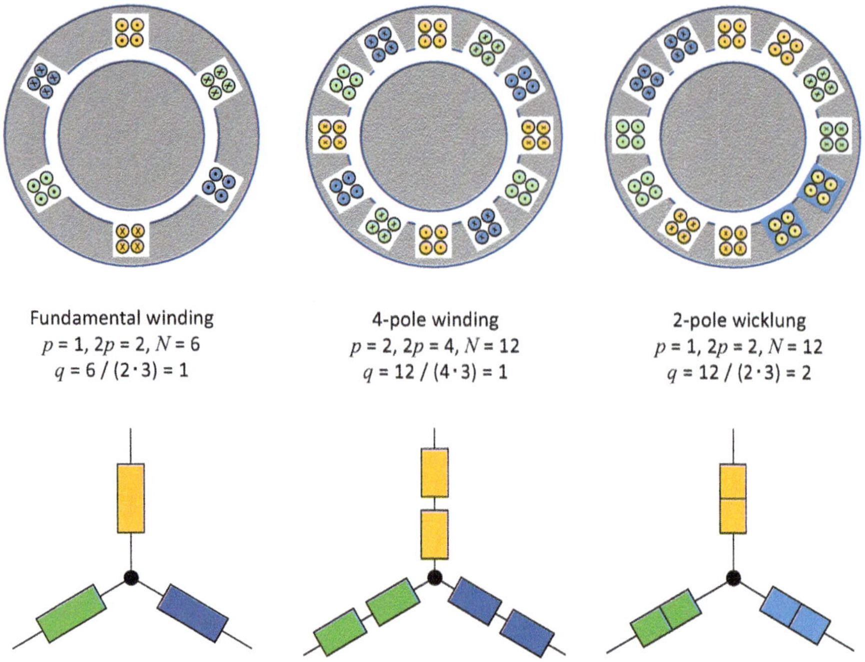

Fig. 7.4 Layout of three-phase windings with a higher number of poles p and slots/pole/phase q

The adjacent number of slots per pole per phase is designated by the letter q. q is decisive for the NVH behavior of windings as it discloses information about the winding factor and its harmonics. For traction motors with low harmonics, q should be at least 2, preferably 3 or even 4.

The calculation rule is:

$$q = \frac{N}{2p \cdot m}. \tag{7.3}$$

where $N =$ number of slots, $2p =$ number of poles, $m =$ number of phases (usually 3).

Figure 7.4 shows the structure of 2- and 4-pole stator windings with $q = 1$ and $q = 2$.

Sometimes two-layer windings are used. In this case, the coil sides of two different coils are placed in each slot. They can be arranged next to each other or on top of each other. For space reasons, each coil can only take up half the space and must therefore have fewer windings or thinner wires. This measure can also improve the NVH behavior of the winding.

Example 7.1—Number of slots, coils, and poles

A common number of slots is 54, so assuming a single-layer winding, the motor has 27 coils.

If we further assume that the number of slots per pole and phase is $q=3$, then there are 3 coil groups per phase, each with 3 adjacent coils. 3 coils per phase result in 6 poles, i.e. $2p=6$ or $p=3$. ◀

Windings with $q>1$ are called distributed windings while windings with $q<1$ are called concentrated windings or single-tooth windings. In addition, windings with q being an integer are called integer-slot windings. Windings with q being a fractional number are called fractional-slot windings.

These different types of windings all have similar properties in terms of noise, vibration, and torque oscillations within their groups but are distinctly different from one group to another.

7.4 Three-Phase Electric Machines

The three most common types of three-phase electric machines are:

- **Permanent Magnet Synchronous Machines** (PMSM) are used in the Toyota Prius, BMW i3, VW ID.3, Daimler e-Smart, Nissan Leaf, Tesla Model 3, Porsche Taycan, cars from BYD, Geely, and many others. Today, PMSM are the most used electric motors in electric vehicles including electric bicycles.
- **Externally Excited Synchronous Machines** (EESM) are used in the Renault Zoe (since 2013) and in many cars from BMW. As EESM do not need permanent magnets and have the highest cycle efficiency, they are favored by many manufacturers and will be included in many cars in the future. However, they are more difficult and costly to produce and have a slightly lower power density than PMSM.
- **Induction Machines** (IM) are used in the Audi e-tron 55, Renault Twizy, Mercedes EQA, EQB, and EQC. Induction machines have a lower power density, a lower peak power, and a lower efficiency compared to synchronous machines, but they are cheaper to make. In future, induction machines will be often used in all-wheel drive electric cars on the boost axle (usually the front axle) while the normal drive axle (usually the rear axle) is equipped with a synchronous machine. This design is, for example, used in cars from Geely (Volvo, Polestar, Zeekr, Smart, Lotus), BYD, NIO, Volkswagen, and Tesla.

These three machine types are presented in detail below.

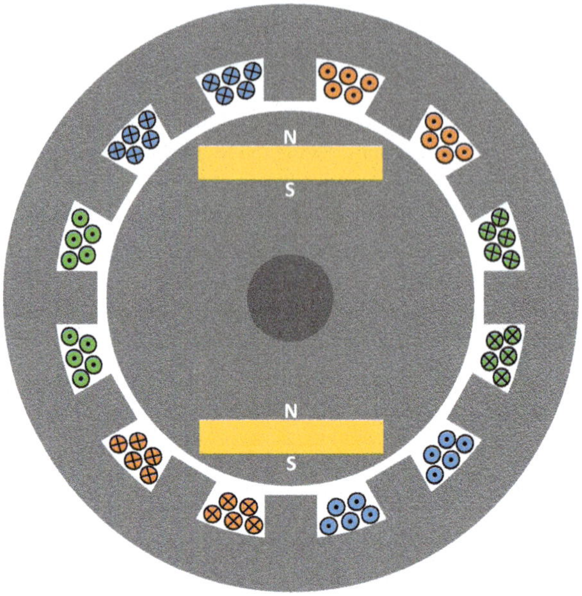

Fig. 7.5 Basic structure of a 2-pole PMSM

7.4.1 Permanent Magnet Synchronous Machine (PMSM)

Figure 7.5 shows an example of a simple PMSM with 2 poles and 12 slots. In this case, two coils per phase, each with 5 turns, are inserted into adjacent slots in the stator ($q=2$).

The individual coils of each phase can be connected in series or in parallel. This depends on the available battery voltage: parallel connection requires less voltage but more current, series connection requires more voltage but less current. Parallel connection is therefore preferred for low system voltages (48 V, 120 V) and series connection for high system voltages (400 V, 800 V). This ensures that the number of turns of the coil is not too small and not too large. If the number of turns is too small, the current per conductor will be high, resulting in thick conductors that are difficult to handle in production. The lowest practical number of turns per coil and slot is about 6.

Excessively large numbers of turns are also undesirable due to the thinner wire and the high manufacturing effort involved in coil production. The highest practical number of turns per coil and slot is around 100.

One end per each phase is brought together internally in the motor to form the star point, the other three ends are connected to the external connections, as shown in the figure above. This is called internal star connection.

Alternatively, the three phases can also be connected in a delta connection. The ends of the coils of each phase are then connected to the beginning of the coils of the next phase. The choice of star or delta also comes down to the system voltage:

Low voltage can be better served with delta connection; higher voltages are better suited for star connection.

However, the delta connection has the distinct disadvantage that internal circular currents may be created due to magnetic asymmetries, which can cause considerable additional losses. This must be considered during the design phase of the motor.

The machine shown in Fig. 7.5 contains magnets (shown in yellow) that are located inside the rotor (buried magnets or embedded magnets) and pre-magnetized in the direction of the letters S $\Rightarrow$ N. Seen from the stator, the lower magnet therefore acts like a south pole, the upper one like a north pole.

The operating principle of the PMSM has already been described in the previous section: A uniformly rotating field of constant amplitude is created in the stator by the three-phase current, which pulls the rotor along with it.

For the machine shown in Fig. 7.5, f1 = 100 Hz and $p = 1$ result in a speed $n_1 = 100/\text{s}$ or 6000/min.

The rotor follows the stator field at the same speed. However, there is an angular displacement between the two.

The angular displacement is not constant but changes depending on the load. When idling, i.e. at low currents and power, the north pole of the stator field and the south pole of the rotor field are exactly aligned with each other. When the machine is successively loaded, i.e. when it delivers more and more mechanical output power, the increasing current of the stator generates an opposing field to the rotor magnets. This leads to an offset in the angular position of the rotor in relation to the stator field. Both fields continue to rotate at the same speed, i.e. synchronous, but the rotor follows the stator field with a load-dependent angle.

Maximum torque and power are reached when the angular displacement approaches 90°. This is called the breakdown torque. When the loading is increased even more, the machine becomes overloaded and stalls: the rotor can no longer follow the stator field and will eventually come to a standstill. However, stalling leads to strong torque pulsations, which can damage the machine and the gearbox. This situation must never occur in real operations. For this reason, torque and angular displacement are closely monitored by the motor controller.

Note

Permanent magnets are used in electric machines to build up the excitation field. The field of the current-carrying coils, the armature field, is superimposed onto the excitation field to form a resulting total magnetic field.

In synchronous motors, the excitation field is generated by the rotor, while the armature field is generated by the stator. In direct current machines, which we will not discuss in this book, it is the other way around. When idling (no load), the currents in the stator coils are zero and only the excitation (rotor) field exists. Under heavy load, the stator currents are at their maximum. The armature field counteracts the excitation field and pushes it sideways out of the poles.

Note
During production, the permanent magnets are magnetized once by briefly exposing them to a strong magnetic field. This creates the remanence flux density and gives the magnets their magnetic effect. A certain amount of energy is required for magnetization, but this is only small compared to the subsequent output of the motor.

During operation, the magnets provide the excitation field that enables the transfer of force, power, and energy between the rotor and stator. However, the magnets contribute nothing to the energy balance.

One can imagine the process like towing a car: The energy is applied by the vehicle in front (= current of the stator coils). The rope (= permanent magnets) merely transfers the power of the car in front to the towed vehicle. Just like a tow rope, the magnets do not wear out during power transmission, they do not become weaker over time, and they do not lose any magnetism.

There are many reports and videos on the Internet that show motors that supposedly work purely based on permanent magnets. All these reports are based on fraud and illusions. A tow rope alone is never able to pull a vehicle if there is no other vehicle in front of it. In the same way, permanent magnets alone cannot keep a rotor in permanent motion if there is no additional external power, i.e. current-carrying coils or another energy source.

Figure 7.6 shows a false color image of the magnetic field of a PMSM. The magnetic flux density is particularly high in light yellow areas and small in dark blue areas.

7.4.2 Electrically Excited Synchronous Machine (EESM)

The stator of an electrically excited synchronous machine (EESM) is no different from that of a permanent-magnet synchronous machine. Here too, a uniform, rotating magnetic field is generated by multiple coils and three phase-shifted currents (Fig. 7.7).

In the rotor of the EESM, the magnets are replaced by excitation coils (shown in yellow). In contrast to the stator, however, a direct current (DC) flows in the excitation winding. The excitation coil generates a magnetic flux that is very similar to that of the magnets in Fig. 7.5.

Compared to the PMSM, the strength of the rotor field of an EESM can be adjusted depending on the load and speed to improve efficiency. However, the rotor coils require a current that causes additional losses and thus reduces the efficiency. Depending on the operating point, one or the other effect will dominate. Overall, however, there are clear efficiency advantages for the EESM.

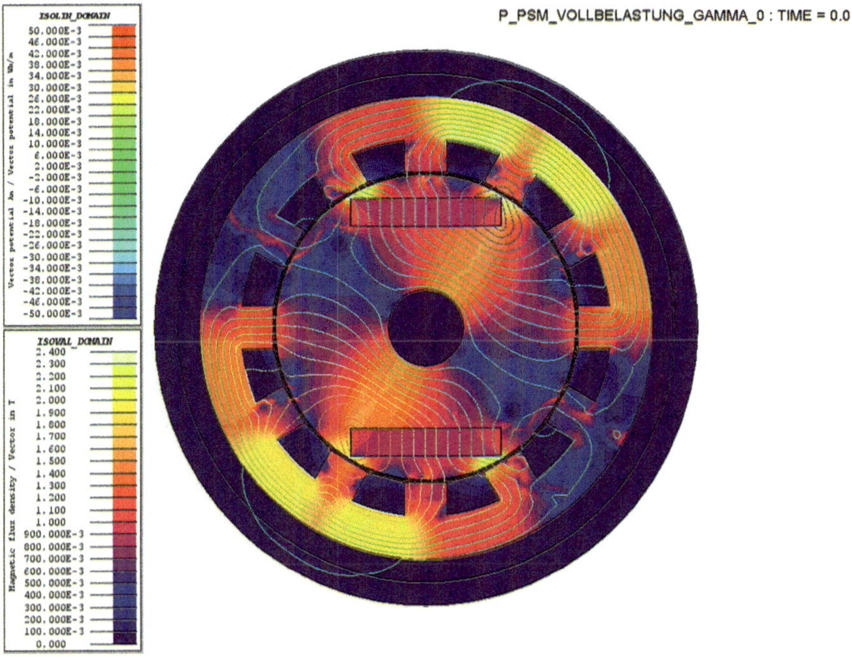

Fig. 7.6 Magnetic flux of a PMSM at full load, current angle = 0°

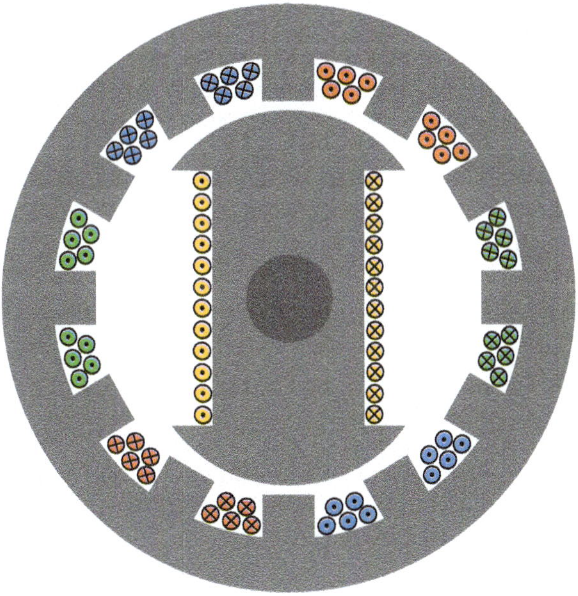

Fig. 7.7 Basic structure of a 2-pole EESM

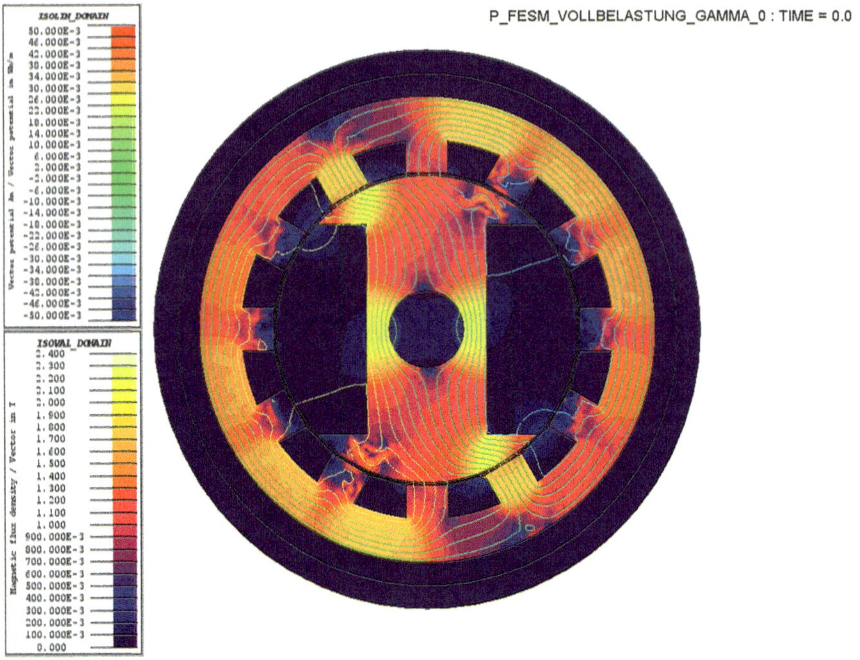

Fig. 7.8 Magnetic flux of an EESM at full load, current angle = 0°

The technical challenge of the EESM is the cost-effective and reliable transfer of electrical energy to the rotating rotor. In the Renault Zoe, slip rings are used for this purpose. Contactless energy transmission based on the transformer principle can also be realized. Some companies (ZF, Mahle) have introduced EESM prototypes with contactless rotor supply and BMW is producing cars with such machines (Fig. 7.8).

7.4.3 Induction Machine (IM)

The design and function of the stator of the induction machine are the same as those of the PMSM and the EESM. However, the functional principle of the rotor is different (Fig. 7.9).

Many conductive rods made of aluminum or copper are embedded in the rotor's laminated core. They are shown in yellow in the picture. At both axial ends, all rods are connected to each other by a short-circuit ring, which is not shown in the picture. This design is also known as squirrel-cage rotor. It was invented and patented by Michael Dolivo-Dobrowolski in 1889.

When the stator field passes the bars of the rotor, an induced voltage is created in each bar. The induced voltages in the bars are different from each other: The highest voltage is induced where the stator field has its greatest rate of change, i.e.

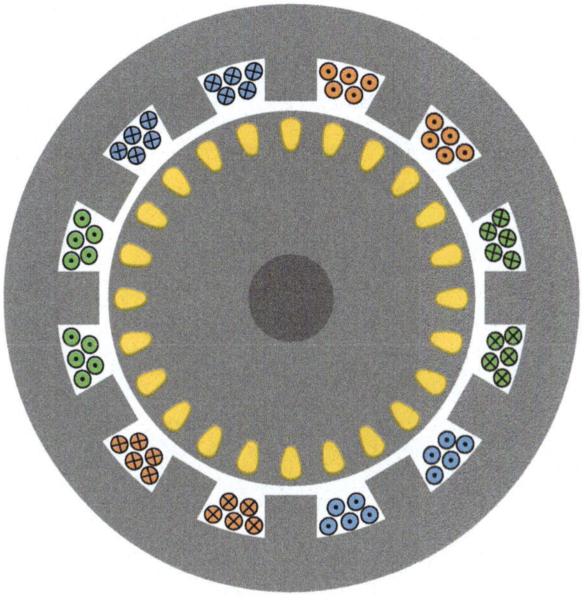

Fig. 7.9 Basic structure of a 2-pole IM

in the zero crossing between two magnetic poles. The lowest voltage is induced where the rate of change is lowest, i.e. exactly under the north and south poles.

As all the rods are connected to each other at the ends, these voltages lead to currents. The currents then generate their own magnetic flux, the rotor flux. The same number of magnetic poles are formed in the rotor as in the stator, as the rotor field is generated by the stator poles.

The principle of magnetic attraction then acts again between the stator flux and the rotor flux, and the rotor is also rotated. Physically, this can be explained as the formation of a Lorentz force between the magnetic field of the stator and the current-carrying conductors of the rotor.

However, voltage induction in the rotor bars can only take place as long as the bars intersect the flux lines of the stator. So, from the perspective of the rotor, there must be a change in the magnetic flux. If the rotor were to rotate at the same speed as the stator, i.e. synchronously, this condition would no longer exist and no voltage induction would take place. In this case, no rotor currents would flow and there would be no force. In that case, the rotor idles at full speed but without producing any torque and power (no-load).

Therefore, the rotor of an asynchronous motor must always rotate slightly slower than the stator field. The so-called slip s indicates the percentage deviation of the mechanical rotor speed n from the speed of the stator flux n_1:

$$s = \frac{n_1 - n}{n_1} \tag{7.4}$$

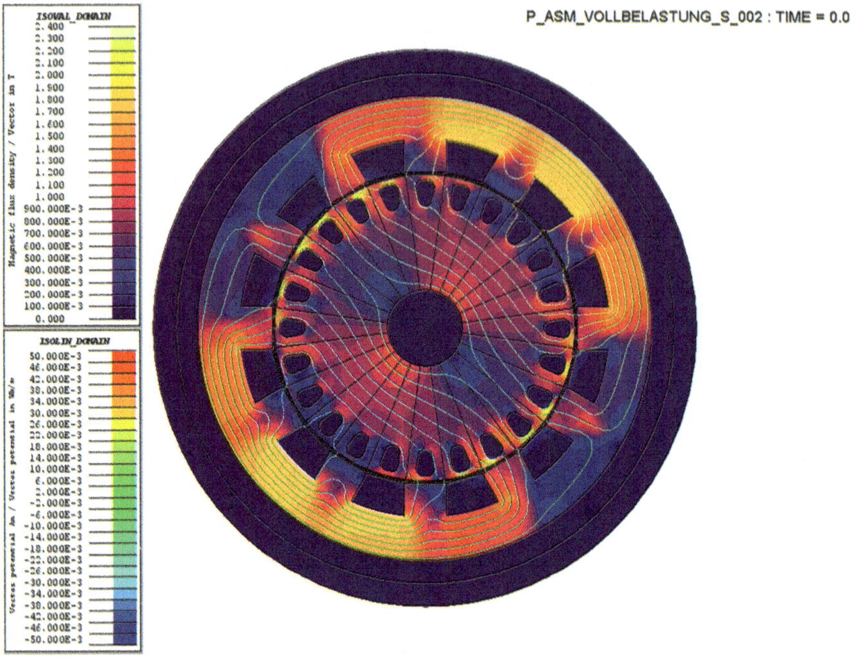

Fig. 7.10 Magnetic flux of an IM at full load, slip = 2%

When idling, the slip is close to zero, i.e. the rotor rotates at the same speed as the stator's magnetic flux.

As the load increases, the slip gradually increases as well. Typical values are 3–10% in continuous operation and around 20% at the highest torque, the so-called breakdown torque. A further increase in load causes the engine to stall, i.e. blockage. The slip is then 100%. However, the motor controller will prevent this situation from happening in a car.

When an induction machine starts up, the frequency and consequently the speed of the stator field is gradually increased from zero to the desired speed by the motor controllers. The rotor follows accordingly with a load-dependent slip of around 5–20% (Fig. 7.10).

In regenerative operation (braking energy recovery), the slip behaves in the opposite way to motor operation: Now the external torque applied by the tires will turn the rotor of the induction machine slightly faster than the stator field moves. The slip becomes negative.

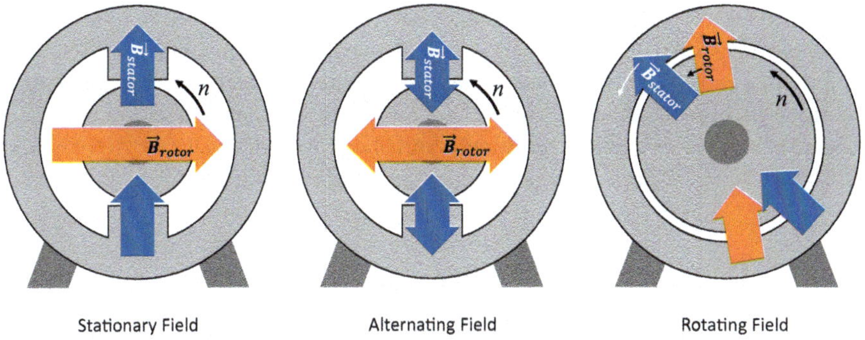

Fig. 7.11 Basic types of rotating electric machines

Note

The rotor of an induction machine does not contain any magnets. Therefore, no voltage is generated at the open terminals when the induction machine is mechanically driven from the outside. Additionally, no braking torques are generated.

Nevertheless, an induction machine can also be used as a generator: as soon as a three-phase voltage is applied externally and the rotor is actively turned, the coils of the stator drive a current back into the battery. Braking energy recovery is therefore only possible with an external voltage source. However, this situation is always present in electric cars due to the battery. The induction machine therefore has no disadvantages in terms of braking energy recovery compared to synchronous machines.

7.5 Types of Electric Machines

All rotating electromagnetic machines can be classified into the fundamental categories of stationary field, alternating field and rotating field machines (rotary field machines).

The drives considered so far fall into the category of rotating field machines (Fig. 7.11).

7.5.1 Rotating Field Machines

Rotating field machines have already been discussed in detail in the previous sections. The stator generates a uniformly rotating magnetic field of constant amplitude, which pulls the field of the rotor along with it, which therefore also rotates.

The speeds of the stator field and rotor field are either the same (synchronous), or the rotor rotates slightly slower (asynchronous motor) or faster (asynchronous generator).

Examples of rotating field machines are the permanent magnet synchronous machine (PMSM), the electrically excited synchronous machine (EESM, also referred to as separately excited SESM), the induction machine (IM, also known as an asynchronous machine ASM), the switched reluctance machine (SRM) or the synchronous reluctance machine (synRM). However, the latter two types of machines have not yet been used in series-produced electric cars.

7.5.2 Stationary Field Machines

In DC machines, the magnetic fields of the stator and rotor are perpendicular to each other, i.e. they are electrically offset by 90°. The fields have a constant amplitude as with rotating field machines, but they are stationary and do not rotate.

The rotor begins to move due to its attraction to the magnetic field of the stator. A mechanical switching device, the commutator, ensures that the currents in the rotor coils are switched after a short rotation of just a few angular degrees so that the rotor field jumps back to its starting position. Depending on the number of commutator laminations, this happens sooner or later, so that the rotor field tilts back and forth more or less around its neutral position.

Today, millions of DC motors are produced and are also being used in cars. Modern luxury-class vehicles can contain up to 100 DC motors, e.g. for seat adjustment, windshield lifters, windshield wipers and as pumps and fan drives.

Due to the low power density and the comparatively high maintenance requirements for continuous operation, DC motors are no longer an option for main drives in motor vehicles.

Note

If a permanent magnet synchronous machine (PMSM) is operated with block-shaped voltage from a frequency inverter with variable DC link voltage, the combination of motor and electronics is sometimes referred to as a brushless DC machine (BLDC). The level of DC voltage in the intermediate circuit then determines the speed, like in a DC machine. Smaller BLDC machines, for example in PC fans, often only have two DC connections and the electronics are concealed in the motor housing.

However, all rotary field machines, even induction machines, can be controlled in this way. The basic operating principle of BLDC is based on rotating magnetic fields in the stator and rotor. It is the same as for all other rotating field machines and differs fundamentally from that of actual DC machines. The term BLDC is therefore misleading and technically incorrect.

Such machines are technically correctly referred to as EC motors (EC = Electronically Commutated).

7.5.3 Alternating Field Machines

Alternating field machines (universal motors) are similar to DC machines, except that a single-phase alternating current is used instead of direct current. The coils of the rotor and stator must be connected in series, so-called series-wound machines.

As with a direct current machine, the magnetic fields of the stator and rotor are spatially fixed and shited by 90° in relation to each other. However, their amplitudes are not constant, but pulsate with the supply frequency.

Alternating field machines used to be very important as drives for electric locomotives as so-called railroad motors. With the widespread introduction of power electronics, however, they were completely replaced in this application by rotary field machines.

Today, simple universal motors are still used in large numbers in household appliances, e.g. in hair dryers or vacuum cleaners.

However, alternating field machines are no longer used in modern electric vehicles and are therefore not discussed further.

7.5.4 Axial Flux Machines

The different electromagnetic operating principles of electric machines can also be applied to other geometric arrangements of stator and rotor.

The magnetic flux that flows across the air gap between the stator and rotor is decisive for the generation of force. In the machines presented so far, this transition took place in a radial direction, i.e. from the inside to the outside and vice versa. These machines are referred to as radial flux machines (RFM).

It has been known for around 200 years that electric machines can also be built in such a way that the stator and rotor are arranged next to each other rather than surrounding each other. The magnetic flux in the air gap then moves parallel to the axis, from left to right and vice versa, resulting in an axial flux machine (AFM). There are axial DC machines (AFDC), axial PMSMs (AFPM), axial induction machines (AFIM), and so on. In fact, the very first machine built by an engineer was an axial DC machine in 1834 (Moritz Hermann Jacobi in Königsberg, Prussia) (Fig. 7.12).

Axial flux machines can be designed in several varieties. In the simplest form, there is just one stator and one axial rotor disk (single-sided AFM). As such designs suffer from strong axial attraction forces, twin-disk AFMs with one stator in the middle and one rotor disk on either side are more common. This design is called YASA and is shown in the figure above. It is also possible to have just one rotor disk in the center, which is surrounded by two stators on either side. Finally, several of these basic arrangements can be mounted next to each other to further increase performance.

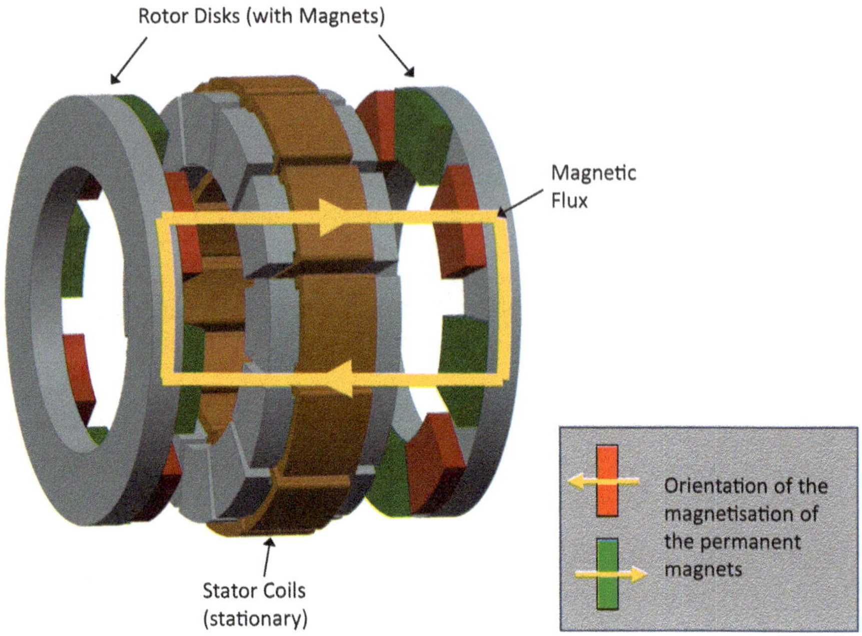

Fig. 7.12 Structure of an axial flux permanent magnet synchronous machine (AFPM)

Today, small axial flux machines are mass-produced for some applications. However, larger axial flux machines have rarely been commercially successful due to their complex and therefore expensive design.

In recent years, there has been some hype surrounding axial flux machines for cars. The hope is for a particularly high-power density, albeit at a high production cost.

Example 7.2—Axial flux machines

English companies like YASA Motors Ltd., Dana TM4 (formerly Ashwoods Electric Motors), and GKN EVO eDrive Systems are actively promoting axial flux machines in electric vehicles—so far with little success. Outside of the United Kingdom, enthusiasm for axial flux machines has been rather subdued.

The most powerful axial flux motors from YASA (types 400 and 750R) reportedly achieve a power density of 4 kW/kg (continuous) and nearly 7 kW/kg (peak). This is slightly better than the performance of the new VW ID.3 motor (type AP550, nearly 5 kW/kg peak). High-performance electric motors in radial flux design for the Formula E racing series achieve around 8 kW/kg (continuous) and 10–14 kW/kg (peak), which is about double the performance of similar axial flux machines. YASA Motors Ltd. was acquired by Mercedes-Benz in 2021. An axial flux machine is expected to be used for the first time in series production in the upcoming AMG sports car from Mercedes-Benz. ◀

A major disadvantage of axial flux machines, apart from the complicated mechanical design and the high production costs, is the limitation of the maximum speed. All currently known designs of axial flux machines suitable for use in motor vehicles are limited to speeds below 8000/min due to the centrifugal forces and the complex rotor design. This is less than half the maximum speed of radial flux machines for cars. However, the power density of electric machines is directly depending on their speed as will be shown later in this chapter.

Developments over the next few years will show whether axial flux machines for cars and trucks have a future and what performance can be achieved. However, it is unlikely that these machines will achieve the same importance on the market as radial flux machines.

7.5.5 Transverse Flux Machines

The third and final option to route the magnetic flux from the stationary part into the rotating part is transverse to the shaft. Consequently, these machines are called transverse flux machines (TFM). They, too, can be built according to all electromagnetic operating principles like transversal permanent magnet (TFPM), transversal induction (TFIM), transversal reluctance (TFRM), and more.

One distinct advantage of transverse flux machines is their simple winding design. Just one large cylindrical coil is required for each phase. The number of phases is created by simply stacking two, three, or more structures in axial direction. Each structure must be rotated by a certain angle according to the desired phase shift. The figure above shows just a single phase of a TF-Permanent Magnet Machine.

A disadvantage of TFM is the high number of poles and therefore the low speed, which is even lower than in axial flux machines. The machine in Fig. 7.13 has 24 pole pairs. If we assume a maximum stator frequency of 1 kHz, which is already very high, the top speed will be $n_1 = f_1/p = 1000\ \text{Hz}/24 = 41{,}7/\text{s} = 2500/\text{min}$. This is 7 times slower than typical electric machines for cars.

The mechanical construction of TFM is another challenge and much more complex than even axial flux machines.

Example 7.3—Transverse flux machines for busses

The company Voith Turbo GmbH & Co. KG from Heidenheim, Germany, built a 2-phase TFM-PMSM with 28 pole pairs for use in the axles of city buses around the turn of the millennium. However, the project was discontinued before the start of series production. ◀

Like axial flux machines, there are various mass applications for small, low-power transverse flux machines. A particularly well-known example is the claw pole generator, which used to be found in all old cars. It is ultimately a transverse flux direct current machine. The author is not aware of any current developments in transverse flux machines as traction motors for vehicles.

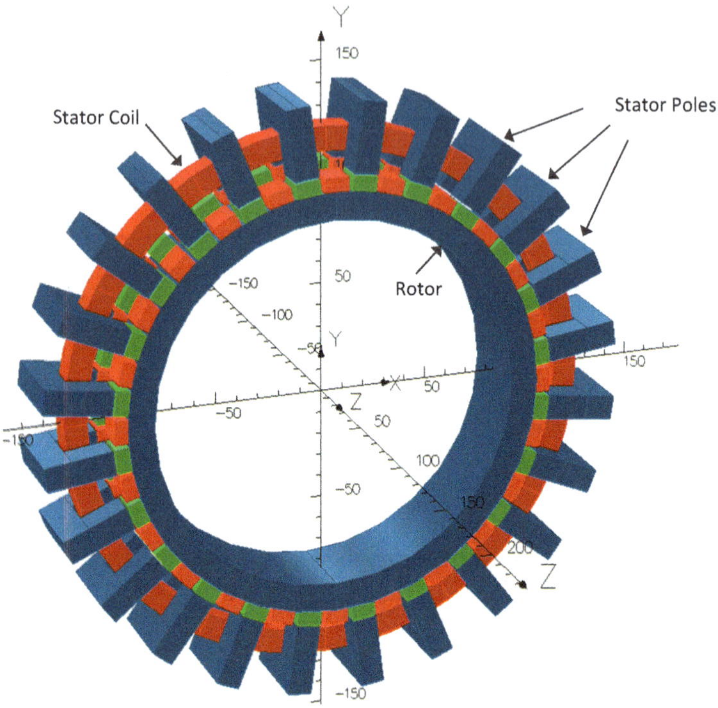

Fig. 7.13 Structure of a single phase of a transverse flux permanent magnet machine (TFPM)

Note

As already explained in Sect. 2.1, wheel torque is important for high-performance sports cars. However, wheel torque results from motor power, not from motor torque. More specifically, a high-power density is needed to get the most wheel torque with a light motor.

The two electric motor concepts discussed in this section, namely axial flux and transverse flux machines, deliver more torque than a conventional electric motor of a similar size but are significantly limited in the achievable speed. Their power and power density are comparatively low. Therefore, these motor designs are not well suited for high-performance cars.

A high motor torque is only important if the motor drives the wheel directly, i.e. without a gearbox. Axial and transverse flux motors may be used as a direct drive in the large wheels of tractors or other commercial vehicles in the future.

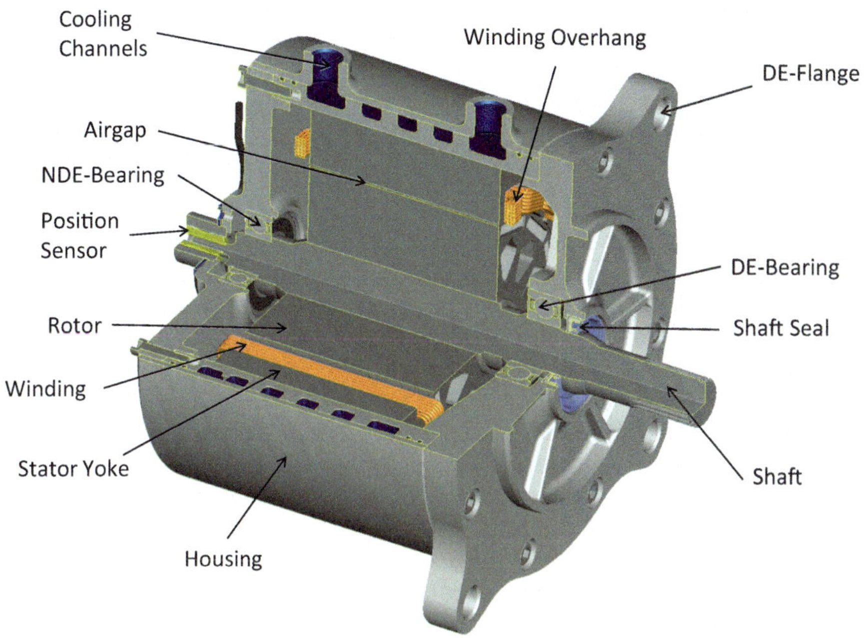

Fig. 7.14 Structure of an electric traction machine

7.6 Design of Electric Machines

7.6.1 Mechanical Design

The design of a modern electric traction machine is shown in Fig. 7.14 using the example of a PMSM.

The stator is practically a hollow cylinder. It consists of electrical steel (stator yoke, also called stator back iron, and teeth), the winding embedded in the slots, and the housing attached around it.

The shaft exits the housing on the right-hand side. This is where the gearbox is mounted (DE = Drive End). A position sensor is fitted on the left-hand side (NDE = Non-Drive End).

The winding consists of multiple round or rectangular copper wires, which are placed into the slots in the stator electrical sheets. The copper wires are coated by several thin layers of enamel and are thus electrically insulated from each other.

To improve the insulation against the electric sheets, which are at earth potential, the slots are also lined with insulating paper (slot insulation). The slot insulation also serves to protect the enameled copper wire from the sharp edges of the sheet stack. The materials used are polyester films, polyamide films, fleece, Mylar (a variant of polyester film) or Nomex (meta aramid fiber), often in a combination of several different layers.

In most electric machines with distributed winding, the individual coils in the winding head are additionally insulated from each other by a layer of insulating paper. This is known as a phase insulation. The materials are similar to those used for slot insulation.

The stator and rotor yoke and the teeth conduct the magnetic field.

The housing transmits the torque, contains all electrical connectors, and includes the cooling system. The cooling of today's high-performance machines is of great importance. Most traction machines are cooled with liquid. The picture shows the standard process: A mixture of water and glycol is placed around the stator core as a cooling jacket. The topic of cooling is discussed in more detail later in the chapter.

The inside of the rotor is often partially hollow to save weight. In rotors with larger diameters, particularly in machines for hybrid cars, there is so much available space inside the rotor that further technical assemblies can be accommodated, for example couplings or rotating axle shafts.

The stator and rotor consist of laminated sheets that are electrically insulated from each other by a thin layer of paint (laminated core). A sheet thickness of 0.35 mm is often used in traction motors. Thinner sheets have fewer losses but are more expensive to produce. More recent developments tend to go in the direction of 0.25 mm or even 0.20 mm.

Depending on the design, a wire winding (EESM) or a bar winding (IM) or permanent magnets (PMSM) are included in the rotor. There are also rotors that contain no magnetization device at all, but only a complex structure of air zones. They are built for reluctance machines (synRM) but have not yet been used in electric cars.

The distance between the stator and rotor is called the air gap. In conventional traction motors, the air gap has a height of around 0.50–1.0 mm, depending on the machine type.

The shaft is held in place by bearings located in the two end shields. A sensor (position encoder) is usually located on the NDE-side end of the shaft, with which the connected electronics can determine the exact angular position of the rotor at any time. Shaft seals are used to protect against dirt or gear oil. In modern high-performance motors, the shaft is often hollow and a coolant flows through it.

Note

In some electric machines, the position of the rotor and stator is reversed: the stator is located inside, and the pot-shaped rotor rotates around it. This type of machine is also known as an external rotor. Millions of such motors are produced for applications in the fan sector, for example. They are also well suited for wheel-hub drives.

Almost all machines with external rotor are PMSM.

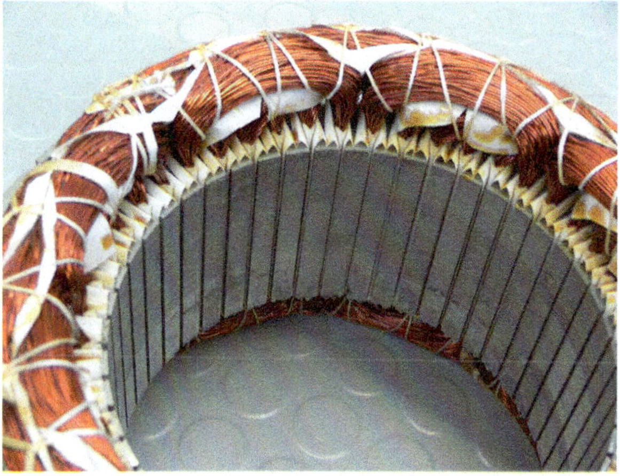

Fig. 7.15 Stator with distributed winding from a Toyota Prius

7.6.2 Windings

7.6.2.1 Distributed Windings

All the windings described above fall into the category of distributed windings. The coil sides of the three phases are distributed evenly over the slots. At the end of the laminated core, the conductors emerge, are bended, and returned a few slots further on.

The stator shown in Fig. 7.15, for example, has a winding with 48 slots, $q=2$, $2p=2$ and a coil width of 6, i.e. the return wires of each coil run 6 slots apart from the forward wires. The structure is basically the same as shown in Fig. 7.5, but for four pole pairs instead of just one.

7.6.2.2 Concentrated Windings

In concentrated windings, also known as single-tooth windings or tooth coil windings, each coil encloses only a single tooth (Fig. 7.16).

Concentrated windings have the following advantages over distributed windings:

- They are compact in axial direction, have a very short winding overhang and thus save space and have a high-power density.
- Production is simpler and can be fully automated. It is therefore cost-effective. Another advantage is the elimination of the insulating papers between the coils in the two end sections, the so-called phase separators, which also leads to lower production cost.

Fig. 7.16 Electric motor for an internal starter generator with tooth coil winding

However, there are also some significant disadvantages:

- As the pole width (pole pitch) must always be about as wide as the coil width (slot spacing) for electromagnetic reasons, machines with concentrated windings require increasingly more magnetic poles with larger diameters and more slots. This results in increasingly slower speeds and hence lower power density, especially for larger machines. If high speeds are required, the frequency of currents and voltages must be increased considerably, resulting in additional iron losses, which in turn have a negative effect on efficiency and power density.
- The magnetic field in the air gap of concentrated windings is highly distorted. Many harmonics and in some cases even subharmonics of the fundamental field curve occur, and the harmonic dispersion is high. This results in strong oscillation torques and radial forces (noise).

The disadvantages of the concentrated winding play a minor role in hybrid motors, which are usually installed between the combustion engine and transmission or directly inside the transmission: Torque oscillations and noise from the combustion engine are dominant and the speed is relatively low anyhow. This is why electric motors for hybrid applications are predominantly built with concentrated windings today.

In the field of traction motors, i.e. as the main drive of electric vehicles, distributed windings are used almost exclusively because noise, smooth torque, and high-power density (= high speed) are critical here.

7.6.2.3 Form Wound Windings

In recent years, form wound windings for traction motors have become increasingly popular. They are a compromise between distributed windings and concentrated windings: Their production can be fully automated, as with the concentrated winding. However, their electromagnetic properties largely correspond to those of the distributed winding. The disadvantage of form wound windings is the high tooling and production machine costs, which only pay off in mass production. Another disadvantage is the low flexibility regarding changes in the number of windings and therefore changes in the rated voltage. This is also only acceptable for mass production in the automotive industry. Form wound windings are hardly suitable for the varied production of industrial motors.[1]

Form wound windings usually have between 2 and 8 conductors per slot. They can be used for up to 800 V system voltage, see for example in the Porsche Taycan.

Form wound windings for traction motors in cars are divided into the following variants: Hairpin, D-pin, I-pin and wave winding (Fig. 7.17).

In hairpin, D-pin, and I-pin windings, individual pieces of wire, usually rectangular conductors (rods), are inserted into the slots of the stator core. The wires protrude by a few centimeters on both sides of the laminated core, the so-called winding overhang. In the next stage of the production process, they are twisted (bent) and welded together at the ends. Obviously, with many slots and up to 8 wires per slot, a very large number of welding points is required.

The three variants, Hairpin, D-Pin, and I-Pin, differ in the shape of the conductors and the wiring complexity. With I-pins, all forward and return conductors of a coil are manufactured separately. I-pins must therefore be welded at both ends. With D-pins and hairpins, the forward and return conductors are bent from a single piece of wire. Welding points are therefore only required on one side of the stator's electrical sheets. Hairpins have a wave-like shape, while the winding head of D-pins is bent in a three-dimensional loop shape. From an electromagnetic point of view, there is no significant difference between the three variants.

In contrast to these three variants, wave windings, also known as continuous hairpin windings, are made from one continuous wire. The rectangular, rigid wire is bent into its final shape by a winding machine outside of the laminated core. The result is a flat mesh of copper wire with a height equal to the length of the iron and a length corresponding to the diameter of the stator, for example, 30 cm × 100 cm. There is no welding required as the winding is produced in one piece. The mesh is then rolled into the slots of the stator laminated core from the inside.

---

[1] Industrial high-voltage motors and generators with powers of dozens or even hundreds of megawatts are always equipped with form wound high voltage windings for technical reasons. The construction of these windings, however, is different from the hairpin form windings used for traction motors.

Fig. 7.17 Hairpin winding (left) and a single hairpin (right)

Example 7.4—Wave windings

The production of the mesh is a complex process that is only mastered by few companies, for example, Aumann AG, Schaeffler Elmotec Statomat GmbH, or Vitesco Technologies Group AG. Such windings are already used by some car manufacturers, for example by Volkswagen. ◀

7.6.3 Rotors

7.6.3.1 Rotors for PMSM

The rotors of permanent magnet synchronous machines must hold the magnets, which can be placed either inside (embedded magnets) or on the surface. Surface magnets are used for most industrial motors, while nearly all traction motors have embedded magnets. The physical background is explained in detail in Sect. 8.5.

Three basic designs of rotors with embedded magnets have become established, of which many variants and combinations exist (Fig. 7.18):

I-type magnets are typical for starter-generators and for gearbox-integrated motors (GEM), i.e. primarily for applications in hybrid vehicles, for example, the Toyota Prius (Fig. 7.19).

This design is inexpensive and leaves plenty of free space inside the rotor. As there is hardly any magnetic field there, the laminated core can also be designed as a hollow ring. This leaves space for the integration of a clutch, for example.

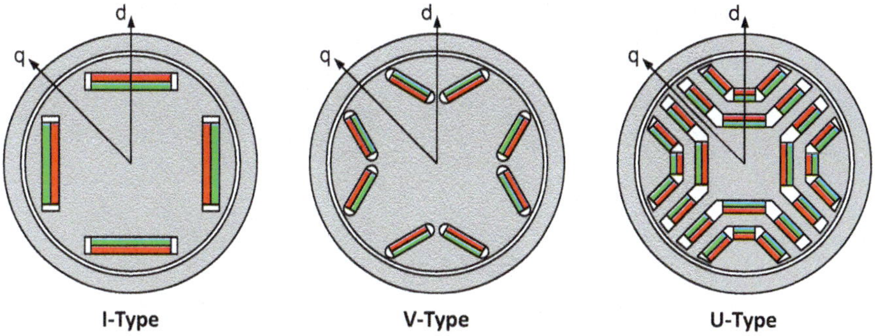

Fig. 7.18 Basic design variants of PMSM rotors with enhanced reluctance

Fig. 7.19: 8-pole I-type PMSM rotor from Toyota Prius

V-type and especially U-type arrangements are preferred in traction drives of battery electric vehicles. There are variants with two or three levels of magnets arranged one above the other. Some variants have pockets that are left empty to increase the reluctance torque.

A combination of I- and V-type is also possible and is known as a delta arrangement. The BMW i3 uses such a rotor.

7.6.3.2 Rotors for EESM

Electrically excited synchronous machines (EESM) require windings and an excitation current in the rotor. The current can be transmitted by slip rings, see the following illustration, or contactless by a high-frequency rotary transformer. Such machines have been developed by Mahle GmbH, ZF Friedrichshafen AG, and BMW AG, for example (Fig. 7.20).

Fig. 7.20: 4-pole EESM rotor with slip rings

Fig. 7.21 Rotor of an induction machine made of die casted copper

7.6.3.3 Rotors for IM

Nowadays, the rotor of an induction machine is predominantly designed as a squirrel-cage rotor: Many conductor bars (rods) are placed around the circumference of the laminated core and are connected at both ends by short-circuit rings. These rotors are simple and inexpensive to manufacture (Fig. 7.21).

The internal rods cannot be seen from the outside. However, the short-circuit rings, which is attached to both ends of the laminated core, are clearly visible.

The cage can be manufactured by die casting with molten metal or by mechanically hammering in copper rods. In the latter case, the end rings must then be fitted and welded. This process was used by Tesla Motors, for example.

Usually, Aluminum is used in the die-casting process, for example in the motors for the Audi e-tron 55. Although aluminum has a higher electrical resistance than copper and the motors therefore have a slightly lower efficiency, the manufacturing process is much simpler. Copper die-casting requires high temperatures of around 1100 °C. As a result, the casting mold is subject to high wear, and the manufacturing costs increase. There are only a few manufacturers in the world who can mass-produce rotors for induction machines using die-casted copper, for example Favi S.A. in France and Wieland-Werke AG in Germany.

7.7 Voltage Induction

A basic knowledge of electrical engineering is required to understand the following sections. The basic principles are briefly summarized in the Appendix of this book.

The voltage induced in a single stator conductor placed in the flux of the rotating rotor magnets can be calculated directly from Eq. (A.73):

$$u = B \cdot l \cdot v \tag{7.5}$$

The rotor speed, i.e. the speed of the rotor surface, is generally linked to the frequency (rotational speed) and the width of two magnetic poles:

$$v = 2 \cdot \tau_p \cdot f \tag{7.6}$$

The magnetic flux of the rotor changes direction from one magnetic pole to the next. The amplitude is approximately sinusoidal, as Fig. 7.22 shows:

The illustration shows only one of several pole pairs of a real machine. However, as they all have the same structure, only one must be calculated and the effects of the other pole pairs are simply considered by multiplication. For simplification, the diagram has been linearized, i.e. the air gap has been straightened. The left end of the diagram is therefore directly connected to the right end.

Also, the stator is greatly simplified for the calculation: The slots are no longer present, and the conductors (the red circles in Fig. 7.22) have been placed directly on the surface of the electrical sheets. One virtual conductor is located at the center of each slot.

Although this simplified model seems to be quite far off from a real motor, the model is well suited for analytical calculation of the operating behavior. It is widely used in practice.

With this model, the peak value of the induced voltage in a single conductor can be calculated:

$$\widehat{U}_{conductor} = \widehat{B} \cdot l \cdot 2\tau_p \cdot f \tag{7.7}$$

Assuming a sinusoidal distribution of the flux density B, the flux of a magnetic pole can be calculated. The mean value of an arc of the sinusoidal function is $2/\pi$. The magnetic flux generated by a magnetic pole results from the flux density multiplied by the area.

$$\phi_{Pol} = \frac{2}{\pi} \cdot \widehat{B} \cdot l \cdot \tau_p \tag{7.8}$$

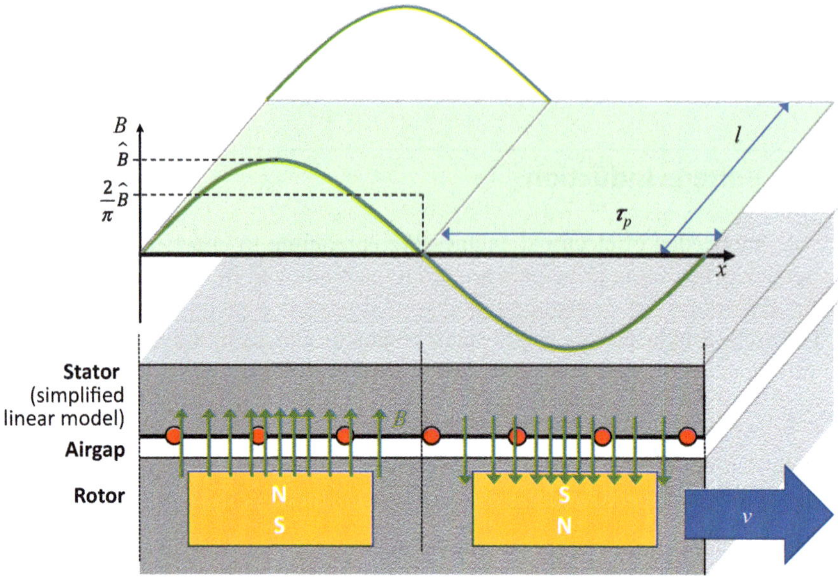

Fig. 7.22 Air gap flux of the rotor magnets

Both equations combined lead to the effective conductor voltage U:

$$U_{conductor} = \frac{\pi}{\sqrt{2}} \cdot f \cdot \phi_{pol} \tag{7.9}$$

This equation can also be applied to motor operation to calculate the flux for a given conductor voltage $U_{\text{conductor}}$:

$$\phi_{pol} = \frac{\sqrt{2}}{\pi} \cdot \frac{U_{conductor}}{f} \tag{7.10}$$

To calculate the terminal voltage, all voltages of the conductors in a coil and all coils of a phase must be added. It must be considered that the voltages are slightly out of phase with each other if several neighboring coils jointly form a pole, i.e. if the number of slots per pole per phase $q > 1$.

This phase shift is considered by the so-called winding factor ξ (xi), which is also known as the group factor. When there is only one coil per pole and phase, i.e. $q = 1$, there is no phase shift and $\xi = 1$.

For $q > 1$, the winding factor becomes increasingly smaller than 1. Typical values are around $\xi = 0.93$–0.96. The effect is not very pronounced.

The slots are often not parallel to the shaft but slightly twisted (skewed) in axial direction. This results in advantages regarding noise and reduces torque oscillations. Skew is considered mathematically by an additional multiplier in the winding factor, the skew factor. In many machines, skewing is performed on the rotor side and not the stator slots. However, in the end, both options are equivalent.

Another measure to improve the electromagnetic properties of a machine is to shorten the step size of the coils. The angular distance between the two sides of a coil is called coil pitch. Normally, the coil pitch should correspond almost exactly to the angular distance of a magnetic pole. This is called a full pitch winding. In that case, maximum voltage induction and the best utilization of the magnetic material are guaranteed. For example, in a 6-pole machine ($2p = 3$) with 48 slots, the coils should have a coil pitch of $48/6 = 8$ slots.

However, it can be advantageous to close the coils one or two slots "too early", for example already after 7 or even 6 slots. The coil then forms a chord. Accordingly, this variant is called chord winding. The effect of a chord winding on induced voltage is considered by an additional factor, the chording factor.

Therefore: Winding factor = group factor · skew factor · chording factor.

There are extensive sets of formulas for the winding factors of the individual winding variants in the literature, but that would go too far at this point.

7.8 Magnetic Forces

Electric and magnetic fields are the cause of a variety of forces.[2]

Magnetic forces:

- Force on current-carrying conductors in a magnetic field (Lorentz force)
- Force on ferromagnetic material (reluctance force)
- Force on hard magnetic material (permanent magnets)
- Forces inside ferromagnetic material (magnetostriction force)

Electric forces:

- Forces on charges (Coulomb force)
- Forces inside non-conductive material (electrostriction force)

In electric machines, only the first three effects are used to generate mechanical force. The magnetostriction force occurs in transformers and causes the well-known "transformer hum".

There are experimental electric motors that work electrostatically using the Coulomb force. However, such machines are limited to low power because, for physical reasons, the power that can be transmitted in an electric field across an air gap is much lower than the power that can be transmitted by a magnetic field.

[2] Physically, all electrical and magnetic force effects can be traced back to the attraction and repulsion of charges (electromagnetic force). Magnetic forces are caused by relativistic effects due to the movement of charges.

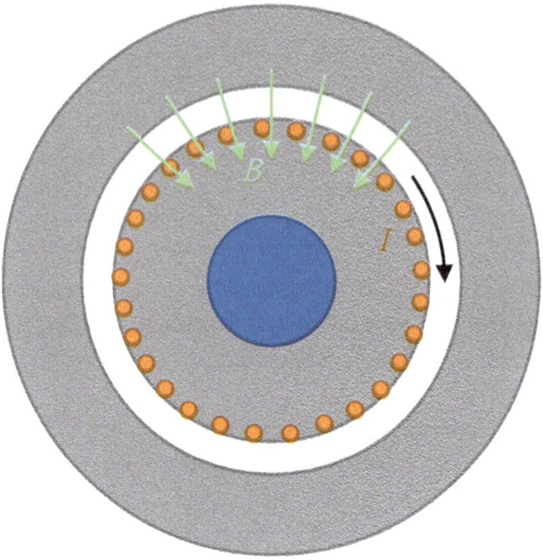

Fig. 7.23 For calculating the force and torque on the rotor of an induction machine

7.8.1 Lorentz Force

The Lorentz force acts on current-carrying conductors in a magnetic field. It amounts to:

$$\vec{F} = i \cdot \left(\vec{l} \times \vec{B} \right) \tag{7.11}$$

where i is the current in the conductor, B is the flux density, and l is the length of the conductor in the magnetic field.

The Lorentz force in electric machines results from the interaction of the excitation field and the current in the coils.

If, for example, the currents (orange circles) in Fig. 7.23 are directed towards the viewer, then the three-finger rule results in a clockwise rotation of the rotor.

To calculate the torque T of an electrical machine, the instantaneous values of all conductor forces F must be added and multiplied by the lever arm, which is approximately half the rotor diameter d:

$$T = \frac{d}{2} \cdot \sum F = \frac{d}{2} \cdot z \cdot I \cdot l \cdot B \tag{7.12}$$

$z \cdot I$ is the effective total current in all conductors z.

The power P then results from the known relationship:

$$P = T \cdot \Omega = T \cdot 2\pi n \tag{7.13}$$

The speed n must be entered in revolutions per second.

The Lorentz force is used to calculate the torque and power of most electric machines. It is irrelevant whether the force effect on the rotor conductors or the force effect on the stator conductors is calculated, as the principle of "action = reaction" applies.

This principle can also be used to calculate the torque of machines with the Lorentz force that has no conductors in the rotor, for example synchronous machines excited by permanent magnets.

Inserting (7.8) into (7.12) leads to a different representation of the torque:

$$T = \frac{d \cdot z}{2\tau_p} \cdot \phi_{pol} \cdot I \tag{7.14}$$

This equation clarifies that torque results from the product of magnetic flux ϕ and current I. This correlation will become important later when analyzing the basic operating characteristics of electric motors.

7.8.2 Power Density

The surface force f_S or shear stress is the total force acting on the rotor surface in relation to the surface area. The summation of the Lorentz forces of all conductors ultimately results in a surface force:

$$f_S = A \cdot B \tag{7.15}$$

where B is the average magnetic flux density in the air gap. In typical electric machines, B has an average value of 0.8–1.1 T in the air gap. The unit of flux density is 1 V·s/m^2 or 1 T (Tesla) for short. The flux density is practically limited by the iron saturation and the power factor. The power factor is the cosine of the phase shift between voltage and current.

A is the current loading, that is the sum of all currents flowing in all conductors related to the rotor circumference:

$$A = \frac{\sum i}{\pi d} \tag{7.16}$$

A is also limited in practice, mostly by the possibility of cooling (heat dissipation) of the losses. Assuming copper conductors and water cooling, this results in a maximum current loading of around 50–100 kA/m in continuous operation and 100–200 kA/m for short-term peak loads.

The technical limit of the maximum surface force f_S that can be achieved with the Lorentz force in electric motors is around 200 kN/m^2.

The utilization factor C (Esson's constant) is traditionally defined as:

$$C = \pi^2 \cdot \xi \cdot A \cdot B \tag{7.17}$$

where ξ is the previously defined winding factor. The winding factor is typically around 0.96. The torque can be specified using the utilization factor:

$$T = \frac{1}{2\pi} \cdot C \cdot d^2 \cdot l \cdot \eta \cdot cos\varphi \tag{7.18}$$

The symbol η (eta) stands for the efficiency of the drive, i.e. the ratio of mechanically delivered power to electrically consumed power. The power factor is denoted by $cos\ \varphi$ (cosine phi). The angle φ is the angle between current and voltage at the terminals of the motor. In traction drives, it is controlled by the power electronics and is usually small, so that $cos\ \varphi = 1$ can be set as a good approximation.

This results in the following equation for the mechanical power of the drive:

$$P = n \cdot C \cdot d^2 \cdot l \cdot \eta \cdot \cos\varphi \tag{7.19}$$

The power density, i.e. power per volume, becomes:

$$p = n \cdot C \cdot \eta \cdot \cos\varphi \tag{7.20}$$

A high-power density is particularly desirable for traction drives, as this reduces volume (installation space), weight and also costs.

Equations (7.20) and (7.17) show that the decisive factors for high power density are

- rotational speed n,
- current loading A, and
- flux density B.

For this reason, electric motors are poorly suited as direct drives, e.g. as in-wheel motors with low speeds up to around 1500/min but high torques. It is better to allow the motor to rotate very quickly with low torque and to produce the desired speed using a gearbox. Fortunately, the gearbox then also ensures a high torque.

The aim of a good drivetrain design must therefore be to operate the electric motor at the highest possible speed. Motor top speed is limited by:

- Centrifugal forces on the rotor surface
- the bearings of the shaft,
- the required frequency of the supply voltage (number of pole pairs), i.e. ultimately the iron losses in the machine and the switching losses in the inverter, and finally by
- the design of the gearbox, which reduces the engine speed to the wheel speed and increases the torque accordingly.

Table 7.1 provides the reference values for the achievable surface velocities with different materials and rotor constructions:

Typical speeds of electric motors for traction applications in series production today are between 16,000 and 20,000/min. Near-series developments exceed 20,000/min.

Table 7.1 Technically feasible surface speeds depending on the rotor design

	Maximum surface velocity
Carbon fiber	Up to 1800 m/s
Solid steel cylinder	Up to 620 m/s
Switched reluctance machine (SRM)	Up to 200 m/s
Induction machine (IM)	Up to 160 m/s
Permanent magnet synchronous machine (PMSM)	Up to 120 m/s (more with a carbon sleeve)
Electrically excited synchronous machine (EESM)	Below 100 m/s

Example 7.5—Xiaomi high-performance motors

By the end of 2023, the Chinese manufacturer Xiaomi revealed its HyperEngine V8s. This PMSM has a max speed of 27,200/min, 425 kW output power, and 635 Nm peak torque. The cooling is performed by a bidirectional oil cooling in the stator and in the rotor. Start of series production is planned for 2025.

Xiaomi also revealed a prototype motor that reaches speeds up to 35,000/min.

Source: https://www.mi.com/global/discover/article?id=3095 ◀

7.8.3 Reluctance Force

The reluctance force arises at the boundary between two materials of different permeability when a magnetic flux passes through this surface. The force acts perpendicular to the surface, i.e. in the direction of the field lines.

The general formula for the reluctance force is:

$$\vec{F} = -\frac{1}{2}\int_{v} \vec{H}^{2} \cdot \operatorname{grad} \mu \cdot dV \tag{7.21}$$

The equation can be greatly simplified when the flux lines exit electrical steel with μ_r close to infinity and enter air with $\mu_r = 1$:

$$F = \frac{B^2 \cdot S}{2\mu_0} \tag{7.22}$$

Here S is the area through which the field lines emerge, and B is the magnetic flux density. μ_0 is a proportionality factor, the so-called magnetic field constant. It has the fixed value $\mu_0 = 4\pi \cdot 10^{-7}$ N/A^2.

In all synchronous machines for traction drives, reluctance and Lorentz forces always occur in combination. Synchronous machines for industrial use are usually designed so that mostly the Lorentz force is utilized.

The reluctance force is also responsible for the radial attraction of the stator core to the rotor. This results in vibrations that are emitted as sound.

Note
In the physical reality of an electrical machine, there are no Lorentz forces at all. The wires of the stator winding are located in slots, but the field lines run through the neighboring teeth. Therefore, the field lines do not meet currents, and no Lorentz forces arise.

However, in the simplified arrangement shown in Fig. 7.22, the slots are missing. Instead, the virtual wires are now arranged on the surface of the laminated core and are indeed influenced by flux lines. In this arrangement, Lorentz forces are indeed created.

It turns out that the calculation with the substitute arrangement and with Lorentz forces leads to the correct results. This procedure has become established in the calculation of electric machines all over the world.

7.9 Base Speed Range (Constant Torque Range)

Two previously derived equations are essential for understanding the operating behavior of electric motors.

From (7.10) follows:

$$\phi = k_1 \cdot \frac{U}{f} \tag{7.23}$$

The magnetic flux in the motor therefore results from the ratio of applied terminal voltage U and frequency f. The constant k_1 is given by the design of the machine. This includes, for example, the number of conductors and the interconnection of the coils.

From (7.14) follows:

$$T = k_2 \cdot \phi \cdot I \tag{7.24}$$

The torque results from the coil current I and the magnetic flux ϕ. The constant k_2 considers design characteristics of the machine, i.e. in particular the number of conductors, the diameter of the rotor, and the pole width.

These basic properties apply to all types of rotating field machines, i.e. IM, PMSM, and EESM and, albeit with different physical backgrounds, even to switched reluctance machines (SRM).

In the base speed range, the magnetic flux is kept constant by adjusting the voltage U precisely in proportion to the frequency f and thus to the speed n of the machine. This adjustment is performed by the motor controller and the power electronics (inverter).

A constant flux ϕ ensures optimum magnetic utilization of the motor. In this case, the torque is proportional to the motor current I. The maximum available torque is also constant. The base speed range is also referred to as the constant torque range, see Sect. 2.4.1.

7.10 Field Weakening Range (Constant Power Range)

The frequency inverter generates the AC voltage for the motor by chopping the DC voltage supplied by the battery. The frequency can be set over a wide range. However, the amplitude of the AC voltage output can at most be as high as the DC voltage of the battery.

The speed limit of the constant torque range is therefore reached when the frequency inverter must set the maximum available output voltage U_{max} to keep the magnetic flux constant (7.23). This is called base speed n_{base}.

A further increase in the speed is easily possible by further increasing the frequency. However, as voltage will remain constant, this leads to a successive drop in the magnetic flux ϕ. This part of the drive performance is called the field weakening range.

We can see from (7.24), that the available torque T will drop with the available flux which in turn drops with the speed.

The I_{max} current limit is set by the inverter and is not depending on speed. Therefore, the maximum torque T_{max} exactly drops inversely proportional with speed. The available power P_{max} remains constant as power is generated from torque times speed.

The continuous current limit I_{th} results from the losses and the cooling system of the motor. As losses increase slightly with speed, I_{th} is continuously being reduced. Therefore, T_{th} drops a little more than inversely proportional with speed.

In reality, the relationships are even more complex, but this will be explained in more detail later.

As the state of charge of the battery (SOC) decreases, the battery voltage also decreases. The limit of the base speed range is then already reached at a slightly lower speed. The drop-in torque therefore begins at lower speeds, see the dashed lines in Fig. 7.24. This effect also reduces the available peak power of the drive.

An acceleration car race should therefore be started with the battery as fully charged as possible. The reduction in power is already slightly noticeable from an SOC of less than 80%, and very noticeable below 20 or 30%.

7.11 Materials

7.11.1 Conductor Materials

Practically only copper and aluminum are suitable materials for current conductors. Their main technical properties are summarized in Table 7.2.

Only silver is a slightly better conductor than copper, but for obvious reasons, it is not an option to produce motors.

Although aluminum is a poorer conductor, it is significantly cheaper than copper. Due to its lower melting point, aluminum can be easily processed in a die-casting process. In industrial induction machines, rotor bars cast from aluminum are therefore usually used.

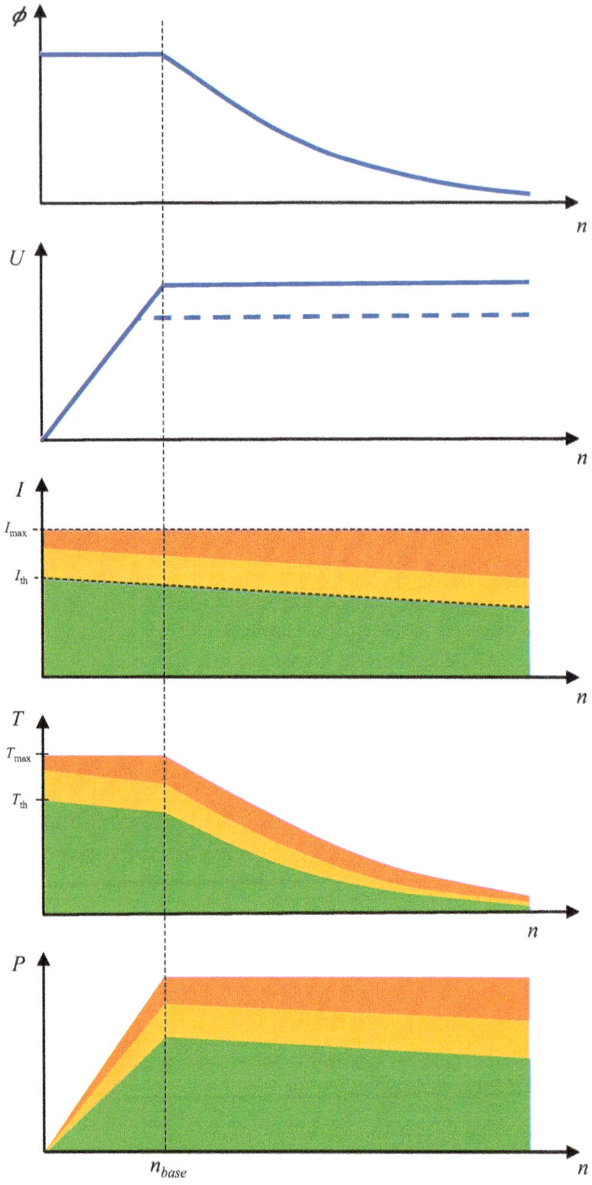

Fig. 7.24 Flux, voltage, current, torque, and power over speed[3]

[3]Green areas: continuous operation, yellow ranges: minute operation, red areas: second operation.

Table 7.2 Technical properties of conductor materials

	Copper	Aluminum
Density	8.92 g/cm^2	2.7 g/cm^2
Electric conductivity (Pure material at 0 °C)	$65 \cdot 10^6$ 1/(Ωm)	$41 \cdot 10^6$ 1/(Ωm)
Electric conductivity κ_{20} (Wire at 20 °C)	$56 \cdot 10^6$ 1/(Ωm)	$33 \cdot 10^6$ 1/(Ωm)
Thermal conductivity	400 W/(m·K)	235 W/(m·K)
Melting point	1085 °C	660 °C
Price/kg (as of 2024)	7.70 €	2.00 €

In the field of traction motors, however, copper has also established itself as a conductor material for the cages of induction machines.

In addition to its low conductivity, aluminum is difficult to solder, which is why it is less suitable as a wire material. This is why copper wire is used for motor windings.

The resistance of a wire R results from the electrical conductivity κ (kappa), the wire length l, and the cross-section S to:

$$R = \frac{l}{\kappa \cdot S} \tag{7.25}$$

The specific resistance ρ (Rho) is obtained from the conductivity at $\rho = 1/\kappa$. It is temperature-dependent:

$$\rho(\kappa) = \rho_{20} \cdot [1 + \alpha \cdot (\vartheta - 20\ °\mathrm{C})] \tag{7.26}$$

Here ϑ (Teta) is the ambient temperature in °C and α is the temperature coefficient. It is $3{,}93 \cdot 10^{-3}\ \mathrm{K}^{-1}$ for copper. Simply put, the resistance of copper increases by about 4% every 10 Kelvins.

For example, the resistance of a copper wire increases by around 50% when heated from 20 to 150 °C. The copper losses I^2R increase to the same extent.

A simple way to determine the average temperature of a copper wire winding in the motor is to use the following method:

The motor rests for several hours until it has completely reached room temperature ϑ_1. The winding resistance R_1 at the terminals is measured.

The motor is then operated with load, which heats up the winding. After thermal equilibrium has been reached, the winding resistance R_2 is measured again.

The average winding temperature ϑ_2 under load then results from the relationship:

$$\frac{R_1}{R_2} = \frac{\vartheta_1 + k}{\vartheta_2 + k} \tag{7.27}$$

According to IEC standard 60034-1, $k=235$ K must be used for copper conductors and $k=225$ K for aluminum conductors. The relationship applies at any temperature ϑ_1 and ϑ_2.

7.11.2 Ferromagnetic Materials

Soft magnetic materials are ferromagnets. They can be magnetized easily in a magnetic field as only low magnetic field strength is required.

Soft magnets are used to reduce the magnetic resistance in the magnetic circuit of electric machines. In this way, the desired magnetic flux density B in the air gap, which is essential for generating force, can be generated with less magnetic field strength H or less current linkage Θ, see Sect. 17.4.

7.11.2.1 Electrical Sheets

In practice, electrical sheets are used for most electric machines. The sheets consist of a type of steel that contains around 2–4% silicon and aluminum as alloy components. This material is known as silicon-iron alloy (FeSi).

There are other types of steel suitable for electric motors, in particular, cobalt-iron alloys (CoFe) and nickel-iron alloys (FeNi), but these are significantly more expensive and are therefore only used in special cases (Table 7.3).

Figure 7.25 shows a realistic B/H curve of a typical electrical sheet. Up to a flux density of around 1.5 T, the increase in flux density is approximately linear. The relative permeability μ_r, which corresponds to the slope of the curve, is approximately constant.

At higher flux densities, the curve bends sharply and flattens out. It is no longer possible to specify a constant μ_r; instead, the permeability must be calculated at each operating point.

Table 7.3 Technical properties of electrical sheets

	Electrical sheet (FeSi)
Density	7.87 g/cm^2
Electric conductivity	$10 \cdot 10^6$ 1/(Ωm)
Thermal conductivity	80 W/(m·K)
Melting point	1538 °C
Saturation polarization	1.25 … 2.2 T depending on the alloy

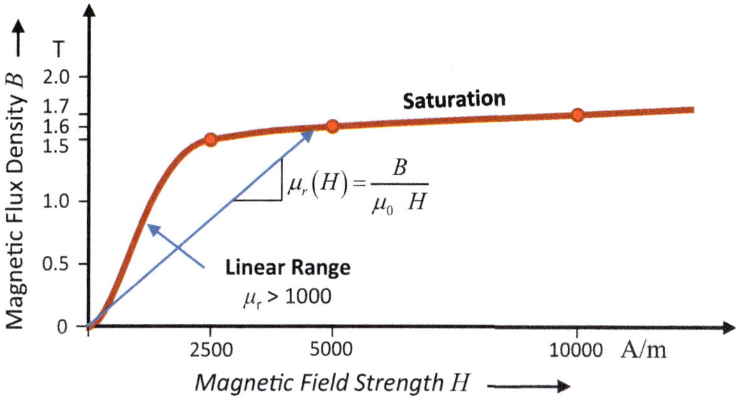

Fig. 7.25 Typical B/H characteristics of non-grain-oriented electrical steel (silicon-iron)

Note

The axes of the *B*/*H* curve can be interpreted as follows: The magnetic field strength H is proportional to the current of the stator coils. The magnetic flux density B is proportional to the generated force and thus to the torque and ultimately also to the power.

The curve therefore should rise as steeply as possible (μ_r as large as possible) because then a high power, i.e. a high magnetic flux density, is generated with a low current, i.e. a low magnetic field.

With plastic or other non-magnetic materials, $\mu_r = 1$ and so the curve has just a very low slope. This is why all electric motors use electrical sheets. Unfortunately, the amplification effect decreases significantly above a certain flux density. This effect is known as iron saturation. Developers of electric machines must design the magnetic circuit so that the operating point lies in the bending curve. If it were much below, power would be wasted; if it were above, too much current is used without a corresponding amount of power in return.

In the case of soft magnetic materials, a distinction is made between grain-oriented and non-grain-oriented grades:

Grain-oriented electrical steel has a preferred direction for the magnetic flux. In this preferred direction, the magnetic properties, in particular the magnetizability, are significantly better. Grain-oriented sheets are used for transformers, as the magnetic flux always goes in the same direction.

Non-grain-oriented electrical steel conducts the magnetic flux equally well in all directions, but slightly worse overall than grain-oriented material in the

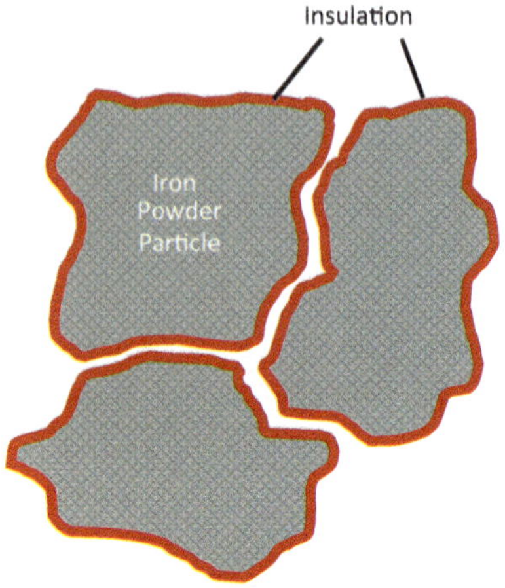

Fig. 7.26 Oxidized iron powder particles

preferred direction. However, in electric motors with rotating magnetic fields, there is no other choice because the magnetic flux is constantly changing direction.

7.11.2.2 SMC

SMC (soft magnetic composite) is an isotropic magnetic compound material. It consists of iron powder particles coated with an electrically insulating layer. SMC can be formed into complex 3D shapes using powder metallurgy.

SMC can be sintered either on a plastic basis or oxide-insulated. In the first case, the iron particles are embedded in a plastic matrix and thus electrically insulated from each other. As a result, the eddy current losses, but the thermal resistance is high, and the soft magnetic properties are poorer because a significant proportion of the finished product consists of plastic. This variant is rather uncommon commercially (Fig. 7.26).

In most applications, oxidized sintered SMC is used. The iron particles are electrically insulated from each other with an oxide coating (rust). The higher iron losses are a disadvantage, but the material is easier to shape using powder metallurgy and is more temperature resistant.

The typical particle size is between 50 and 200 μm. SMCs with larger particles have more magnetic material per volume and hence a higher permeability. But there are also higher eddy current losses, particularly at higher frequencies (Fig. 7.27).

The figure shows two applications of SMC. On the left is a single-tooth coil with stacked electrical sheets—the winding overhands are not covered. In the center with SMC—the winding overhangs contribute to the formation of magnetic flux as they are surrounded by ferromagnetic material. On the right, a three-dimensionally shaped SMC cover (gray) is placed around a toroidal coil (red).

Compared to electric sheets, SMC parts are more expensive and have a lower magnetic permeability. However, some variants of electric machines have such

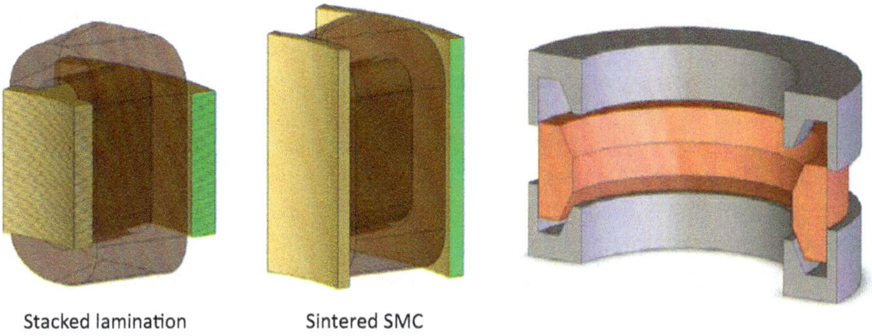

Fig. 7.27 Examples of SMC parts[4]

complex magnetic structures that the use of SMC parts is beneficial. These include, for example, many variants of axial and transverse flux machines.[5]

7.11.2.3 Amorphous Metals

Amorphous metals (metallic glasses) are produced by very rapid cooling of a melt. The molten metal is poured through a ceramic nozzle onto a cooled copper wheel (approximately 0.5–1 m in diameter) and cooled from around 1400–400 °C in milliseconds. The resulting sheets are extremely thin, typically 0.025 mm, and are therefore difficult to punch and stack. Instead, they are wound into ring cores. Iron–boron–silicon (FeBSi) alloys with 1–5% boron and 5–10% silicon are used for soft magnetic amorphous metals.

Amorphous metals are already widely used in electric transformers, especially in Asian countries. They have rarely been used for electric motors because they are difficult to shape into the required geometries.

The advantages are the low iron losses and the extremely high magnetic permeability. The high mechanical tensile strength is also advantageous. Amorphous metals are well-suited as a material for high-speed rotors.[6]

7.11.3 Iron Losses

Continuous re-magnetization of the laminated core within an electric motor produces losses that are referred to as re-magnetization losses or iron losses. They have two technical causes, namely hysteresis losses and eddy current losses.

[4] Source: Höganäs AB.

[5] *Evaluation of the Application of Soft Magnetic Composites in Novel Designs of Electrical Traction Machine*, Bo Zhang, Dissertation, Karlsruhe Institute of Technology, 2016.

[6] *Improving High-speed Electrical Machines by Amorphous Metals,* Jing Ou, Dissertation, Karlsruhe Institute of Technology, 2019.

In practical electric machines, hysteresis and eddy current losses are of similar importance. The magnetic losses in the iron, i.e. the sum of hysteresis and eddy current losses, typically account for 10–50% of the total losses depending on the load point.

7.11.3.1 Hysteresis Losses

All electric machines used for traction drives (rotary field machines) are operated with multiphase alternating current. The currents cycle between the negative and positive peak value and consequently the magnetic field strength H exhibits the same behavior. The magnetizing curve shown in Fig. 7.25 is actually point symmetrical: It continues into the third quadrant towards negative field strengths and flux densities.

On closer inspection, however, the flux density B is always slightly lower with increasing H than with decreasing H and a hysteresis curve is created, see Fig. 7.28.

When the field strength H is switched off, residual magnetism remains in the iron. This is known as the remanence flux density B_R. The effect is well known from steel screwdrivers, where it can be used to magnetically lift screws.

The magnetic field only disappears at a certain negative field strength $-H_C$, the coercive field strength.

Figure 7.28 shows the hysteresis curve in a highly exaggerated form for illustration purposes. In reality, the hysteresis is narrow and hardly recognizable at a normal scale.

Due to the hysteresis loop, a little more energy must always be applied to build up the magnetic field than is recovered during decay. This results in a loss of magnetizing energy, which is referred to as hysteresis loss. In physical terms, this can

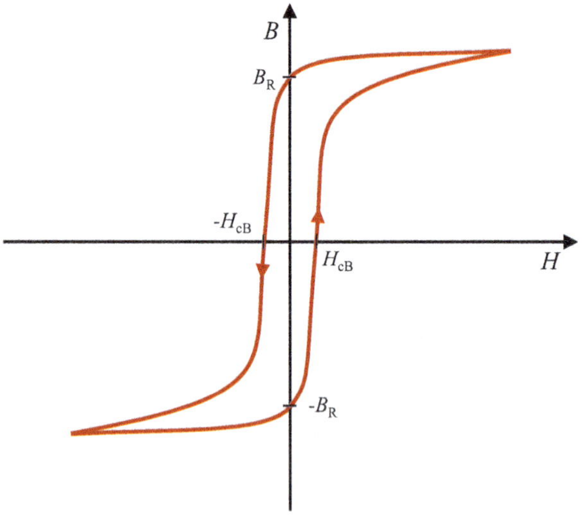

Fig. 7.28 Hysteresis curve of soft magnetic material

be thought of as the frictional work involved in aligning the elementary magnets in the iron, the so-called Weiss districts.

Manufacturers specify a material characteristic value σ_{hyst}, which indicates the specific hysteresis losses per kg of electrical steel at a peak flux density of 1.5 T and a frequency of 50 Hz.

The actual hysteresis losses of a motor are determined using the laminated core mass m to:

$$P_{hyst} = \sigma_{hyst} \cdot m \cdot \left(\frac{f}{50\ \text{Hz}} \right) \cdot \left(\frac{B_{\max}}{1.5\ \text{T}} \right)^2 \tag{7.28}$$

The hysteresis losses increase proportionally to the frequency f and quadratically with the peak flux density B. They can be reduced by selecting a higher quality steel grade with lower specific losses.

7.11.3.2 Eddy Current Losses

The law of induction, Eq. (A.67), does not only apply to coils but generally to all electrically conductive materials in pulsating magnetic fields.

Induction voltages also arise in the laminated core of electric motors, which is constantly exposed to a changing magnetic field. These voltages occur mostly in the axial direction of the motor.

If the core was made of solid iron with a low electrical resistance, the axial voltages would lead to axial currents that close at the ends of the core. These currents are therefore also known as eddy currents.

Like all other currents, eddy currents generate I^2R losses and these can be quite considerable in electric machines.

As early as the 1850s, it was recognized that eddy currents in the iron core are undesirable and must be suppressed. The solution commonly used today is to build the iron core from a stack of thin sheets of metal (electrical sheets), each of which is electrically insulated from the other by a layer of paint (laminated core). The circuit is now interrupted and the axial voltages cannot create axial currents.

However, axial currents are still generated within each sheet. The thinner the sheet, the more effectively eddy currents are suppressed. Typical sheet thicknesses are 0.2–1.0 mm. However, the costs of rolling, insulating, punching, and stacking increase with thinner sheets.

Another way to reduce eddy currents is to increase the electrical resistance R. According to $U = R \cdot I$, less current will flow for the same amount of induced voltage and due to $P = I^2R$ less power will be lost.

The steel normally used for electrical sheets is alloyed with silicon for this purpose as explained in the previous chapter. However, silicon reduces the thermal conductivity and magnetic properties, so only a relatively weak alloy can be used.

Like the hysteresis losses, the steel manufacturers publish a material characteristic value σ_{wb} for the specific eddy current losses. The total eddy current losses are calculated accordingly from:

$$P_{wb} = \sigma_{wb} \cdot m \cdot \left(\frac{f}{50\text{ Hz}}\right)^2 \cdot \left(\frac{B_{\max}}{1.5\text{ T}}\right)^2 \quad (7.29)$$

Note that the eddy current losses increase quadratically with the frequency, while the hysteresis losses increase linearly with the frequency.

Example 7.6—M230-35A sheet metal
M230-35A is a typical sheet metal used in electric traction motors. It has a thickness of 0.35 mm and specific losses of 2.30 W/kg at 50 Hz and 1.5 T. ◀

7.11.4 Hard Magnetic Materials

Hard magnetic materials (permanent magnets) have a high remanence flux density B_r and a high negative coercive field strength H_{cB}. Unlike soft magnets, hard magnets cannot be exposed to a frequently changing magnetic flux. Their hysteresis loop is large and would lead to enormous losses due to remagnetization. Therefore, hard magnets are only used in areas of the motor where the flux is constantly flowing in a particular direction, which is the rotor.

Note
The remanence flux density B_r determines the surface area of the magnet required to achieve a desired magnetic flux density in the air gap of the electrical machine.
The coercive field strength H_{cB} determines the required thickness of the magnet. The lower H_{cB}, the thicker the magnet must be made.

7.11.4.1 Variants of Permanent Magnets

The best permanent magnets known today are made of rare-earth materials (NdFeB = neodymium–iron–boron) and have a coercive field strength of more than −1,000,000A/m and a remanence flux density of up to 1.5 T.

Normal steel also has weak hard magnetic properties. This is known, for example, from magnetized screwdrivers that hold small screws in place. However, the coercive field strength H_{cB} of steel is only around −10 A/m (Table 7.4).

Almost all hard magnets are sintered materials with a ceramic structure and are therefore quite brittle. A good coating is important for NdFeB, as the raw magnet consists of around two thirds iron, which oxidizes in a normal environment (Fig. 7.29).

Table 7.4 Characteristics of the most important hard magnetic materials

	AlNiCo	Hard ferrite	Samarium–Cobalt	Neodym–Iron–Bor
			$SmCo_5$ Sm_2Co_{17}	NdFeB
In use since	1940	1960	1970	1990
Coercive field strength H_{cB}	−40 … −170 kA/m	−100 … −350 kA/m	−600 … −850 kA/m	−500 … −1100 kA/m
Remanence flux density B_r	0.6 … 1.3 T	0.2 … 0.4 T	0.9 … 1.2 T	0.7 … 1.5 T
Energy density	10 … 50 kJ/m³	10 … 40 kJ/m³	140 … 300 kJ/m³	300 … 450 kJ/m³
Use in electric machines	Not suitable for motors due to low H_{cB}	Simple and cost-effective drives	Drives for high temperatures, expensive	Standard material for high-performance motors

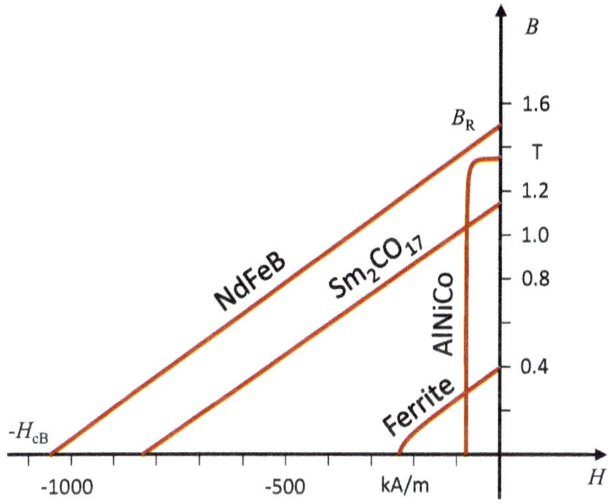

Fig. 7.29 Typical characteristics of hard magnetic materials

There are several methods of manufacturing magnets, but the most common method is powder metallurgy.

Suitable raw materials are melted down under vacuum or inert gas in an induction melting furnace. The molten alloy is either poured into a mold or onto a cooling plate or processed in a strip caster, a device that produces a thin, continuous metal strip. These hardened "chunks" of metal are crushed and pulverized to produce a fine powder with a grain diameter of 3–7 microns.

In the presence of an external magnetic field, the crystalline powder is heated and pressed into the desired magnet shape to achieve densification by “liquid phase sintering”. For this reason, the magnets are often referred to as sintered magnets. During this process, the crystals align themselves with their preferred axis of magnetization in the direction of the external magnetic field.

After sintering, the magnet initially no longer has its own magnetization. It must be magnetized once before being installed in the motor. This is done in a special magnetizing machine where the magnet is exposed to an enormously strong external field for fractions of a second.

With certain motor geometries, it is also possible to carry out magnetization only after the magnets have been installed on the complete rotor. This simplifies production as the handling of magnetized magnets is difficult, but the costs of the magnetization device are significantly higher.

After magnetization, a small amount of magnetic energy is stored in the permanent magnet. This energy is not consumed during operation and the service life of the magnets is basically unlimited. However, magnets can be demagnetized by excessively high opposing fields, high temperatures and strong vibrations. The high proportion of iron in NdFeB magnets can lead to oxidation if the passivating cover layer is damaged, which can pulverize the magnet.

As with soft magnets, a relative permeability μ_r, i.e. the gradient of the *B*/*H* curve, can also be specified for hard magnets. Because the curve is very flat, the relative permeability is small, only about 1.05, like vacuum ($\mu_r = 1.0$). In terms of the transmission of external magnetic fields, a permanent magnet therefore behaves like air. It has a high magnetic resistance.

7.11.4.2 Operating Point and Demagnetization

The operating point of a permanent magnet in the magnetic circuit of an electrical machine is determined by the intersection of the magnet’s characteristic curve with the air gap line.

The characteristic curves of the different magnet materials are published in the manufacturers’ catalogs. The red lines in Fig. 7.30 are examples of a typical NdFeB magnet at different operating temperatures.

The gradient of the air gap straight line, the blue line 1 in Fig. 7.30, is determined by the geometry of the motor. An ideal iron circuit without an air gap would result in a vertical line. The larger the air gap is designed, the more the air gap line is inclined to the left, i.e. it becomes flatter.

At no-load and no stator currents, the airgap line passes through the origin. In Fig. 7.30, for example, for a cold motor and 20 °C, the operating point 1 is resulting from the intersection of the air gap line and the magnet’s characteristic. A field strength of around −300 kA/m and a flux density of around 1 T is created by the magnet.

As the stator current successively increases with the mechanical load, an opposing stator field builds up. This field shifts the air gap line parallel to the *H*-axis to the left, for example to operating point 2. The offset can be calculated from the

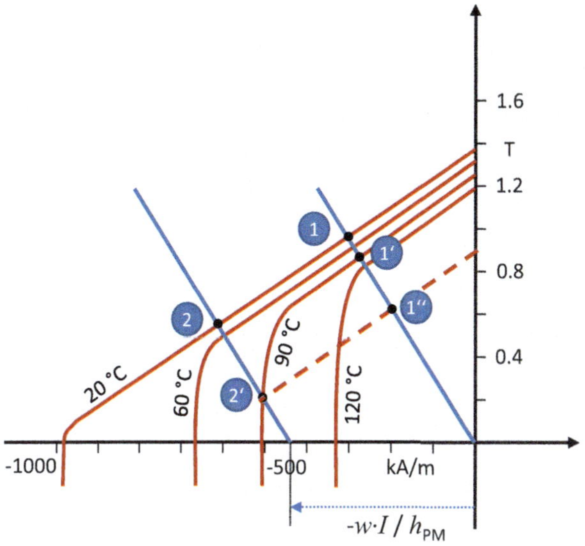

Fig. 7.30 Temperature-dependent characteristics of a typical NdFeB magnet (red) and air gap curves (blue)

current loading ($w \cdot I$) divided by the height of the permanent magnets h_{PM}. The superimposed magnet and stator fields create a field strength of around −600 A/m and a flux density of around 0.6 T in the permanent magnet.

When the load disappears again later ($I=0$), the air gap line returns to the origin and operating point 1 is set again. The process is entirely reversible.

Now consider the same situation for a hot motor, for example a magnet temperature of 90 °C. Due to the higher temperature, the magnet's characteristic curve drops and the no-load operating point shifts from 1 to 1′. This process is also reversible, i.e. after cooling down to 20 °C, the operating point 1 is set again.

However, the motor needs more current (more magnetic field $-H$) to achieve the same flux density B and hence the same Lorentz force, torque, and power as before. When the motor becomes hot, efficiency is reduced not only due to increased ohmic resistance in the wires but also due to the need for more current to generate the same magnetic flux.

The situation becomes critical when the operating point exceeds the bending point of the magnetic characteristic curve, for example when it reaches point 2′. If the motor is later unloaded from this operating point, the magnet no longer follows the path along the original characteristic curve. Instead, it forms a new linear characteristic curve, which is shown as a dashed line in the diagram. As a result, the new operating point 1″ and not 1′ or 1 is reached in no-load. The magnet has become significantly weaker and thus partially demagnetized.

This process is not reversible. The partial demagnetization of a permanent magnet under high counterfields is permanent. The motor can no longer achieve its full power. It has a significantly poorer efficiency in the remaining operating range.

Theoretically, it would be possible to remove the magnet from the motor for re-magnetizing. In practice, however, this is generally not possible: Removing a motor from a car and dismantling the motor itself is very time-consuming. The magnets are usually glued into the rotor pockets and cannot simply be dismantled. Also, the magnetizing device is only available at the original manufacturing plant of the motor.

When designing a permanent magnet synchronous motor, care must therefore be taken to ensure that the operating point never exceeds the bending point of the magnet's characteristic curve.

Possible design measures are:

- Selection of a magnet material with higher temperature resistance (this can be achieved by adding dysprosium to the material mix),
- reduction of the air gap,
- increasing the magnet height h_{PM},
- limiting the electric currents in the power electronics, and
- limiting the permissible operating temperature (temperature sensor).

However, dysprosium is a very expensive, heavy rare earth material. For this reason, various methods are currently being investigated to improve the thermal resistance of magnets with minimal use of dysprosium.

7.12 Losses and Efficiency

The ratio of power output to power input is referred to as efficiency η (eta). Depending on whether the machine is operating in motor or generator mode, the role of electrical power and mechanical shaft power is reversed:

$$\eta = \frac{P_{out}}{P_{in}}, \eta_{mot} = \frac{P_{mech}}{P_{el}}, \eta_{gen} = \frac{P_{el}}{P_{mech}} \tag{7.30}$$

The electrically consumed power P_{el} is calculated from the input voltage, i.e. the phase voltage, the phase current and the power factor as follows (Fig. 7.31):

$$P_{el} = \sqrt{3} \cdot U_1 \cdot I_1 \cdot cos\varphi \tag{7.31}$$

The losses in the winding of the stator are obtained from the phase current. In the case of the usual star connection, the phase current is equal to the phase current and the following applies:

$$P_{Cu1} = 3 \cdot I_1^2 \cdot R_1 \tag{7.32}$$

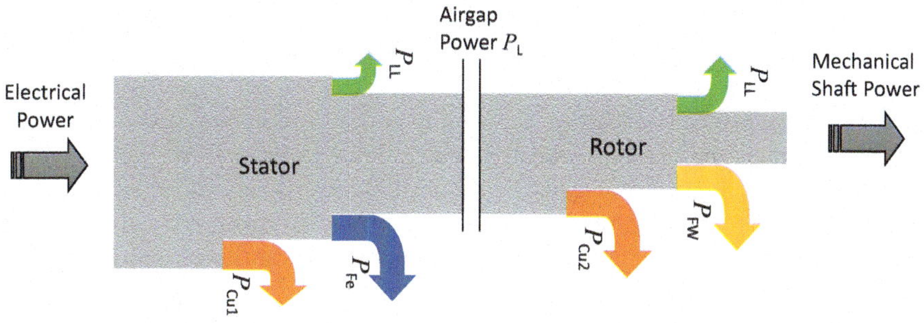

Fig. 7.31 Power flow in an electric motor

The iron losses P_{Fe} (remagnetization losses) result from hysteresis and eddy current losses as described above. They change depending on the frequency and the flux density.

Further loss mechanisms are summarized under the term load-dependent additional losses P_{LL}, which only account for around 2–5% of the total losses. These include, for example, eddy current losses in the housing or currents on the surfaces of the stator and rotor electrical sheets.

When the electrical input power is reduced by the losses of the stator, the result is the air gap power:

$$P_L = P_1 - P_{Cu1} - P_{Fe} - P_{LL} \tag{7.33}$$

This power is transferred magnetically to the rotor.

In all electric machines, the rotor rotates either exactly or at least approximately at the speed of the stator field. Within the rotor, the frequency of the magnetic fields is therefore either zero (synchronous machines) or at least very low (asynchronous machines). Because the iron losses increase with the frequency, they are also small and can be neglected in the rotor.

However, higher-frequency magnetic fields can penetrate the rotor surface due to slot harmonics, causing losses. In addition, the material for the rotor is produced free of charge when the stator laminated core is punched. For this reason, rotors today are normally also constructed from laminated sheet metal.

In an induction machine, the current heat losses of the rotor cage must also be considered. They can be calculated using a virtual rotor current I_2':

$$P_{Cu2} = 3 \cdot I_2^2 \cdot R_2 = P_L \cdot s \tag{7.34}$$

where s is the slip and can be calculated from:

$$s = \frac{n_1 - n}{n_1} \tag{7.35}$$

where n_1 is the synchronous speed $n_1 = f_1/p$ and n is the mechanical speed of the rotating shaft. In induction machines, the rotor losses are identical to the product of air gap power and slip.

There may also be some additional losses P_{LL} in the rotor. However, they are small and are either neglected or added to the additional losses of the stator in the calculation.

In PM-synchronous machines, some eddy current losses may occur in the permanent magnets, but they are usually small. Although these losses hardly play a role in the overall motor efficiency, they can still lead to a noticeable heating of the magnets. As shown in the previous chapter, this heating can lead to critical operating states. Therefore, the rotors of high-performance machines with permanent magnets are also cooled at great expense today.

In the case of externally excited synchronous machines, the winding losses of the excitation winding result from the excitation current $I_E^2 \cdot R_E$.

The result of input power and electromagnetic losses is the internal mechanical power and the internal torque of an electric machine:

$$P_{mechi} = P_1 - P_{Cu1} - P_{Fe} - P_{LL} - P_{Cu2} \tag{7.36}$$

$$T_i = \frac{P_{mechi}}{\Omega} = \frac{P_{mechi}}{2\pi \cdot \mathrm{n}} \tag{7.37}$$

where $n = n_1 \cdot (1 - \mathrm{s})$ is the mechanical speed of the shaft ($s = 0$ in synchronous machines).

In the typical efficiency maps of electric machines, see for example Fig. 2.3, only the internal losses and the internal torques are specified.

Mechanical friction and windage losses P_{FW} can be quite high, particularly in high-speed machines. They result from friction losses of the bearings and losses due to air turbulence in the air gap. Friction and windage losses are difficult to calculate.

Finally, the mechanical shaft power and torque are obtained from:

$$P_{mech} = P_{\mathrm{mechi}} - P_{FW} \tag{7.38}$$

$$T = \frac{P_{mech}}{2\pi \cdot \mathrm{n}} \tag{7.39}$$

Note

It is known from combustion engines that losses and efficiency tend to deteriorate with larger engines with more displacement and more cylinders, assuming the same driving profile, i.e. the same power output.

The additional volume and the additional cylinders must be filled with additional gas–air mixture. Friction losses increase, for example at the cylinder piston seals and the camshaft. All in all, the fuel economy of large, powerful combustion engines reduces significantly.

However, the situation is different with electric motors.

A larger electric motor has more and longer wire windings. The coil resistance R_1 increases roughly in proportion to the diameter and length. At the same time, however, the rotor diameter and surface become larger.

Therefore, the stator current I_1 is lower in bigger machines when the same torque is required. The reduction of current is also roughly proportional to the diameter and length of the motor.

The main losses of the motor are the current heat losses, i.e. $I^2 \cdot R$ losses. It is immediately apparent that the reduction of the current I has a greater influence on the losses than the increase in resistance R.

This leads to the surprising fact that larger electric motors are more efficient than smaller electric motors!

This concept has long been known in the field of industrial motors. It has been used for decades to build highly efficient motors in the higher IE classes (IE = International Efficiency, see the IEC 60034-31 standard).

The actual relations are of course more complex than the simple explanation given above. The larger motor has more friction and air turbulence losses. In addition, the iron mass increases, which increases the iron losses. On the other hand, the magnetic flux density B and thus also the iron losses can be reduced due to the larger rotor surface. Ultimately, the magnetic design and the energy-optimal motor control is a highly complex, multi-layered optimization task.

But in the end, larger motors have an advantage in terms of efficiency. You can see this clearly in today's electric cars. Many cars are offered with high peak power outputs, often 400 hp and more, and still have exceptionally low energy consumption under normal driving conditions.

7.13 Cooling

Due to the high-power density of electric machines, the density of losses is also high, even though the machines have a very good efficiency over a wide operating range.

Example 7.7—Losses in electric motors

A modern electric motor with a mechanical output of around 150 kW has a size of approximately 10 L. Assuming an efficiency of 95%, around 7.5 kW losses occur. This corresponds to the heat output of 4 medium-sized cooking plates, which must be actively dissipated continuously. ◀

In today's high-performance machines, cooling is a key success factor and highly complex in detail (Fig. 7.32).

The following cooling methods are known and mostly already implemented in series engines, often in different combinations:

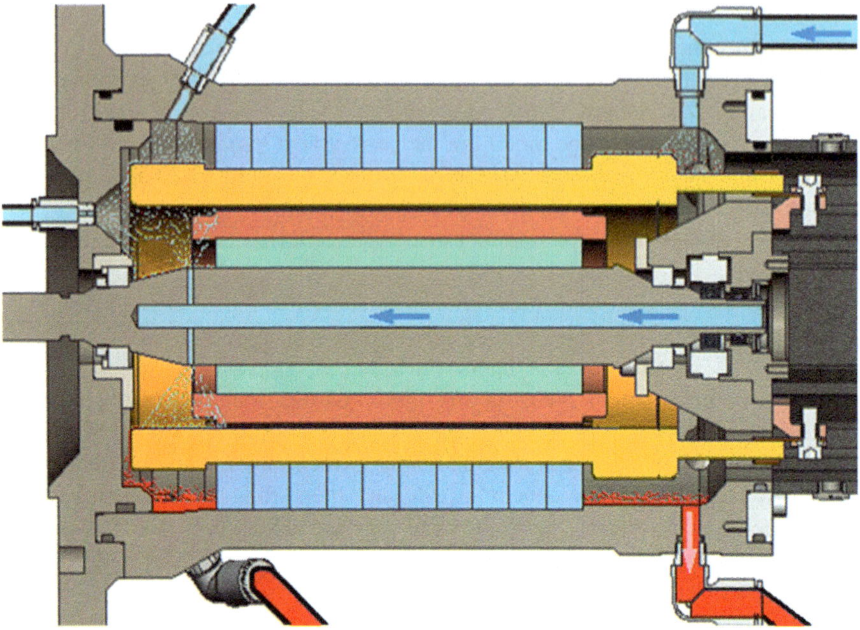

Fig. 7.32 Cross-section of an electric motor with various cooling concepts

- Coil housing cooling (Coolant (usually a water/glycol mixture) flows through channels around the outside of the stator housing.)
- Laminate package cooling (Coolant (typically non-conductive oil) flows through channels inside the laminated stack.)
- Slot interior cooling (Channels made of plastic are located in the slots between copper conductors, through which coolant (usually a water/glycol mixture) flows.)
- Conductor cooling The conductors are designed as hollow conductors, with coolant (usually a water/glycol mixture) flowing inside.)
- Winding head cooling (Coolant (usually non-conductive oil) is directed to the winding heads. There is spray mist cooling (active oil atomization through nozzles, creating a spray mist) and passive spray cooling (oil is directed through openings to the winding head). The cooling can occur radially from the inside through holes in the shaft, radially from the outside through nozzles or holes in the laminated stack, or axially from the front through nozzles.)
- Rotor shaft cooling Coolant (either oil or a water/glycol mixture) is fed axially through the hollow rotor shaft. The shaft is typically only accessible from one side, with a tube inserted in the hollow section through which the coolant is injected.)
- Internal air cooling Air is blown into the motor from one side, flowing over the winding heads and through the air gap. The air is then sucked out on the opposite side.)

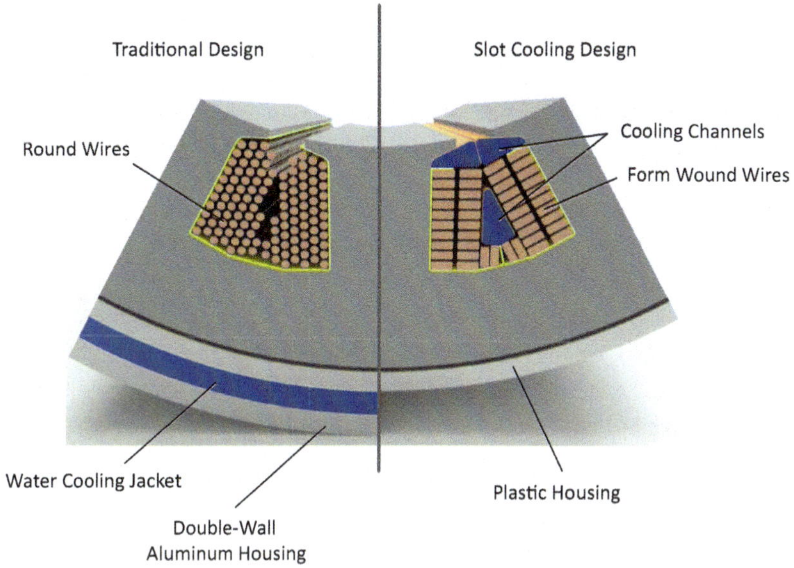

Fig. 7.33 New cooling concept (indirect slot cooling) without cooling jacket

- External air cooling (Cooling air flows over the machine surface, with the housing usually equipped with fins. The air can be forced by a separate fan or flow passively, as in the case of motors in e-bikes.)

The illustration demonstrates different cooling methods. The outside of the stator core is enclosed by a double-walled housing. Spiral channels run through the housing, transporting a liquid coolant, usually a water-glycol mixture. This design has been the standard for traction motors for many years.

The illustration also shows a hollow shaft. Coolant is supplied from the right via a lance (straw). Such rotor cooling systems can now be found in many vehicles, for example the Audi eTron 55 and Tesla Model 3.

In the illustration, the coolant is not sucked back through the rotor but rather sprayed out at the end onto the inside of the stator winding heads by centrifugal force. In this case, a non-conductive fluid must be used, usually oil. This cooling concept is used in the Tesla, for example. The cooling oil can also be injected centrally into channels in the laminated core and then fed outwards to the left and right over the winding heads (not shown in the illustration). This is done in the Lucid Air, for example.

Even better cooling can be achieved by spraying cooling oil directly onto the winding heads with the aid of nozzles, either axially or radially (Fig. 7.33).[7]

[7] *Thermische Modellierung permanenterregter Synchronmaschinen mit wellenseitiger Öl-Sprühkühlung*, Felix Hoffmann, Dissertation, Karlsruhe Institute of Technology, 2024.

The next logical step is the integration of the cooling medium directly into the slots of the winding.[8] This not only significantly improves the cooling effect, but also eliminates the need for a metal housing. Costs can be saved by using a plastic housing. During production, one of the end shields can be cast in a single shot together with the housing. Add-on parts, such as electric connector housings, can also be integrated.

Such concepts are currently being investigated by OEMs and Tier 1 suppliers and are likely to be found in series products in the next few years.

[8] *Indirekte Wicklungskühlung von hochausgenutzten permanenterregten Synchronmaschinen mit Zahnspulenwicklung*, Markus Schiefer, Dissertation, Karlsruhe Institute of Technology, 2017.

Potentiale durch Kunststoffeinsatz in Elektromotoren, Steffen Reuter, Dissertation, Karlsruhe Institute of Technology, 2024.

Synchronous Machines 8

Abstract

This chapter is dedicated to the detailed description of the function and operating characteristics of synchronous machines in their various forms. In particular, the description of the machine behavior and the control in d–q coordinates are explained. This results in the equations for describing the generation of torque from synchronous torque and reluctance torque. The parameter plane is introduced in a short section. It enables a systematic rotor design to achieve the desired torque–speed behavior. Finally, critical operating states such as short circuit and sudden loss of load are discussed, which must be mastered in the event of a fault.

8.1 The d/q Coordinate System

The physical operation of three-phase windings and synchronous machines has already been presented. In practice, it proves useful to use a more abstract representation instead of the momentary currents and voltages in the three phases. This simplifies the equations considerably and is the basis for the operation of a machine controller that sits inside each frequency inverter.

A new coordinate system is introduced that rotates along with the rotor: the d/q system.

The d-axis is called the direct axis. The d-axis is oriented with the axis of the main rotor flux, caused by permanent magnets (PMSM) or excitation coils (EESM).

The q-axis is the transverse axis ("Querachse" in German). It is oriented in the next pole gap directly following the d-axis in the counterclockwise direction. The rotor does not drive any flux through the q-axis, but the stator may.

M. Doppelbauer, *Introduction to Electromobility*,
https://doi.org/10.1007/978-3-658-45482-1_8

Note

If a rotor carries multiple pairs of magnets, we will consider only one of them as they are all acting alike. This is called electrical coordinates: 0–360° (0 to 2π) is one pole pair. The other pole pairs are introduced in the calculations by the factor p.

In electrical coordinates, d- and q-axis are always perpendicular to each other at an angle of 90°.

In actual, mechanical coordinates, the angle between d- and q-axis is 360°/(4p).

The direction of the magnetic flux of the stator depends on the mechanical loading of the machine. In no-load, the stator flux is oriented in parallel to the rotor flux. Both are in the direction of the d-axis and no torque is created. When the torque is gradually increased, the stator current also increases and the axis of the stator flux is moved away from the d-axis towards the q-axis.

As the machine turns, the magnetic flux of the rotor ϕ_{Rotor} is pulled along by the stator flux ϕ_{Stator}. In stationary operation, both rotate at the same speed ω, i.e. synchronously. As the load increases, the stator flux turns more and more in the direction of the q-axis—the angular displacement in synchronous machines ϑ (teta) is created.

In physical reality, there is only a single magnetic flux that results from the superposition of rotor and stator flux.

Note that the d–q coordinate system and thus also the stator and rotor fluxes constantly rotate with the speed ω. As the coordinate system is fixed at the rotor, all three-phase alternating variables, i.e. magnetic fluxes, voltages, currents, and inductances, become constant after the d–q transformation. This simplifies all calculations considerably.

Furthermore, the d- and q-parts of each quantity are combined by using complex numbers, which also happen to have two parts. The real part corresponds to the magnitude of the variable in the d direction and the imaginary part corresponds to the magnitude of the variable in the q direction.[1]

Mathematically, the d/q transformation consists of two steps: First, the real variables (currents, voltages, magnetic fluxes, inductances, resistances) are converted from three phases (u, v, w) with a 120° displacement angle to two virtual stator phases (α, β) with a 90° displacement angle. This transformation is also known as the Clarke transformation after its inventor Edith Clarke.

[1] Note that this assignment of the d- and q- axes to real and imaginary axes of complex numbers is common and useful in the field of variable drive technology. In the field of large generators, grids, and electrical power supply, a different assignment is more beneficial. Here, the terminal voltage of the generators $\underline{U}_1$ is assigned to the real axis.

In the second step, the coordinate system is transformed from the stator (α, β—stationary) to the rotor (d, q—rotating). The combination of both transformations is also known as the Park transformation after its inventor Robert H. Park. It was first described in 1929.

The Park transformation is easy to describe mathematically by using just a few trigonometric functions:

$$\begin{bmatrix} i_d \\ i_q \\ i_0 \end{bmatrix} = \frac{2}{3} \cdot \begin{bmatrix} cos\gamma & cos\left(\gamma - \frac{2\pi}{3}\right) & cos\left(\gamma + \frac{2\pi}{3}\right) \\ -sin\gamma & -sin\left(\gamma - \frac{2\pi}{3}\right) & -sin\left(\gamma + \frac{2\pi}{3}\right) \\ \frac{1}{2} & \frac{1}{2} & \frac{1}{2} \end{bmatrix} \cdot \begin{bmatrix} i_U \\ i_V \\ i_W \end{bmatrix} \tag{8.1}$$

This equation is written for currents, but similar transformations apply to voltages, fluxes, inductances, etc. The component i_0 is zero in symmetrical systems and can be neglected.

The angle γ is the momentarily position of the rotor (d-axis) relative to the stator (α-axis). This angle must be known to carry out the transformation. For this reason, all synchronous machines are equipped with an angle encoder.[2] This is a sensor attached to the shaft that can be queried by the motor controller and reports the current rotor angle position at any moment.

The reverse transformation is as follows:

$$\begin{bmatrix} i_U \\ i_V \\ i_W \end{bmatrix} = \begin{bmatrix} cos\gamma & -sin\gamma & 1 \\ cos\left(\gamma - \frac{2\pi}{3}\right) & -sin\left(\gamma - \frac{2\pi}{3}\right) & 1 \\ cos\left(\gamma + \frac{2\pi}{3}\right) & -sin\left(\gamma + \frac{2\pi}{3}\right) & 1 \end{bmatrix} \cdot \begin{bmatrix} i_d \\ i_q \\ i_0 \end{bmatrix} \tag{8.2}$$

Here again, knowledge of the momentary rotor position is required to carry out the transformation.

The rotor flux induces a voltage in the stator winding, the synchronous generated voltage U_p. This voltage is shifted by 90° relative to the rotor flux, i.e. it is in the direction of the q-axis.

The reason for this shift is the law of induction $u = -d\phi/dt$. The derivative of a sine function results in a cosine function and therefore a phase shift of 90°. The induced voltage is highest at the highest rate of change, which occurs at zero crossing.

The synchronous generated voltage is always purely imaginary, as it is always oriented in the q direction.

[2] There are model-based methods for estimating the rotor angle from certain characteristics of the current. In this way, the rotor angle can be obtained without using an angle sensor. However, these methods are not always reliable and are therefore not yet used in electric motors for cars for safety reasons.

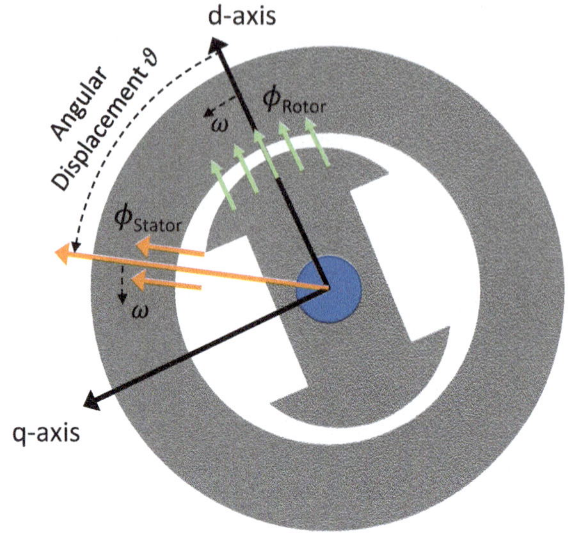

Fig. 8.1 d/q coordinate system of the synchronous machine, illustrative representation of the pole wheel angle (not mathematically exact)

In machines with permanent magnets (PMSM), the rotor flux is constant and therefore the synchronous generated voltage is purely linearly dependent on the speed. In electrically excited machines (EESM), the rotor current I_E generates the rotor flux $\underline{\phi}_{Rotor}$. It can be varied as required. Per definition, the rotor flux is always oriented in the d-axis.

The stator current $\underline{I}_1$ generates the stator flux $\underline{\phi}_{Stator}$ and is therefore oriented in its direction. It can vary between the d- and q-axis.

The terminal voltage (stator voltage) $\underline{U}_1$ is the vectorial sum of the synchronous generated voltage $\underline{U}_p$ and the ohmic voltage drop $R_1 \cdot \underline{I}_1$ and the voltage across all stator inductances $j\omega L_1 \cdot \underline{I}_1$.

The synchronous generated voltage is mathematically defined as the angle between $\underline{U}_1$ and $\underline{U}_p$. If the electrical resistance of the stator winding is small, this corresponds to the angle between the stator and rotor flux as shown in Fig. 8.1.

The angle between terminal voltage $\underline{U}_1$ and stator current $\underline{I}_1$ is referred to as the phase angle φ. This angle is needed to compute the power factor cos φ. With EESM, the power factor can be influenced over a wide range by varying the excitation current and the amplitude of the synchronous generated voltage.

The self-inductances and mutual inductances L between the stator and rotor windings are also separated in the direction of the two coordinate axes:

- L_d is the self-inductance of the stator winding in the direct axis,
- L_q is the self-inductance of the stator winding in the transverse axis,
- L_{Fd} is the mutual inductance between the stator and rotor windings, which only occurs in the d-axis and only exists with EESM.

Note
In classical electrical machine theory, a distinction is made between salient-pole and cylindrical rotor (turbo type) synchronous machines.

Salient-pole machines are constructed as shown in Fig. 8.1. The air gaps in the d- and q-directions are quite different. Therefore, the inductances L_d and L_q, which ultimately describe the magnetic resistance, are also different.

Turbo-type machines have a circular rotor so that the magnetic resistances are nearly the same in all directions, so $L_d = L_q$ approximately.

Cylindrical rotor machines have the highest torque in the base speed range. For this reason, industrial servomotors, e.g. robot drives, are preferably designed as cylindrical rotor machines albeit using permanent magnets instead of excitation coils.

On the other hand, cylindrical rotor machines require a much higher current in the field weakening range than salient-pole machines. They are therefore less suitable as traction drives and are not used in this application.

All following equations are derived for the general case of salient-pole machines. Cylindrical rotor machines are included in the equations as a special case by setting $L_d = L_q$.

8.2 Mathematical Model

Based on Eq. (A.67), the following can be written for the terminal voltage of a coil in general terms:

$$u = R \cdot i + \frac{d\psi}{dt} \tag{8.3}$$

In it, ψ is the linked flux of all internal and external fluxes. The sign of ψ is dependent on the counting arrow and was chosen to be positive here in accordance with the usual literature. Ultimately, the flux linkage can be imagined as the magnetic flux ϕ introduced earlier. However, the number of turns of the coil z must be considered: $\psi = \phi \cdot z$.

By introducing complex numbers, the equation is simplified to:

$$\underline{U} = R \cdot \underline{I} + j\omega\, \underline{\Psi} \tag{8.4}$$

Please note that the time derivate $d\psi/dt$ simplifies to a factor $(j\omega)$ when using sinusoidal quantities and complex numbers. This is one of the benefits of complex numbers in electric circuit analysis.

We now set:

$$\begin{aligned} \underline{U} &= U_d + jU_q \\ \underline{I} &= I_d + jI_q \\ \underline{\Psi} &= \Psi_d + j\Psi_q \end{aligned} \tag{8.5}$$

and get:

$$\begin{aligned} U_d &= \Re\{\underline{U}\} = R_1 \cdot I_d - \omega \cdot \Psi_q \\ U_q &= \Im\{\underline{U}\} = R_1 \cdot I_q - \omega \cdot \Psi_d \\ U_F &= R_2 \cdot I_2 \end{aligned} \tag{8.6}$$

The fluxes result from:

$$\begin{aligned} \Psi_d &= L_d \cdot I_d + \Psi_{Rotor} \\ \Psi_q &= L_q \cdot I_q \end{aligned} \tag{8.7}$$

Depending on the machine type, the rotor flux is generated in different ways.

$$\text{In PMSM} : \Psi_{Rotor} = \psi_{PM} \tag{8.8}$$

$$\text{In EESM} : \Psi_{Rotor} = L_{Fd} \cdot I_2 \tag{8.9}$$

$$\text{In SynRM} : \Psi_{Rotor} = 0 \tag{8.10}$$

If required, the inductances can be further subdivided into a part originating from the main flux L_h, which is the flux between the stator and rotor, and a part resulting from the leakage flux L_σ, which is the flux that closes purely in the stator and does not reach the rotor:

$$\begin{aligned} L_d &= L_\sigma + L_{hd} \\ L_q &= L_\sigma + L_{hq} \end{aligned} \tag{8.11}$$

The leakage flux L_σ is the same in both axial directions d and q, as it remains entirely in the stator and is therefore independent of the rotor position. In the following, the subdivision into main and leakage flux is not used any further. However, it is often found in the literature.

With these basic principles, the equation for the terminal voltage can be fully given:

$$\underline{U}_1 = U_d + j\mathrm{U}_q = R_1 \cdot \underline{I}_1 + \mathrm{j}\omega \cdot \left(L_d \cdot I_d + \mathrm{j}L_q \cdot I_q\right) + \mathrm{j}\omega \cdot \Psi_{Rotor} \tag{8.12}$$

The last term is the synchronous generated voltage, i.e. the voltage induced in the stator by the rotor's magnetic field:

$$\underline{U}_p = \mathrm{j}\omega \cdot \Psi_{Rotor} \tag{8.13}$$

The torque generally results in complex representation from:

$$T = \frac{3}{2} p \cdot \Im\{\underline{\psi}^* \cdot \underline{I}\} = \frac{3}{2} p \cdot \left(\psi_d \cdot I_q - \psi_q \cdot I_d\right) \tag{8.14}$$

The equation can be derived from the mathematical model of two-axis theory with power-variant space vectors. However, this would go too far at this point.

Active and reactive power are calculated from:

$$\begin{aligned} P &= \frac{3}{2} \cdot \Re\{\underline{U}_1 \cdot \underline{I}_1\} = \frac{3}{2}(U_d \cdot I_d - U_q \cdot I_q) \\ Q &= \frac{3}{2} \cdot \Im\{\underline{U}_1 \cdot \underline{I}_1\} = \frac{3}{2}(U_q \cdot I_d - U_d \cdot I_q) \end{aligned} \tag{8.15}$$

We have now derived a set of equations from which we can determine the currents in the d- and q-axis if we know the speed and the terminal voltage. We also need to know the machine parameters, i.e. the inductances and resistances of the coils. The currents can be used to calculate the torque as well as the active and reactive power. Later, we will get to know a control method to adjust the voltage so that the desired currents and thus the desired torque are obtained.

8.3 Torque Creation

8.3.1 Synchronous Torque and Reluctance Torque

By transforming Eq. (8.14) using (8.7), the important torque equation of the synchronous machine is obtained:

$$T = \frac{3}{2}p \cdot \left(\Psi_{Rotor} \cdot I_q + \left(L_d - L_q\right) \cdot I_d \cdot I_q\right) \tag{8.16}$$

The first summand is referred to as the synchronous torque. It is present in every synchronous machine with magnets (PMSM) or excitation coils (EESM) and can ultimately be attributed to Lorentz forces.

The second summand is referred to as reluctance torque. It only appears when $L_d \neq L_q$ or otherwise the parenthesis would become zero.

Surprisingly, reluctance also provides a torque if the machine does not generate any rotor flux at all, i.e. if $\Psi_{Rotor} = 0$. There are indeed synchronous machines without windings and without magnets in the rotor, the so-called synchronous reluctance machines (synRM). This second summand is also present in all EESM and in some PMSM, but only in those PMSM who have magnets that are placed in pockets inside the rotor. Other PMSM have magnets on the rotor surface and in this case, L_d becomes equal to L_q and the second summand is zero.

The difference of L_d and L_q should be as large as possible so that the reluctance component is large. In practice, however, it can only be around 2–5 times for design reasons.

All synchronous machines for traction drives are designed so that both components, i.e. synchronous torque and reluctance torque, are present. This is also referred to as permanent magnet synchronous machines (PMSM) with an increased reluctance component. The same effect occurs in electrically excited machines (EESM). This combination of synchronous and reluctance torque is required to utilize a wide field weakening range, as will be shown later in Sect. 8.5.

Technically, the reluctance part is achieved by different inductances in the d- and q-axis by embedding the magnets in pockets in the rotor core (so-called buried magnets or internal permanent magnets). This machine variant is called PMSM-IPM or simply IPM.

The currents in d- and q-axis are set by the motor controller inside the inverter depending on the operating point to optimize the torque generation in terms of efficiency.

The design and control of PMSM-IPM machines were already elaborated in the early 1990s in detail by Soong and Miller.[3] All PMSM traction motors in electric cars starting with the Toyota Prius in 1997 (and maybe even before that) including the latest motors from Telsa Inc. use this design.

There is a different type of synchronous machine, which is based purely on the reluctance principle, i.e. it has no electrical or permanent magnets in the rotor. These machines can either be designed for switched voltages (SRM = switched reluctance machine) or for sinusoidal currents (synRM = synchronous reluctance machine). synRM machines are built by companies like ABB or KSB for industrial applications. So far, no mass-produced electric vehicle uses such motors. However, several manufacturers are looking into the topic.

Note

Unfortunately, an internet cult has developed in recent years around the long-known technology of internal permanent magnet synchronous machines.

A machine-type designation such as "IPMSynRM", which can be found in internet forums, is an oxymoron (a contradiction in itself) and simply nonsense. The first part (IPM) indicates the existence of magnets inside of the rotor, while the second part (synRM) indicates their absence.

What is meant is simply a PMSM-IPM machine. It goes without saying that this machine has an increased reluctance torque, which is automatically produced by the internal rotor structure. Such a design is required for all traction drives that use permanent magnets to reduce the currents in field-weakening operation. This knowledge dates back to the 1990s and has been used in all corresponding electric motors ever since, including that of the Toyota Prius, BMW i3, Tesla Model 3 and many more.

All in all, there are four different types of synchronous machines depending on L_d, L_q, and Ψ_{Rotor}, as shown in the following table.

[3] *Theoretical limitations to the field-weakening performance of the five classes of brushless synchronous AC motor drives*; Soong, W.L.; Miller, T.J.E.; Sixth International Conference on Electric Machines and Drives, 1993, pp. 127–132.

	Rotor excitation Ψ_{Rotor}	Inductances	
EESM	$L_{Fd} \cdot I_2$ (variable)	$L_d > L_q$	Electrically excited synchronous machine
PMSM-SPM	Ψ_{PM} (fixed)	$L_d = L_q$	Surface PM synchronous machine
PMSM-IPM	Ψ_{PM} (fixed)	$L_d < L_q$	Internal PM synchronous machine
synRM	0	$L_d \ll L_q$	Synchronous reluctance machine

8.3.2 Rotor Design

Figure 8.2 shows examples of four basic types of synchronous motors. For traction drives, 6–8 poles are common. Gearbox-integrated motors or starter generators often have significantly more poles.

In the PMSM with reluctance component (PMSM-IPM = Internal Permanent Magnets), the magnetic pockets act like an additional air gap and thus increase the magnetic resistance in the d-direction: L_d becomes smaller than L_q. The same effect occurs with the synchronous reluctance machine (synRM).

The opposite is true for the separately excited synchronous machine. Here, the largest air gap is in the q-direction: L_q becomes smaller than L_d.

The PMSM without reluctance component (PMSM-SPM = Surface-Permanent-Magnets) consists of a cylindrical rotor with glued-on magnets. Because the magnets have a relative permeability coefficient μ_r of approximately 1, they behave magnetically almost like air. The two inductances in the d- and q-directions are therefore the same.

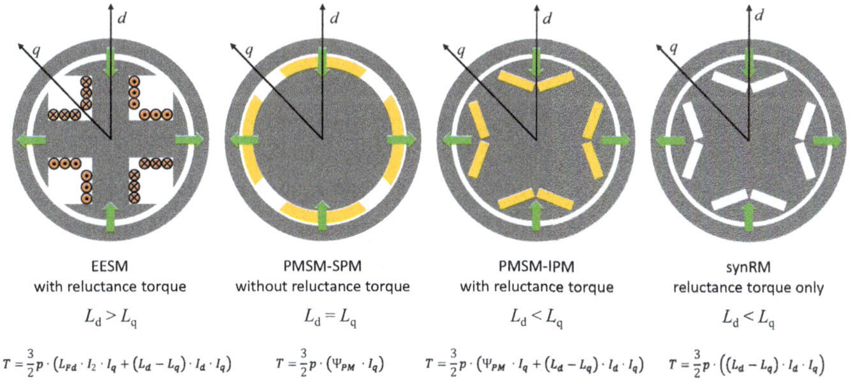

Fig. 8.2 The four basic types of synchronous machines

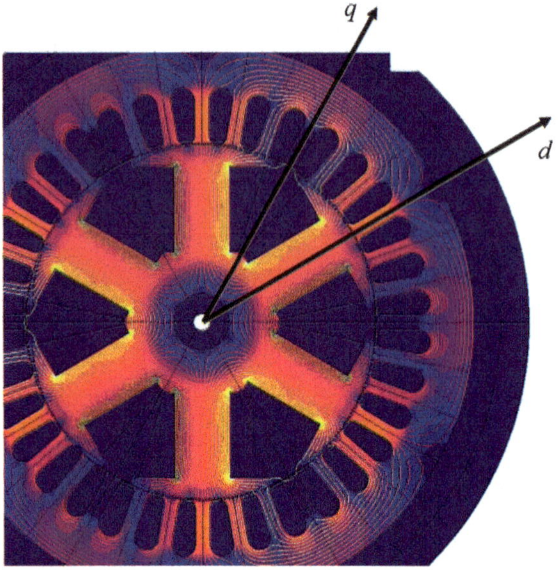

Fig. 8.3 Numerical field calculation of a 6-pole (p = 3) EESM with distributed winding

Current electric vehicles use motors as shown in Fig. 8.2 (a) (EESM, e.g. Renault Zoe) or as shown in Fig. 8.2 (c) (PMSM-IPM, e.g. BMW i3, Tesla Model 3, Toyota Prius and many more). The other two variants are only built as industrial motors because they either have disadvantages in the field-weakening range (PMSM-SPM) or because their power density is lower (synRM).

Section 7.6.3.1 presented different variants of permanent magnet rotors. It can now be seen that these variants differ only in terms of the inductances in the d- and q-directions and in the amount of magnetic material used, that is the amount of generated magnetic flux Ψ_{PM}. Other features like mechanical stability, cooling, etc., are not relevant for the torque–speed characteristic of the motor.

Figures 8.3 and 8.4 show numerical calculations of the magnetic fluxes in an EESM and a PMSM corresponding to Fig. 8.2 (a) and (c). The light-yellow areas are regions with high flux density B. Dark-blue areas are areas with low flux density.

In the PMSM shown in Fig. 8.4, the magnetic flux reaches its maximum quite exactly in the middle between the d- and q-axis. The calculated angular displacement is therefore approximately 45°. The load on the machine cannot be estimated from the angular displacement because, as mentioned, the torque and angular displacement are controlled independently by the power electronics within certain limits.

In the EESM shown in Fig. 8.3, the flux is at its maximum exactly in the d-axis, i.e. the angular displacement is approximately 0°. In this case, the machine was simulated in no-load mode.

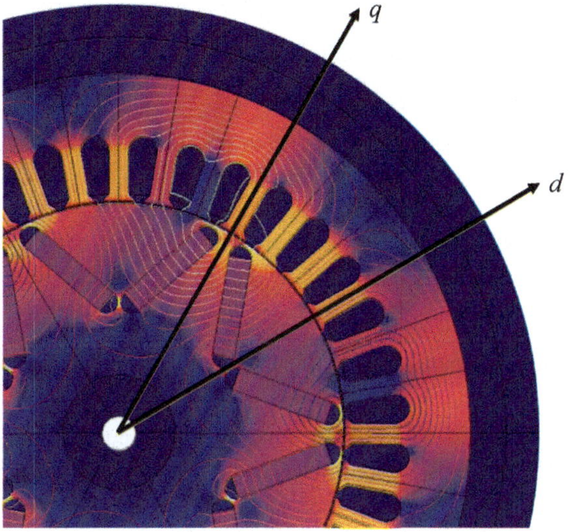

Fig. 8.4 Numerical field calculation of a 6-pole (p = 3) PMSM-IPM with distributed winding

8.4 Current Control Angle

The angle from the q-axis (synchronous generated voltage $\underline{U}_P$) to the stator current $\underline{I}_1$ or stator flux is referred to as the current control angle γ (gamma). It results from the correct addition of the phase angle φ (from $\underline{I}_1$ to $\underline{U}_1$) and the angular displacement ϑ (from $\underline{U}_1$ to $\underline{U}_P$): $\gamma = -(\varphi + \vartheta)$.

The d and q components of the stator current can be easily specified using the current control angle:

$$\begin{aligned} I_d &= \left|\underline{I}_1\right| \cdot \sin(\gamma) \\ I_q &= \left|\underline{I}_1\right| \cdot \cos(\gamma) \end{aligned} \tag{8.17}$$

This simplifies the torque Eq. (8.16) as follows:

$$T = \widehat{T}_{syn} \cdot \cos(\gamma) + \widehat{T}_{rel} \cdot \sin(2\gamma) \tag{8.18}$$

The doubled angular displacement (2γ) in the reluctance torque results from the transformation of the product $\sin(\gamma)\cdot\cos(\gamma)$ in the second summand.

The amplitudes of the synchronous and reluctance torques are:

$$\widehat{T}_{syn} = \frac{3}{2}p \cdot \Psi_{Rotor} \cdot \left|\underline{I}_1\right| \tag{8.19}$$

$$\widehat{T}_{rel} = \frac{3}{2}p \cdot \left(L_d - L_q\right) \cdot \frac{1}{2}\left|\underline{I}_1\right|^2 \tag{8.20}$$

In machines with a cylindrical rotor and surface magnets (SPM, i.e. $L_d = L_q \Rightarrow T_{rel} = 0$), the maximum torque T is reached at $\gamma = 0$, i.e. $i_d = 0$.

For machines with pure reluctance torque (no magnets $\Rightarrow T_{syn} = 0$), the optimum angle is $\gamma = 45°$.

With conventional traction drives (IPM), both T_{syn} and T_{rel} play a relevant role. The optimum current control angle is, therefore, between 0 and 45°.

All the previously introduced formulas were based on a purely linear machine behavior. Due to iron saturation, however, the behavior becomes non-linear, and the analytically determined angles may shift slightly.

8.5 Parameter Plane

The design of the rotor regarding T_{syn} and T_{rel} is decisive for the driving behavior of the machine and must be adapted precisely to the desired application.

The parameter plane described below enables a systematic preselection of suitable rotor designs (Fig. 8.5).

Every conceivable rotor design of a reluctance or PM synchronous machine can be represented in this plane. The horizontal axis of the plane represents the normalized permanent magnet flux, i.e. T_{syn}. The vertical axis represents the ratio of L_d/L_q, i.e. T_{rel}.

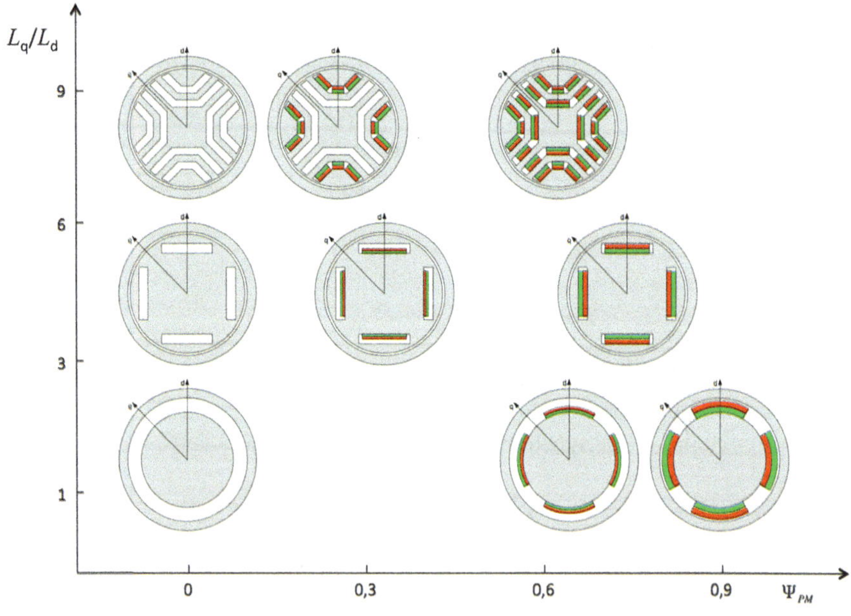

Fig. 8.5 Systematic classification of the rotor designs of synchronous machines at the parameter plane

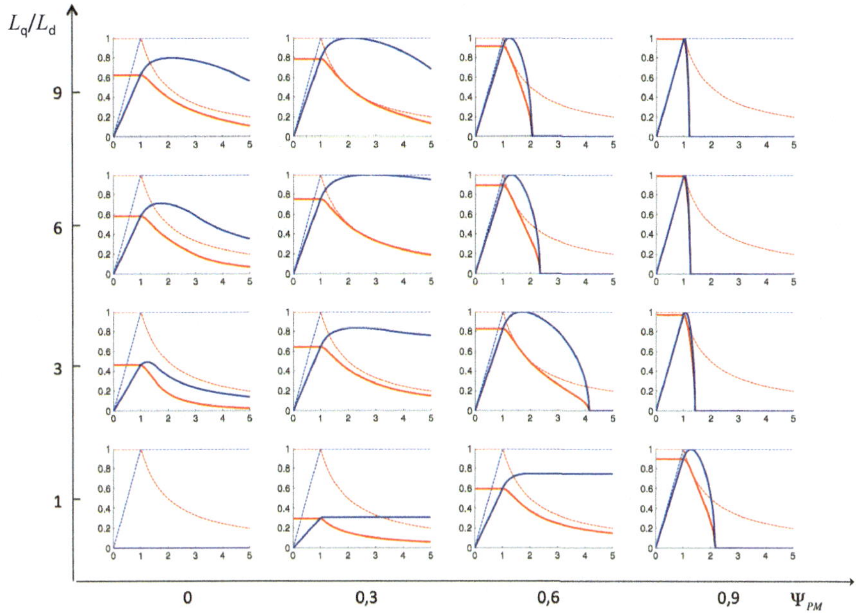

Fig. 8.6 Standardized torque (red) and power curve (blue) of the rotor designs from the previous figure

At the bottom left is a machine without magnets and without reluctance. Such a machine has $T_{syn} = T_{rel} = 0$ and will not produce any torque and hence is not useful.

Further upwards the x-axis, the reluctance torque of the rotors is increased. Further to the right, the synchronous torque generated by the magnets is increased.

Figure 8.6 shows the normalized curves of torque and power developed by the rotors. It can be seen, for example, that the rotor design in the lower right area of the diagram has a lot of torque and power in the base speed range. It is therefore the preferred design for industrial servomotors. However, this design is hardly usable in the field weakening range.

For traction drives, motor designs in the upper area of the second column of the diagram are particularly suitable because the base speed range and field weakening range are both high here.

Further analyses of the parameter level provide basic information about the induced voltage, the short-circuit current, and the possible constant power range of the respective rotors.

The parameter level does not take losses or iron saturation into account. The rotor design selected in this way can therefore by no means be regarded as final, but rather represents the starting point of an extensive optimization process that must subsequently be carried out using analytical and numerical calculation programs.

8.6 Special Operating Conditions

8.6.1 Open Circuit

If the current flow is interrupted in the event of a fault while the vehicle is rolling, the PMSM continues to generate synchronous voltage. This voltage is now fully present at the external terminals and thus also reaches the power semiconductors in the inverter. In machines with surface magnets (SPM), the synchronous generated voltage at maximum speed can be a multiple of the rated voltage and thus destroy the power semiconductors of the inverter if no suitable protective circuits are installed. In machines with embedded magnets (IPM), the synchronous generated voltage is considerably lower and usually below the rated voltage even at maximum speed. This is another reason why IPM-rotor design is preferred for traction drives.

8.6.2 Active Short Circuit

Due to the synchronous generated voltage problem, it has become common practice in the automotive industry to short-circuit the terminals of the machine by the inverter in the event of a fault (active short-circuit = safe state). The motor voltage U_1 thus becomes zero.

However, the machine must be designed accordingly. The synchronous generated voltage that is still present during rotation drives a short-circuit current I_1 that must not be higher than the permissible maximum current I_{max} or otherwise it will destroy the semiconductors in the inverter. This must be ensured by a suitable winding and rotor design.

The short-circuit current also generates a braking torque. It must not be too high, otherwise there is a risk of skidding. The braking torque can become critical if several synchronous motors are installed in the vehicle, which act on individual wheels. If just one of them fails, a yaw torque is created that can cause the vehicle to skid. In this case, the control system must ensure that both motors of an axle are always short-circuited simultaneously, even if a fault occurs in only one of them.

Induction Machines (IM) 9

Abstract

In the short chapter on asynchronous machines, the equations for the single-phase equivalent circuit diagram are explained. The torque and operating behavior can be derived from this.

The electrical operating behavior of the induction machine can be described with one equation for the rotor and one for the stator.

In contrast to the synchronous machine, it is not possible to carry out a rotor-fixed d-q transformation, as the rotor of the induction machine rotates at a different speed to the stator field. For the motor control, a coordinate system is therefore used that is oriented with the rotor flux. The rotor then rotates slowly under this coordinate system. This will be explained in a later chapter.

The rotor speed is

$$n = n_1 \cdot (1 - s) \tag{9.1}$$

And the frequency of the rotor's fundamental currents is

$$f_2 = s \cdot f_1 \tag{9.2}$$

In this equation, n_1 is the synchronous speed, i.e. the speed of the stator flux and the rotor flux, and f_1 is the synchronous stator fundamental frequency, i.e. the frequency of the rotary voltage supplied by the inverter to the stator.

n is the mechanical speed of the rotor. In motor operation, the rotor is slower than the flux, in generator operation it is faster. s is the slip that is the relative difference between the rotor speed and the speed of the flux (synchronous speed), see Eq. (7.4).

M. Doppelbauer, *Introduction to Electromobility*,
https://doi.org/10.1007/978-3-658-45482-1_9

It has become common practice to specify the equation system of the induction machine for only one phase of the stator (single-phase equivalent circuit). The other two phases can be treated accordingly and only appear as a factor of 3 for power and torque.

9.1 Single-Phase Equivalent Circuit Diagram

The system of equations can be derived in a similar way to the transformer because the induction machine also ultimately has two magnetically coupled winding systems. This results in the following system of equations:

$$\begin{aligned} \underline{U}_1 &= R_1 \cdot \underline{I}_1 + \mathrm{j}X_{1\sigma} \cdot \underline{I}_1 + \mathrm{j}X_h \cdot (\underline{I}_1 + \underline{I}'_2) \\ 0 &= R'_2 \cdot \underline{I}'_2 + R'_2 \cdot \left(\frac{1-s}{s}\right) \cdot \underline{I}'_2 + \mathrm{j}X'_{2\sigma} \cdot \underline{I}'_2 + \mathrm{j}X_h \cdot (\underline{I}_1 + \underline{I}'_2) \end{aligned} \tag{9.3}$$

The equivalent circuit diagram can be specified for this system of equations which is shown in Fig. 9.1.

In the single-phase equivalent circuit diagram, the stator and rotor circuits are electrically connected to each other. This simplifies the calculation considerably. A part of the stator current I_1 therefore flows directly into the rotor $\underline{I}'_2$. In the real machine, however, there is only magnetic coupling and no electrical coupling.

To enable electrical coupling, a trick must be used: the rotor variables in the single-phase equivalent circuit diagram, i.e. currents, resistances, and inductances, are not the values that can be measured in the real rotor but are derived from them using a transformation factor. This is indicated by the apostrophe, for example, $\underline{I}'_2$.

The variables in the equivalent circuit diagram have the following meaning:

$\underline{U}_1$ is the phase voltage, i.e. the voltage between one of the three-phase conductors (L1, L2 or L3) and the neutral point. $\underline{I}_1$ is accordingly the current of this phase.

R_1 denotes the ohmic resistance of all coils of a phase in the stator, R'_2 accordingly the resistance of the rotor. The rotor usually consists of short-circuited bars, the so-called cage rotor. This construction can also be regarded as a winding and there are formulas for calculating the appropriate phase resistance. The virtual

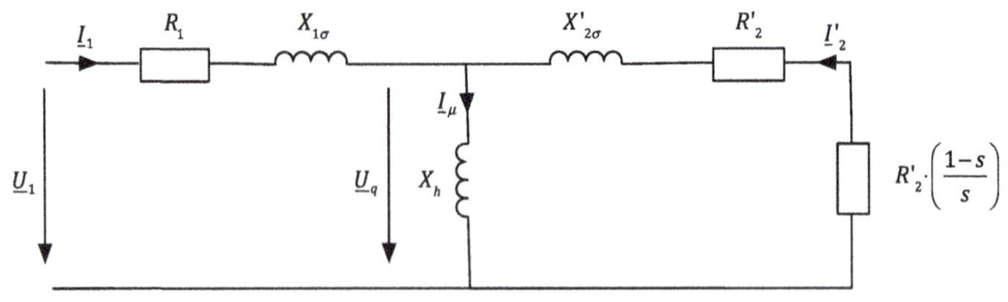

Fig. 9.1 Single-phase equivalent circuit diagram of the induction machine

rotor phase current $\underline{I}'_2$ can be converted to the actual currents in the individual bars using suitable formulas.

$X_{1\sigma}$ and $X'_{2\sigma}$ are the leakage reactances of the stator and rotor. They represent the fluxes that close purely in the stator or purely in the rotor and do not make it across the air gap to the other side. Depending on the design, the associated leakage inductances $L_{1\sigma}$ and $L'_{2\sigma}$ are around 2–5% of the main inductance L_h (remember: $X = \omega L$).

X_h is the main reactance. It represents the magnetic flux that passes between the stator and rotor. In the equivalent circuit diagram, X_h is caused by the magnetizing current $\underline{I}_\mu$. In the real machine, the main flux results from the superposition of the stator and rotor current fields. $\underline{I}_\mu$ is therefore a virtual current that is not actually present in the machine.

What remains is the terminating resistor $R'_2 \cdot (1 - s)/s$. This virtual resistance represents the mechanical power emitted (or absorbed) by the machine. Power in a resistance is calculated by the formula for current heat loss (I^2R). The following therefore applies:

$$P_2 = 3\left(\underline{I}'_2\right)^2 \cdot R'_2 \cdot \left(\frac{1-s}{s}\right) \tag{9.4}$$

In this formula, s is the slip introduced earlier. The factor 3 results from the three phases.

The two rotor resistances are often summarized in R'_2/s in the literature. This is mathematically sound, as both resistances are in a series connection. Physically, however, this combination is useless, as the two resistors have completely different meanings.

The main flux is obtained from:

$$\phi_h = k_3 \cdot \frac{U_q}{f_1} \tag{9.5}$$

When stator resistance R_1 and leakage inductance $L_{1\sigma}$ are small, the voltage across the main inductance U_q corresponds approximately to the terminal voltage U_1.

Equation (9.5) is then similar to Eq. (7.23), which applies to synchronous machines.

The internal torque of the induction machine is obtained from:

$$T_i = k_4 \cdot \phi_h \cdot I_1 \cdot \sin\left(\measuredangle\left(\underline{I}_\mu, \underline{I}_1\right)\right) \tag{9.6}$$

When the stator resistance R_1 is small, the magnetizing current $\underline{I}_\mu$ and the stator current $\underline{I}_1$ are perpendicular to each other, and the sine term becomes 1. In that case (9.6) is identical to (7.24). Therefore, induction and synchronous machines can be operated in a similar way using power electronics.

The two constants k_3 and k_4 contain the specifics of the stator winding, in particular, the number of phases, number of poles, and winding factor.

The maximum torque T_k of an induction machine (breakdown torque) is magnetically limited. It is

$$T_k = k_5 \cdot \left(\frac{U_1}{\omega}\right)^2 \tag{9.7}$$

This is also an analogy to the synchronous machine. The diagrams in Fig. 7.24 therefore apply equally to both types of machines.

9.2 Operating Characteristics

The torque curve over the speed can be approximated using the so-called Kloß equation:

$$T = T_k \cdot \frac{2}{\frac{s}{s_k} + \frac{s_k}{s}} \tag{9.8}$$

T_k is the breakdown torque, that is, the maximum torque the machine can produce at any given voltage and frequency.

s_k is the breakdown slip, i.e. the slip at which the breakdown torque is delivered by the machine. It is calculated approximately as

$$S_k \cong \frac{R_2'}{2\pi f_1 \cdot (L_{2\sigma}' + L_h)} \tag{9.9}$$

Please note that the stator resistance R_1 has been neglected in these formulas. This is only permissible for large machines where $R_1 \ll X_{1\sigma}$.

Equation (9.8) can be used to specify the torque–speed map of the induction machine (Fig. 9.2).

Each individual torque–speed curve is given for a fixed frequency and a fixed voltage. In the base speed range, the voltage is set in proportion to the frequency, for example, 400 V/100 Hz, 200 V/50 Hz, 100 V/25 Hz, and so on. In this way, the breakdown torque remains constant, and the curves are simply shifted to the left with the frequency.

In the field weakening range, the voltage is constant and only the frequency is increased further, for example, 400 V/100 Hz, 400 V/150 Hz, 400 V/200 Hz, and so on. The breakdown torque then decreases quadratically with increasing frequency and the shape of the curves changes accordingly.

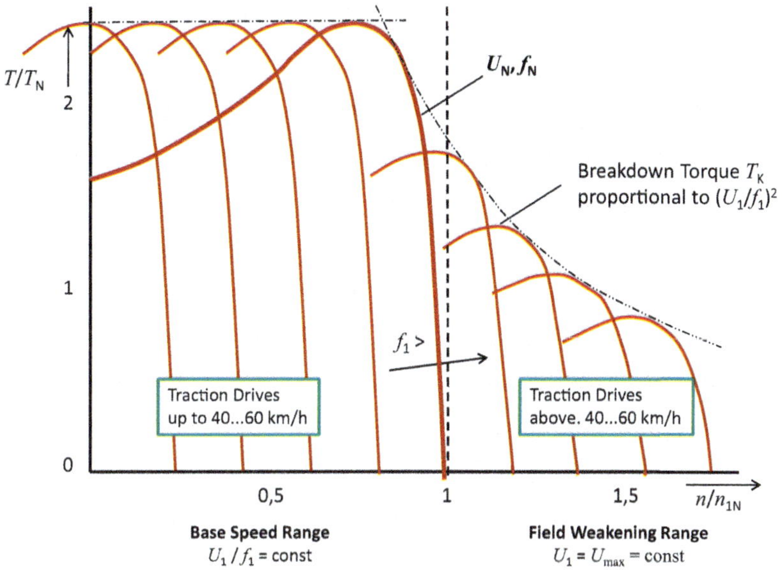

Fig. 9.2 Speed–torque characteristic diagram of an induction machine when operating on an inverter with variable frequency f_1 and frequency-dependent voltage U_1

Power Electronic Components 10

Abstract

This chapter presents the most important semiconductors and power electronic components that make up the assemblies required in hybrid and electric vehicles, in particular, DC/DC and DC/AC converters. These are primarily semiconductor valves (diodes) and power electronic switches (MOSFET, IGBT).

10.1 Semiconductors

Semiconductors have an electrical conductivity that lies between that of conductors (metals) and insulators. Most semiconductor devices used today are silicon based. Other semiconductor materials such as GaN (gallium nitride) and SiC (silicon carbide) are also used in large quantities in certain applications.

The conductivity of metals is only determined by the number of free electrons (concentration n). Some semiconductors also contain holes (electron-empty areas), also known as defect electrons (concentration p), which can also conduct electricity.

The current in bipolar components, i.e. diodes, bipolar transistors, and IGBTs, is made up of n-type and p-type conduction. MOSFETs are unipolar components that only use the n-type conduction.

Semiconductors as such do not conduct electricity and are therefore initially useless. Impurities are specifically added to the semiconductor by doping. There are two variants: n-doping (addition of electrons) and p-doping (addition of holes). This makes it possible to control the electrical properties of the semiconductor.

The charge carrier density, i.e. the number of electrons or holes per volume, is determined by doping the semiconductor crystal with foreign atoms. Silicon has an electron valence of 4, i.e. four valence electrons. The n-doping is carried out with 5-valent substances (phosphorus, arsenic, …). The excess electron can easily

M. Doppelbauer, *Introduction to Electromobility*,
https://doi.org/10.1007/978-3-658-45482-1_10

be removed from the compound. The p-doping is realized with trivalent substances (boron, aluminum, gallium, …).

By itself, a p- or n-doped region is electrically neutral. It acts like an ohmic resistor. If two differently doped semiconductor layers are brought together, a pn junction is created (diode). By combining more layers, much more complex electronic components can be produced.

Semiconductor components are available in various forms, namely, for signal processing applications or as power semiconductors. The former are only suitable for small currents and voltages but are very fast. Depending on the application, the switching frequency can be well into the GHz range. Such components are used, for example, in radio devices (cell phones) and form the basis of microprocessors and memory components.

Semiconductors can switch voltages up to well over 1000 V and electric power up to over 100 kW. The switching frequencies that can be achieved are lower than those of signal transistors and are up to around 10 kHz (IGBT) to 100 kHz (MOSFET). Much higher frequencies can be achieved with wide-bandgap semiconductors, which will be discussed in a later chapter.

10.1.1 Diodes

The combination of p- and n-doped zones is crucial for the function of all semiconductor components. In the boundary region, the excess electrons in the n-region are balanced out by an exchange of charge carriers (diffusion) with the holes in the p-region (diffusion current).

Within a very short time after production of the component and later when the external voltage is switched off, all freely moving charge carriers disappear from the boundary layer and it becomes an insulator (depletion zone). This insulating layer is called the space-charge region (SCR) or junction.

The width of the space-charge region depends on the doping. An electric field exists within the junction due to the separation of charge carriers, see Fig. 10.1, while the semiconductors are otherwise electrically neutral.

When an electrical voltage U_0 is applied to the outer edges of the semiconductor, the field strength E across the space-charge zone is either strengthened or weakened depending on the polarity. The width of the space-charge zone changes accordingly.

When the external voltage ($+U_0$) is opposite to the diffusion voltage and is sufficiently large, the space-charge region disappears completely. Its insulating effect is thus eliminated and a current flow i_F (forward current) becomes possible. This polarity is referred to as the forward direction. The internal resistance of the semiconductor becomes very small, so the current is determined only by the external resistance R.

When the external voltage is applied in the same direction as the diffusion voltage ($-U_0$), it further increases the space-charge region, and a current flow is not possible.

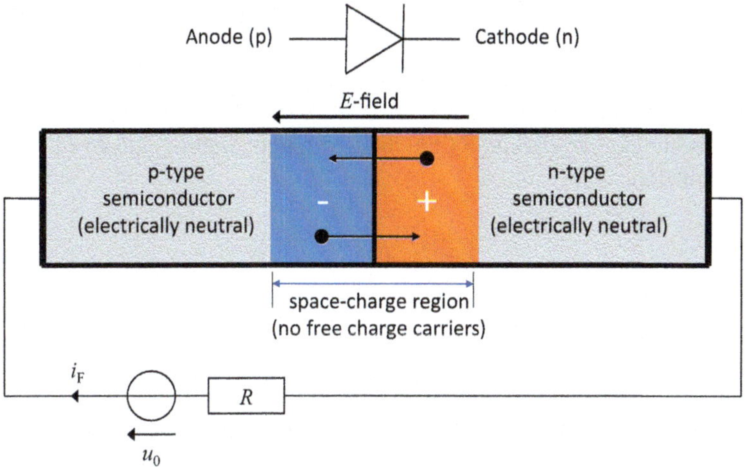

Fig. 10.1 pn junction (space-charge region) and external connection

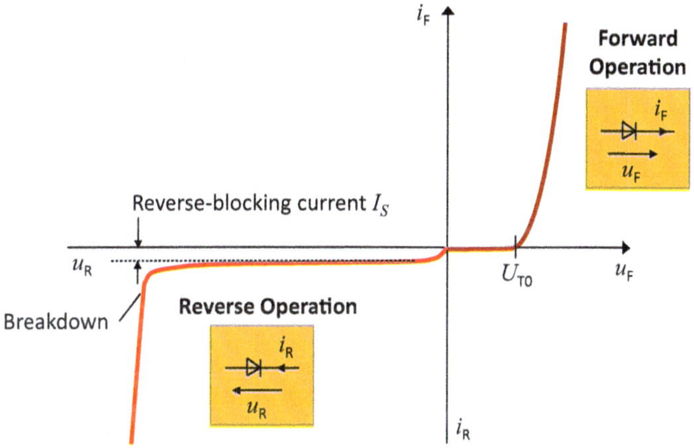

Fig. 10.2 Diode characteristic in blocking and pass-through mode

There is a small reverse-blocking current (i_R). This reverse saturation current is typically in the order of $i_S = 10^{-10}$ A. Therefore, it can be neglected in practical applications.

This pn junction results in a diode whose characteristic curve is shown in Fig. 10.2.

Note that the scale of the y-axis is very different in the forward (i_F) and reverse (i_R) directions. The reverse saturation current is very small.

The diode is a non-controllable semiconductor component that only allows current to flow in one direction. It acts like an electrical valve. The threshold voltage U_{TO} depends on the size of the space-charge region and thus on the amount of doping. It is typically 0.5–1.0 V.

The threshold characteristic of the diode (first quadrant) is idealized by the following equation:

$$I = I_s \cdot \left(e^{\frac{U}{N \cdot U_T}} - 1 \right) \tag{10.1}$$

with

I_S = reverse-blocking saturation current = app. 10^{-10} A
U_T = thermal voltage, app. 26 mV at 300 K (room temperature)
N = diode factor = 1 … 3

The permissible reverse current i_R is up to 10 million times smaller than the permissible forward current i_F. The full reverse saturation current is reached from blocking voltages of around 0.1 V. In addition, a temperature-dependent reverse current flows, which doubles approximately every 6 K.

The differential resistance in the forward range r_F is the ratio of forward voltage to forward current:

$$r_F = \frac{du_F}{di_F} = \frac{N \cdot U_T}{i_F} + R_s \tag{10.2}$$

R_S is the so-called ohmic path resistance.

If the maximum reverse voltage is exceeded ($u_R > U_Z$), a breakdown occurs, and the diode is destroyed. An exception is the so-called Z diode (Zener diode), which may be operated permanently in the reverse direction in the breakdown voltage range. However, Z diodes are only manufactured for low power ratings.

10.1.2 Power Electronic Switches

Unlike a diode, which always lets an electric current pass in one direction, controllable semiconductor valves act as electronic switches. All power electronic switches have three connections: two for conducting the motor current and a third for activation (on/off). Depending on the component, switching can be controlled either with a small current or with a small voltage.

The historically first power electronic switches were thyristors and triacs. These devices can be turned on by supplying a small ignition current. However, they cannot be turned off. Instead, when the main AC current reaches a zero crossing, the devices turn off by themselves. Afterwards, they must be switched on again if required.

Today, thyristors are only used in very few motor applications, mainly for controlling direct current machines, for example, in electric trams. However, they are still important in power electronics for high-voltage direct current networks.

Bipolar junction transistors (BJT) can be switched on and off at any time by the controller. They have three terminals, namely, C (Collector), E (Emitter), and B (Base). The main current flows from the collector to the emitter. Like a diode, the transistor does not allow a current flow in opposite direction.

The flow of the main current can be controlled by a small current into the base connection. Because of the control per current, compared to voltage-controlled elements, the activation of bipolar transistors is more complex.

To increase the capability to withstand voltage, bipolar power transistors (BPT) are applied with an additional, weak doped n-zone in the n-collector. This results in a reduced current amplification, which in turns yields to a further increase of activation effort compared to a signal transistor.

By connecting multiple transistors in series, the current amplification is increased. This so-called Darlington connection is often used in power transistors.

A disadvantage of the BPT is the relatively large, frequency-proportional switching losses, which must be dissipated as heat.

Today, bipolar transistors are in billions as signal processors and for linear amplifiers, for example, in audio circuits. Due to their high switching losses, bipolar power transistors play no role in the field of electrical drive technology.

The two components that are important for electrical drives are MOSFETs and IGBTs.

10.1.2.1 MOSFETs

A MOSFET is a unipolar component. In the typical n-type components, the electric current is only conducted via electrons and not via holes.

Figure 10.3 shows the basic semiconductor structure of a MOSFET (metal-oxide field-effect transistor).

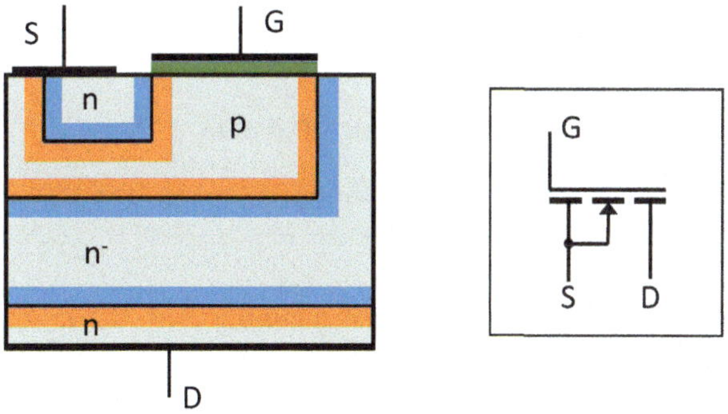

Fig. 10.3 Structure and circuit symbol of a MOSFET

AOK MOSFET has three connections, which are labeled Source (S), Drain (D), and Gate (G). The current is conducted between the source and drain. In practical devices, MOSFETs consist of many identical elements which are connected in parallel.

The gate is separated from the p-doped substrate by an insulating layer of metal oxide. The conductivity of the p-channel under the gate is controlled by an electric field that is built up by applying a voltage between the gate and source.

When a positive voltage U_{GS} is applied, the structure behaves like a plate capacitor and a negative induced charge, i.e. electrons, is formed in the substrate. These cause the channel to become electrically conductive and the current can flow from source to drain. In that case, the voltage U_{DS} must be>0. When the control voltage U_{GS} is switched off, there are no more induced charges and the channel from source to drain becomes non-conductive.

However, when a negative voltage is applied between the source and drain ($U_{DS}<0$), the channel is always conductive (reverse conductivity). What initially appears to be a disadvantage often proves to be beneficial in practice because this reverse conductivity is required in many circuits, see the following section.

A MOSFET has only low switching losses due to the voltage control instead of current control. High switching frequencies can therefore be achieved.

The on-resistance of a MOSFET becomes relatively high if the component is designed for high voltages, as the n-zone must then be made wider. The typical application range of Si-MOSFETs is therefore up to around 200 V, although individual components are already available up to 650 V (super-junction MOSFETs).

Silicon carbide SiC-MOSFETs are available up to much higher voltages (2200 V). They are now becoming quite popular, see Sect. 10.2.2

Table 10.1 summarizes the most important properties of MOSFETs.

In power electronic applications, the device is used either in its on-state (OP2) or in its off-state (OP1) but never in the linear range, see the following illustration. Even though a MOSFET is a voltage-controlled element by design, a high gate current is required for a short time during switching. This is to quickly recharge the internal capacities and to build up the U_{GS} voltage to switch between the two states as quickly as possible.

Table 10.1 Properties of commercially available MOSFETs

n-MOSFET
Voltage-controlled switch
Unipolar charge transport by electrons only
No charge storage, therefore frequency-independent losses
Positive temperature coefficient
High on-resistance at high currents due to small charge injection
Blocking voltages up to 600 V, up to 2200 V for SiC-MOSFETs
Currents up to 300 A

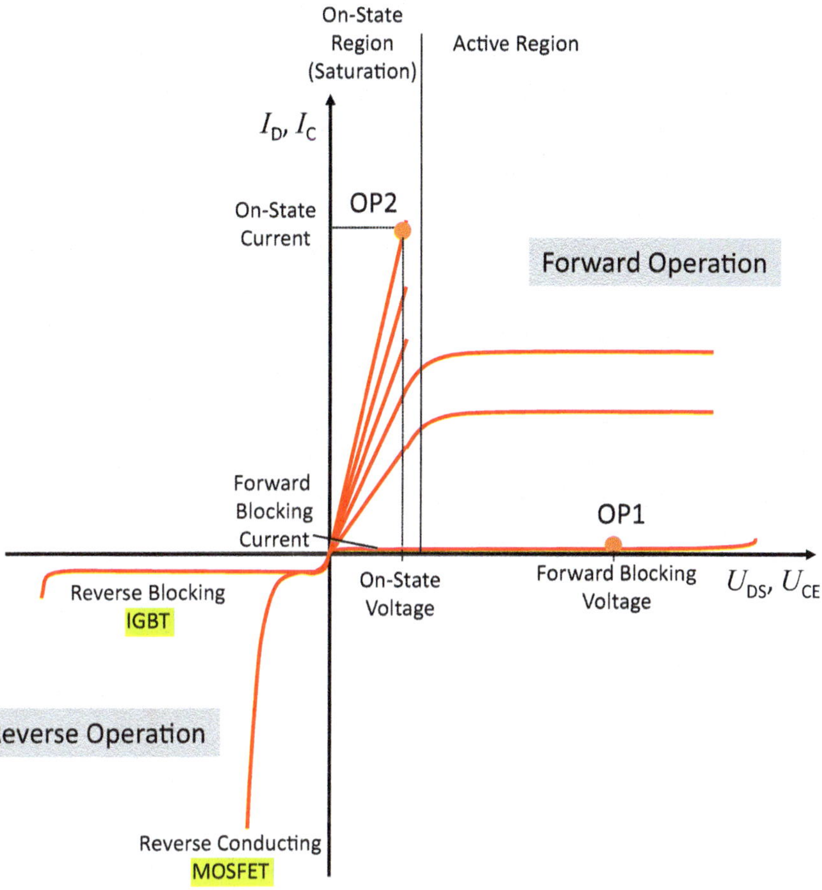

Fig. 10.4 Operating characteristics of MOSFETs and IGBT with antiparallel diode

$U_{GS(th)}$ is the threshold voltage of the MOSFET. Depending on the design it is typically between 1 and 3 V (Fig. 10.4 and Table 10.2).

10.1.2.2 Freewheeling

When switching off inductive loads like electric motors it is essential that the circuit is conductive in the reverse direction. Otherwise, the magnetic energy stored in the coil will lead to a high voltage across the contacts of the switch or the semiconductor. This voltage must be avoided so that the semiconductor switches and the insulation of the coils are not destroyed by a voltage breakdown.

Freewheeling diodes are arranged antiparallel to the semiconductor switch. When the switch turns off, the current is commutated and taken over by the diode. As a result, no high voltages can build up across the semiconductor. As no new energy is supplied to the coil after the current has been commutated to the diode,

Table 10.2 Operating modes of power MOSFETs

Operation type	Control voltage	Operational behavior
Forward operation $U_{DS}>0$ $I_D>0$	Switched-off $U_{GS}=0$	Blocking-state (OP1) I_D = nearly zero (small blocking current)
	Low voltage $U_{GS(th)}>U_{GS}>0$	Active region No continuous operation due to high losses
	Switched-on $U_{GS}>U_{GS(th)}>0$	On-state (OP2) Low ohmic resistance; I_D determined by external circuits
Reverse operation $U_{DS}<0$ $I_D<0$	Switched-off $U_{GS}<U_{GS(th)}$	Reverse conduction The intrinsic body diode acts as a freewheeling diode and passes currents (app. 0.8 V drop)
	Switched-on $U_{GS}>U_{GS(th)}>0$	Reverse conduction Small ohmic losses depending on the design

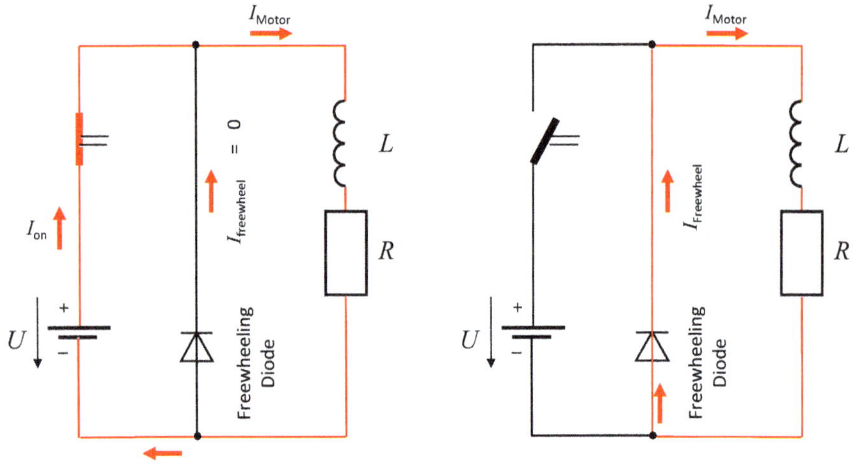

Fig. 10.5 Inductive load fed by a battery in on-state (left) and in freewheeling off-state (right)

the freewheeling current decreases rapidly due to the ohmic resistances and the inductances in the circuit (Fig. 10.5).

Every MOSFET has an integrated (intrinsic) built-in reverse diode due to its design. In reverse operation ($U_{DS}<0$), this diode becomes conductive and the MOSFET does not block the current. However, the intrinsic diode is about three times worse than an external, separate diode. Due to the semiconductor structure, it is difficult to optimize both the forward and reverse diodes simultaneously. An external diode with improved switching behavior is therefore usually connected in parallel with the body diode (Fig. 10.6).

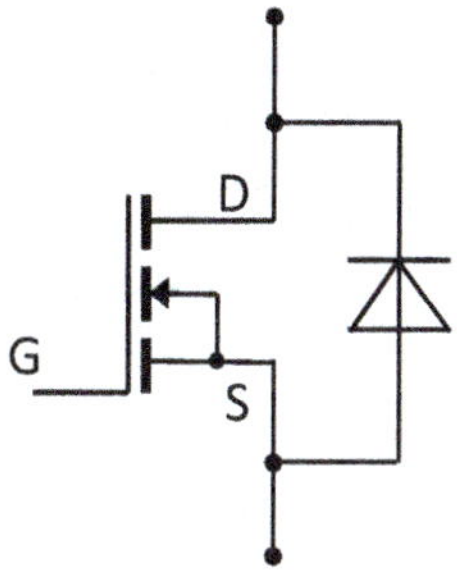

Fig. 10.6 Basic circuit of an n-MOSFET with external freewheeling diode

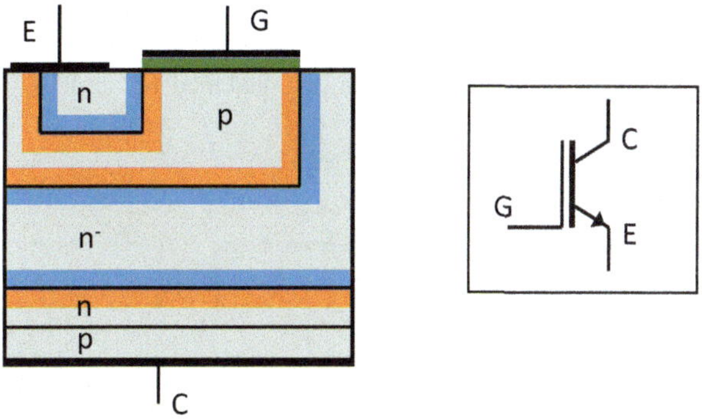

Fig. 10.7 Structure and circuit symbol of an IGBT

10.1.2.3 IGBTs

Although the unipolar MOSFET can be controlled almost powerless and achieves high switching frequencies, it has comparatively high conduction losses. Conduction losses are significantly lower in bipolar components, but current-controlled activation is only possible at relatively low frequencies and higher switching losses occur.

IGBTs (insulated-gate bipolar transistors) are a mixture of bipolar transistors (good conduction behavior) and MOSFETs (powerless and fast voltage control). The structure is like that of a MOSFET, but there is an additional p-layer in front of the collector.

For operating voltages above 600 V, IGBTs are most important components in power electronics today. Their structure is shown in Fig. 10.7.

The connections of the IGBT are labeled collector (C), emitter (E), and gate (G). This already suggests that it is a combination of MOSFET (gate) and bipolar transistor (collector, emitter).

In simplified terms, the IGBT can be imagined as a pnp bipolar transistor whose base current is controlled via a MOSFET, as illustrated in Fig. 10.8.

The conduction behavior of the IGBT is determined by the bipolar transistor. In the reverse direction, the base–emitter (BE) diode blocks the current. Therefore,

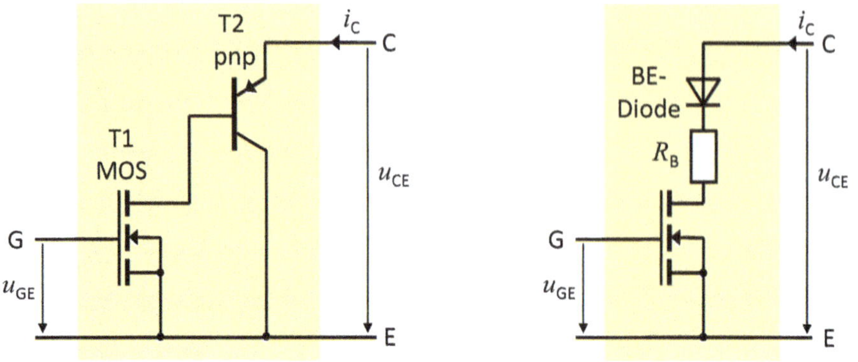

Fig. 10.8 Simplified equivalent circuit diagram (left) and model of an IGBT for obtaining the characteristic curve

Fig. 10.9 Section of a semiconductor module with three chips connected in parallel, each consisting of an IGBT (large rectangle) and a freewheeling diode (small rectangle)

unlike a MOSFET, an external freewheeling diode is always required parallel to IGBTs. MOSFETs and IGBTs can both conduct a short-circuit current for a short time and switch it off safely.

In practice, power semiconductors for higher currents are constructed internally from several IGBT chips that are all connected in parallel (Fig. 10.9).

The base–emitter diode of the IGBT leads to a higher voltage drop at low current loads than in a MOSFET. MOSFETs are therefore the preferred component for lower voltages (Fig. 10.10).

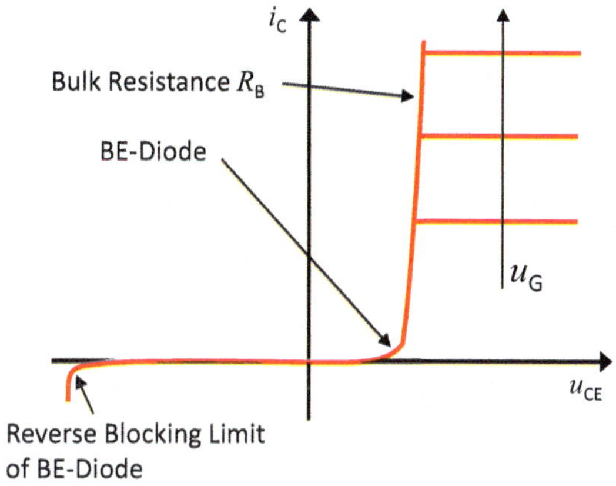

Fig. 10.10 Characteristic curve of an IGBT

As the current load increases, charge carrier flooding occurs in the IGBT in the weakly doped n-region, whose conductivity thus increases. As a result, the on-state resistance (and also the voltage drop) of the IGBT becomes increasingly favorable. Similar to the MOSFET, the IGBT is only used when fully switched on or off, not in the active region.

10.1.2.4 Conduction and Switching Losses

As already mentioned, semiconductors are not ideal switches. When IGBTs are switched on, some time passes (t_d=delay time) before the collector current has reached its final value. The switch-on time is determined between the 10 and 90% mark of the collector current.

During this time, a voltage is already building up between the collector and the emitter u_{CE}. The product of i_c and u_{CE} results in an instantaneous value of the power loss p_V. The switch-on loss energy W is obtained by integration:

$$p_v = u_{CE} \cdot i_c \tag{10.3}$$

$$W = \int_t p_v \, \mathrm{d}t \tag{10.4}$$

When switched on, a small saturation voltage u_{CEsat} remains between collector and emitter. This voltage leads to conduction losses. They cannot be avoided and result from the electrical resistance of the doped semiconductor material.

To initiate the switch-off process, the polarity of the gate emitter voltage is first reversed. After a certain delay time, the collector–emitter voltage u_{CE} builds up gradually. The collector current only begins to drop when the collector–emitter voltage has reached the value of the DC link voltage (battery voltage), because only then can the opposite diode take over the output current.

The switch-off loss energy is obtained by integrating the instantaneous values of the switching power.

$$P_s = (W_{on} + W_{off}) \cdot f_s \tag{10.5}$$

The total losses, which are the sum of the average conduction and switching losses, are important for the thermal design of an IGBT:

$$P_v = P_D + P_s \tag{10.6}$$

In practice, the switching frequency is selected so that both loss components are approximately the same value. As the switching losses increase proportionally with the number of switching operations and therefore proportionally to the switching frequency, this results in a typical switching frequency of 4–8 kHz for IGBTs.

Compared to the MOSFET, the decay process of the current is several times longer in an IGBT and the switching losses are correspondingly higher. The permissible switching frequency of the IGBT is therefore lower than that of the MOSFET.

Figure 10.11 shows the switch-on and switch-off behaviors of an IGBT and its freewheeling diode.

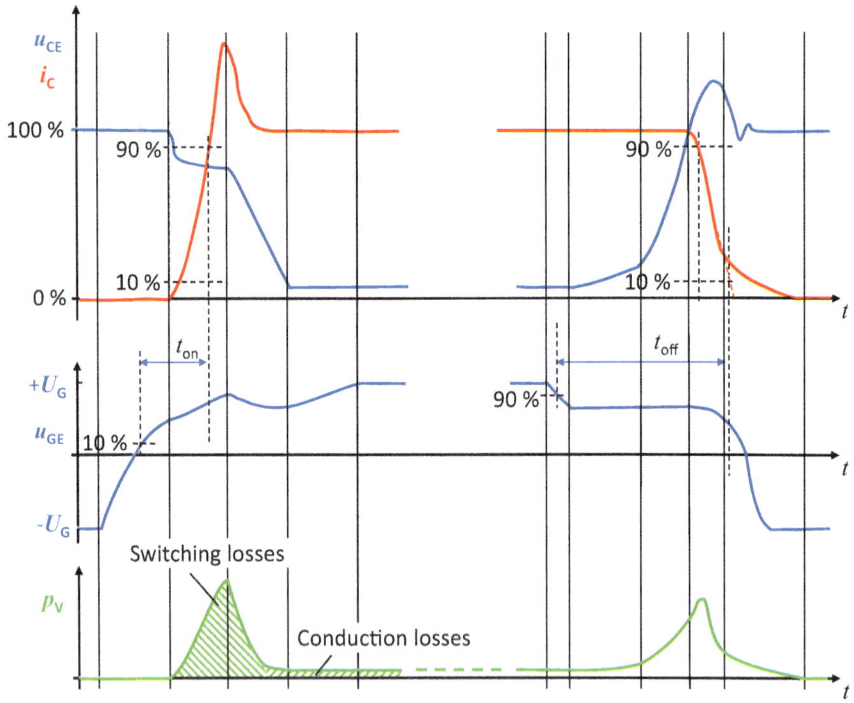

Fig. 10.11 Switch-on and switch-off behaviors of an IGBT and the freewheeling diode

The upper diagram shows the forward voltage u_{CE} and the forward current i_C. The diagram in the middle shows the corresponding control voltage u_G (gate voltage). The switching and conduction losses p_V are plotted at the bottom.

10.1.2.5 Standard IGBT Components

A wide variety of IGBTs are available today. The voltage classes 600, 1200, and 1700 V have become established for industrial applications. IGBTs are also available in higher voltage classes, but with much lower production quantities.

The voltage class corresponds to the highest permissible blocking voltage. A certain safety margin is required in relation to the DC link voltage. As can be seen in Fig. 10.11, there is a voltage peak during the switch-off process that must be safely below the blocking voltage. Therefore, a system voltage of 800 V requires 1200 V components. This corresponds to an AC motor voltage of around 570 V_{eff}. Accordingly, a system voltage of up to 400 V can be operated with 600 V components, which corresponds to an AC motor voltage of around 280 V_{eff}.

The ratio of voltage to current is referred to as R_{DSon} = resistance between drain and source when switched on. An IGBT has a significantly lower forward voltage drop compared to a conventional MOSFET for components with a higher reverse voltage. However, MOSFETs have a lower forward voltage at lower current densities. The slope of the forward voltage drop curve can be changed with the chip area of the semiconductor. In the partial load range, the unipolar MOSFET is advantageous because the voltage drop caused by the pn junction, i.e. the buck voltage of around 0.7 V, of bipolar IGBTs is eliminated.

The saturation voltage for 600 V IGBTs is typically 1.5 V, for 1200 V IGBTs typically approx. 2.5 V.

10.1.2.6 Power Semiconductor Modules

As will be explained later, six IGBTs and the corresponding antiparallel connected freewheeling diodes are required for a three-phase bridge circuit (DC/AC converter).

Due to the wide range of applications for this circuit, modules have become established on the market that combine all the required components in one assembly. These power semiconductor modules contain all the necessary chips in a common housing, mounted on a heat sink. They are interconnected by bonding wires, usually made of aluminum or copper.

In the lower power range, up to several 100 kW, the modules also contain other components, for example, for current measurement and potential-free control directly from the digital output of a microprocessor.

Modules for higher outputs only contain the semiconductors, i.e. six IGBTs and six diodes. All other components must be connected externally.

Figure 10.12 shows a cross section of a power semiconductor module.

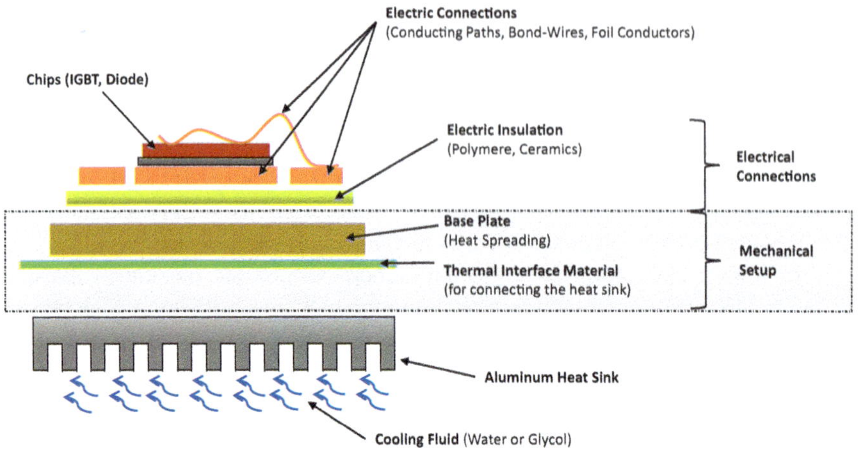

Fig. 10.12 Schematic structure of a power semiconductor module

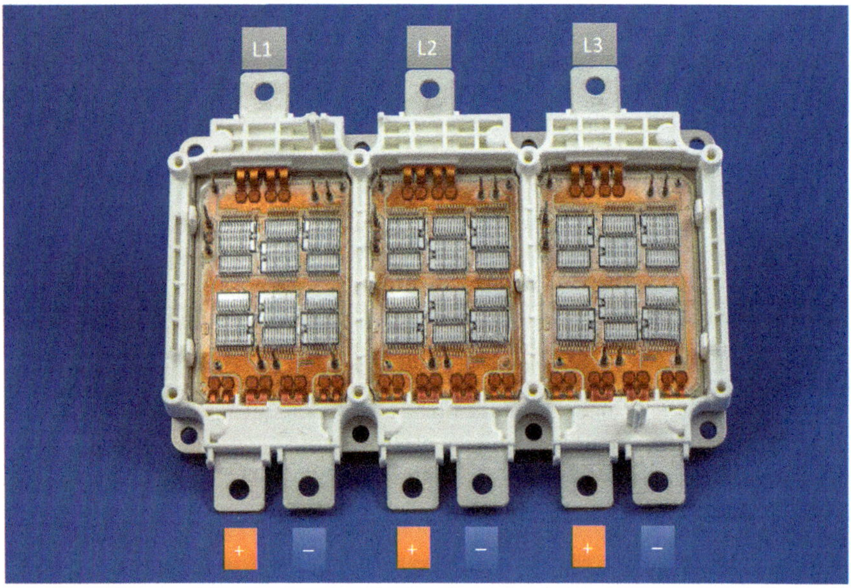

Fig. 10.13 Opened power semiconductor module (Infineon HybridPack) with the motor and battery connections

Figure 10.13 shows the inside of a power semiconductor module with three half bridges. Each bridge consists of three IGBT and diode units connected in parallel to increase the current-carrying capacity. The battery connections are separate for each half bridge and must be connected externally. The three-phase connections for the motor are located at the top. The remaining, smaller contacts are used to control signals and the built-in temperature sensors.

Example 10.1—Semiconductor modules

The semiconductor chips of the module in Fig. 10.14 are specified for a maximum reverse voltage $U_{CES} = 750$ V and a rated current $I_{CN} = 820$ A. The maximum current that can be switched off is $I_{CRM} = 2 \cdot I_{CN} = 1640$ A. This is the maximum reliable value of the motor current per phase including any current ripples. In practice, a sufficient safety margin must be maintained from this value.

The "Maximum RMS module terminal current" (highest effective AC phase current of the module) is specified as 500 A_{eff}, i.e. 707 A_{peak}. The value is limited by the module connections and is slightly below the rated current of the semiconductor chips.

When operated on a battery with a nominal voltage of 400 V, the module can therefore provide a continuous power of $P = \sqrt{3} \cdot 400\ \text{V}/\sqrt{2} \cdot 500\ \text{A} = 245$ kVA. With a standard power factor of $\cos \varphi = 0.8$, this is just under 200 kW motor power. When the battery voltage drops with its state of charge, the maximum possible output power is reduced accordingly.

However, as the data from the semiconductor chips show, the module can be overloaded for a few seconds and, for example, provide significantly more current and therefore more torque for quick starts.

The module would also allow for higher battery voltages of up to around 500 V possible. However, such voltages are unusual in practice.

Note:

The maximum possible conductor-to-conductor voltage is the DC voltage of the battery divided by $\sqrt{2}$, as shown in the estimate above. This is quite a bit more than would be expected in theory. However, the modulation process in the inverter does not generate a pure sine wave but is adding a third harmonic. The sine wave is thus flattened in the upper range (Fig. 10.14). ◀

The different thermal expansions of the individual layers of the module cause thermal stresses during operation, which ultimately lead to material fatigue and wear. The service life of power semiconductor modules is for the most part limited by the number of temperature cycles.

With typical temperature cycles of around 100 K, the nominal service life of a module is around 1000 cycles. Continuous operation at a constant temperature, i.e. long, steady highway driving, has hardly any impact on the module's lifetime.

For example, when the module is heated up from 20 °C ambient temperature to 120 °C module temperature once per day, then the calculated service life is reached after around 3 years.

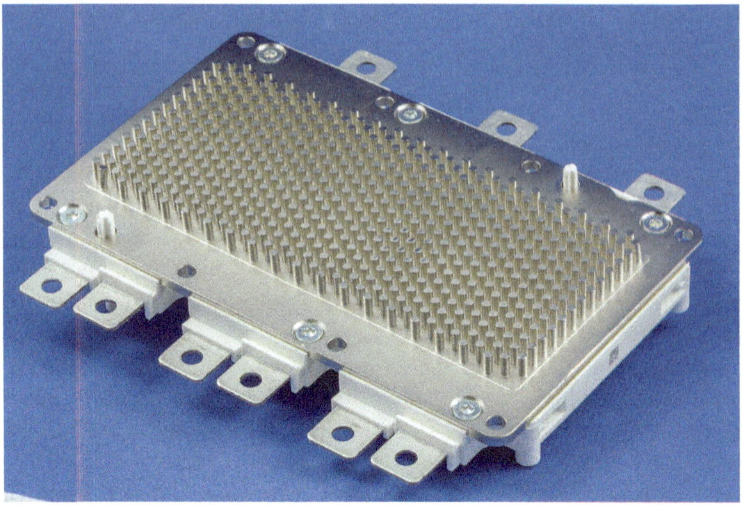

Fig. 10.14 Flip side of a power semiconductor module (Infineon HybridPack) with direct cooling via pin–fin technology

Oversizing modules can reduce the temperature cycle and extend the service life. At a temperature cycle of 70 K, for example, the nominal service life increases to around 10,000 cycles.

10.2 Wide Bandgap Semiconductors

Two new other semiconductor materials have become established in recent years and are increasingly being used in series products. These are gallium nitride (GaN) and silicon carbide (SiC). They are known as wide-bandgap (WBG) semiconductors because the energy gap between the valence band and the conduction band is greater than that of silicon.

The increased bandgap increases the breakdown field strength and allows thinner components for higher voltages. This in turn reduces switching losses, improves thermal conductivity, and enables higher switching frequencies.

SiC semiconductors enable switching frequencies around 100 kHz and GaN semiconductors even into the MHz range. This has advantages in terms of size, cost, and durability of the passive components, especially the inductors. Due to $X_L = \omega L$, when the frequency ω is increased, the inductance L can be reduced. Smaller inductances are cheaper and reduce EMC problems (EMC = electromagnetic compatibility). Switching frequencies above 20 kHz are particularly important in the automotive sector because audible noises are then eliminated (Table 10.3).

Table 10.3 Selected properties of silicon (Si), silicon carbide (SiC), and gallium nitride (GaN) semiconductors

	Bandgap in eV	Breakthrough field strength in MV/cm	Thermal conductivity in W/(m·K)
Si	1.12	0.3	150
SiC	3.26	3.5	490
GaN	3.4	2.6	130

Note

According to the band model of quantum mechanics, electrons are only permitted on certain intervals of the energy scale.

The gap between the highest band still fully occupied by electrons (valence band) and the next highest band (conduction band) is important.

Only excited electrons in the conduction band can move practically freely through a solid and contribute to electrical conductivity. At finite temperatures there are always some electrons in the conduction band due to thermal excitation, but their number varies greatly with the size of the bandgap.

The bandgap is therefore used to classify conductors, semiconductors, and isolators. The exact limits are fuzzy, but the following can be used as a rule of thumb: conductors have no bandgap. Semiconductors have a bandgap in the range from 0.1 to about 4 eV. Non-conductors have a bandgap greater than 4 eV.

10.2.1 Gallium Nitride (GaN)

Gallium nitride is a semiconductor consisting of gallium and nitrogen. The voltages of today's GaN components are typically around 100 V, at most up to 650 V. So-called HEMTs (high electron mobility transistors) are usually made from GaN.

In contrast to all other semiconductor components, the structure of GaN HEMTs is lateral, meaning that the current flows across the component, i.e. parallel to the circuit board. Nowadays, these components are used for low power and switching frequencies in the MHz range, for example, for switching power supplies. They are also used millions of times to control lamps with light-emitting diodes (LEDs).

The use of GaN semiconductors in drives with higher power, as required for the automotive sector, is not expected in the foreseeable future.

10.2.2 Silicon Carbide (SiC)

SiC is a chemical compound of silicon and carbon. It has a good thermal resistance.

Silicon components for switching voltages above 650 V are always bipolar components, i.e. IGBTs, because the on-state resistance (R_{DSon}) increases with approximately the 2.5th power of the on-state voltage. MOSFET silicon components are only available up to around 650 V. The losses of silicon MOSFETs for higher voltages would be too high.

However, SiC-MOSFETs can be manufactured with higher voltages up to 2200 V (and in future possibly up to 3300 V and more), because the on-state resistance of SiC is lower and the blocking voltage is higher.

The advantage of MOSFETs over IGBTs is the elimination of the buck voltage (0.7 V pn junction). This results in lower losses, particularly at low currents in the partial load range. This is important for automotive applications, as peak power is only required in a few operating states and partial load operation is decisive for driving cycle efficiency.

In principle, SiC semiconductors allow higher junction temperatures than silicon semiconductors. The junction temperature of Si components can be up to 175 °C and more for SiC it can be around 220 °C.

A problem, however, is the assembly and connection technology, i.e. the connection of the semiconductor chip to the heat sink and the bonding of the connecting wires, see Fig. 10.13. Because different materials with different thermal expansion coefficients are in direct contact in the semiconductor chip, different expansions occur during temperature load changes, which lead to the delamination of soldered connections and the detachment of the bonding wires. Due to the high load cycling stress caused by the driving cycle, the maximum semiconductor temperature can hardly be utilized even with silicon semiconductors.

The challenges in using the new semiconductor materials lie in production and costs. SiC components are also built on wafers, but they are more prone to defects (impurities in the crystal structure) and are technically more complex than silicon wafers. Also, SiC wafers are smaller: up to 200 mm compared to 300 mm for the largest Si wafer. A smaller wafer causes almost the same processing costs but delivers fewer components.

SiC components (diodes and MOSFETs) are now commonly available up to 2200 V commercially. They are suitable for the construction of high-power inverters (DC/AC converters) with high switching frequencies in the 100 kHz range. Several car manufacturers are already using SiC-based inverters, especially in 800 V cars. It is expected that the power semiconductor market for cars will have largely switched to SiC in a few years' time (Fig. 10.15).

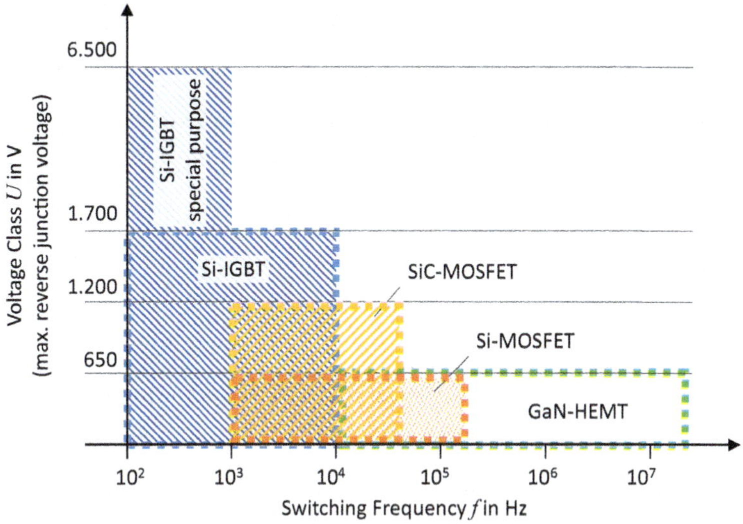

Fig. 10.15 Overview of the application areas of commercially available MOSFETs and IGBTs

10.2.3 Gallium Arsenide (GaAs)

Gallium arsenide (GaAs) is not a wide-bandgap semiconductor in the true sense of the word, as the bandgap is only 1.42 V. This III–V semiconductor material is quite expensive due to the high cost of the raw materials and the relatively complex process for manufacturing the single crystals required for chip production.

There are some applications for gallium arsenide semiconductors, but they are all outside the field of energy technology. High electron mobility transistors and GaAs gun diodes are used, for example, for low-noise high-frequency amplifiers in mobile telephones, in satellite communications, or in radar systems. There are light-emitting and laser diodes made of gallium arsenide. GaAs can also be used as a base material for certain types of photovoltaic cells.

Topology of Electric Drivetrains 11

Abstract

In this chapter, the basic structure and function of the main components of the electric drivetrain, namely, power electronic components are presented in a brief overview. The connection between the battery (DC) and the motor (AC) is established via power electronic energy converters. A distinction is made between DC/AC and DC/DC converters. This chapter presents the various switching and control concepts. Further sections are devoted to the control of current, voltage, torque, and speed of the two types of electrical machine (synchronous and induction motor).

11.1 Drivetrain Components

Figure 11.1 gives an overview of the basic structure of the electric drive system of battery electric and fuel cell vehicles. Hybrid vehicles are similar in design but have lower voltages. As an example, two electric motors were assumed, although most vehicles only have one motor. But there are also cars with three or four electric motors.

In general, the HV voltage of the battery (DC) is converted via an inverter (DC/AC converter) into a three-phase alternating voltage with adjustable amplitude and frequency (AC) for the motor. Typical operating ranges are 0–400/800 V AC and 0–1000 Hz, sometimes also higher voltages and higher frequencies. If the vehicle

M. Doppelbauer, *Introduction to Electromobility*,
https://doi.org/10.1007/978-3-658-45482-1_11

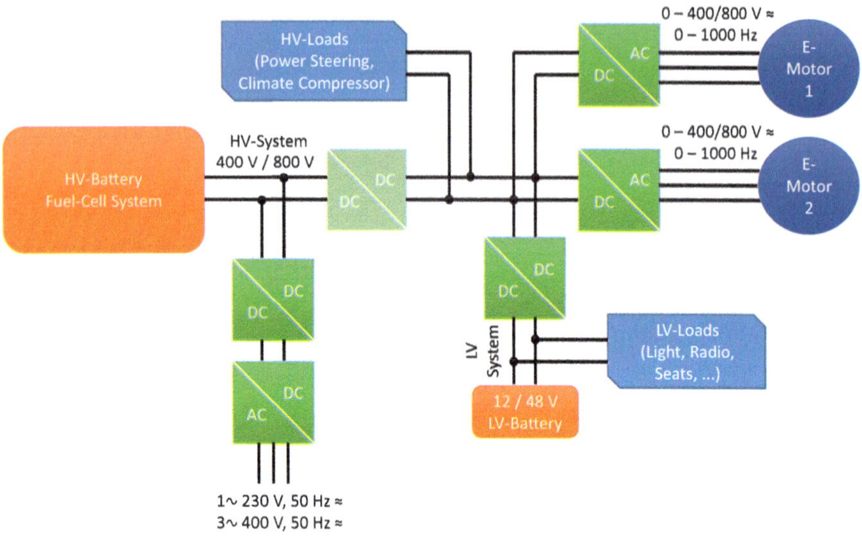

Fig. 11.1 Topology of an electric drive system

has more than one traction motor, each one must be equipped with its own DC/AC converter.[1]

The energy for the 12 V/24 V or 48 V LV system is provided by a separate battery. For charging, a DC/DC converter is used that connects the LV battery to the HV system. This converter is usually unidirectional, which means it can only convert energy from the HV level to charge the LV battery.

In recent years, there has been an increasing trend towards connecting particularly powerful consumers directly to the high-voltage grid. These can be, for example, the drives of the air conditioning compressor or power steering.

The topology of the electric drive system is completed by the charging circuit. Batteries are always charged with a regulated direct current (DC). Depending on where the rectification of the AC mains voltage takes place, a distinction is made between AC charging, in which the AC/DC converter is in the vehicle, and DC charging, in which the AC/DC converter is located in the charging station.

The voltage of electrochemical storage systems such as lithium–ion batteries or nickel–metal hydride batteries depends on the state of charge. It can drop to around 80% of the nominal value when the battery is fully discharged. In addition, a load-dependent voltage loss occurs, but this is only small due to the low internal

[1] In the past, it was common for electric trains to supply several motors with one inverter (group drive). However, this is only possible with asynchronous motors and worsens the control behavior. For this reason, group drives are not used in electric cars.

resistance of the batteries. Therefore, the unregulated voltage of the battery is connected directly to the HV system.

The entire HV power electronics are bidirectional, meaning that they can also be used for energy recovery when braking the electric motor.

The structure of an electrical system for fuel cell vehicles is similar to that for battery electric vehicles. However, an additional HV-DC/DC converter is required to stabilize the voltage, which fluctuates greatly depending on the load. HV-DC/DC converters have a high efficiency, typically 95–98%, but generate continuous losses and are also complex, i.e. large and expensive. In addition to the fuel cell, an additional HV battery is required (not shown in the picture) to absorb the recuperated energy, as the fuel cell system is not bidirectional. This battery is also used to start up the fuel cell system (cold start) and to bridge the dead time of the fuel cell during rapid acceleration.

The following illustration shows an inverter (AC/DC converter) for battery electric cars with up to 450 V DC voltage.

11.2 DC/AC Converter (Frequency Inverter)

The inverter is the heart and brain of an electric drive, the electric motor is its muscle. It provides the translation between the DC voltage of the battery and the AC voltage at the motor terminals. The inverter also contains the motor controller, which generates suitable voltage curves from the torque setpoint (accelerator pedal position) so that the motor accelerates or brakes the vehicle as desired.

A DC/AC converter (three-phase bridge circuit) can set the desired alternating voltages in a wide frequency and amplitude range. It works both ways, i.e. it can supply energy to the motor during driving (DC to AC) and feed energy back into the battery during braking energy recovery (AC to DC).

11.2.1 Fundamentals of PWM

The function of the bridge circuit is based on the principle of chopping: a DC voltage is broken down into on and off phases by rapid switching. The longer the switch-off phases are, the lower the effective value of the output voltage. The longer the switch-on phases, the higher the output voltage. The desired sinusoidal output voltage is produced in the moving average.

Chopping occurs at a fixed time interval (clock rate). Typical clock frequencies are 4–16 kHz, typical frequencies of the AC voltage to the motor terminals (base frequencies) are 0–1000 Hz, or higher depending on the number of poles of the motor and the desired speed. Inverters based on SiC semiconductors achieve clock frequencies up to 100 kHz. The higher the switching frequency, the lower the harmonics in the current and the associated additional losses, torque oscillations, and vibrations (noise) in the motor.

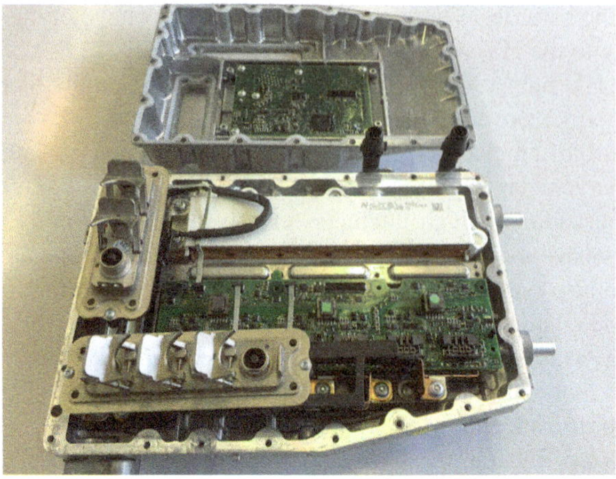

Fig. 11.2 Inverter for a battery electric vehicle

Figure 11.2 shows the basic principle. The desired voltage curve (green) is approximated by switching the DC voltage (red) between +1 and −1. If the desired voltage is positive, wide pulses with positive polarity (+1) are generated and only narrow pulses with negative polarity (−1). At zero crossing of the target voltage, the positive and negative pulse widths are the same. If the setpoint voltage is negative, the pulse width with negative polarity is wider than that with positive polarity. This procedure is known as pulse width modulation (PWM) (Fig. 11.3).

In this example, a sinusoidal voltage with 1.5 Hz and an amplitude of 0.7 was approximated by a PWM with a clock frequency of 20 Hz, i.e. 40 switching operations per second.

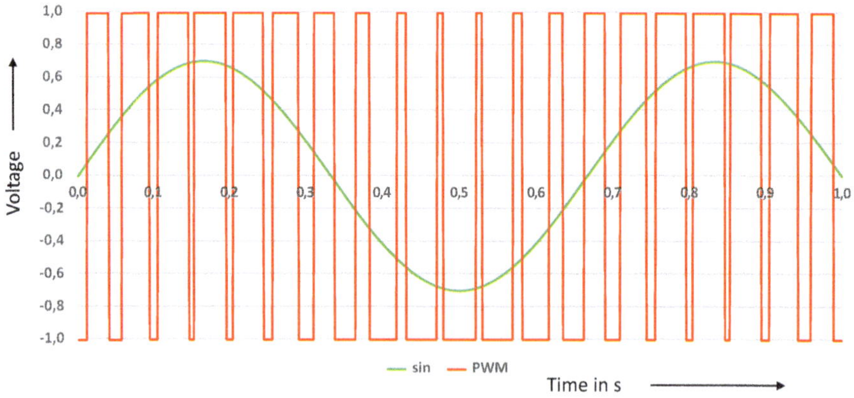

Fig. 11.3 Approximation of the desired sinusoidal voltage (green) by chopping a DC voltage (red)

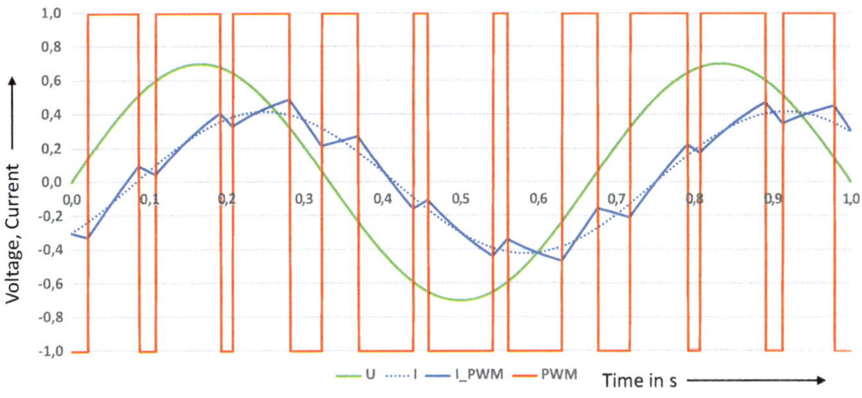

Fig. 11.4 Current build-up and reduction (blue) with PWM clocking of the terminal voltage (red)

Even though the effective values are similar, the red voltage looks nothing like the target voltage (green) as it contains many harmonics. To achieve a clean, sinusoidal output voltage, powerful low-pass filters would have to be added. However, a sinusoidal current is much more important than a sinusoidal voltage as the current ultimately generates the torque and mechanical power in the motor. Fortunately, the coils in the motor already act as a low-pass filter for currents. Therefore, additional filtering of the voltage is not required to achieve a smooth current curve, as shown in Fig. 11.4.

The current rises or falls with the difference between the internal voltage (sinusoidal synchronous generated voltage) and the external voltage (PWM-modulated terminal voltage). The slope of the current is determined by the inductance of the motor winding. A high inductance leads to a slower gradient and small inductances generate faster current changes. The illustration shows how the actual current (blue) follows the sinusoidal target curve (blue dotted line) very closely. The more cycles per sinusoidal period are performed, the better this is achieved. As a rule of thumb, the clock frequency should be at least 10 times the highest fundamental frequency. If the motor is to be fed with up to 1200 Hz, for example, then the PWM clock must be set to at least 12 kHz.

11.2.2 Single-Phase Circuit

The basic concept of switching a battery voltage to a motor winding can be seen in Fig. 11.5.

This circuit has four operating states that always alternate (Fig. 11.6).

At first, the upper semiconductor is turned on and the voltage across the inductive load becomes +200 V (top left). Then, the switch is turned off and the current is commutated to the freewheeling diode of the lower circuit (top right). The voltage

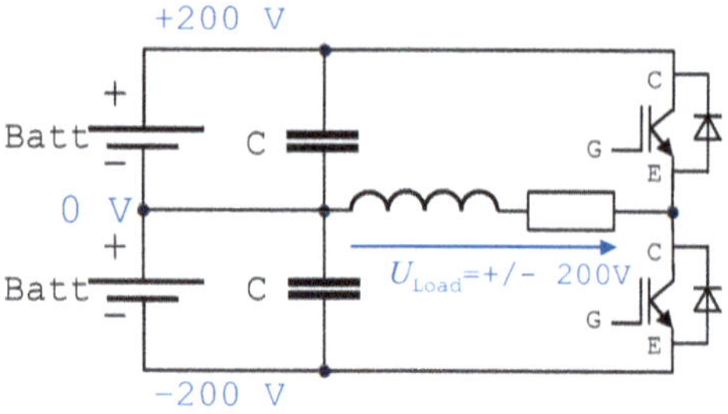

Fig. 11.5 Circuit of a single-phase DC/AC converter

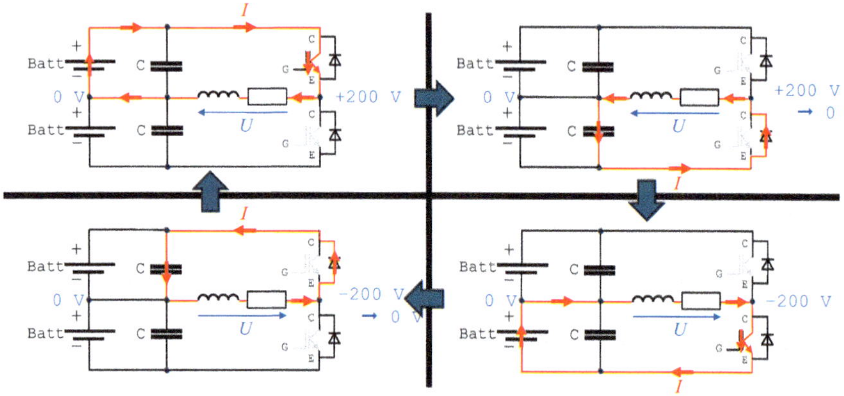

Fig. 11.6 The four operating states of the single-phase DC/AC converter

across the load returns to zero. Next, the lower switch is turned on, thereby reversing the voltage to −200 V (lower right). Finally, the lower switch is turned off and the current is commutated to the freewheeling diode of the upper circuit (lower left). The process then repeats.

The semiconductor switches are controlled by a microprocessor. However, the gate connection is at high voltage and requires much more switching energy than a standard microprocessor output can supply. Therefore, an additional control circuit is required, the so-called gate driver. It provides the necessary switching energy and implements galvanic isolation, usually via optocouplers.

The two switches (top/bottom) must never be activated simultaneously, as this would result in a short circuit of the battery, i.e. a direct connection of the two terminals. This is referred to as interlocking. In commercially available power semiconductor modules, interlocking is ensured by the gate driver. The shorter the interlocking time is selected, the more power an inverter can produce. However, short interlocking also increases the risk of an excessively high internal current flow and thus the risk of thermal failure of the semiconductors.

The two capacitors are primarily used to stabilize the input DC voltage. They are required if the inverter is fed by an upstream rectifier from sinusoidal mains voltage. This is common in industrial inverters. When used in an electric car, the DC voltage is provided by the battery. It already has a perfectly even curve. The two capacitors can therefore be reduced in size or omitted altogether.

11.2.3 Three-Phase Circuit

The single-phase circuit is simply repeated for each of the three phases (Fig. 11.7).

The battery voltage U_d is stabilized in an intermediate circuit by one or two capacitors. Each end of the three phases (R, S, T) of the motor winding can be connected to the positive voltage of the intermediate circuit via an upper switch (T1, T3, T5) and to the negative terminal via a lower switch (T2, T4, T6). By clocking these switches appropriately, it is possible to apply a sinusoidal current with adjustable amplitude and frequency to the motor.

Each switch is fitted with a freewheeling diode to short circuit the high voltages after switching off and to carry the decaying coil current.

The voltages between any two of the motor terminals (motor line-to-line voltages u_{RS}, u_{ST}, u_{TR}) can be obtained from the two voltages between the motor terminals and ground (switching voltages u_{R0}, u_{S0}, u_{T0}):

$$\begin{aligned} u_{RS} &= u_{R0} - u_{S0} \\ u_{ST} &= u_{S0} - u_{T0} \\ u_{TR} &= u_{T0} - u_{R0} \end{aligned} \tag{11.1}$$

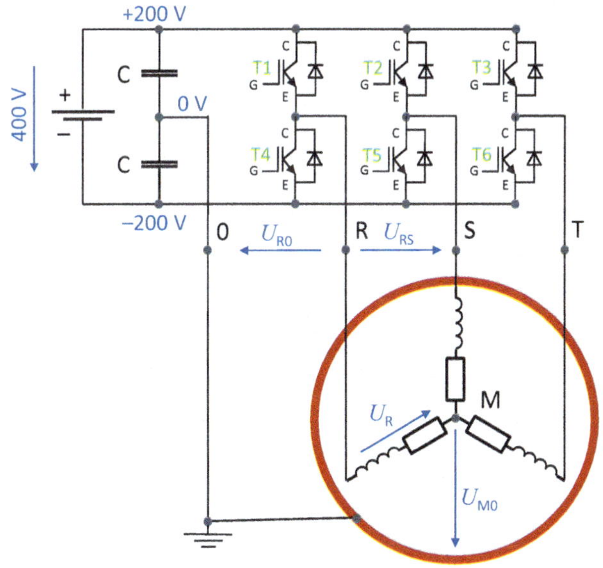

Fig. 11.7 Circuit of a three-phase DC/AC converter and motor

The voltages along the phases (motor phase voltages u_R, u_S, u_T) are

$$u_R = \frac{1}{3}(u_{RS} - u_{TR})$$
$$u_R = \frac{1}{3}(u_{RS} - u_{TR}) \quad (11.2)$$
$$u_T = \frac{1}{3}(u_{TR} - u_{ST})$$

The voltage between the star point and ground u_{M0} is

$$u_{M0} = \frac{1}{3}(u_{R0} + u_{S0} + u_{T0}) \quad (11.3)$$

11.2.4 Sine-Triangle Modulation

The switches are controlled via a microcomputer according to the principle of pulse width modulation already shown above: Depending on the desired amplitude and frequency of the AC voltage, the duty cycle of the pulses is regulated.

The question is now, how to determine the switching pulses of the sinusoidal three-phase set voltages. A variety of modulation methods are known from literature.

The simplest variant is the so-called sine-delta modulation: The shape of the desired output voltage, the set voltage, is computed by the microprocessor based on the given voltage amplitude, frequency, and phase angle. It is then stored as a table with sufficiently many grid points. In addition, the microprocessor calculates the course of a triangular voltage with full amplitude and the desired switching frequency, e.g. at 10 kHz. This triangular curve never changes and only must be computed once. Only the set voltage table is adjusted according to the motor controller requirements.

Whenever the set voltage (green in Fig. 11.8) and the delta voltage (blue) intersect, the switching states of the two power semiconductors of the affected half bridge are reversed (on/off to off/on to on/off, etc.). This results in a PWM switching pattern (red) whose fundamental wave corresponds exactly to the desired setpoint voltage.

With this simple algorithm, AC line-to-line voltages of up to $\sqrt{3}/2\ U_{DC}/\sqrt{2} \cong 0.612\ U_{DC}$ (rms) can be provided to the motor terminals.

The achievable AC line-to-line voltage can be increased to up to $U_{DC}/\sqrt{2} \cong 0.707\ U_{DC}$ (rms) by using advanced modulation methods such as space vector modulation and the injection of a third harmonic.

Example 11.1—Line-to-line voltage

For a given battery voltage of 400 VDC, the line-to-line voltage at the motor terminals can reach up to 283 VAC_{rms} or 400 VAC peak. ◀

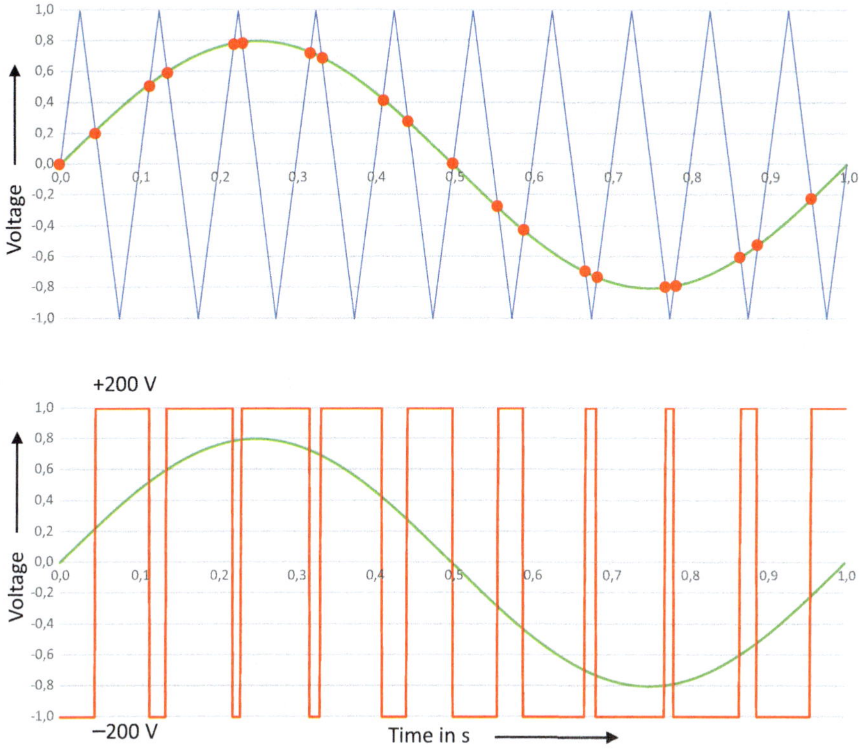

Fig. 11.8 Single-phase sine-delta PWM generation

Three-phase voltages can be created by using three time shifted set curves of sinusoidal tables against the same triangular curve.

The following three figures show examples of the line-to-line voltage, the phase voltage, and the common-mode voltage. All of them are given for a fundamental motor frequency of 1,7 Hz, battery voltage $U_{DC} = 400$ V, a switching frequency of 30 Hz, and a modulation index 1 (maximum voltage without overmodulation) (Fig. 11.9).

11.2.5 Common-Mode Voltage

In three-phase, purely sinusoidal systems, the sum of the three-phase voltages against ground potential u_{M0} is always equal to zero at any given moment in time.

If, as described above, the terminal voltages are generated by chopping a DC voltage, this of course no longer applies, because u_R, u_S, and u_T no longer change sinusoidally, but only assume the five discrete values $0, \pm U_{DC}/3$ or $\pm 2U_{DC}/3$, see Fig. 11.10.

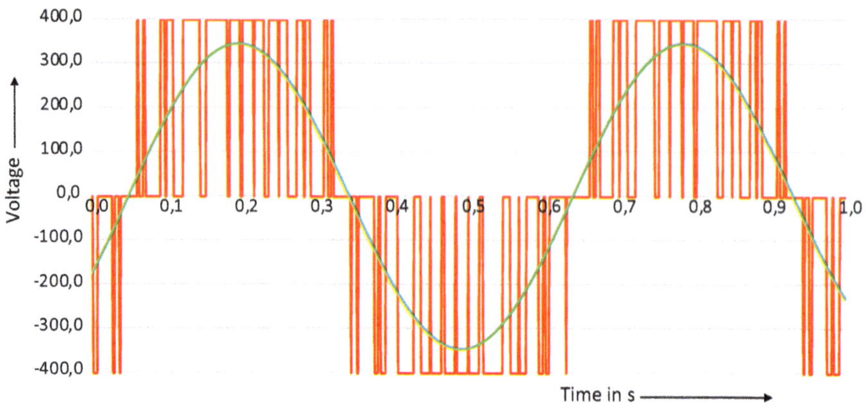

Fig. 11.9 Line-to-line voltage u_{RS}

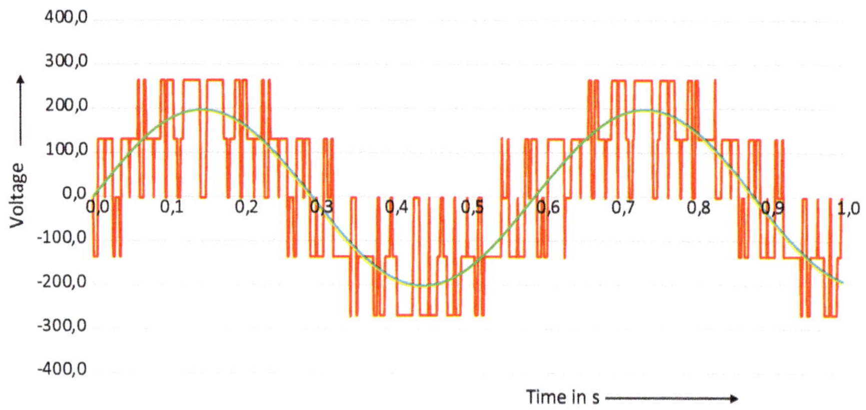

Fig. 11.10 Phase voltage u_R

The sum of the three-phase voltages results in the so-called common-mode voltage, which oscillates at three times the basic frequency and has a peak value of $U_D/2$ (Fig. 11.11).

If the star point of the motor has a connection to earth potential, then the common-mode voltage drives a current. However, this earth connection does not have to be a direct line. At the typical switching frequencies of a few kHz, undesirable coupling inductances and coupling capacitances are formed between the stator, winding, and rotor, which can transmit an alternating current. This current leads to several undesirable effects and can, among other things, damage the motor bearings due to current passage and sparking. See also Sect. 11.5.

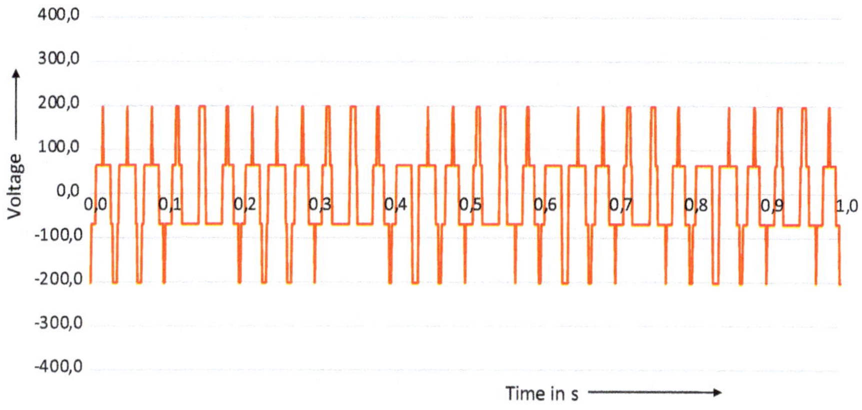

Fig. 11.11 Common-mode voltage u_{M0}

11.3 Motor Control

11.3.1 Rotor-Flux-Oriented Control of Synchronous Machines

The basic procedure for controlling electric machines is as follows: The frequency is specified by the vehicle's driving speed and thus the current rotational speed of the motor (ω). The accelerator pedal position provides the target value for the torque (T_{set}).

The control is carried out using so-called rotor-flux-oriented control. It is based on the coordinate transformation of voltage and current quantities from their original three-phase values in d/q coordinates. In the rotor-based coordinate system, all AC quantities become DC values, so the current controller can be designed quite simply.

In the first step, the current speed and the set torque are used by the motor control unit to determine the two target currents I_{dset} and I_{qset}. The target values for the currents are determined with the aim of achieving an optimum motor operation, i.e. minimum losses. Suitable methods for obtaining the best currents are presented later (Fig. 11.12).

The two set currents I_{dset} and I_{qset} are converted into voltages U_d and U_q via a PI controller, considering the deviation from the currently measured I_d and I_q currents.[2] The calculated voltage values must then be converted back into the three-phase system by a sequence of Park and Clarke transformations before they can be used. Finally, a PWM modulator generates the six control signals for the three-phase bridge circuits.

[2] Only two of the three-phase currents need to be measured, as the third is automatically given by the equation $i_U + i_V + i_W = 0$.

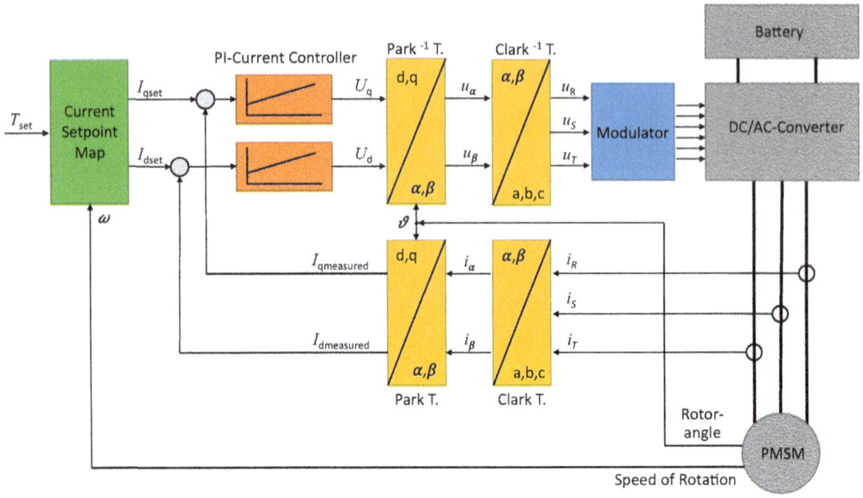

Fig. 11.12 Block diagram of the rotor-oriented control of a synchronous machine

This control loop is repeatedly run at the clock frequency of the PWM of the inverter, i.e. many thousands of times per second.

The rotor position is supplied by a measuring system, for example, a resolver, which is connected to the rotor shaft. There are several ways to determine the optimum target values of the currents $I_{dset} = f(T, \omega)$ and $I_{qset} = f(T, \omega)$. They can be determined either online, i.e. during operation by the motor controller, or offline, i.e. calculated in advance during motor development.

Generally, the goal is to minimize losses at each operating point (T, ω). Online procedures are rather limited and will usually optimizing the target currents for heat losses only. That means, the total current is minimized. This procedure is called MTPA[3] (maximum-torque-per-amp).

Particularly in the field weakening range, it can also be important not to fully utilize the available voltage so that a voltage reserve remains in the event of a sudden overload. This procedure is called MTPV (maximum-torque-per-volt).

11.3.2 Offline Determination of Set Currents

However, online determination procedures are rather limited by the available details of the magnetic circuit of the motor and by the speed of the calculation. After all, an optimum pair of target currents needs to be found for each iteration of the current controller, which is many thousand times per second.

[3] Alternate designation: MTPC (maximum-torque-per-current).

In offline optimization, a non-linear motor behavior due to iron saturation effects and iron losses can be considered. It turns out that in some operating points a slight increase in heat losses (i.e. more current) can be beneficial if this leads to a reduction in iron losses (i.e. less voltage). This can all be considered in offline determination methods for the target currents. The optimal I_{dset}, I_{qset} currents are determined by computer simulations for a close-meshed grid of operating points (T, ω) and stored in a table inside the motor controller. This procedure is called MTPL (maximum-torque-per-loss).

In principle, the same procedure is used for EESM. However, there is an additional, third degree of freedom with the excitation current I_2. Three target values for the currents must therefore be optimized and stored at each operating point. Figures 11.13, 11.14, and 11.15 show an example of such an optimization for an EESM.

The currents in the images are calculated so that the efficiency is optimal at every operating point. The d-current is largely zero in the base speed range. It only increases significantly in the field weakening range. Both q-current and excitation current are mostly proportional to the torque. In the calculations, iron saturation was considered realistically, that is, non-linearly.

To determine the target values of the three currents over the whole plane, typically 8–16 torques must be calculated at every 1000 revolutions. The calculations can be based on the equivalent circuit diagram (fast, but somewhat inaccurate) or

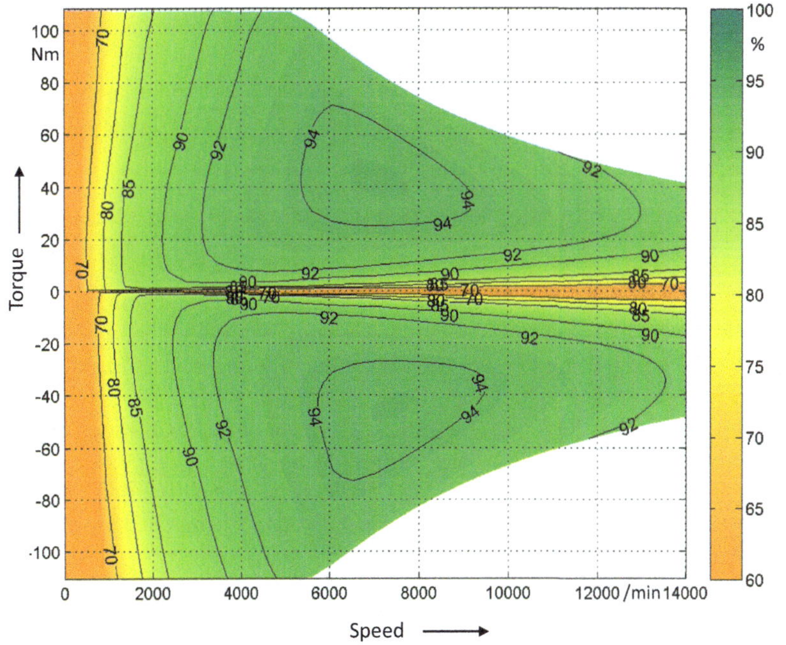

Fig. 11.13 Efficiency map of a typical EESM traction motor for cars

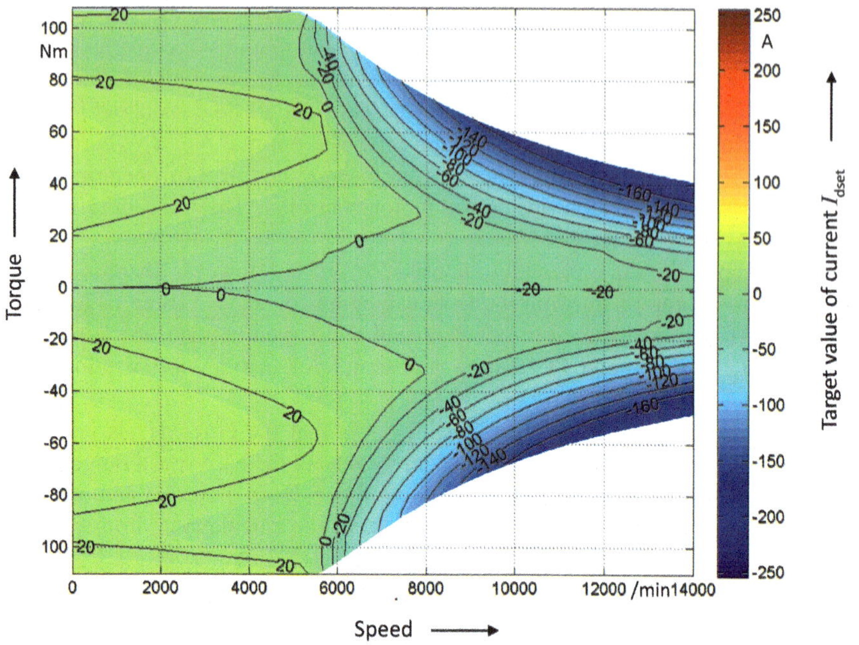

Fig. 11.14 Calculated target values of the d-current

based on numerical field calculations (accurate, but slow). Suitable interpolation is used between the grid points.

Depending on the position of the accelerator pedal (desired torque) and the current vehicle and motor speed, the motor controller can then read the target values of the currents from the predetermined maps.

Note

For most speed–torque operating points, there are various combinations of current (I_d, I_q) that lead to the desired outcome. Normally, the combination that results in the least losses and, consequently, the best efficiency is selected.

During a rapid acceleration process, it can be beneficial to change the strategy and choose currents that allow the next operating point to be reached as quickly as possible. To do this, the current in the direction of the lowest inductance must be changed at first. For PMSM, this is the d-axis, and for FESM, it is the q-axis.

Especially with a cold motor and wintry temperatures, there is the option to intentionally increase losses by choosing unfavorable currents to raise the temperature. Tesla Motors employs this method to preheat the battery, which is connected to the same cooling circuit as the motor.

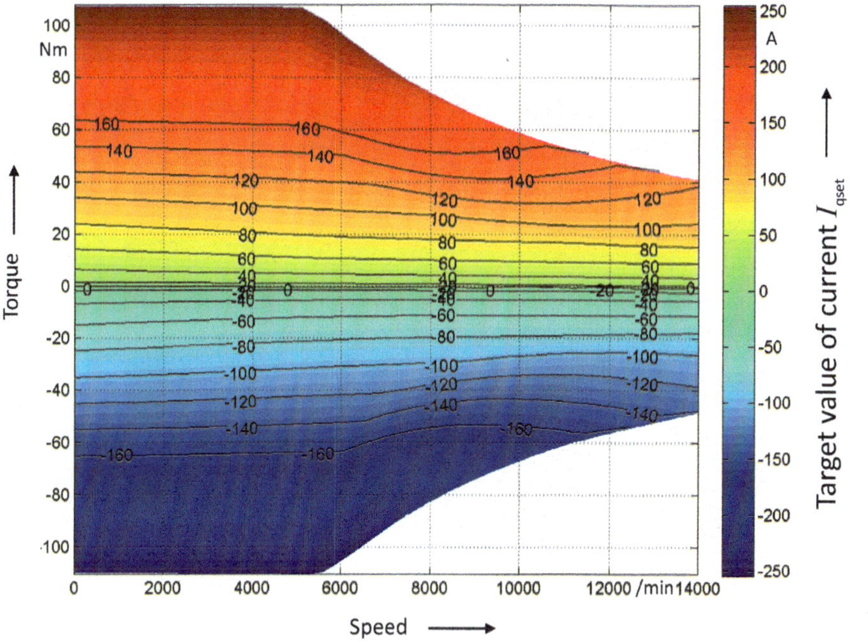

Fig. 11.15 Calculated target values of the q-current

11.3.3 Id/Iq Current Plane

The Id/Iq-plane is a great way to visualize the relationships between currents, torque, voltage, voltage limits, and current limits. It is used frequently in motor and inverter development. The MTPA and MTPV procedures mentioned above can also be visualized well in the Id/Iq-plane.

The plane shows the current in the d-direction on the x-axis and the current in the q-direction on the y-axis. Each combination of (I_d, I_q)-currents that can be set by the inverter therefore corresponds to a point in the diagram.

I_{max} is the maximum available current, limited by the semiconductors in the frequency inverter or by the battery state of charge or temperature. We can easily plot the I_{max} limit into the Id/Iq-plane by drawing a circle around the origin:

$$|I|^2 = I_d^2 + I_q^2 \tag{11.4}$$

All permissible combinations of (I_d, I_q) must lie inside this circle or otherwise the current limit is exceeded. See, for example, the gray shaded circle in Fig. 11.16.

For PMSMs, the following formulas are known from Chap. 8 (stator resistance R_1 neglected):

$$U_d = -\omega \cdot L_q \cdot L_q \tag{11.5}$$

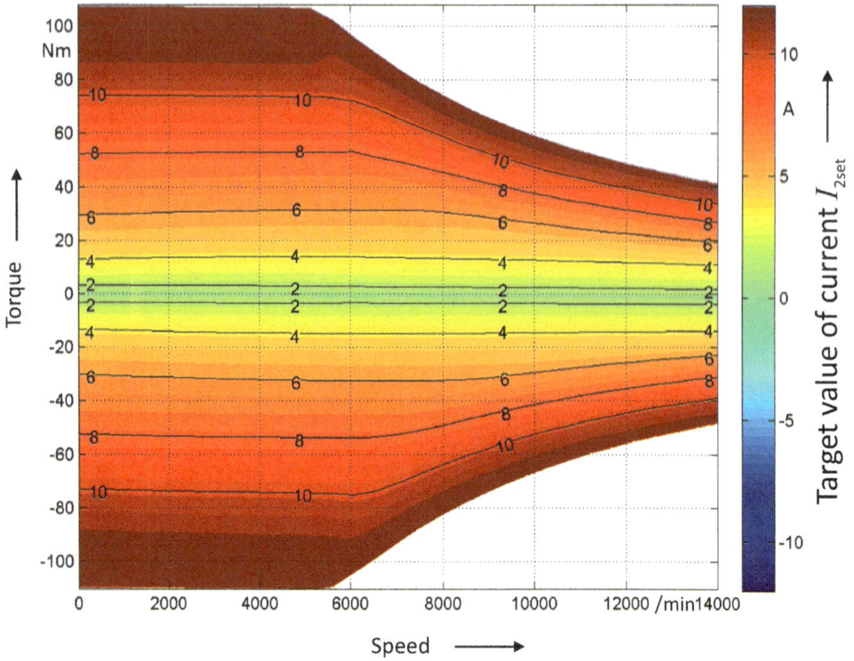

Fig. 11.16 Calculated target values of the excitation current

$$U_q = \omega \cdot (L_d \cdot I_d + \psi_{Rotor}) \tag{11.6}$$

The value of the terminal voltage is obtained from:

$$|U|^2 = U_d^2 + U_q^2 = \omega^2 \cdot \left[(L_d \cdot I_d + \Psi_{Rotor})^2 + \left(L_q \cdot I_q\right)^2\right] \tag{11.7}$$

This equation can be converted into an elliptic equation of the following form:

$$\frac{x^2}{a^2} + \frac{y^2}{b^2} = 1 \tag{11.8}$$

where x and y are the coordinates and a and b are the lengths of the major and minor semi-axes, i.e. the diameters of the ellipse.

An algebraic transformation of (11.7) results in

$$1 = \frac{\left(I_d - \left[\frac{-\Psi_{Rotor}}{L_d}\right]\right)^2}{\left(\frac{U}{\omega \cdot L_d}\right)^2} + \frac{\left(I_q - 0\right)^2}{\left(\frac{U}{\omega \cdot L_q}\right)^2} \tag{11.9}$$

A comparison with Eq. (11.8) shows that (11.9) describes an ellipse in the Id/Iq-plane. The center of the ellipse lies at the coordinates $I_d = -\Psi_{Rotor}/L_d$ and $I_q = 0$.

The lengths of the semi-axes are given by the terms under the fractions. The ellipse shrinks with increasing frequency ω, i.e. at higher rotational speeds.

Just like before, the voltage is also limited to U_{max}, namely, by the available battery voltage and the modulation method.

If we set $U = U_{max}$ in Eq. (11.9), we obtain an ellipsis that encloses all permissible voltages. This is shown in the blue shaded area in Fig. 11.17. Outside this ellipsis, a voltage would be required that is higher than the available battery voltage.

Naturally, both voltage and current limit must be observed at any time. Therefore, all permissible current combinations (I_d, I_q) are in the intersection area of the voltage ellipse U_{max} and the current circle I_{max}.

Finally, we consider the already known torque equation:

$$T = \frac{3}{2} p \cdot \left(\Psi_{Rotor} \cdot I_q + \left(L_d - L_q \right) \cdot I_d \cdot I_q \right) \tag{11.10}$$

In the Id/Iq-plane, this equation represents a hyperbola for constant torque T whose asymptote ($T = 0$) is described by $I_d = -\Psi_{Rotor}/(L_d - L_q)$ and $I_q = 0$.

Figure 11.17 illustrates all relationships.

For example:

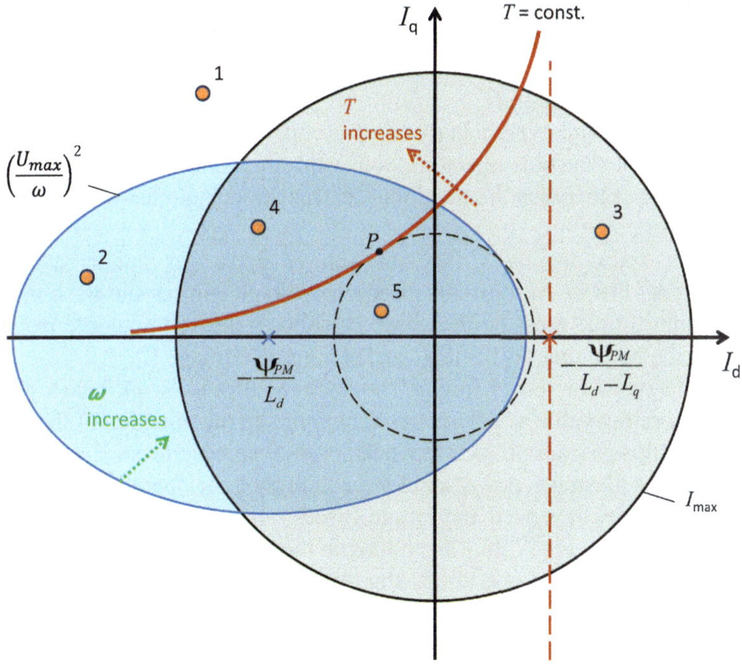

Fig. 11.17 Id/Iq-plane for a PMSM with reluctance component ($L_d < L_q$)

1 = impermissible current combination (current and voltage limit exceeded),
2 = impermissible current combination (current limit exceeded),
3 = impermissible current combination (voltage limit exceeded),
4 and 5 = permissible current combinations.

As the frequency increases, the voltage ellipse contracts. In the example shown, the center of the ellipse lies inside the I_{max} circle. In this case, there are always permissible operating points even at the highest frequencies (i.e. very small voltage ellipsis).

Depending on the motor design, however, the center of the ellipse can also be located outside of the I_{max} circle to the left. In this case, there is a maximum frequency above which the motor can no longer be operated because there are no more currents that meet the U_{max} and I_{max} conditions at the same time.

The hyperbola of a certain torque is also shown in the figure as an example. All (I_d, I_q)-combinations along the red line result in the same motor torque.

The motor control will select the specific combination (I_d, I_q) that causes the lowest losses for a given torque. As the current heat losses have the greatest influence, this best point is largely identical to the point of the lowest absolute current (P).

The MTPA method is based on these considerations. It contains analytical formulas for determining the point P as a function of the torque.

Figure 11.18 shows a realistic example of a PM synchronous. The black circle represents the maximum permissible current and the blue ellipses enclose all currents that can be set with the maximum permissible voltage. As the speed increases, the ellipses contract.

The leftmost torque hyperbola (red) represents the maximum achievable torque T_{max} of the motor. The sole operating point with this torque is P1. Higher torques cannot be set as otherwise the maximum current (black circle) would have to be exceeded.

When the speed increases, the blue voltage limits get smaller and smaller (range I in Fig. 11.19). At a certain speed, the voltage ellipses shrink so much that the maximum torque must be decreased. It follows the yellow curve from P1 to P2. This is the beginning of the field weakening range (range II).

When the speed increases further, the voltage ellipses shrink below point P2. Then it is more favorable to select operating points in the direction of the center of the voltage ellipse because the current heat losses then become smaller (range III).

Figure 11.19 illustrates power and torque in these three operating ranges.

In this example, the speed and torque curves largely correspond to the simplified theory shown in Fig. 7.24. Depending on the design of the motor, however, it is also possible that the power drops sharply in range III and that operation is no longer possible above a certain speed.

It is the task of the motor designer to design the electromagnetics of the motor according to the desired torque and power in the field weakening range.

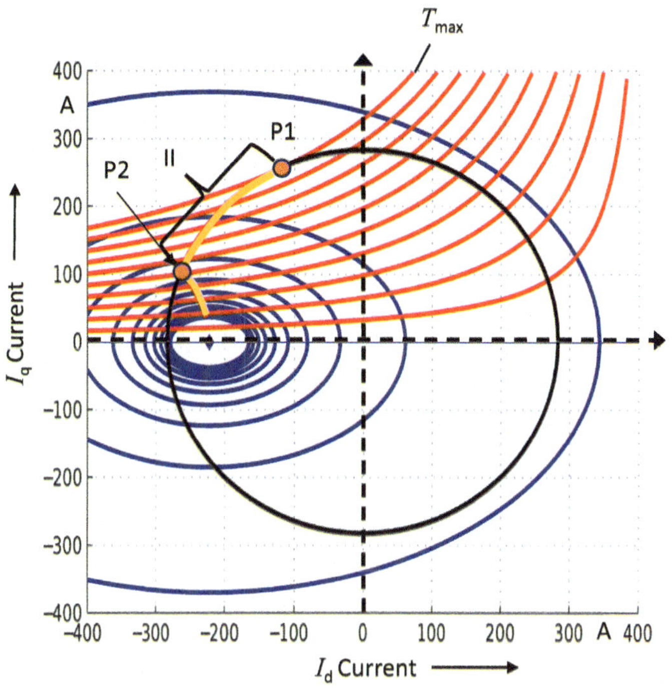

Fig. 11.18 Practical example of the MTPA target currents of a PMSM

11.3.4 Rotor-Flux-Oriented Control of Induction Machines

An induction machine can be controlled in a similar way to a PMSM as shown in the previous chapter. The transformed coordinate system is also aligned in the direction of the rotor flux.

However, the mechanical position of the rotor is no longer suitable as a reference for the coordinate system, as the rotor rotates slower than its flux (or faster in case of generator operation). As the angle of the rotor flux cannot be obtained from a position sensor it is calculated by means of a mathematical machine model using the equivalent circuit diagram. The calculation is based on the measured voltages and currents. To improve the model, the measured speed can also be used (Fig. 11.20).

The rotor resistance is critical in the calculation. It is temperature dependent, but the rotor temperature is not known. Intensive efforts are being made in research and development to automatically adjust the parameters of the models. This allows the induction machine to be operated with the same high control quality as a synchronous machine.

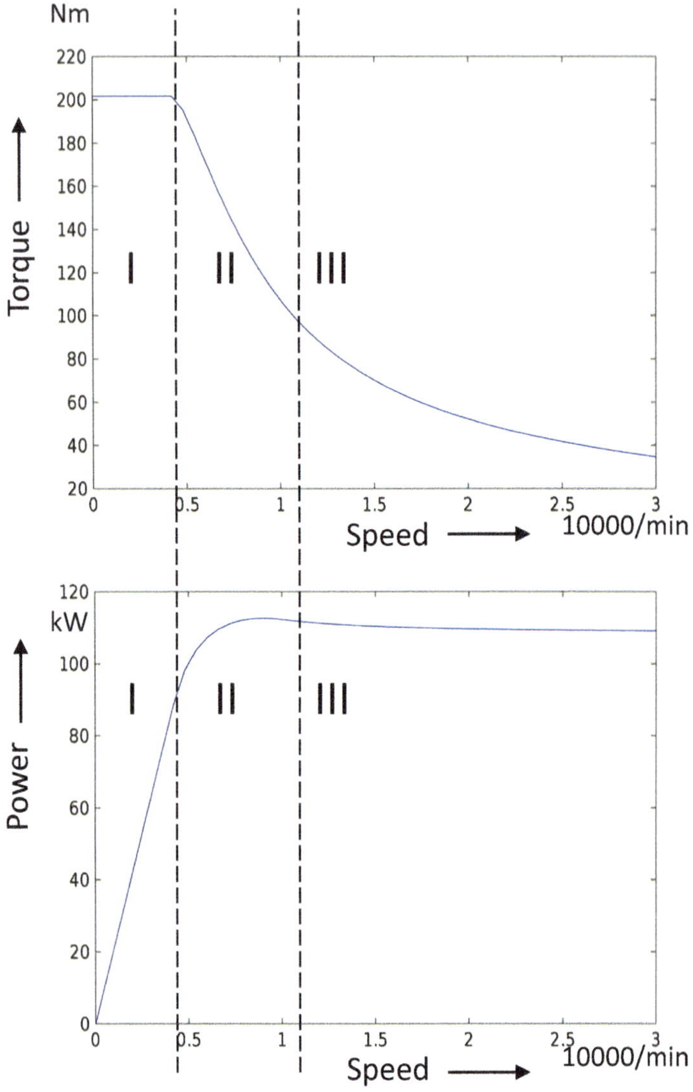

Fig. 11.19 Maximum torque and maximum power of the example motor

11.4 DC/DC Converter (DC Voltage Controller)

The DC/DC converter is an electronic circuit that acts like a DC transformer: it converts DC electricity from one voltage level to another.

Every electric car requires at least one DC/DC converter between the HV level (400/800 V) and the low voltage level (12/48 V), see Fig. 11.1. A DC/DC

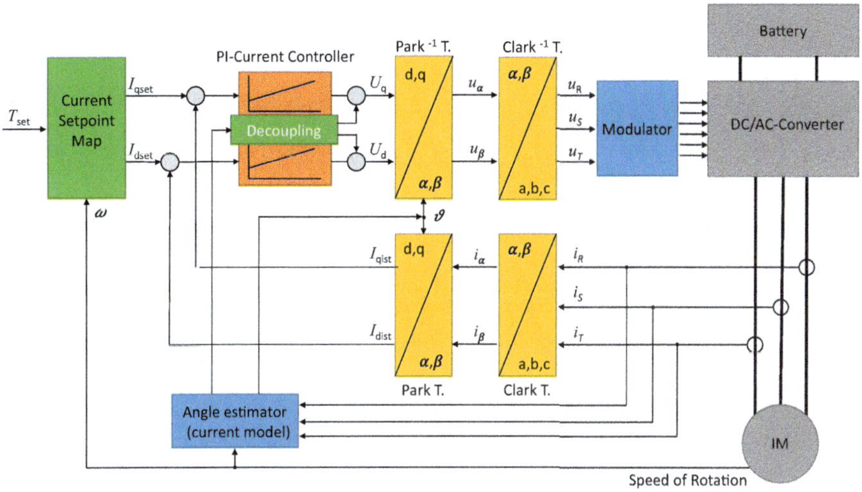

Fig. 11.20 Block diagram of the rotor-flux-oriented control of an induction machine

converter is also used for adjusting the battery voltage during charging thereby controlling the charging current.

11.4.1 Electric Circuit

Figure 11.21 schematically shows the concept of a DC/DC converter circuit.

The circuit combines a low voltage level U_L and current I_L with a higher voltage level U_H and current I_H. Apart from the internal losses, the power on the input and output side is the same ($P_L = U_L \cdot I_L = P_H = U_H \cdot I_H$).

Energy can flow in both directions. Energy flow from U_L to U_H is referred to as a step-up converter or boost converter and energy flow from U_H to U_L as a step-down converter or buck converter:

This circuit is not able to raise a lower voltage on the right-hand side to the left-hand side. If both sides of the circuit can be at a lower or higher voltage level than the other side, then a second DC/DC converter must be connected in parallel in the opposite direction.

11.4.2 Boost Converter

In step-up operation, U_L is the input and the energy flows to the high voltage level U_H. An inductive energy storage device L is used to provide an additional voltage, which is added to the input voltage to create the higher output voltage U_H.

Switch T1 is periodically opened and closed, switch T2 is always open. There is a work cycle for charging L (T1 closed) and for discharging L (T1 open).

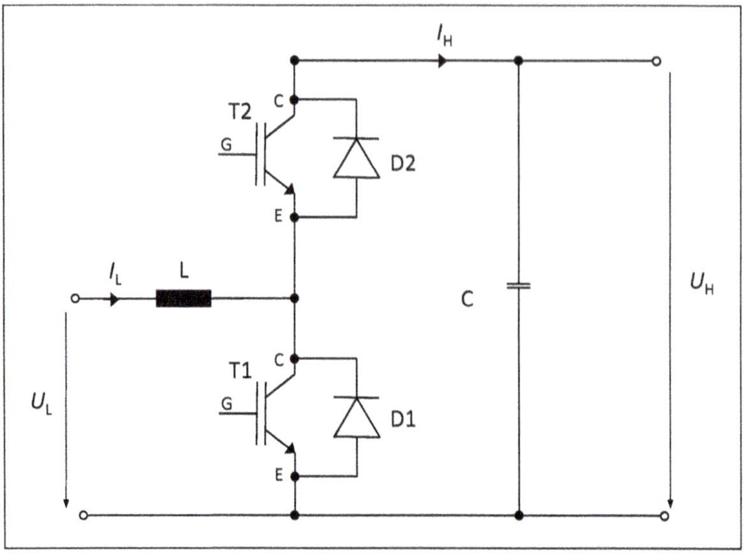

Fig. 11.21 Electric circuit of a DC/DC converter

When T1 is closed, the current I_L increases up to a specified limit value while the inductance is being charged. The switch T1 is then opened and the current continues to flow via the diode D2 to the high voltage side U_H.

The capacitor C stabilizes the output voltage during the recharging phases of the inductance (T1 closed). During this time the diode D2 ensures that the current of the capacitor can only flow to the output U_H and not back into the circuit.

11.4.3 Buck Converter

In step-down operation, U_H is the input voltage and the energy flows to the low voltage level U_L. Transistor T2 is switched periodically and transistor T1 remains permanently open.

When transistor T2 is switched on, a current $I_H = I_L$ flows directly from the high voltage side U_H to the low voltage U_L. The voltage across transistor T1 is at the same level as U_H.

When transistor T2 is opened, the current I_L on the low voltage side is driven further by the energy stored in the inductance. It now flows via the freewheeling diode D1, i.e. the voltage across transistor T1 is zero.

T2 is now switched on and off cyclically. A pulsating voltage with a lower average value than U_H is produced at output U_L.

If the inductor L is discharged before the switch T2 is closed again, phases without secondary-side current flow occur (the current is said to "gap").

The capacitor C is used to compensate line inductances on the input side and thus facilitates the switching of T2 (reduction of switching losses and avoidance of overvoltages).

In both operating modes, the voltage on the output side is heavily loaded with a switching triple. If necessary, it must be smoothed by additional filters.

11.5 Effects in the Motor Due to Switching Voltages

The hard switching of AC/DC converters with frequencies in the range between 10 and 100 kHz and the steep switching edges (short voltage rise times) cause several undesirable effects in the motor:

- High-frequency voltages can cause radio interference. This is why most electric cars today do not have AM receivers, not even in the USA, where AM radio is still available.
- Steep voltage switching edges can lead to an uneven distribution of the terminal voltage over the individual coils of a phase thereby increasing the voltage of the first coil considerably. Also, voltage reflections can lead to a doubling of the effective terminal voltage. In both cases, the wire-to-wire or wire-to-ground insulation can be overloaded, resulting in a quick failure of the insulation system.
- High-frequency capacitive currents can flow through the motor bearings and destroy them after a relatively short operating time. For example, the lubricating grease can turn black due to micro-sparks, or craters can form on the running surface of the balls.
- The high-frequency voltages lead to harmonic currents, which in turn lead to additional losses that can reduce efficiency and heat up permanent magnets.
- Finally, the high-frequency voltages and currents lead to attraction forces between stator and rotor causing stator vibrations and noise. In some cars, this noise can be heard as a high pitch beeping during driving or charging.

Various measures are taken to combat these effects.

The obvious solution seems to be to flatten the steep switching edges with RLC low-pass filters to such an extent that they no longer cause any damage. However, filters have many disadvantages: they require large, heavy, and expensive components. Filter capacitors have a limited lifetime and generate additional losses. They also introduce a time delay that makes motor control difficult and the motor performance sluggish. For these reasons, filters are avoided as much as possible in traction drives today.

High-frequency effects are combated by using metallic housings or by direct shielding of the motor winding. Developments of shielded windings have been carried out by various manufacturers but have not yet been transferred to series production.

Damage to the insulation is largely prevented by improved and thicker layers of wire and slot insulation. However, this has the disadvantage of higher costs and a reduced filling factor of the slots (less copper per slot).

Bearing currents are prevented by insulated bearings (ceramic ball bearings) or by earthing brushes. They short circuit the bearing voltage by connecting the shaft to the same potential as the housing so that no currents can flow through the bearings.

Charging Systems 12

Abstract

The chapter provides an overview of charging methods for EVs. This is followed by a detailed discussion of the different plug variants and their performance. The focus is on conductive charging for cars, but systems for commercial vehicles and contactless systems are also briefly discussed. The chapter ends with an overview of the technical characteristics of hydrogen refueling stations.

12.1 Introduction

A battery is always charged with a controlled direct current (DC).

In case of AC charging, the car is connected to an AC socket and generates and controls the DC current by a built-in charger. Integrated chargers are usually developed for outputs of 11 kW three phase or 7.4 kW single phase. Some vehicles also offer 22 kW AC charging.

When more charging power is required, it is economically feasible to include the charging electronics into the charging station rather than in the car. The connection to the car is made via a DC cable and is referred to as DC charging.

For AC charging, a separate cable must be carried in the car because the charging station only provides a socket. DC charging stations always have a permanently connected cable, so a separate charging cable is not required.

The maximum current, given by the C-rate and the maximum voltage depending on the end-of-charge voltage of the individual cells, must always be observed while charging. Otherwise, the battery will degrade with a corresponding loss of lifetime and, in extreme cases, may even catch fire.

A charging location is a place where you can charge your car. There is one or often several charging stations at each charging location. These can be commercial

M. Doppelbauer, *Introduction to Electromobility*,
https://doi.org/10.1007/978-3-658-45482-1_12

charging stations or just a private or semi-public wallboxes. Charging stations can support AC charging, DC charging, or both. Each charging station has individual cables and sockets to connect the cars. They are called charging points. A charging station has at least one charging point, but often two and rarely even three. Sometimes the charging points support different charging standards, for example, CCS, IEC Type 2, CHAdeMO, or NACS. With many fast DC charging stations, the available power is shared between all the charging points.

12.2 Structure of the On-Board Charging System

The purpose of the charging circuit is to provide a regulated DC voltage to control the charging current of the battery. The permissible current is determined by the battery management system (BMS) and is depending on many parameters (SOC, temperature, state of health).

12.2.1 On-Board AC Charging

With AC charging, the vehicle is connected directly to the AC network. The charger in the vehicle first converts the AC voltage into a DC voltage (AC/DC converter) and then regulates the DC voltage via a DC/DC converter (Fig. 12.1).

The on-board charger (OBC) typically consists of three identical single-phase modules. In some cars, the modules can be reconfigured by the charging controller depending on the charging situation (single or three phase):

- 1 module on one phase (1∽16 A, 230 V = 3.7 kW);
- 2 modules in parallel on one phase (1∽32 A, 230 V = 7.2 kW);

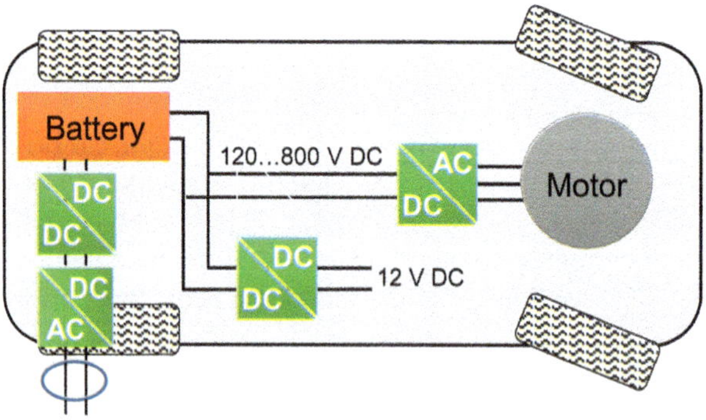

Fig. 12.1 Circuit topology AC charging

- 2 modules, each on one phase (2∽16 A, 230 V = 7.2 kW) (rarely implemented);
- 3 modules, each on one phase (3∽16 A, 230 V = 11 kW) (Fig. 12.2).

The figure shows a typical configuration for 11 kW three-phase charging. Each module is connected to a different phase.

12.2.2 DC Charging

With DC charging, the charging circuit is in the charging station. The vehicle has a DC connection that circumvents the on-board charger and leads directly to the battery. There must be a frequent exchange of information between the car's battery management system and the charging point through the car's charging controller as the station adjusts the charging current (Fig. 12.3).

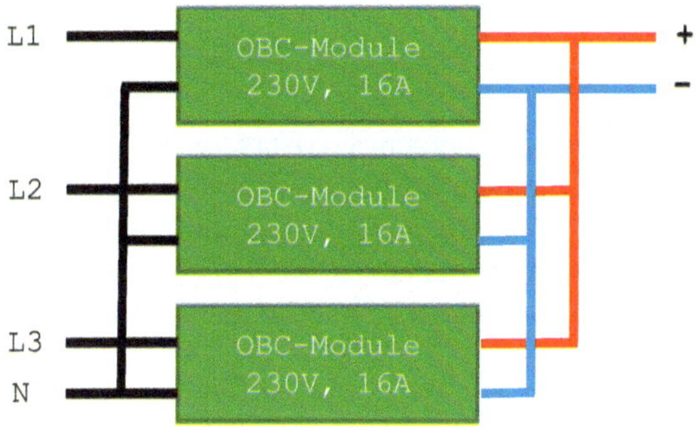

Fig. 12.2 On-board charger with three single-phase modules

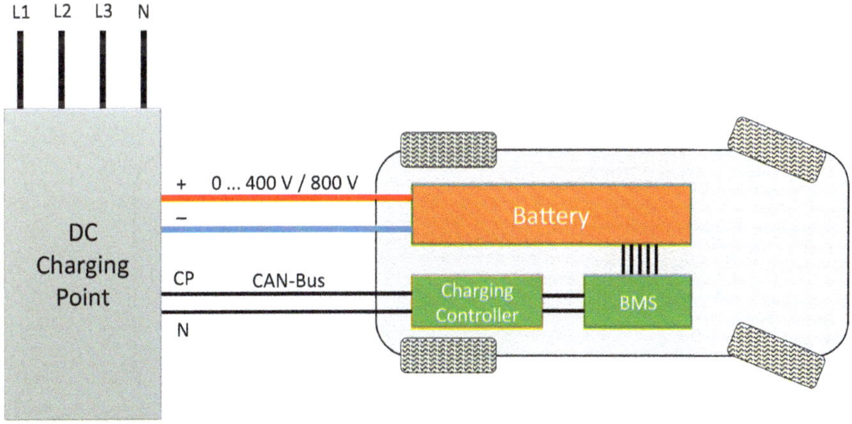

Fig. 12.3 Circuit topology DC charging

12.2.3 On-Board AC Inverter Charging

The two charging methods presented above require additional charging electronics. However, every electric car already has built-in charging electronics, namely, the inverter that supplies the drive motor. Charging is ultimately the same as recovering braking power. However, the electric motor must not rotate during stationary charging. So the motor is usually disconnected from the inverter using contactors. With electrically excited synchronous machines, however, it is sufficient to switch off the power supply of the rotor (Fig. 12.4).

The concept of AC inverter charging is being pursued by Renault (Zoe) and AC-Propulsion (reductive charger), for example. However, the lifetime of the inverter is reduced by charging due to the thermal load. Most car manufacturers therefore favor a separate charging circuit.

12.2.4 Electrical Safety

For safety reasons, the entire HV system in the car is marked with an orange signal color.

All housings, plugs, service openings, covers, and other HV access points, i.e. ultimately all detachable components of the HV system, are connected in series with a 12 V safety line (HV interlock). If the circuit of this pilot line is interrupted by disconnecting one of the plug connections or by opening a part of the housing, the HV system is immediately de-energized by the HV controller by switching off the main relays.

The HV system of electric cars is insulated from the chassis and all other components (voltage class B system). During charging there is also a separation from the mains by an isolating transformer. Insulation faults, i.e. low-resistance electrical connections to the car's ground (chassis), can occur due to soiling, for example, moisture, defective plugs, or damaged cables. Faults must be detected as quickly as possible, and the system must be switched off safely.

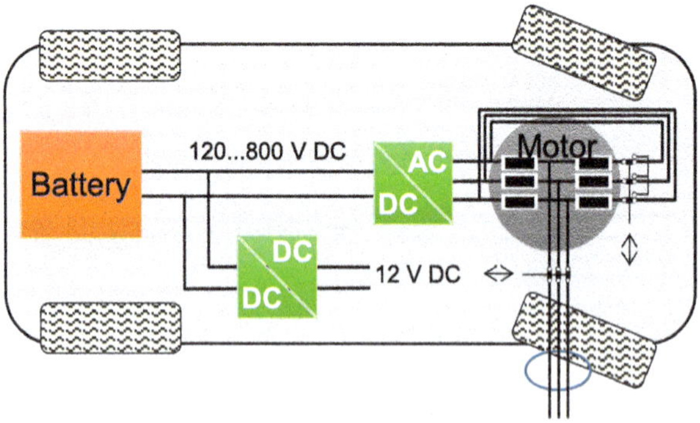

Fig. 12.4 Circuit topology AC inverter charging

For this purpose, the insulation resistance, i.e. the resistance between the power lines of the high-voltage system and the chassis of the vehicle, is continuously monitored by a device in the car (IMD = Insulation Monitoring Device). In normal operation, the resistance is several megohms.

DC fast charging stations are designed as unearthed power supply systems (Mode 4 IT system) with isolating transformers and insulation monitoring. The monitoring of the charging station protects the entire unearthed charging circuit. The insulation monitoring system in the car is switched off during DC charging.

When AC charging from a household socket with a cable box (Mode 2) or at a charging station/home charger (Mode 3), fault currents are monitored. The currents in all connection cables are measured continuously. As soon as these deviate noticeably from one another, there must be a parasitic current path resulting from an insulation fault. If such a residual current is detected, the charging process is switched off immediately for safety reasons.

Residual current protective devices (RCDs) are available in various designs. When charging electric cars, not only AC but also DC residual currents can occur. Therefore, only the AC/DC sensitive type B RCD should be used. These RCDs reliably switch off DC residual currents $I_{\Delta n}$ of around 6 mA (milliamperes) and AC residual currents of around 30 mA.

After locking the car, the HV battery is disconnected from the rest of the HV system. This de-energizes all other HV components. The car then relies solely on the 12 V battery to detect the presence of the car key and reconnect the HV battery.

Note—Loss of 12 V power

It occasionally happens that one of the many 12 V car control units does not switch to idle mode after the vehicle has been switched off and locked but rather continues to run and consumes energy. If this situation continues for a long time, the 12 V battery may discharge completely. The car can then no longer be started, even if the HV battery is at a high state of charge.

In that case, the 12 V battery must be recharged externally for several minutes, for example, with a jumper cable. As soon as 12 V power is restored, the HV battery is reconnected and the 12 V battery is charged from the HV battery through the DC/DC converter. After such an incident it is advisable to keep the car activated for a long enough time to allow the 12 V battery to recover. This can be done either by driving, by leaving the car unlocked, or by recharging the car from a charging point.

12.3 Charging Power and Charging Time

Typical high-energy lithium–ion batteries for BEVs and PHEVs enable charging rates of 2–3 C, that is, a charge from below 10–80% or 90% in 20–30 min, see Sect. 6.11. High-power lithium–ion batteries for hybrid vehicles have a lower

Table 12.1 Common charging methods and charging times for electric cars[1]

	# of phases	Voltage and current	Power	Charging system	Charging time for 300 km range
Slow charging Level 1 <10 kW	1~AC	200–240 V 10 A	2.3 kW	Household plug	30 h
	2~AC	200–240 V 32 A	7.4 kW	Commercial plug	10 h
Rapid charging Level 2 11 or 22 kW	3~AC	400–460 V 16 A	11 kW	IEC Type 1 or 2	6,5 h
	3~AC	400–460 V 32 A	22 kW	IEC Type 1 or 2	3 h
Fast charging DC >50 kW	DC	400 V/125 A	50 kW	IEC CCS NACS CHAdeMO	90 min
	DC	400 V/500 A	200 kW		30 min
	DC	800 V/500 A	350 kW		10 min

energy density and can be charged at higher C rates. Faster electrochemical storage systems are being investigated but have not yet reached market readiness.

The charging capacity that can be achieved at a charging point depends on the available current and the system voltage. The current is limited by the plug, the cable, and the power electronics. Table 12.1 provides an overview of common charging systems.

Charging with 150 kW and more is known as high-power charging (HPC). The current is 400 A or more and the charging cable between the charging point and the car is usually liquid cooled.

Example 12.1—Average charging power

According to a recent study,[2] the typical average charging power from 10 to 80% SOC for compact-class cars is between 50 and 80 kW (peak 100 kW), for middle class cars between 100 and 200 kW (peak 250 kW), and for luxury cars between 150 and 230 kW (peak 280 kW). ◀

From the user's point of view, it is not the charging power but the recharged range per time that is important. The P3 Group has introduced the P3 charging index for this purpose.[3] It indicates how much range can be recharged in 20 min compared

[1] With an assumed consumption of 20 kWh per 100 km range.

[2] https://www.p3-group.com/p3-charging-index-vergleich-der-schnellladefaehigkeit-verschiedener-elektrofahrzeuge-aus-nutzerperspektive_07-22/

[3] https://www.p3-group.com/

to a reference distance of 300 km. For example, an index of 0.5 means that 150 km of range can be charged in 20 min.

Example 12.2—Charging index

By mid-2022, only the KIA EV6 has achieved a P3 charging index just above 1.0.

Luxury vehicles from Mercedes-Benz, BMW, and Porsche are close behind with values above 0.9. The Tesla Model 3 LR (1st gen) comes in at 0.74 and Model Y LR at 0.58.

All values are based on real test drives carried out as part of the study, not on manufacturer specifications. ◀

Example 12.3—Charging stations versus filling stations

With chemical energy (diesel, petrol), around 10 MW of refueling power can be delivered. This corresponds to a range of around 300 km per minute of refueling time.

We assume that up to six petrol or diesel cars per hour can be served at a filling station, i.e. 10 min refueling including arrival, departure, and payment. We further assume that an average range of 500 km is refueled per car. This results in a fuel pump capacity of 3000 km per hour.

Electric cars typically recharge less range, for example, 300 km, to shorten the waiting time. Around 100 kWh can be recharged per 150 kW charger per hour on average. So around six electric fast charging points with 150 kW each are required to enable the same throughput with electric cars: 10 cars $\cdot$ 300 km = 3000 km range per hour = 600 kWh of required net energy in the car. Charging losses for the most part occur in the fast charger and must be added.

A large freeway service station with, for example, 10 fuel pumps will therefore have to install around 60 fast charging points in the parking lot to accommodate the same number of electric vehicles in the future as petrol cars today.

The largest EV fast charging stations in the world already today (2024) have between 50 and almost 100 charging points.[4] ◀

Note—Utilization of high charging power

A high charging power is only required for batteries with a high capacity due to the limitation of the C-rate. For example, a battery must have a size of at least 75 kWh to utilize 150 kW charging power when limited to 2 C charging rate. Charging capacities of around 300 kW are only useful

[4] The Harris Ranch Tesla Supercharger in California, USA, was expanded to 98 charging points by the end of 2023.

with even bigger batteries, for example, a 100 kWh with a charging rate of around 3 C.
However, the permissible charging currents are depending on the temperature and the SOC. Therefore, a high charging power is only needed at the beginning of the charging process.

The batteries of commercial vehicles usually have a larger capacity than those of passenger cars and, as a result, the charging power is correspondingly higher. New charging standards for charging of heavy-duty trucks enable up to 4.5 MW charging power at 1500 V and 3000 A. This will allow battery capacities of over 500 kWh to be recharged in 20 min.

Urban buses do not require nearly as much energy as long-distance buses. The daily distance traveled is relatively short and braking energy recovery is relatively high due to the many stops, so the charging capacity does not need to be particularly high.

Example 12.4—Electric buses and taxis in China

By the end of 2022, more than three-quarters of all urban buses in China were "new energy vehicles" (414,000 BEV and PHEV plus 4000 FVEC = 542,600 total).

With approximately 16,000 electric buses, Shenzhen was the first city in China that replaced all diesel buses with battery electric buses already in 2020. Shenzhen is a major Chinese city that borders Hong Kong to the north and has a population of 12 million. A total of 40,000 charging points have been installed in the city's 180 bus depots. The charging process only takes 2 h per bus and takes place at night, as the typical daily mileage is only 200 km.

All 22,000 taxis in the city have also been converted to battery electric vehicles. ◀

Note—Single-phase versus three-phase on-board chargers

An on-board charger is required to charge the vehicle at home or at AC charging points.

It is important that the OBC can operate on three phases with at least 16 A to achieve a charging power of 11 kW. In this way, even large batteries can be fully charged overnight. Suitable 11 kW home chargers are available at reasonable prices. Most inner-city charging stations offer 11 or even 22 kW AC charging power.
Cars with on-board chargers that are limited to single-phase charging are not recommendable. In many countries, for example, Germany, Austria, and Switzerland, single-phase charging with more than 16 or 20 A is not permitted and in that case the charging power is limited to a measly 3.7 kW.

12.4 Charging Losses

AC charging of electric cars causes heat losses in various places:

- In the domestic distribution board, typically around 1%.
- In the supply line to the home charger or station, typically less than 1%.
- In the home charger or charging station, typically less than 1%.
- In the charging cable, typically less than 1%.
- In the on-board charger, typically 5–10%.
- In the HV cable inside the car, typically less than 1%.
- In the battery, typically less than 1%.

The German automobile club ADAC has performed many measurements of electric cars that included the determination of charging losses. The charging losses of a typical 11 kW AC home charger or charging station are between 5 and 10% of the electrical input power. When charging from a household socket using an in-cable control box (ICCB) with a power output of around 2.3 kW, the losses are higher, around 10–25%.

Charging at high-power DC charging points causes similar losses to charging from an AC home charger, around 10%. The liquid cooling of the cable may give the impression that the losses are higher. However, this is not the case, the charging process is just much shorter and therefore the momentarily generated heat is greater. It can generally be found that electrical systems with higher power tend to have a better efficiency.

12.5 Charging Systems for Cars and Light Commercial Vehicles

See Table 12.2.

12.5.1 Europe

In Europe, the IEC type 2 plug is used for AC charging and CCS combo type 2 for DC charging. In Germany, Austria, and Switzerland, almost all residential buildings (single and multi-family homes) are connected to the three-phase low-voltage grid. Therefore, most private AC home chargers are operated with three-phase 16 A (11 kW). In most other European countries, however, houses are usually only connected to the single-phase grid. In some cases, e.g. in the UK, the connection supports single-phase charging with 32 A (7.4 kW), while in others, e.g. Italy, the connection is often so weak that even single-phase charging with 10 A (2.3 kW) can lead to disruptions in the domestic supply. Charging an electric car at a normal household socket is possible, but only with a special adapter and at a maximum of 10 A (2.3 kW).

Table 12.2 Overview of worldwide charging standards and plugs

	AC charging		DC charging	
Europe	Type 2 IEC 62196-3	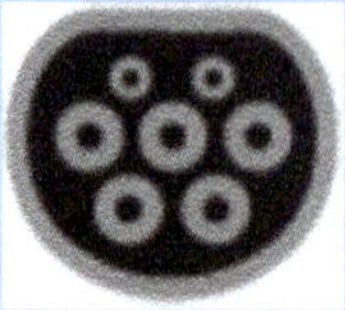	CCS2 IEC 62196-3 FF	
North America	Type 1 IEC 62196-3 SAE J1772	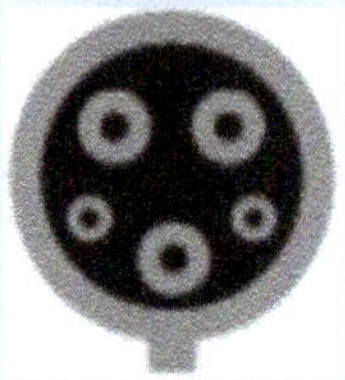	CCS1 IEC 62196-3 EE	
North America (NACS)	NACS Tesla-Plug SAE J3400		NACS Tesla-Plug SAE J3400	
Japan Korea	Type 1 IEC 62196-3 SAE J1772		CHAdeMO IEC 62196-3 AA	
China	Type 2 (male) IEC 62196-3 GB/T 20234	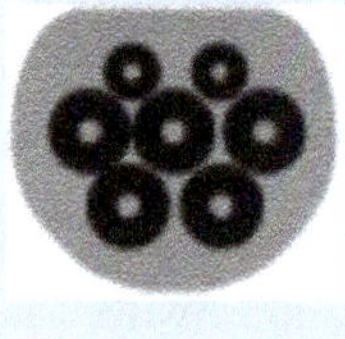	GB/T 20234 IEC 62196-3 BB	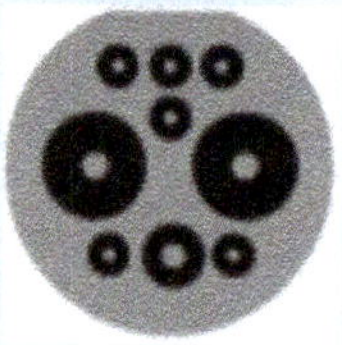

Public AC charging points in Europe are usually three phase and offer a 16 A (11 kW) or 32 A (22 kW) connection.

DC fast charging in Europe is carried out with the CCS plug, which is an extension of the type 2 plug. Since the market launch of the Tesla Model 3 in 2019,

Tesla has converted its European Superchargers to CCS and is now fully compatible with all other European DC charging stations. Therefore, all Tesla cars in Europe can charge at all public DC charging stations and all other electric cars can charge at Tesla superchargers.

By the end of 2023, the number of charging points in the EU was about 600,000 including about 80,000 DC fast chargers.

At the beginning of modern electromobility, some Japanese manufacturers (Nissan, Mitsubishi) introduced the CHAdeMO plug in Europe. There is still a growing number of chargers that offer two or even three different plugs including one CHAdeMO plug, but these chargers are mostly limited to 50 kW. All in all, CHAdeMO did not catch on in Europe and no longer plays a role.

12.5.2 USA

In the USA, private households and many small industrial companies only have a single-phase connection to the electricity supply. Partly, however, with two voltages (split-phase three-wire system). AC charging takes place via the SAE J1772/IEC type 1 plug.

Level 1 charging uses a separate fuse box to charge directly from a household socket at 120 V single phase. Charging currents of up to 10 or 12 A and outputs of between 1.3 and 2.4 kW are common.

With level 2 charging, a home charger is permanently installed. The voltage is either 208 V (households) or 240 V (industry). Charging currents between 12 and 64 A and outputs of up to around 15 kW are possible.

For DC fast charging, the CCS Combo Type 1 has been the standard for non-Tesla cars up to the end of 2023.

However, more and more manufacturers are now turning away from this rather large plug and using the Tesla plug, whose specification was released in November 2022 (NACS = North American Charging Standard).[5] It can be assumed that the CCS system will gradually disappear in North America.[6]

With CCS, outputs between 24 and 200 kW at 400 V and up to 350 kW at 800 V are possible. The Tesla plug supports 500 A at 500 V or 1000 V system voltage. It has a maximum charging power of 250 or 350 kW.

By the end of 2022, there were about 140,000 publicly available charging ports in the United States. Of those, just 28,000 were DC fast charging ports. Just as in

[5] There are official adapters from Tesla that allow the cars to be connected to CHAdeMO and CCS Combo 1 stations.

[6] The NACS plug is limited to single-phase AC charging. It is therefore not suitable for the European market. In Europe, Tesla converted its entire vehicle fleet and all superchargers to IEC Type 2 CCS since 2019.

Europe, Japanese manufacturers have tried to establish the CHAdeMO plug in the USA. There are still some medium-speed chargers with this plug, but the standard is practically irrelevant today.

12.5.3 Japan

AC charging systems in Japan use the same plugs as in North America, i.e. SAE J1772/IEC Type 1. Households are mostly connected to single phase with two voltages (split-phase three-wire system), while commercial areas and industrial companies are supplied with three-phase voltage.

In the southwest of the island, the charging voltages are 100/200 V single phase or 400 V three phase with 50 Hz while in the northeast the voltages are 110/220 V single phase or 440 V three phase with 60 Hz.

The national CHAdeMO standard has been established for DC charging in Japan since 2010. The first version was limited to 50 kW power. The second generation can charge up to 320 kW (800 V, 400 A).

So far, the market of electric cars in Japan is lagging far behind that of other industrialized nations. The charging network has around one-third of the charging points compared to Germany, even though Japan's population is 50% larger.

The Japanese manufacturers are only slowly switching to battery electric cars. By the end of 2022, there were just 30,000 EV charging points. The Japanese government aims for 300,000 chargers installed by 2030.

12.5.4 Korea

Korea uses the same AC charging system as North America and Japan, that is, SAE J1772/IEC Type 1. Domestic chargers are operated at 220 V and can reach charging speeds up to 3.2 kW. Public AC chargers are typically rated at 7 kW.

For DC charging, Korea has adopted the CCS combo type 1 but also uses the Japanese CHAdeMO system. Rapid chargers used to be rated around 50 kW, but more and more faster chargers with much more power are being opened.

By the end of 2022, Korea had more than 200,000 public EV chargers. The government's aim is to install 1.23 million public chargers by 2030, including 12,000 DC fast chargers.

12.5.5 China

In China, charging plugs according to GB/T 20234 series standards are used. The plug for AC charging is like the European IEC Type 2 plug. However, the pins (male side) and sockets (female side) are reversed. In Europe, the cable has pin contacts for connection to the charging point and sockets for connection to the vehicle. The Chinese standard GB/T 20234.2 is exactly the opposite.[7] If needed, adapters can be used to charge original Chinese cars in Europe.[8]

Typical charging powers in China are 7 kW single phase and 40 kW three phase.

A separate plug is used for DC fast charging, which is not compatible with CCS, CHAdeMO, or Tesla.[9] It supports up to 1000 V system voltage and 300 A without water cooling and 800 A with water cooling. Common DC charging powers are 60, 120, and 150 kW, sometimes even more up to 480 kW.

As of 2022, China had nearly 1.8 million public EV charging points. Almost half of these were DC fast chargers. The annual increase in charging points is around 50%. China is currently the country with the best infrastructure for electric cars, far ahead of Europe, Korea, and the USA, while Japan is in last place of all industrial nations by some distance.

12.5.6 Other Countries

Brazil has by far the largest fleet of flexible-fuel vehicles in the world. In mid-2018, over 30 million such cars were on the road, representing a market share of around 89%. Flexible-fuel vehicles can run on any mix of E20–E25 petrol and up to 100% hydrous ethanol fuel (E100). Brazil is the world's second largest producer of ethanol fuel. The production is based on efficient agricultural technology for cheap sugarcane cultivation. Electromobility has not yet gained momentum in Brazil, partly due to the widespread use of environmentally friendly biofuels. In mid-2022, there were only 300 type 2 AC charging stations and a few fast chargers.

Electromobility is slowly growing in Malaysia. As of 2023, there were around 1,500 public EV charging stations including some high-performance chargers. The European plug system (IEC Type 2 for AC and CCS2 for DC) is mostly used in Malaysia. The country operates on a 240 V/415 V supply voltage and 50 Hz.

---

[7] In general, the wear on sockets (metal springs) is higher than that on plugs (pin contacts). Whenever possible, cables that are easy to replace should therefore be fitted with sockets and vehicles that are more difficult to repair should be fitted with plug contacts. This approach was chosen in Europe.

[8] Almost all manufacturers adapt the charging ports of their cars to the respective market. It only happens occasionally that low-volume vehicles are sold with connections for the "wrong" continent.

[9] There are adapters for charging cars with the Chinese plug at CHAdeMO and CCS charging stations. However, as the charging protocols are different, these adapters must be equipped with their own microcontrollers. As of 2024, the reliability of these adapters still leaves a lot to be desired.

India uses the European voltages (230/400 V, 50 Hz) and plugs (IEC Type 2, CCS) for charging their EVs. The number of electric cars is growing slowly. At the beginning of 2023, there were around 5000 public EV charging stations. However, there are many three-wheeled electrical rickshaw taxi services, the so-called e-tuk-tuks. As of 2022 India had about 2.4 million of these on its roads. Most e-tuk-tuks are still equipped with lead batteries and charged via household sockets.

12.6 Battery Swapping

To shorten the charging time, some vehicles allow for removing the empty battery and replacing it with a full battery within a few minutes. These cars are equipped with a battery in the underbody that can be easily removed mechanically. The empty battery remains in the charging station and is slowly recharged while the car is already on its way with the filled-up replacement battery.

The company Better Place was founded in the USA already in 2007 for this purpose. There have been several pilot projects in Denmark and Israel with Renault vehicles. Ultimately, however, the concept was not successful and Better Place went bankrupt in 2013.

Since 2018, the Chinese car manufacturer NIO is reviving the concept with considerably more success: at the end of 2023, there were over 2000 battery swap stations in China and 30 stations in Europe, mostly in the Netherlands and Norway. However, the stations are only compatible with vehicles from the manufacturer NIO itself. Other cars cannot be served.

12.7 IEC Type 1 and 2 and CCS Charging Standards

The IEC 62196 series standards define three plug types in different variants for AC, DC, and combined use (CCS = Combined Charging System) (Table 12.3).

Type 1 plugs are common in the USA, Japan, and Korea. They were also occasionally used in Europe, for example, in the Opel Ampera.

Table 12.3 Common standards for charging procedures and plugs

	IEC type 1 (USA, Japan, Korea)	IEC type 2 (China, Europe)	IEC type 3 (obsolete)
AC charging	SAE J1772 IEC 62196-2	IEC 62196-2	IEC 62196-2
DC charging (obsolete)	IEC 62196-3	IEC 62196-3	IEC 62196-3
Combined charging system CCS	SAE J1772 IEC 62196-3	IEC 62196-3	IEC 62196-3

In January 2013, the EU officially committed to the type 2 plug. Type 2 plugs are available at practically every public charging station in Europe. They are also used in China.

The type 3 plug was originally favored by France and Italy but has since been abandoned and is no longer used (Fig. 12.5).

An IEC type 2 AC plug contains five power contacts (N, PE, L1, L2, L3) and two signal contacts (PP = proximity pilot, CP = control pilot), see the following illustration (Fig. 12.6).

The power contacts can each carry 500 V and up to 63 A. With the usual 400 V AC voltage, up to 43 kW AC charging power can be transmitted.

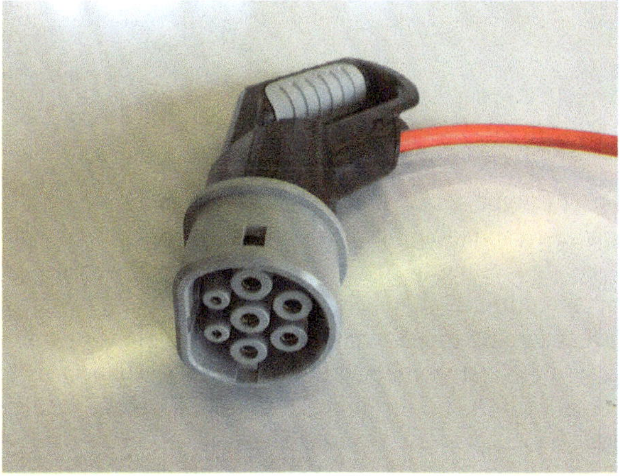

Fig. 12.5 IEC type 2 AC plug

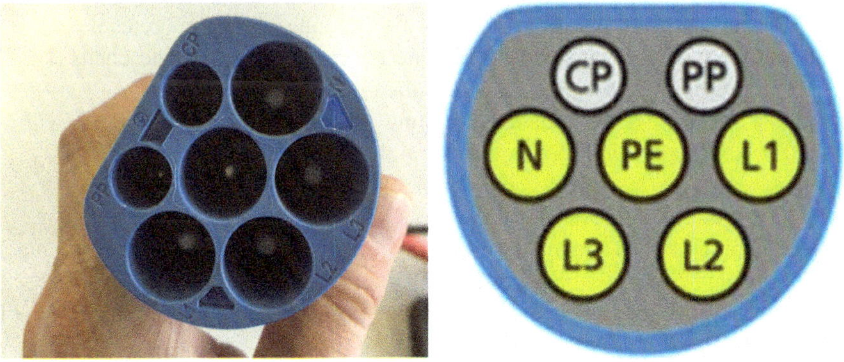

Fig. 12.6 Contact assignment of an IEC type 2 AC plug

Fig. 12.7 IEC type 2 CCS plug

Most public IEC type 2 AC charging stations in Europe limit the available power to 22 kW, corresponding to 32 A current. Some stations only support 16 A (11 kW) and very few allow 63 A (43 kW). However, there are no more cars on the market that can utilize 63 A AC charging.[10]

There is a DC charging mode for the type 1 and 2 plugs that utilizes the N and L3 connectors for the positive pole and the L1 and L2 connectors for the negative pole. This mode allows DC power up to 70 kW but has never really been used and is now obsolete.

Note—Compatibility of IEC Type 1 and 2

The permissible charging current depends on the maximum current of the car, the charging point, and the cable.

Before the charging process starts, the charging point and the cable communicate with the vehicle to determine the weakest link in the chain. The current during the charging process is then limited accordingly.

This means that all vehicles, charging points and cables equipped with either IEC Type 1 or Type 2 are compatible with each other.

The CCS plug is an extension of the IEC type 1 or 2 plug. Two additional contacts for DC power are added beneath the AC plug. A vehicle equipped with CCS can therefore also use all type 1 or 2 AC charging points if a suitable cable is available (Fig. 12.7).

[10]The first-generation Renault Zoe, sold from 2013 to 2019, had 43 kW AC charging capability.

12.7.1 AC Charging Modes

The IEC 61851-1 standard defines three charging modes for the type 1 and 2 plugs in accordance with IEC 62196.

In mode 1, there is no communication. The vehicle is connected directly to a domestic socket using a suitable adapter cable. Protection is provided by the RCD (residual current device) in the household connection. There is no communication or monitoring between the vehicle and the charging connection. For this reason, Mode 1 is practically not used in modern vehicles.

Mode 2 also charges from a domestic socket. However, the cable contains a control box (ICCB = In-Cable Control Box), which performs safety checks and communicates with the vehicle's charging circuit. The temperature of the plug is continuously monitored to ensure that no overloading occurs in the household socket. Corresponding adapter cables are available from all car manufacturers and many third parties, for example, Mennekes or Juice Booster.

Mode 2 allows a charging current up to 32 A and charging power of up to 7.4 kW single phase or 22 kW three phase. Currents higher than 10 A usually require industrial plugs, for example, red and blue CEE plugs. Household plugs, for example, Schuko, can only carry up to 10 A or 2.3 kW and are used for emergency charging.

Mode 3 is the standard case: The vehicle is connected to a dedicated charging facility, that is a home charger or a public station. There is full communication between the vehicle and the charging facility (Fig. 12.8).

12.7.2 AC Charging Communication

The plugs of a charging point are always de-energized when no charging takes place. Charging only begins after all safety checks are finished, after the cable

Fig. 12.8 Mode 2 in-cable control box for domestic charging at 230 V 10 A (2.3 kW)

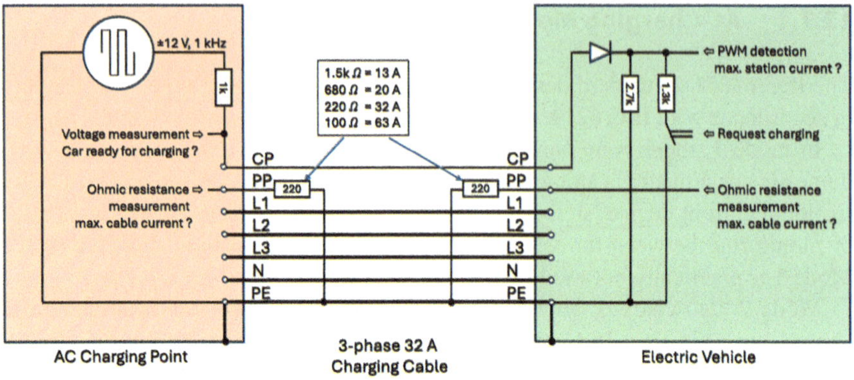

Fig. 12.9 AC charging communication through PWM voltage and resistors

plugs have been mechanically locked into the sockets on both ends and the maximum permissible charging current has been determined.

In modes 2 and 3, communication via pulse width modulation (PWM) between the home charger and the charging station or, in mode 2, the control unit in the charging cable (ICCB) and the vehicle takes place before and during the charging process. Several parameters are transmitted and synchronized via the CP (control pilot) and PP (proximity pilot/plug present) signal lines, see Fig. 12.9.

The charging standards IEC 61851, ISO 15118, DIN 70121, and VDV 261 define the interchange of information between the vehicle and the infrastructure. The communication is simple and fool-proof. It is based on voltage and resistance measurements and a PWM, which can be generated by a small micro-controller.

The charging station and the electric car use the PP contact to detect how much load the connected charging cable may carry. The plugs on both ends have a fixed resistor installed between PP and PE, which indicates the cross section of the charging cable, see Fig. 12.9.

The detailed AC charging procedure and communication is listed in Table 12.4.

The maximum admissible charging current of the charging point is determined by the car by measuring the duty cycle of the PWM. The duty cycle is the ratio of the high phase ("on" time) to the total cycle length (at 1 kHz this is equal to 1 ms). The coding of the duty cycle is given in Table 12.5. Between 10 and 85%, intermediate values are also permitted. Duty cycles below 3% or above 97% are not allowed and will stop the charging process (Fig. 12.10).

Note—Excess PV charging (solar mode)

Smart chargers can dynamically adjust the charging current via the CP signal to enable charging purely from the surplus energy of a photovoltaic system. The charging current is controlled depending on the momentarily PV generation and the consumption of the house.

However, most cars cannot charge with less than 6 A and some cars are unable to restart a charging process once it has been stopped, therefore smart chargers usually do not go below 6 A and do not stop charging even in the event of heavy cloud covers.

12.7.3 Combined Charging System (CCS)

CCS 1.0 supports DC charging currents up to a maximum of 200 A and voltages up to 500 V. This enables an output power of up to 100 kW at 400 V system voltage.

With CCS 2.0, the system limits are raised to 500 A and 920 V voltage. The power limit is 350 kW. This version of the standard is referred to as HPC350 (HPC = High-Power Charging) (Fig. 12.11).

12.7.4 High-Level Charging Communication

With DC charging, the energy flow from the charging station to the EV must be controlled instantaneously by the charging station (reactive load balancing). For this reason, far more complex communication with the vehicle's battery management system is required (HLC = High-Level Communication) than with AC charging, when the vehicle controls the charging current.

While initiating the charging process, the charging point signals the vehicle to switch to HLC by placing a PWM signal on the CP contact with a duty cycle between 3 and 7%.

Before the communication is established, however, a check of the communication line quality is performed by a signal level attenuation characterization (SLAC). SLAC is a protocol to measure the attenuation between two power line communication (PLC) modules. This is done to prevent crosstalk if there are several EVs connected to nearby charging stations.

After a successful SLAC procedure, the PLC modules of the charging station and the car set up a network. The network is based on CAN–bus over power line communication (PLC) on the CP signal line. Compatible Home Plug Green PHY modems (IEEE1901) and the established frequency band of 1,8 to 30 MHz are used.

The communication protocol is defined in the OCPP (open charge point protocol) as provided in the international ISO 15118 (road vehicles—Vehicle-to-grid communication interface) series of standards.

HLC is used for exchange of charging parameters, for example, voltage and current as well as other information, for example, state of charge, remaining charging time, and next maintenance. There is also the possibility to enable and operate a payment system.

Table 12.4 AC charging process for IEC type 1 and 2 plugs

State	CP	
A No car connected	+12 V	The socket is de-energized (N, L1, L2, and L3 are open) and the PWM is deactivated; a voltage of +12 V is permanently applied to CP via the 1 kΩ resistor.
B1 Car connected but not ready for charging	+9 V	When an electric car is plugged, it connects the CP line to PE via a diode and a 2.7 kΩ resistor. This reduces the CP voltage from +12 V to +9 V, which is recognized by the charging point.
B2 Exchange of max. current	PWM (+9 V/−12 V)	The PWM signal is activated with a pulse width corresponding to the available charging current, see table below. Due to the 1 kΩ resistor in the charging point, the diode and the 2.7 kΩ resistor in the electric car, the square wave signal at CP oscillates between +9 V and −12 V. The electric car measures the pulse ratio of the PWM signal to determine how much charging current is available.
C Connected and ready for charging	PWM (+6 V/−12 V)	When the electric car is ready to charge, it switches an additional 1.3 kΩ resistor between the diode and PE. This pulls the upper voltage of the square wave signal from +9 V to +6 V The charging point now activates the power supply of all contacts (N, L1, L2, and L3) and the charging process starts During charging, the car continuously monitors the PWM signal and dynamically adjusts the charging current whenever the charging point signals a new limit If the PWM signal stops, the electric car must stop charging immediately Once the electric car has finished charging or the driver interrupts the charging process, it deactivates the 1.3 kΩ resistor, causing the upper limit voltage of the square wave signal to return to +9 V. The charging station then de-energizes the socket again
D Connected and ready for charging, ventilation is required	PWM (+3 V/−12 V)	Same as state D but with additional ventilation. Instead of the 1.3 kΩ resistor, 270 Ω is used thereby reducing the CP signal to +3 V. This mode is rarely used.
E Internal fault	0 V	In the event of a connection error or other internal errors, CP and PE are connected, and the CP voltage becomes zero. The charging process stops immediately, and the plugs are unlocked.
F External fault, charging point unavailable	−12 V	If the charging station is no longer available, for example, in the event of a power failure, the CP signal is set to −12 V.

Table 12.5 Coding of the permissible current of the charging point in the PWM signal

High phase of duty cycle	Charging current limit
3 … 7%	Switch to high-level communication protocol ISO 15118
10%	6 A
16%	10 A
22%	13 A
26%	16 A
40%	24 A
53%	32 A
90%	64 A (single-phase only)
96%	80 A (single-phase only)

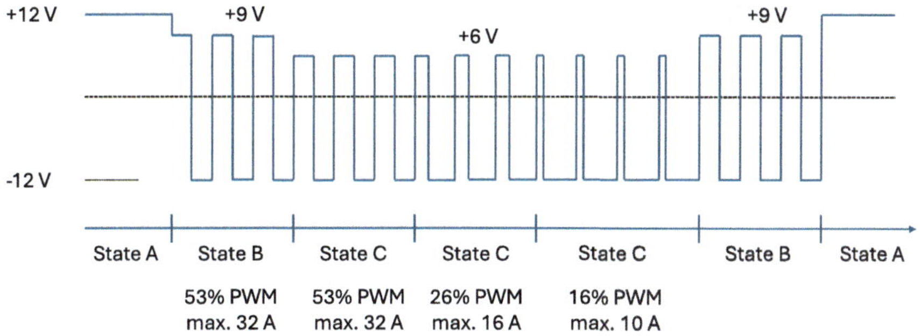

Fig. 12.10 Example of a CP signal during a charging process

12.7.5 Charging Authorization

The user must be authorized before the charging process can begin. In particular, the payment information for subsequent billing must be provided.

12.7.5.1 External Identification Means (EIM)

EIM is an identification method that requires additional user interaction or other identification processes. This can be a smartphone app, a credit card, an RFID card, or scanning the license plate at a charging station. Most fast charging stations currently only support EIM.

12.7.5.2 AutoCharge

The electric vehicle is automatically authorized at the charging station via its vehicle ID. When the vehicle is connected to the charging station with the charging cable, the vehicle sends its MAC address (media access control) to the charging

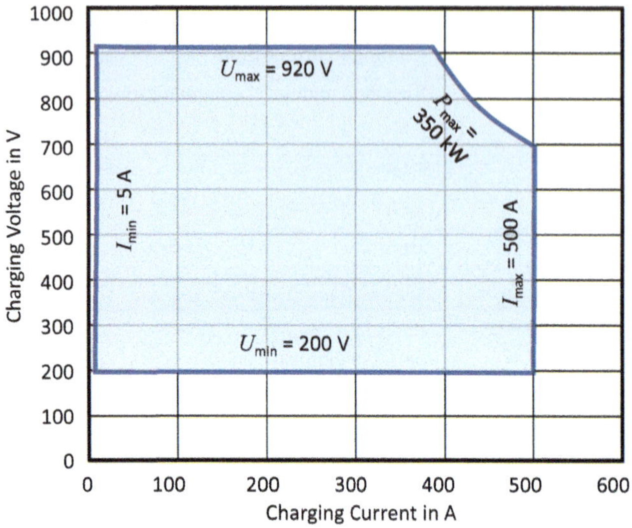

Fig. 12.11 Voltage, current, and power limits of the CCS 2.0 HPC350 standard

station. The charging station uses an OCPP request to send the data to the charging station management system (CSMS), where the address is matched against an internal or external address list. If the match is successful, an authorization response is sent back to the charging station; if the match is unsuccessful, an error status is transmitted. AutoCharge is mostly used at private charging stations and rarely at public charging stations due to security concerns.

12.7.5.3 Plug & Charge (P&C)

With P&C authorization technology, the vehicle is automatically identified after plugging in and authorized for charging. P&C works with an asymmetric key algorithm that uses a key pair with a private and a public key. These are linked so that coded messages can only be read with the corresponding key.

More and more manufacturers in Europe are supporting P&C. However, the pioneer in this field was Tesla, which has always offered P&C at all its superchargers.

Example 12.5—Charging at Tesla superchargers

Tesla Motors vehicles have been using a plug & charge process in the manufacturer's own charging network, the so-called superchargers, for many years. The charging stations have neither a display nor a keyboard.

The vehicle's charging port is opened by pressing a button on the plug of the cable. Billing information is exchanged via wireless communication between the vehicle and the charging station. The amount owed is shown on the vehicle display together with the SOC and the amount of energy charged. Billing is done automatically via the credit card stored in the Tesla account.

Today, many superchargers also support charging by non-Tesla cars. In that case, the charging process must be activated via the Tesla smartphone app. ◀

12.8 CHAdeMO

The CHAdeMO standard was developed already at the beginning of electromobility in 2010 by the Tokyo Electric Power Cooperation (TEPCO) and five Japanese automakers.

CHAdeMO is available in all markets worldwide but has largely been superseded by IEC CCS in North America, Europe, and India. China uses its own high-power charging standard. Tesla cars in the USA can be connected to a CHAdeMO charging station by a special adapter. The car automatically switches to the CHAdeMO communication protocol when the adapter is plugged in.

The CHADEMO plug has two power contacts (DC+and DC−) but no protective earth (PE). The ground contact is only intended for the signal lines. Communication takes place via two CAN–bus lines. Additional lines enable direct start/stop signaling between the charging station and the car. There is also an interlock line.

The first version of CHAdeMO was established in 2012 and offered a maximum charging power of 62.5 kW (125 A, 500 V DC) or 50 kW with 400 V. This is still the version found on most CHAdeMO chargers today.

CHAdeMO 2.0 was introduced in 2018 and allowed up to 400 kW (400 A, 1000 V DC) charging power. However, this standard has not become widespread outside Japan and some Asian countries.

The CHAdeMO standard has integrated vehicle-to-load capability since 2014. In this respect, it is ahead of CCS, which has still not fully integrated V2L by 2024.

12.9 ChaoJi (CHAdeMO 3.0)

The new ChaoJi charging plug has been under development since 2020. It was developed for near megawatt charging of large cars and light commercial vehicles.

ChaoJi allows up to 900 kW charging power (1500 V, 600 A with liquid cooling) (Fig. 12.12).

ChaoJi-1 is operating under the GB/T protocol, for primary deployment in mainland China. ChaoJi-2 is operating under the CHAdeMO 3.0 protocol, for primary deployment in Japan and other parts of the world. Both variants were harmonized in 2020/21.

The plug has three main connectors for DC+, DC−, and PE (protective earth). In addition, there are two pairs of signaling lines for CAN–bus communication (S+and S−) and for connection confirmation (CC1 and CC2).

Fig. 12.12 ChaoJi pinout

ChaoJi is V2L and P&C ready.

However, ChaoJi is incompatible with all previous charging systems. The charging power is far too high for cars but still too low for heavy commercial vehicles. Therefore, the success of the system is quite doubtful. As of 2024, there are no correspondingly equipped series vehicles.

12.10 Tesla Charging Standard

Tesla developed a proprietary charging connector for the Tesla Model S already in 2012 and used it on all their subsequent EVs in North America, Europe, and Asia.[11]

In 2022, Tesla opened the specification of their charging standard to all manufacturers in North America (NACS = North American Charging Standard). NACS is now being standardized as SAE J3400 and supported by a growing number of car makers. It is expected that NACS will supersede IEC Type 1 CCS in North America soon.

The Tesla plug uses a five-pin layout: two large pins for power transfer, one smaller pin for ground, and two signaling lines (CP = control pilot and PP = proximity pilot). The two large pins can be used to transfer high-power DC or single-phase AC for slow charging.

The communication uses a CAN–Bus interface. Tesla offers a short adapter cable that enables compatibility with CHAdeMO, so that Tesla cars can charge on CHAdeMO stations. However, this requires an adaptation of the communication protocol, which is automatically carried out by the vehicle when the adapter is plugged in.

Originally, the Tesla plug was rated for 500 V. In November 2022, a revision was made that created a backwards compatible configuration that is rated for up to

[11] In 2019, with the introduction of the Model 3 in Europe, the IEC Type 2 CCS standard was generally adopted by Tesla in the European market.

1000 V. There is no specification of the maximum current, but the temperature of the connector's interface must not exceed 105 °C. Tesla uses the connector up to 250 kW at 400 V in their version 3 superchargers. It is expected that higher charging power will be available at 800 V.

Up to version 3, the cable and plug are air cooled, very different from high-power CCS, which uses liquid cooling. With version 4, which was announced by Tesla by the end of 2022, liquid cooled cables were introduced for higher currents and megawatt charging.

The Tesla plug also includes a button on the top of the handle. When depressed, the plug emits a UHF signal. Before charging, this indicates the vehicle to open the charge port door electrically. When pressed while charging, the signal stops the charging process and unlocks the plug.

12.11 Charging Systems for Medium- and Heavy-Duty Vehicles

12.11.1 OppCharge

In 2016, the SAE (Society of Automotive Engineers = U.S. American standardization institute) formed the "medium- and heavy-duty vehicle conductive charging task force" to develop a recommended practice for heavy-duty electric vehicle conductive charging. This led to the OppCharge (opportunity charging) system, which was first introduced in 2016. The consortium was led by Volvo Buses and is supported, for example, by ABB and Siemens. The charging is performed via an infrastructure-mounted cross-rail connection, that is, a curbside charging station with an overhead structure overhanging the street. After the vehicle pulls up to the charging station, a pantograph is lowered from the overhead charger to connect to rails mounted on the forward roof of the vehicle.

Test installations of OppCharge have been made in Geneva, Switzerland, at Schipol Airport in Amsterdam, Netherlands and for the New York City Transit Authority along its M42 route.

Power ratings are currently 150, 300, and 450 kW. The charging current is controlled according to IEC 61851 part 1 and part 23 (electric vehicle conductive charging system).

12.11.2 Catenary Systems

Various systems for charging electrically powered hybrid trucks via overhead lines (catenary system—eHighway) are being tested in Germany and other countries. In contrast to OppCharge, which can only be used when the vehicle is stationary, the routes are several kilometers long and can be used at normal driving speeds.

For example, two test tracks with lengths of 12 and 5 km were built near Frankfurt, Germany. Another 2.6 km track is located near Karlsruhe, Germany.

Table 12.6 MCS charging levels

Case	Charge current	Connector	Inlet
Level 3+	3.5 kA	Liquid cooled	Liquid cooled
Level 3	3 kA	Liquid cooled	Liquid cooled
Level 2+	1.5 kA	Liquid cooled	Passive cooling
Level 2	1 kA	Liquid cooled	Passive cooling
Level 1	350 A	Passive cooling	Passive cooling

The energy to the vehicles is supplied by a two-pole 600 V DC overhead contact line with two parallel catenary systems, each consisting of a high-strength copper contact wire with magnesium alloy and a suspension cable. The vehicles are equipped with two pantographs mounted side by side.

However, catenary systems are not supported by any vehicle manufacturer. It is therefore questionable whether this technology will become established in the future.

12.11.3 Megawatt Charging System (MCS)

The megawatt charging system (MCS) for heavy commercial vehicles is a further development of the CCS system and was presented at the EVS trade fair in Oslo, Norway, in 2022.

The maximum voltage is 1250 V and the maximum current is 3000 A resulting in a maximum power of 3750 kW. The plug is intended for installation in trucks and is always fitted on the left-hand side, approximately where the tank inlet is located in conventional vehicles. From the beginning, the MCS system is designed to feed energy back into the grid (V2x) (Table 12.6 and Fig. 12.13).

In addition to the two power contacts and the protective earth (PE) contact, the plug has the two familiar CP and PP connections for high-level communication.

12.12 Contactless Inductive Charging

With inductive charging, the energy is transferred contactless by a magnetic field, like a transformer. The first patent for a contactless electric charging of a vehicle dates to 1903 and was granted to the Westinghouse company.

Contactless charging requires no driver intervention. There is no need to plug and unplug the cable and charging takes place automatically as required, resulting in a high level of convenience and simple operation. Inductive charging is protected against vandalism and the effects of the weather as the coil is invisibly integrated into the pavement. Finally, the aesthetics are advantageous, as neither pillars, wallboxes nor cables are visible.

Fig. 12.13 The MCS plug

Contactless charging can be used in a private garage when the car is parked for a longer period of time. Contactless chargers can also be installed at frequently used fixed points, such as taxi stations at airports or at train stations. There are even developments to enable inductive energy transfer while driving.

Example 12.6—Inductive taxi charging

In 2020, the EnergiCity project has started the test phase for contactless charging of electric cabs in the Norwegian capital Oslo. 25 Jaguar I-Pace will be inductively charged in taxi use. The charging power exceeds that of standard AC charging (Fig. 12.14). ◀

Example 12.7—Inductive roads

Sweden is building a permanent electrified road for EVs to charge while driving by 2025. The road is between the cities Hallsberg and Örebro, in the middle of the country's three major cities, Stockholm, Gothenburg, and Malmö.

In 2023, North America's first wireless charging road debuted in Detroit, Michigan. The quarter-mile stretch of road comes from a collaboration between the state of Michigan and the tech company Electreon. A one-mile stretch of a four-lane highway in Florida will be equipped with inductive charging in 2024/25.

Already in 2018, China built a 1 km long stretch of inductive road in the city of Jian, 300 km south of Beijing, that even has solar panels integrated into the pavement. ◀

Fig. 12.14 Taxi stand in Oslo, Norway, with inductive charger, 2022

However, the costs of inductive charging systems, i.e. charger, charging coil, receiver coil, and charging regulator in the car, are significantly higher than with conventional, plugged-in cable solutions. Also, the efficiency of contactless charging is lower. Therefore, contactless charging is not a replacement but a useful addition to charging via cable.

12.12.1 Technology

A stationary and a mobile system are required for energy transmission.

Depending on the application, the stationary systems can utilize line conductors, which are permanently installed, elongated conductor loops for stationary charging on public parking spots, or compact coils, e.g. for installation in the garage floor as the primary coil of a transformer.

Inductive charging with line conductors is also possible at high speeds of up to 100 km/h.

The mobile system, i.e. the secondary coil, is installed in a transformer head in the vehicle. It can move along the line conductor or is positioned as precisely as possible above the coil of the stationary system. Multiple inductive heads can also be used. The air gap is approx. 15–30 cm. Typical positioning accuracies are in the range of around 10–15 cm, both laterally and longitudinally.

Due to the small main inductance, the transformer has unfavorable transmission characteristics. The large air gap and the correspondingly large leakage as well as the low main inductance of the line conductor or coil with only a few windings lead to the generation of only a low voltage in the secondary coil. For compensation, capacitors are used to form a resonant circuit, which produces approximately the original voltage.

This capacitor can be connected either in parallel (C_P) or in series (C_s) with the coil. If a high voltage is required in the secondary part, a capacitor connected in parallel is used. If a high current is required, the capacitor connected in series is advantageous.

Inductive transmission elements can be designed to be particularly small at high frequencies, but the cost of distance compensation increases with increasing frequency. Frequencies in the range of 25–150 kHz (long-wave radio) are therefore usually used. The current standard for contactless charging of electric vehicles provides for a transmitter frequency of 140 kHz.

To make the voltage and frequency of the transformer heads usable, a so-called matching converter is inserted. It converts the alternating voltage from the kHz range into a DC voltage.

Information is exchanged between the vehicle and the energy transmission system by wireless network (WLAN).

The maximum theoretically achievable efficiency depends solely on the coupling k and the quality of the coils Q_1, Q_2. Coil quality is the ratio of reactance (X_L) to ohmic resistance (R): $Q = X_L/R$. The higher the Q, the better the coil. A lower coupling can be compensated for by a higher coil quality and vice versa. Added to this are the losses in the power semiconductors and passive components such as coils and capacitors.

Coil qualities of up to around 100–1000 are economically feasible. This allows air gaps in the order of the coil diameter to be realized with good efficiencies.

Depending on the distance, geometry, and quality of the coils, transmission losses of between 10 and 20% occur.

12.13 Hydrogen Filling Stations

Recharging hydrogen vehicles requires filling stations that are only at first glance similar to today's filling stations for gasoline or diesel fuels.

The downstream sector of the fuel industry comprises the processing chain from crude oil to mineral oil in refineries, as well as the distribution and transportation of mineral oil products. Some oil companies have the downstream chain completely in their own hands. But independent importers and traders also operate in the wholesale and retail sectors.

Example 12.8—Downstream sector for the fuel industry in Germany

In Germany, most oil-based fuels are currently produced in 14 domestic refineries.

In addition, there is a nationwide network of 279 fuel depots with a total capacity of around 68 million m^3, which are supplied from the refineries by ship; rail; or, in some cases, by pipeline. The fuel is delivered to wholesalers by inland tankers, rail tank cars, road tankers, or through the approximately 3300 km of product pipelines, which in turn supply the retail trade, i.e. public road filling stations.[12] ◀

All of this would have to be rebuilt to supply millions of fuel cell vehicles with hydrogen.

12.13.1 Technology

The refueling process for 700 or 350 bar compressed hydrogen (CH_2) is complex. Today's hydrogen filling stations have one or two fuel dispensers that receive the high-pressure hydrogen from a nearby technical area with a size of several hundred square meters.

Within that technical area, a cylindrical gas tank around 10 m high supplies the hydrogen. When full, it has a pressure of around 250 bar. The pressure reduces according to the fill level. The tank's capacity is sufficient to refuel around 200 vehicles.

The vehicle tanks are initially filled with an overpressure of 220 bar. The pressure is increased to 700–900 bar during the refueling process. The hydrogen is refueled at a temperature of −40 °C, so the nozzle cools down noticeably. The low temperature is required because the hydrogen gas heats up during expansion in the tank. There is a risk of explosion above temperatures of around 85 °C.

A fairly large amount of energy is required for the compression of the hydrogen to 750 bar and for the cooling process. The thermodynamic loss of the compression is around 8%, but the pressure must be generated via a high-pressure pump system with relatively poor efficiency. The cooling is also inefficient. This results in total fueling losses of 12–15%.

Table 12.7 gives an overview of different hydrogen filling stations, their sizes and capabilities. As of today, nearly all existing filling stations are "S" size.

The "required space" listed in the last line of the table is the additional space for the technical area that is needed to prepare the hydrogen before fueling. This space is in addition to the normal space requirements of a filling station (refueling area, parking space, buildings, etc.).

Example 12.9—Supplying cars with hydrogen filling stations in Germany

The annual driving range of all cars in Germany is about 580 billion kilometers. There are 14,000 gas filling stations. Assuming 400 km distance between refueling, each filling station serves around 280 cars per day.

[12] Source: *Sektoruntersuchung Kraftstoffe (B8-100/07)*, Bundeskartellamt, Zwischenbericht Juni 2009.

Table 12.7 Size and performance of hydrogen fuel stations[13]

Size	S	M	L
Vehicles	Car, Light commercial vehicles (LCV)	(Cars, LCV, Busses), Medium-sized commercial vehicles (MCV)	(Cars, LCV, Busses), MCV, Heavy-duty vehciles (HDV)
Max H_2 throughput per day	200 kg	500 kg	1000 kg
Average H_2 throughput per day	150 kg	350 kg	700 kg
Average number of vehicles refu-eled per day	6 cars	70 cars	10 trucks or 140 cars
Yearly H_2 throughput	1–10 t	>100 t	>500 t
Number of fuel points	1	2	2–3
Required space	80–250 m^2	200–350 m^2	250–800 m^2

An "L"-type hydrogen filling station can serve half as many cars, namely, 140, and that doesn't even include refueling commercial vehicles (trucks).

Therefore, the additional technical area needed to convert an average gas filling station entirely to hydrogen would be more than a square kilometer. ◀

Example 12.10—Current utilization of hydrogen filling stations in Germany

In mid-2023, Germany had about 90 hydrogen filling stations and about 1800 FCEVs. With a typical annual mileage of 12,000 km and a refueling stop every 400 km, each hydrogen filling station carries out an average of 1.6 refueling processes per day. ◀

The hydrogen refueling process takes around 5 min plus the time required for starting and stopping and the payment process. This is similar to the time required to refuel a conventional vehicle.

However, the tank pressure is provided by an intermediate high-pressure storage at the filling station, which must be refilled and cooled during the breaks between two refueling processes. Depending on the technology used, this process takes around half an hour, which limits the refueling capacity to around 40–50 vehicles per day and pump. In future generations of hydrogen filling stations, the aim is to shorten the waiting times between refueling by using additional high-pressure storage tanks and more powerful pumps. However, this will increase the cost even more and require even more space for the technical area.

[13] Source: *Wasserstoffbetankung von Schwerlastfahrzeugen – die Optionen im Überblick*, H2Mobility, 08/2021; LCV = Light commercial vehicles, MCV = Medium-sized commercial vehicles, HDV = Heavy-duty vehicles.

Example 12.11—Hydrogen filling stations and EV chargers in Europe

At the beginning of 2024, there were around 160 small public 700 bar hydrogen filling stations for cars in Europe. Over half of these are in Germany (88), a further 16 in the Netherlands, 15 in Switzerland, and 11 in France. All other countries have only a few stations.

For comparison, the number of gas filling stations in Europe is about 140,000 and the number of DC fast chargers is about 70,000. Each fast charger can recharge almost the same range per hour as a small hydrogen filling station (up to two BEVs, each with 150–250 kW peak power = 100–150 kW average power = 300–400 km recharged range per vehicle). ◀

Framework of Individual Mobility 13

Abstract

Based on the concept of sustainability, this chapter motivates and explains regional and global climate protection efforts and targets. This is followed by an overview of the global automotive landscape: vehicle fleets, consumption, and mileage. Further sections shed light on global emissions in connection with electricity production, the switch to renewable energies, and the status of electromobility. The chapter concludes with a discussion of the various raw materials required for electromobility. Their range, production conditions, and possible environmental damage. And finally, a section on second use and recycling of batteries.

13.1 Introduction

It is a sound engineering practice to initially assess new ideas through realistic estimates before undertaking comprehensive studies or commencing detailed designs. This approach facilitates a visualization of structural dependencies, key influencing factors, and the anticipated magnitudes of results. Moreover, it offers the advantage of swift outcomes. If these preliminary estimates suggest that the desired results are unattainable, in-depth studies may often prove unwarranted.

Studies frequently encounter challenges in terms of comprehension due to their extensive scope and detailed content, particularly when multiple research institutes are involved, each contributing a small portion to the overall picture. From the author's experience, this scenario often results in coordination issues, potentially rendering the results unrealistic and impractical. Additionally, studies necessitate various assumptions about future technological developments and political framework conditions, which may not always be realistic and openly communicated.

M. Doppelbauer, *Introduction to Electromobility*,
https://doi.org/10.1007/978-3-658-45482-1_13

Notably, many studies may have an underlying agenda, influencing the message they aim to convey.

The subsequent chapters of this book present methods and procedures for obtaining realistic estimates in the realm of electromobility in a relatively straightforward manner. This is not intended to replace thorough studies but aims to empower readers to better comprehend and evaluate the outcomes of such studies. Background information and figures are also supplied, serving as a foundation for further considerations.

The environmental and cost considerations surrounding individual mobility are dynamic and ever-evolving, shaped by political, social, and technological factors. Ongoing adjustments in framework conditions, exemplified by events like the diesel scandal, continue to impact the discourse. Additionally, advancements in technology, especially in battery capacity and cost, play a pivotal role. The public debate, as amplified through the press and the Internet, continually accentuates different focal points.

However, some facts are not debatable.

Electrically powered vehicles stand out for their significantly higher efficiency when compared to combustion engine counterparts. In a direct comparison, the electric drive achieves an efficiency of around 80–90% over a broad operating range, contrasting with the best diesel units that barely reach 35% under optimal conditions and only about 25% in normal operation (including an automatic transmission). In the end, the best way to avoid CO_2 emissions is to stop burning stuff.

Assessing the carbon footprint of electric vehicles involves a multifaceted analysis. Key considerations include the source of electrical energy used to charge these vehicles, losses during transmission and charging, the CO_2 footprint of battery production, and the potential for mitigating emissions through recycling and dual use.

The production of liquid fuels also contributes to CO_2 emissions, raising questions about the corresponding electrical energy requirements.

A substantial reduction of CO_2 emissions in road-based individual and commercial traffic requires comprehensive efforts. The question arises about the costs associated with establishing a charging network and whether alternatives like e-fuels or hydrogen could be practical options.

The subsequent chapters delve into these complex considerations, using current examples and figures to elucidate the key aspects of creating a comprehensive environmental and cost balance for various motor vehicle drive technologies. While the specific examples and figures may become outdated over time, the methodologies introduced in these chapters retain lasting value, offering a framework for ongoing assessment and understanding of the evolving landscape of sustainable mobility.

13.2 Climate Protection

13.2.1 Sustainability

To ensure the long-term existence of humanity on this planet, a fundamental shift towards consuming renewable (e.g. plants), inexhaustible (e.g. wind and sunlight), and replaceable resources obtained through recycling is imperative. This guiding principle is commonly known as sustainability. It extends far beyond energy considerations and encompasses food, water consumption, land use, raw materials for production, and more.

Historically, sustainability was inherent until the onset of the industrial revolution in the mid-nineteenth century, ushering in a lifestyle marked by resource wastefulness. With a few exceptions, such as deforestation in ancient Roman times, societies operated within sustainable parameters.

Examining the estimates for energy resources (primarily crude oil and natural gas) and raw materials critical for key technologies (primarily copper and lithium), it becomes apparent that, at current consumption levels, these resources are projected to last well over 100 years. However, when viewed in the context of humanity's age—well over 100,000 years—this timeframe is notably brief. Future generations, perhaps the next or the one after that, are likely to grapple with a severe shortage of crucial raw materials.

Addressing this challenge becomes humanity's paramount task to secure its future, requiring a return to complete sustainability within the next 100 years. This transformative approach is commonly referred to as the circular economy—a model that emphasizes the cyclical use of resources, reducing waste, and promoting recycling to minimize environmental impact and ensure the long-term viability of essential resources.

13.2.2 Global Climate

A fundamental aspect of sustainability involves achieving climate neutrality, meaning the avoidance of emissions, particularly greenhouse gases, that pose a threat to the global climate. The consensus among the vast majority of climate researchers is that the significant warming observed over the last century, as illustrated in the previous figure depicting the average world's temperature curve, is primarily attributable to human activities (Fig. 13.1).

There is a widespread understanding that a global temperature increase of 2 °C or more would result in severe and unmanageable consequences for the environment. These consequences include rising sea levels, crop losses leading to famines, and an increased frequency of severe storms, even in temperate climate zones. Recent occurrences, such as intense hailstorms in Europe, serve as tangible examples of the impact of climate change. Concerns are mounting that the 2 °C limit may have been partially surpassed already in 2023. Consequently, climate

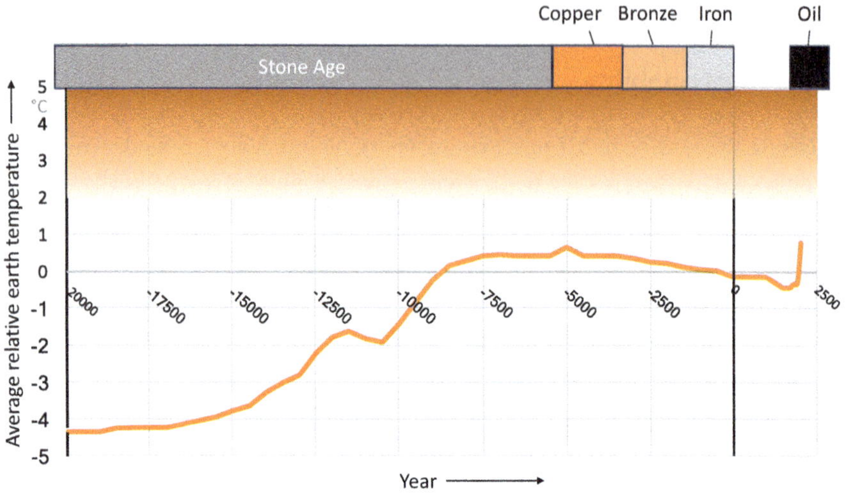

Fig. 13.1 Development of the mean earth temperature from 20,000 BC to the present day

change has transcended the realm of purely scientific discourse and has become a matter of broad public concern.

Another critical issue is the potential transformation of rainforests into savannahs and the thawing of permafrost soils in Siberia and Alaska due to global warming. This process could release substantial amounts of methane, triggering a potential avalanche effect that could jeopardize life on our planet in the medium term.

Recognizing the urgency of these challenges, 195 countries came together for the Paris Climate Change Conference in December 2015. In a landmark move, they collectively agreed on a comprehensive, legally binding, and global climate protection agreement. The primary objective of this agreement is to curb the rise in the global average temperature to well below 2 °C compared to pre-industrial levels. This agreement represents a crucial step in addressing the complex and interconnected issues surrounding climate change on a global scale.

13.2.3 Final Energy Consumption in Germany

Looking at Germany's total energy consumption, the following picture emerges (Table 13.1).

At just under 480 TWh, electrical energy is only in third place behind gas (580 TWh) and oil (850 TWh). However, it is by far the most efficient final energy source and the only one that can be produced in a largely CO_2-neutral way.

The table also clearly shows why it is so important to switch to electromobility as quickly as possible, at least for passenger cars.

Table 13.1 Final energy consumption by energy form and sector in Germany in TWh, 2022[1]

	Industry	Households	Business, trade, and services	Traffic	Total
Stone and brown coal	109	–	–	–	109
Oil and liquid fuel	23	124	53	648	848
Gas	227	249	103	2	581
Electricity	203	139	123	12	477
District heating	45	51	–	–	96
Other	60	115	46	36	257
Total	667	678	325	698	2368

13.2.4 Anthropogenic Emissions of Pollutants

Anthropogenic emissions refer to the release of pollutants into the atmosphere resulting from human activities. While anthropogenic emissions are relatively modest when contrasted with natural emissions originating from oceans, soils, or terrestrial vegetation, a delicate balance exists between natural emissions and natural absorption processes, including those occurring in oceans, soils, and vegetation.

The critical disruption occurs when additional anthropogenic pollutants are introduced into the environment. Approximately half of the carbon dioxide (CO_2) emissions generated by human activities cannot be effectively absorbed by the natural environment. This imbalance leads to a net increase in the atmospheric CO_2 content, contributing to the ongoing challenge of climate change. Recognizing and mitigating these anthropogenic emissions is essential to restoring a more sustainable equilibrium in the Earth's climate system.

In absolute terms, China is the inglorious leader of anthropogenic CO_2 emissions with over 11 billion tons in 2020, followed by the USA (5.1 billion tons), India (2.6 billion tons), Russia (1.8 billion tons), and Japan (1.2 billion tons). Germany follows in sixth place with 700 million tons.

Australia and the USA emitted around 15 tons of CO_2 per capita in 2021, while Germany and China were well below this at 8 tons each. However, the Arab countries are the frontrunners (Qatar 36 t CO_2 per inhabitant per year, UAE 22 t).

Figure 13.2 shows the breakdown of domestic greenhouse gas emissions (including inland and coastal shipping) for Germany. In addition, international shipping (container ships) and air traffic account for a significant proportion of total emissions (not shown).

[1] Data source: Umweltbundesamt, Endenergieverbrauch nach Energieträgern und Sektoren, Januar 2024.

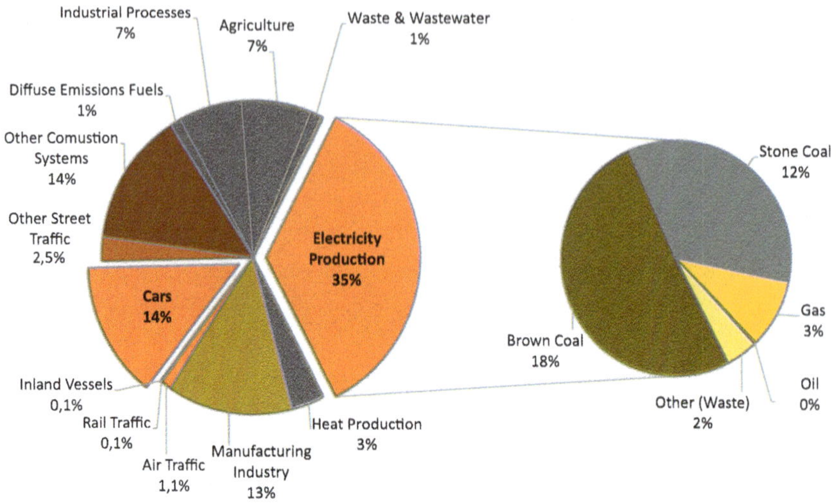

Fig. 13.2 Anthropogenic greenhouse gas emissions in Germany, 2014/15[2]

In this example, electricity production is responsible for around one-third of total greenhouse gas emissions. Half of this is caused by lignite-fired power plants and a third by hard coal-fired power plants.

The second major contributor to emissions is the transport sector, and here predominantly car traffic, which is responsible for over 86% of all greenhouse gas emissions from domestic transport.

In total, car traffic is responsible for around 14% of all anthropogenic greenhouse gas emissions in Germany.

13.2.5 Climate Protection Targets

13.2.5.1 Europe

Already in 2008, the EU set the target to cut emissions by 20% compared to 1990 levels by 2020. In part due to the COVID-19 pandemic, this goal was achieved. Emissions had dropped to 24% by 2019 and to 31% by 2020.

Under the Paris agreement in 2015, the EU committed to reducing greenhouse gas emissions by at least 55% by 2030 and to reach climate neutrality by 2050. The EU is also working to achieve a circular economy by 2050, create a sustainable food system, and protect biodiversity and pollinators.

In June 2022, the European Parliament backed a proposal to reach zero CO_2 emissions from new cars and vans in the EU by 2035.

[2] Data source: Umweltbundesamt, Germany.

13.2.5.2 Germany

The Paris climate protection targets were broken down into a national energy policy by the federal government (Federal Climate Protection Act of 2019). The main goals of the so-called energy transition (Energiewende) are as follows:

- Greenhouse gas emissions must be reduced to 440 million tons of CO_{2equiv} by 2030 (65% less than in 1990), to 150 million tons of CO_{2equiv} by 2040 (88% less than in 1990), and to zero net emission in 2045. From 2045, there must be a balance between greenhouse gas emissions and their reduction. The target for 2050 is even negative emissions, i.e. more greenhouse gases will be absorbed into natural sinks than emitted.
- The share of renewable energy sources (solar, wind, etc.) in final energy consumption is to increase to 60% by 2045.
- The share of renewable energy sources in the electricity supply should increase to over 80% by 2045.
- Energy consumption in the transport sector must be reduced by 40% by 2045 compared to 2005.

To achieve these climate targets, practically all coal-fired power plants will have to be shut down by 2045. In private transport, most vehicles must be electrically powered. Vehicles with combustion engines may only represent a small minority and must use synthetic e-fuels.

The German Expert Council for Climate Issues presents a report every 2 years on the targets, measures, and trends achieved to date.

13.2.5.3 China

Since the Paris Agreement in 2015, China had reduced its carbon intensity (CO_2 emissions per GDP) by 51% from 2005 levels. 50.9% of the country's power generation mix was already non-fossil fuel based.

China's current climate plan targets peak carbon dioxide emissions before 2030, and carbon neutrality before 2060. Meanwhile the country has committed to lower its carbon intensity by over 65% in 2030 from 2005 levels.

The plan also includes increasing the share of non-fossil fuels in primary energy consumption to around 25% in 2030 and increase installed capacity of wind and solar power to over 1200 GW by 2030.

China plans to release new 2030 and 2035 national climate targets under the Paris Agreement's framework in the year 2025.[3]

13.2.5.4 USA

Climate policy in the USA is heavily dependent on the party that holds the presidency. President Biden sets the following goals after taking office in 2021:

[3] Source: S&P Global, COP28: China to release 2030, 2035 national climate targets in 2025, December 02, 2023.

- Reducing U.S. greenhouse gas emissions by 50–52% below 2005 levels in 2030.
- Reaching 100% carbon pollution-free electricity by 2035.
- Achieving a net-zero emissions economy by 2050.
- Delivering 40% of the benefits from federal investments in climate and clean energy to disadvantaged communities.

However, there are fears that the climate targets will be abandoned if the Republican Party takes over the presidency in 2025.

13.3 Megacities

One prevalent trend that all forecasts predict will intensify in the coming years is the rise and expansion of densely populated urban areas, commonly referred to as megacities. Since 2007, most of the global population has been residing in cities, marking a profound shift in demographic patterns.

Currently, there are over 30 megacities worldwide, each hosting more than 10 million inhabitants. Topping this list is the Tokyo metropolitan area with approximately 39 million residents, followed by populous cities such as Jakarta, Delhi, Manila, Sao Paulo, Seoul, Mumbai, and Shanghai. In Germany, the Ruhr region stands as the largest conurbation, boasting over 6 million inhabitants. The epicenter of urbanization is notably in Asia, including India.

Public transportation by train or bus is evidently the most efficient means of mobility in megacities. Nevertheless, the electric car remains an ideal vehicle for individual transportation within these vast urban landscapes due to its absence of local pollutant emissions and user-friendly features, such as one-pedal driving. In this context, the limitations of range and extended charging times do not pose a significant drawback.

However, the increasing density of cars per capita has surpassed acceptable levels in numerous large cities. It is plausible to anticipate that private transport within urban centers will decline in the medium term, giving way to alternative mobility models, including local public transport, car sharing initiatives, and autonomous driving technologies. This shift reflects a broader trend towards more sustainable and efficient modes of transportation in the evolving landscape of megacities.

13.4 Car Fleets

13.4.1 Fuel Economy of Cars in Germany

The carbon dioxide (CO_2) emissions produced during combustion are in a direct correlation with the quantity of fuel consumed. Consequently, mitigating emissions from combustion engines is inherently intertwined with the imperative to reduce fuel consumption. Despite concerted efforts, particularly in the introduction of hybrid cars, achieving success in this realm has proven elusive over the past 15 years. To the contrary, the average fuel economy of passenger cars in Germany has experienced a gradual increase in recent years and currently hovers at 7,4 L per 100 km.

The preceding figure shows the average real-world fuel consumption per car in Germany, calculated from the total amount of fuel sold per year in relation to the annual mileage of all cars.

13.4.2 Global Fuel Economy of Cars

The International Energy Agency (IEA) has published a report on vehicle fuel economy in global markets for 2005–2019.[4]

The fuel consumption in this publication is based on national test cycles that have been converted to WLTP. The reported values are therefore significantly lower than the real-world consumption shown in Fig. 13.3. For example, the IEA reported a fuel consumption of 6.2 L petrol equivalent (lge) per 100 km in Germany in 2019 when the real-world value was 7.4 l/100 km. Nevertheless, the study can be used for comparisons between regions and to identify trends.

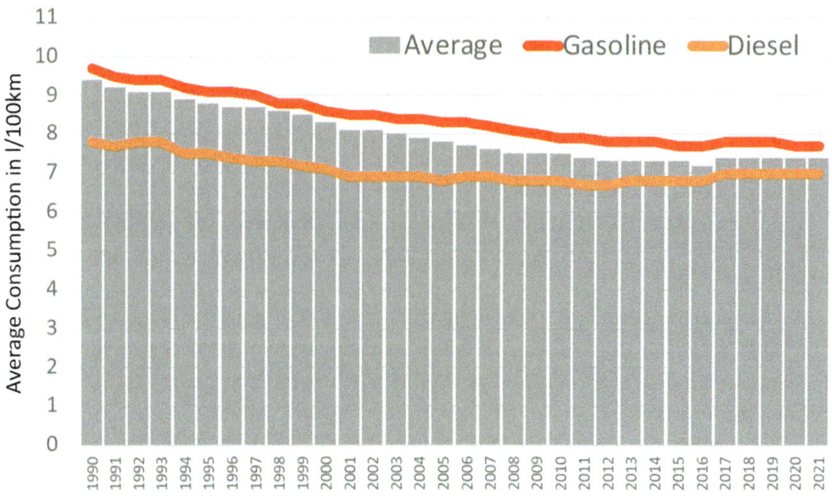

Fig. 13.3 Average real-world fuel consumption per car in Germany[5]

[4] Source: *Vehicle fuel economy in major markets 2005–2019*, working paper 22, IEA, 2021.

[5] Data source: *Verkehr in Zahlen 2023/2024*, 52. Jahrgang, Herausgeber: Bundesministerium für Verkehr und digitale Infrastruktur, Deutschland.

According to the IEA paper, the fuel economy in 2019 was best in Europe and Japan with about 5.8 lge/100 km. The world average is 7 lge/100 km. That was also China's level. Passenger vehicles in the United States, however, have an average consumption of more than 8.5 lge/100 km.

A comparison of global values over the last 15 years shows a downward trend from around 9 to around 7 lge/100 km. However, most of the improvements occurred between 2005 and 2015. In Europe, in particular, average fuel consumption has risen since 2015 from around 5.5–5.8 lge/100 km. An upward trend can also be observed in the United States.

Other sources also confirm that total global emissions from cars and vans have been on an almost constant upward trend for the past decades, only interrupted in 2020 by the coronavirus pandemic. In 2022, they amounted to 3530 million tons of CO_2.[6]

It is frustrating to see that fuel consumption has not fallen in recent years. Obviously, it is not possible to compensate for the increased consumption of large, heavy SUVs through further developments in combustion engine technology. Even the introduction of hybrid vehicles, which took place on a large scale a few years ago, has not been able to reverse the trend.

13.4.3 Stock of Cars

In 2014, the total number of passenger cars worldwide was around 1.2 billion vehicles with an average annual growth rate of 3.7%.

The number of passenger cars in Germany has grown steadily at an average annual rate of 1.2% over the last 10 years. It currently stands at just under 49 million vehicles.

Especially in the emerging BRICS countries, which account for around 40% of the world's population, the number of vehicles is still comparatively low: in Brazil, 250 out of every 1000 inhabitants have a four-wheeled road vehicle (car, SUV, truck, van, bus, etc.), in Russia 290, in India only 32, in China 205, and in South Africa 165.

By comparison, the larger European countries (Germany, France, UK, Spain, Italy) and Japan are in the range of 520–680 vehicles per 1000 inhabitants. In the USA, just under 800 road vehicles per 1000 inhabitants are registered.

It can therefore be assumed that the number of new vehicles sold in many countries will increase in the future. Due to environmental issues (local air pollution is a major health problem), strong political pressure for electromobility is to be expected.

While the vehicle density per capita in emerging and developing countries is therefore likely to increase, the question can be asked as to whether it might not

[6] Source: *Carbon dioxide emissions from cars and vans worldwide from 2010 to 2022*; statista, 2024.

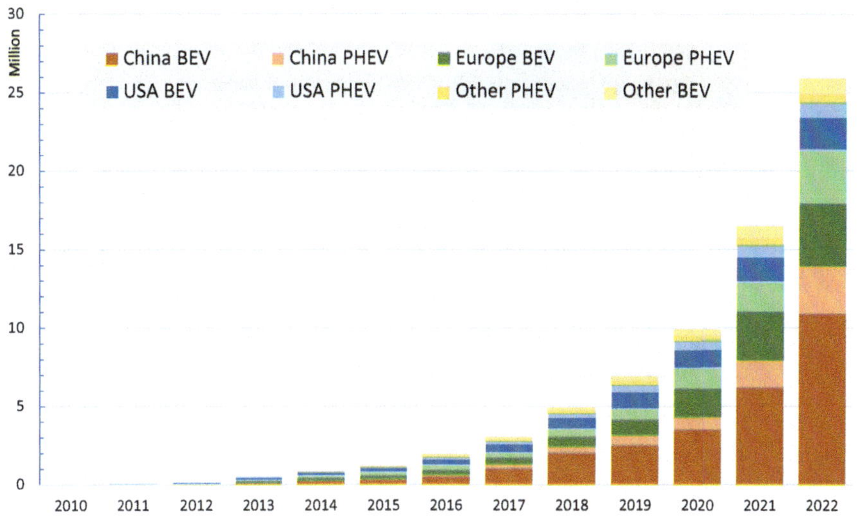

Fig. 13.4 Global stock of electric cars (BEV, PHEV)[7]

decrease again in industrialized countries in the future. This will be facilitated by new business models for meeting individual mobility requirements (car sharing, autonomous driving, etc.).

13.4.3.1 Battery Electric Vehicles

At the end of 2022, there were around 26 million electric passenger cars and light commercial vehicles worldwide. However, this figure includes plug-in hybrids. Most of them are driven in China, with the USA in second place, followed by Germany, France, and the UK. A global stock of over 200 million BEVs is expected by 2030.

To achieve the climate targets of the Paris Agreement, the International Energy Agency (IEA) assumes that the number of electric vehicles worldwide will have to reach 600 million by 2040. That is around half of the current total of 1.2 billion cars (Fig. 13.4).

The worldwide annual growth rate of battery electric vehicles has been 77% in average over the last 10 years from 2012 to 2022.

By the end of 2022, there were about 1,8 million slow charging points and 900,000 fast charging points in the world (760,000 in China, 70,000 in Europe, 28,000 in the United States, and more than 40,000 in the rest of Asia).

[7] Data source: *Global EV Outlook 2023*, IEA, 2023.

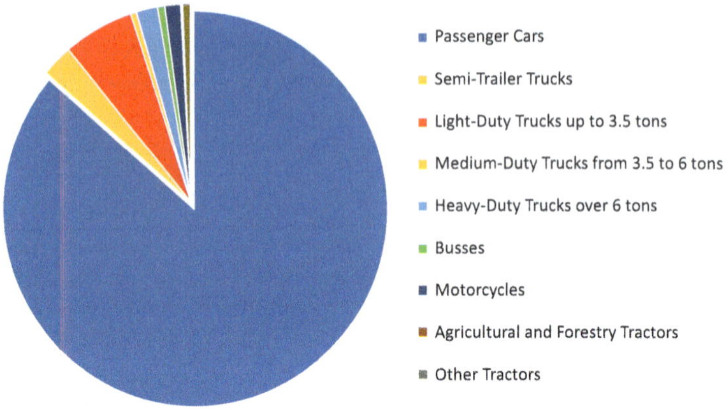

Fig. 13.5 Share of traffic volume (annual mileage) of the various types of motor vehicles in Germany in 2015

13.4.3.2 Fuel Cell-Based Vehicles

Just like electric cars, the number of fuel cell vehicles is also increasing, albeit at a much lower level.

At the end of 2022, there were 72,000 fuel cell vehicles worldwide, of which 58,000 were passenger cars.[8] Most of the fuel cell passenger cars are in South Korea (29,500) followed by California, USA (15,000), Japan (7500), and Germany (2200). China is not a market for fuel cell passenger cars (less than 500).

However, China has the most fuel cell-based light commercial vehicles in the world (800) followed by Germany (43). China is the only country in the world with a notable number of fuel cell medium-duty trucks (3900) and heavy-duty trucks (3300).

All in all, the number of fuel cell vehicles is very low. There are only 2.2 fuel cell passenger cars per thousand battery electric cars in the world and they are concentrated in just four countries (South Korea, California, Japan, and Germany). The share of commercial vehicles is even lower, and they are mostly located in just one country (China).

In the future, the gap between battery electric passenger cars and hydrogen cars will widen even further, as the annual growth rate of battery cars is more than 50% higher (BEV: 77% year-to-year growth compared to FCEV: 50%).

The total number of hydrogen filling stations in the world was just over 1000 by the end of 2022.[61] However, this includes both public and private stations and it includes fueling stations for passenger cars (700 bar) and for commercial vehicles,

[8] Source: *Deployment of Fuel Cell Vehicles in Road Transport and the Expansion of the Hydrogen Refueling Station Network: 2023 Update*; Remi Can Samsun, Michael Rex, Energie & Umwelt, Volume 611, Forschungszentrum Jülich, Germany.

which are mostly 350 bar and therefore incompatible with cars. There were 320 stations in China (some still in construction), 213 in Korea, and 164 in Japan. Germany was reported with 95 stations followed by the United States with 71 stations and France with 58 stations. There are now reports from large oil companies that are permanently closing their hydrogen filling stations for cars worldwide due to high costs and technical problems.[9]

Compared to the number of fast chargers for electric cars, which can recharge at roughly the same number of kilometers per hour as a typical hydrogen station, the proportion of hydrogen filling stations is only 1.1 per thousand.

13.4.4 Traffic Volume

According to the Federal Motor Transport Authority, car traffic in Germany is responsible for more than 86% of total street mileage, or 618.7 billion km per year. Nevertheless, the average car only covers 13,000–14,000 km per year, that is, an average of around 35 km per day. The figures in the neighboring German-speaking countries of Austria and Switzerland are similar.

In addition to cars, 5.9% of the total mileage is covered by trucks up to 3.5 t and 2.6% by articulated trucks. The remaining 5% is made up of all other types of vehicles (motorcycles, buses, agriculture, forestry, etc.) (Fig. 13.5).

13.4.5 Daily Mileage of Cars

In more than 100 years of individual road transportation with internal combustion engines, users have become accustomed to being able to spontaneously travel any distance with their car at practically any time. This option is taken for granted and considered necessary, even though most daily journeys are much shorter (Fig. 13.6).

With a range of 100 km, 87% of all private and 70% of all commercial journeys are already covered. A range of 150 km is sufficient for 93% of all private journeys and 81% of all commercial journeys.

If additional recharging is possible during the day (employee parking lot, cab rank, etc.), a range of 150 km is sufficient for over 98% of all commercial trips.

A study by PriceWaterhouseCoopers from 2010 arrived at the following figures (based on Europe EU-27, cumulative annual car mileage):

- Vacation travelers: 633 billion km, mainly single trips, long distance.
- Leisure travelers: 3293 billion km, single and periodic trips, short and medium distances.

---

[9] Source: *Shell macht abermals einige Wasserstoff-Tankstellen dicht*, Daniel Krenzler, Elektroauto-News, 15.02.2024.

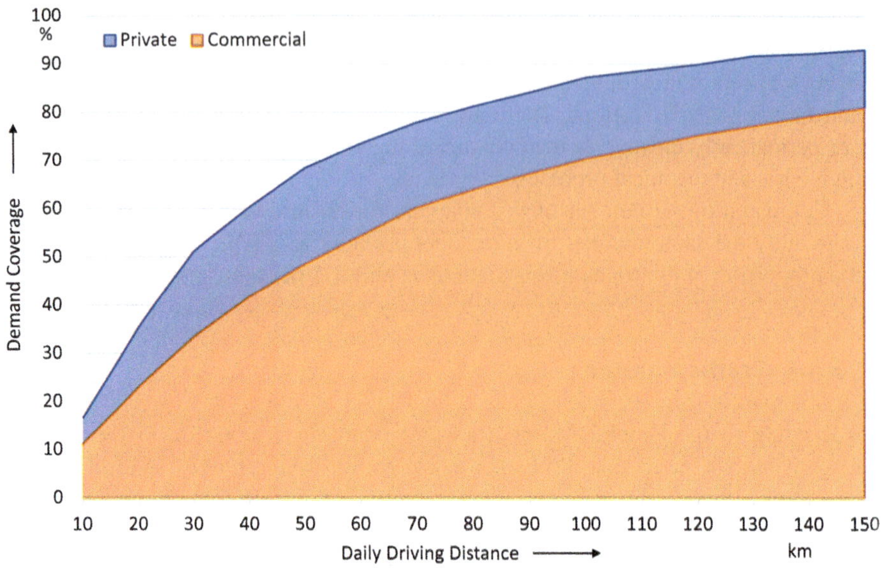

Fig. 13.6 Cumulative daily mileage per car in Germany[10]

- Business travelers: 915 billion km, predominantly single trips, medium and long distance.
- Commuters: 1331 billion km, periodic journeys, mainly short and medium distances.

An important psychological aspect is "range anxiety", i.e. the vague feeling of buying or driving a car that cannot cover the desired distance.

For most practical purposes, a reliable range of approximately 200 miles (320 km) is generally considered sufficient. Achieving this requires electric cars with batteries from approximately 70–80 kWh. When considering typical daily commutes, this range requires one or at most two charging sessions per week.

For longer journeys, the driving time between breaks is typically around 2–4 h, depending on traffic conditions. Modern electric vehicles with high charging power can be recharged in a comfortable 20–30 min before the journey can be continued. Long driving ranges of approx. 400–500 km are only advantageous for frequent long-distance journeys. Fortunately, there are now also several electric vehicles on the market that are tailored to these requirements.

[10] Data Source: *VDE-Studie Elektrofahrzeuge*, 2010.

13.5 Electrical Energy

13.5.1 The Global Picture

The generation of electrical energy accounts for a large proportion of global CO_2 emissions.

In 2022, China emitted around 4700 million tons of CO_2 through the operation of its coal-fired power plants. The USA follows in second place with 1600 million tons. Next are India (1160 Mt), Japan (467 Mt), Russia (409 Mt), and South Korea (264 Mt). In seventh place is Germany with 224 million tons.

The carbon intensity of the production of electric energy is summarized in the CO_2 emission factor, which is given in g_{CO2}/kWh. It is calculated from the direct CO_2 emissions generated during electricity production and the net amount of electricity available to the end consumer.

However, some important influencing factors are missing from the classic emission factor. These are

- the electricity trade balance, i.e. the amount of electricity purchased from or sold to neighboring countries over the year;
- the indirect emissions from the upstream chain, e.g. from the extraction and transportation of coal or gas; and
- the weighted (equivalent) emissions of other greenhouse gases, in particular, CH_4 and N_2O.

For example, the classic emission factor for German electricity production in 2022 is estimated at 434 g_{CO2}/kWh according to the Federal Environment Agency. However, if all other influences are considered, the factor is 498 g_{CO2eqv}/kWh.[11]

In this chapter, the classic emission factor is used in the comparison between countries as it is commonly used internationally. In the well-to-wheel and life-cycle analyses in the following chapters, however, the full emission factor is used (Table 13.2).

Over the past 10 years, most countries have made successful efforts to improve electricity production through hydro, wind, and solar power. The progress was particularly visible in China, the United States, and Europe. In these regions, emissions were reduced by around 150 g_{CO2}/kWh. Only India has not succeeded in reducing its emissions from electricity production over the past 20 years.

Figure 13.7 shows the development of carbon intensity in electricity production in the three main regions in the world and in Germany from 2000 until 2022 and an extrapolated outlook for the next 10 years based on a linear extrapolation of the preceding 10 years.

[11] Source: Entwicklung der spezifischen Treibhausgas-Emissionen des deutschen Strommix in den Jahren 1990–2022, Petra Icha, Thomas Lauf, Umweltbundesamt, 20/23.

Table 13.2 Carbon intensity of electricity production per country in 2022[12]

Country	Emissions g_{CO2}/kWh
India	633
Poland	633
Indonesia	619
Saudi Arabia	588
Malaysia	544
China	534
Australia	502
Mexico	424
Turkey	414
Germany	385
Italy	373
USA	368
Russia	364
EU-27	261
United Kingdom	261
Spain	217
Austria	158
Canada	126
Brazil	102
France	85
Sweden	45
Switzerland	41
Norway	29

In China, in particular, the expansion of renewable power generation is progressing very rapidly. In 2023, China's addition of wind and solar capacity reached 230 GW, Europe added 75 GW, and the USA 40 GW. In addition to improving the CO_2 footprint, this also leads to lower electricity costs for consumers. End-user prices for electricity in China are on average less than half as in Europe.[13]

[12] Source: Our World in Data, *Carbon Intensity of Electricity Generation*, December 12, 2023.

[13] Source: *Renewables helping China to halve power prices compared to US*, Europe; pv magazine, Emiliano Bellini, November 28, 2023.

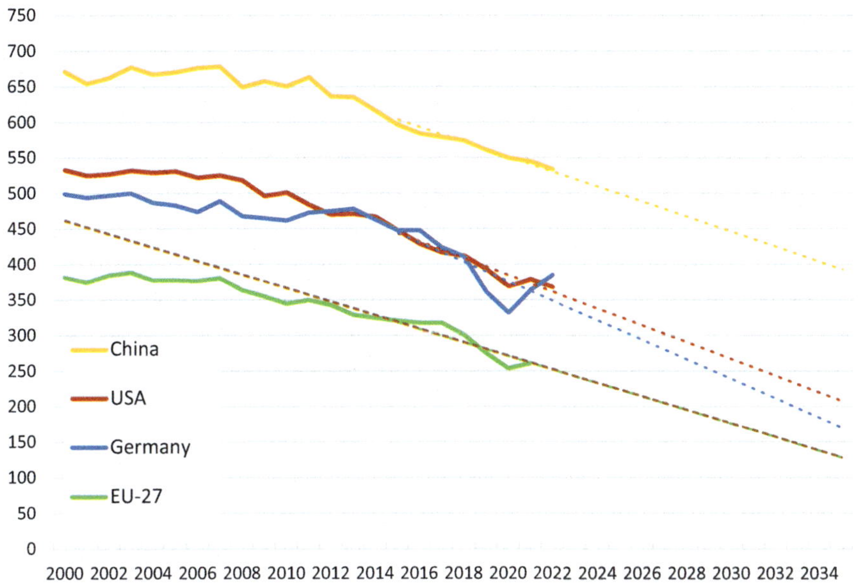

Fig. 13.7 Carbon intensity of electricity production in g_{CO2} per kWh since 2000 and extrapolated outlook

Note
It is sometimes argued that emissions from the electricity generation sector are higher than from road transport and that the switch to electric cars should therefore only begin after CO_2 emissions in the electricity generation have been reduced significantly.

Some fans of combustion engines are even demanding that the emissions from electricity production for electric cars should be based on those from coal-fired power plants only, because electric cars place an additional burden on the grid and additional loads cannot be met with wind and solar power plants (according to the principle of marginal cost).

These arguments are countered by

Humanity needs to significantly reduce CO_2 emissions in both sectors, transportation and electrical energy, and that takes a lot of time. We must proceed in parallel.

While an enormous amount has been achieved in the electricity sector over the last 30 years, little has happened in the transport sector and emissions per vehicle have even increased again in recent years.

Even though the absolute emissions in the electrical energy sector are higher, the emissions of an electric car are still much lower than those of a combustion engine car due to the high efficiency of electric cars and the small amount of electricity used. Furthermore, electric cars will continuously benefit from future improvements in electric power production.

The increased consumption of electrical energy due to the introduction of electric cars has not stopped the significant downward trend in CO_2 emission factors worldwide, nor has it even slowed it down noticeably.

And finally, the argument of marginal cost does not apply because new wind and PV systems are being created much faster than the increase in electricity consumption of electric cars. One can even observe that electric mobility is driving the expansion of renewable electricity generation. For instance, a private photovoltaic (PV) system is only economically viable when there is a high level of self-consumption, for example, by using an electric car and/or a heat pump. Electric vehicles preferentially charge over night when there is abundant wind power in the grid and electricity prices are low. In this way, electric mobility actively contributes to improving the emissions profile of electricity production.

To summarize, we must increase both renewable electricity production and the share of electric cars as quickly as possible. The well-to-wheel emissions from electric cars must be calculated with the same CO_2 emission factor that is used for all other electricity consumers.

13.5.2 Trends in Green Electricity Production

In recent year, the size and power of offshore wind turbines has been rapidly increasing. As of July 2023, the most powerful wind turbine in operation is the 16 MW Mingyang MySE 16-260 in Fujian in China. It has a rotor diameter of over 310 m. Even larger turbines up to 22 MW are already in development.[14]

Example 13.1 Cost and energy production of offshore wind turbines

An offshore wind turbine typically has 4000 full load hours per year with investment costs of 2.5 M€ per MW of installed capacity. ◀

Developments in the PV sector are less spectacular but just as fast. Today, more powerful cells with peak outputs of up to 400 W per module (1 m by 1.6 m) and inexpensive home storage systems are available.

This is almost double the level of 10 years ago and offers great potential for retrofitting old systems. Today's commercially available standard modules made of silicon yield around 22–23% of solar radiation. In future, values of up to around 27% are possible, which could increase the module output towards 480 W.

[14] Source: *List of most powerful wind turbines*, Wikipedia, 2024.

Researchers are working on completely new generations of PV cells that can significantly increase the system output per square meter. In concentrated cells (CPV = concentrated photovoltaics), the light is bundled by lenses, so more powerful, smaller, and cheaper semiconductors can be used. Tandem cells use a sandwich structure consisting of two layers of different semiconductor materials, usually silicon and perovskite, each of which reacts to different spectral wavelengths. In research, efficiencies of up to 50% are achieved.

The electricity for electric cars can easily be generated by private homeowners. With a PV system and battery electric home storage units, self-sufficiency levels of over 50% can be achieved, significantly reducing operating costs. The cars are charged during periods of strong sunlight and will in future even supply electricity to the house for the evening and into the morning hours. Low-cost home storage systems are also making private PV systems considerably more financially attractive despite falling feed-in tariffs.

Many manufacturers, including Tesla Motors, now offer such systems with capacities of between 4 and 16 kWh. Most modern home storage systems feed in energy on all three phases and offer continuous electricity output of between 3.3 and 5 kW. In the past, single-phase home storage systems were also used. They are less recommendable because they place an unbalanced load on the grid and only allow low feed-in power.

Due to the safety requirements, the complex plant technology, and the high primary energy requirement, hydrogen is not suitable for production by private individuals—just as private individuals do not refine crude oil or operate private filling stations today. With a broad-based hydrogen supply for millions of fuel cell cars, society would therefore continue to be dependent on the large energy companies and their nationwide network of filling stations (quasi-monopolization). For example, most public hydrogen filling stations in Germany are operated by H2 Mobility Deutschland GmbH, whose shareholders include two large hydrogen companies and well-known oil companies.

13.5.3 Electricity Production in Germany

The following example from Germany shows how the generation of electrical energy has become increasingly sustainable over the last 20 years (Fig. 13.8).

Gross electricity generation includes the power plants' own consumption (particularly for pumps), which amounts to around 10%.

Since the turn of the millennium, Germany has produced a considerable net electricity surplus. In recent years, it has amounted to between 20 and 40 billion kWh annually. This additional electricity was exported abroad, primarily to Austria and France, but put a strain on the German grids and reduced the price of electricity on the exchange.

The break-even point of regenerative electricity production was already reached in Germany in 2019, i.e. since then, more public electricity has been produced

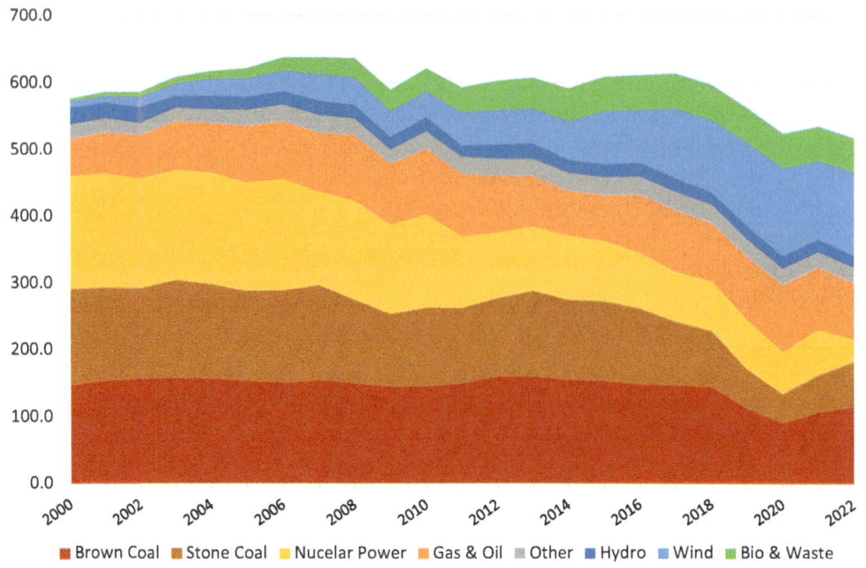

Fig. 13.8 Gross electricity generation in Germany in billion kWh[15]

from renewable sources than from fossil fuels. For the total European electricity supply, this happened 1 year later in 2020.

In 2022, around 244 billion kWh of public electricity was produced from renewable energy sources in Germany. This corresponds to just under 50% of net electricity generation (excluding generation for own industrial use). The share of public electricity production from fossil sources amounted to just under 44%, with the remainder being nuclear power. By 2024, all nuclear power plants in Germany have been shut down.

The rate of increase in renewable electricity generation remains high, as the extrapolation shown in Fig. 13.9. In recent years, the expansion of solar power has increased sharply due to the high electricity prices and the favorable conditions for self-consumption. In contrast, the expansion of wind power in Germany is lagging behind the targets. While the northern and eastern federal states have in some cases exceeded their expansion targets, the expansion of wind power in southern Germany has practically been halted for several years.

Nevertheless, a doubling of renewable electricity generation in Germany over the next 10–15 years seems realistic overall.

However, additional grid expansion will be necessary, particularly for the distribution of electricity from offshore wind turbines, to transport the electricity from the North and Baltic Seas to central and southern Germany.

[15] Source of figures: AG-Energiebilanzen, 2023.

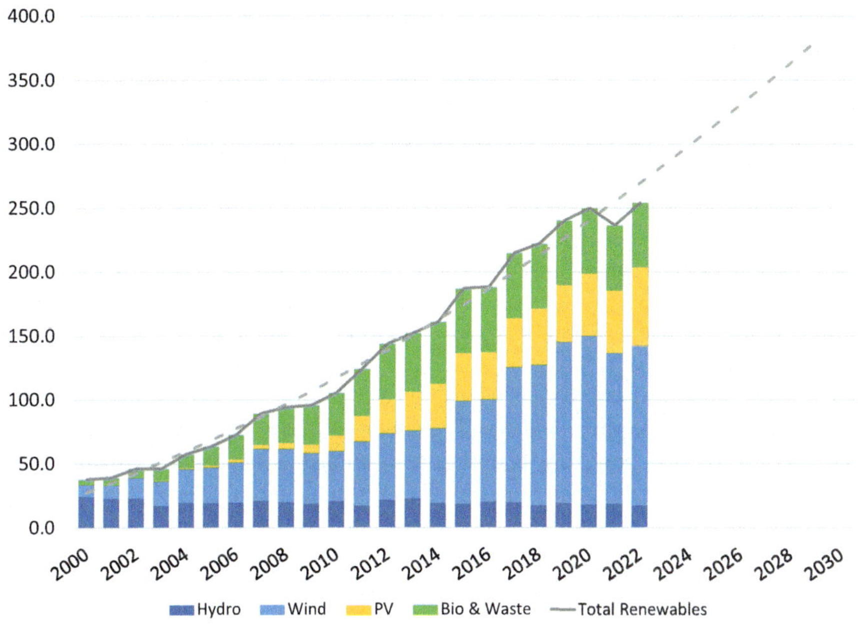

Fig. 13.9 Electricity generation from renewable energy sources in Germany in billion kWh

13.5.4 Cost and Price of Electricity

The cost of renewable energy is on a steady downwards trend since decades. As of 2023, onshore wind is at the low end with around 50 $/MWh (spread $24–$75) followed by solar photovoltaic with 60 $/kWh (spread $49–$185) and offshore wind with 80 $/MWh (spread $72–$140). Concentrated solar power generation is highest but no longer plays a role today. The last recorded price was 141 $/MWh in 2019.

By comparison: the cost of conventional power generation is generally higher than for renewables: gas (combined cycle) at 70 $/MWh, coal at 117 $/MWh (spread 68$–166$), gas (peaking) at 170 $/MWh, and nuclear at 180 $/MWh (spread 115$–221$).[16]

The average end consumer price is subject to political strategies and strong fluctuations. In mid-2023, the highest prices worldwide are generally paid in Europe, ranging from 0.52 US$/kWh in Ireland to 0.40 US$/kWh in Germany to 0.23 US$/kWh in Spain.[17]

The prices in other industrialized countries are generally lower, for example, in Japan (0.23 US$/kWh), United States (0.17 US$/kWh), China (0.08 US$/kWh),

[16] Source: *2023 Levelized Cost of Energy+*, Lazard Research & Insights, April 12, 2023.

[17] Source: *Household electricity prices worldwide in June 2023, by select country*, statista, 2024.

Turkey (0.08 US$/kWh), and Russia (0.07 US$/kWh). However, the price per state varies greatly in the USA. In California and Massachusetts, in particular, it reaches European levels.

Example 13.2 composition of final energy prices in Germany (2023)

In Germany, the end-customer price for electricity has fluctuated between 0.30 and 0.40 €t/kWh in recent years.

About half the price, 55% is spent on the procurement and sale of electricity including the profits of the electricity producers. 18% goes towards the grid fee including the profit of the grid companies. The remaining 27% are taxes and levies.

The composition of the electricity price is not too different from the composition of the petrol price, which only includes 33% for the crude oil and around 10% for the oil company (extraction, refining, transportation, distribution). More than half of the price of petrol (58%) is made up of taxes and levies. ◀

13.5.5 Electricity Demand for Electromobility

The production of the electricity required for electromobility does not pose a particular hurdle in general in most countries, as the following examples show.

Example 13.3 electricity demand (energy) for electromobility in Germany

Assuming an average consumption of 20 kWh per 100 km per car and a charging efficiency of 90%, around 140 billion kWh of electrical energy is required to produce enough electricity to cover the total distance driven by cars in Germany (around 620 billion km). This corresponds to around 20% of the amount of electricity currently produced (including industrial production).

Even today, electricity production from renewable sources (244 billion kWh in 2022) significantly exceeds this demand.

The forecast growth in renewable electricity generation alone over the next 6–7 years will be enough to power Germany's entire car fleet electrically.

In the medium term, all other road vehicles must also be converted to electric drives. This results in an energy requirement of 69 billion kWh for trucks and a further 7 billion kWh for buses. The total requirement will then be around 218 billion kWh. ◀

Example 13.4 electricity demand (power) for electromobility in Germany

Based on 140 billion kWh of annual energy for all cars, this results in an average daily demand of 380 million kWh.

Even if there will be temporal focal points, most vehicles only need to be available with a full battery in the early morning. Variable electricity prices can create an additional incentive for load balancing. As a conservative estimate, we assume that charging only takes place on average over 10 of the 24 h of the day.

This results in an average charging capacity of 38 GW. This is around 16% of the net capacity installed in the German electricity grid in 2023 (around 238 GW). ◀

The main challenge with the electricity supply of electric cars is not the generation but the distribution of electrical energy. Local grids can be overloaded if many electric cars are charged at the same time. In areas with a high housing density, there are not enough opportunities to install charging points. These issues are examined in more detail in the following chapter.

Some people are asking whether we should not first make the electricity grids cleaner before adding many additional consumers through electromobility.

The following example uses the situation in Germany to illustrate the two borderline scenarios.

Example 13.5 How to use electricity from renewable sources in Germany

In the next 7 years, an increase in electricity from renewable sources in the order of 140 billion kWh can be expected (see Fig. 13.9).

Two borderline scenarios can be compared for the use of this electricity:

Scenario A:

The 140 billion kWh are used solely to electrify all passenger cars. The existing electricity consumers are supplied from the existing old plants (including coal-fired power plants).

This approach would not change the emissions from existing electricity production, but it would largely eliminate the emissions from today's car traffic with combustion engines. The savings would amount to around 127 million tons of greenhouse gases per year (not including the CO_2 emissions for the construction of the power plants and the production of the electric vehicles).

Scenario B:

The 140 billion kWh are used solely to shut down coal-fired power plants. The total amount of electricity produced remains the same as of today. No new electric cars are added. The CO_2 emission factor of electricity production improves significantly.

This approach would almost eliminate the emissions of today's lignite-fired power plants (current annual production of around 150 billion kWh) and thus save 145 million tons of greenhouse gases per year (again, not including the CO_2 emissions for the construction of the plants). ◀

This highly simplified example shows, at least as a first approximation, that both borderline scenarios ("introduce electric cars" or "shut down coal-fired power plants") result in similar CO_2 savings. As Germany's climate targets can only be achieved by 2050 when both electric cars are introduced and coal-fired power plants are shut down, the increase in renewable electricity generation should therefore be used to promote both measures equally.

This example also clearly shows the CO_2 savings potential of electric cars and thus answers the question of the ecological benefits of these vehicles compared to cars with internal combustion engines, at least in qualitative terms.

13.6 Raw Materials

The following section examines the reserves and resources available for the most important raw materials that are essential to produce battery electric cars. This applies to lithium, copper, nickel, manganese, phosphor, cobalt, and rare earths (neodymium and dysprosium). Not all the materials are used in all batteries or in all electric motors. For example, cobalt is not used in LFP or sodium batteries. Rare earth magnets are not used in induction machines nor in electrically excited synchronous machines. Batteries also do not use rare earth materials.

13.6.1 Lithium

The key technology for modern electric cars is without doubt the lithium-ion battery.

However, lithium is not only used in batteries. Other major fields of application are glasses and ceramics as well as lithium grease. In 2020, around half of the lithium mined was used for battery production.

The currently known mining reserves of lithium are around 26 million tons. The Earth's identified lithium deposits (resources) are estimated at around 98 million tons (as of 2023). However, these figures increase continuously as new deposits are discovered.

As prices rise, the extraction of lithium from used batteries (secondary lithium), which is hardly practiced today, will become economically lucrative in the future.

In 2025, around 2/3–3/4 of the lithium mined will be used in batteries. 18% will be used in the production of ceramics and 5% in lubricants. The rest will be used for polymers, metal casting, and air treatment, for example.

Note

Reserves are deposits that can be mined economically using current technology.
Resources are deposits that are proven to exist but cannot yet be extracted economically using current technology.

Table 13.3 Total worldwide lithium potential, as of 2023[18]

Lithium	Annual output 2022 in kt	Reserves in kt	Resources in kt
Argentina	6.2	2700	30,000
Australia	61	6200	7900
Bolivia	–	–	21,000
Brazil	2.2	250	730
Canada	0.5	930	2900
Chile	39	9300	11,000
China	19	2000	6800
Congo	–	–	3000
Germany	–	?	3200
USA	–	1000	12,000
Other	2.1	3600	10,200
Worldwide	**130**	**26,000**	**98,000**

Lithium is also found in salt lakes, groundwater, and even in ocean water, albeit in low concentrations (only around 0.18 ppm). The quantity of this water-bound lithium amounts to a further 230 billion tons. At present, however, its extraction is not yet worthwhile as it is very energy intensive (Table 13.3).

The broad distribution of lithium deposits across all continents is remarkable. There are even large quantities in Europe (Germany, Spain).

Example 13.6 Lithium from geothermal power plants

At the end of 2021, a geothermal power plant was put into operation in Bruchsal near Karlsruhe in the Rhine Valley. When the extracted hot water was examined, a high concentration of lithium was found, around 200 mg per liter. At a flow rate of 25 L per second, this results in over 430 kg of lithium per day—enough for over 30 car batteries. It is estimated that enough lithium for 200,000 car batteries per year can be extracted in the Upper Rhine Graben between Basel and Karlsruhe. ◀

The extraction of lithium at the required high purity levels (around 95%) is critical. It requires production capacities that are not yet available to a sufficient extent. The current mining capacity will also not be sufficient to cover the rapidly increasing demand in the short term. This may result in temporary supply bottlenecks.

[18] Source: U.S. Geological Survey, Mineral Commodity Summaries, 2023.

The use of water in lithium extraction has been and continues to be the subject of much critical discussion. As the above explanations show, there are many different deposits and mining options for lithium, and their water requirements vary greatly. In addition, processes have been developed in recent years that significantly reduce the amount of groundwater required for lithium extraction in Latin America.

Around 2010, around 2 million liters of brine were needed to extract 1000 kg of lithium in Chile. Today, new processes have reduced the amount of water required to 200,000–1 million liters. The average lithium requirement for a mid-range electric car is within a range from 12 to 19 kg. Therefore, one car is responsible for a brine consumption of 2400–19,000 L.[19]

By comparison, the water required to produce hydrogen from electrolysis is just under 10 L per kg of H_2. Calculated over the typical lifetime of a car (160 tkm), this adds up to around 17,000 L of water. However, this is pure drinking water, whereas the lithium brine consists of wastewater. Around 0.6 L of water are used to produce 1 L of petrol. Over the life of a car, it comes to around 7700 L, which is in the same ballpark as lithium production from South America.

Furthermore, over half of the worldwide production of lithium is extracted from solid rock by mining, crushing, and grinding the rock. This process does not require a significant amount of groundwater.

Example 13.7 Estimation of lithium availability for battery production

According to the current state of the art, around 1.5–1.7 kg of lithium is required per 10 kWh of battery capacity. This corresponds to around 13 kg of lithium for a 75 kWh battery, which can be regarded as the standard for mid-range electric vehicles (400 km WLTP range).

The known reserves (17 million tons) are therefore sufficient for around 1.3 billion car batteries. This is more than enough to replace the current global car fleet by electric vehicles.

The known resources (80 million tons) are sufficient for a further 6.2 billion car batteries and new deposits are constantly being discovered.

Recycling will significantly increase the amount of available lithium in the future. ◀

13.6.2 Copper

Another key raw material for electrically powered vehicles is copper. Contrary to popular believe, copper is not a critical raw material; the EU describes the supply risk as "very low".

[19] Source: *Lithium aus Lateinamerika*, Edison Handelsblatt, March 07, 2019.

After iron and aluminum, copper is the most mined metallic raw material in terms of volume. Asia currently accounts for almost 70% of global demand for refined copper, Europe for 15%, and North America for 10%. Primary copper is extracted in mineral form in mines and then processed. Secondary copper is obtained through recycling.

Around 25 kg of copper is used in an average mid-range petrol vehicle. This amount will increase in the future as more and more comfort functions are realized with electric motors. Simple hybrid vehicles require around twice as much copper, plug-in hybrids and battery electric vehicles even require around three times as much copper, around 70 kg. It is needed primarily for the HV cables, the large drive motor, the power electronics, and the battery (around 18% of a lithium-ion battery is copper foil).[20]

Unlike lithium, which is indispensable to produce high-performance batteries, copper can in principle be replaced by aluminum as a conductor material. However, this is associated with many disadvantages (size, efficiency) and not a recognizable trend.

Both metals, copper and aluminum, are recovered in large quantities from scrap. Secondary copper has accounted for a constant share of around one-third of global copper production since the turn of the millennium. The remaining two-thirds come from the extraction of deposits (mining and open-cast mining).

The copper reserves known today (870 million tons, as of 2020) will last for around 40 years if consumption continues at the current rate. In addition, extensive resources are known (2100 million tons), of which around 1800 million tons are in porphyry deposits with only a low copper content (0.4–1%). Open-cast mining, which is already being carried out in the Andes (Chile), produces large quantities of overburden. All in all, the copper supply is secured for at least 150 years.

The U.S. Geological Survey estimates that a further 3500 million tons of copper will be found in yet unidentified deposits (Table 13.4).

Example 13.8 Estimation of copper availability for motor production

Just under 70 kg of copper is required to manufacture a typical BEV with all electrical components, including the battery. To equip all 1.2 billion cars in the world with an electric drive, 84 million tons of copper would be needed. That is around one-tenth of the known reserves.

Assuming the usual one-third share of secondary copper, this corresponds to the current copper production volume of 3 years. ◀

[20] Source: Deutsches Kupferinstitut, 2023.

Table 13.4 Total copper potential, as of 2023[103]

Copper	Annual output 2022 in Mt	Reserves in Mt	Resources in Mt
Australia	0.83	97	
Chile	5.2	190	
China	1.9	27	
Congo	2.2	31	
Kazakhstan	0.58	20	
Mexico	0.74	53	
Peru	2.2	81	
Poland	0.39	31	
Russia	1.0	62	
USA	1.3	44	
Zambia	0.77	19	
Other	3.4	200	
Worldwide	**22**	**890**	**2100**

13.6.3 Rare Earths

Rare earth metals are indispensable to produce powerful permanent magnets. These magnets are required for PM synchronous motors, which are the preferred motors in electric cars today. They are also used in many generators in wind turbines.

A distinction is made between LREE (light rare earth elements) and HREE (heavy rare earth elements). Neodymium, samarium (LREE), and dysprosium (HREE) are particularly relevant for permanent magnet motors.

NdFeB permanent magnets consist of around one-third neodymium. The proportion of dysprosium is around 1–5%. Other rare earth materials do not play a significant role in magnet production. The older SmCo (samarium–cobalt) magnets have a lower energy density than NdFeB and are also more expensive. They are no longer used in series-produced vehicles.

Rare earth materials are quite common on earth, some are more common than lead. But the deposits are unevenly distributed. There are large deposits mainly in China, and some in Australia and California.

In 2022, around 300,000 tons of rare earth oxides (REO) were produced worldwide. The Chinese market share was 70%, and 85% for rare earth refining. China drastically reduced illegal extraction from 120,000 tons of rare earth oxide in 2011 to less than 10,000 tons in 2020.

The concentration of rare earths in the oxide is very low. The greatest environmental impact is caused during the separation by the complex, multi-stage solvent extraction process, which produces radioactive residues, among other things (Table 13.5).

Table 13.5 Total rare earth oxide potential, as of 2023[103]

Rare earth oxides	Annual output 2022 in kt	Reserves in kt	Resources in kt
Australia	18	4200	
Brazil	0.08	21,000	
Burma	12	?	
China	210	44,000	
Greenland	–	1500	
India	2.9	6900	
Russia	2.6	21,000	
Thailand	7.1	?	
Vietnam	4.3	22,000	
USA	43	2300	
Other	–	280	
Worldwide	**300**	**130,000**	

The recycling of rare earths currently plays no role in global supply due to low collection rates and complex processing methods. Currently, only 3% of the total demand for LREE and 8% for HREE in Europe is covered by recycling.[21]

There are several electric motor technologies that do not need permanent magnets, in particular, induction machines (used in some Tesla, Mercedes-Benz, Volkswagen, and Renault vehicles, for example), separately excited synchronous machines (used in Renault and BMW vehicles), and reluctance machines (not yet in series production in cars, but used in industrial applications). Externally excited synchronous machines are becoming increasingly popular because they have the best cycle efficiency of all electric motor variants.

In wind turbines, three different types of generators are currently being used. Only one of them, the permanent magnet synchronous generator, requires rare earths. The other variants, namely, induction machines with wound rotors and externally excited synchronous machines, are only made of iron, copper, and some plastic.

With resources of rare earth metals becoming scarce, a wide range of alternative technologies with almost the same performance are available. Rare earths are therefore not a critical raw material for electromobility and the energy transition.

[21] Source: *Seltene Erden – Informationen zur Nachhaltigkeit*, BGR, März 2021.

Table 13.6 Total cobalt potential, as of 2023[103]

Cobalt	Annual output in kt	Reserves in kt	Resources in kt
Australia	5.9	1500	
Canada	3.9	220	
China	2.2	140	
DR Congo	130	4000	
Cuba	3.8	500	
Indonesia	10	600	
Madagascar	3.0	100	
Morocco	2.3	13	
Papua New Guinea	3.0	47	
Philippines	3.8	260	
Russia	8.9	250	
Turkey	2.7	36	
USA	0.8	69	1000
Other	5.2	610	
Worldwide	**190**	**8300**	**25,000**

13.6.4 Cobalt

Cobalt is an important raw material for many technical products. In addition to batteries, cobalt chemicals are also used in ceramics and dyes. Cobalt metals are used in superalloys, carbides, and diamond tools. Cobalt is also used as a catalyst material in fuel cells, where it replaces the otherwise required platinum, which is even more critical.

In 2017, around 8% of the global cobalt production was used for lithium-ion batteries for electromobility. It is estimated that this share will increase to over a quarter by 2026.

China is the world's most important producer of cobalt chemicals. At the same time, it is a key country for the expansion of cobalt recycling. Most of the cobalt processed in China comes from the DR Congo (Table 13.6).

More than 120 million tons of cobalt resources have been identified in polymetallic nodules and crusts on the floor of the Atlantic, Indian, and Pacific Oceans.[101]

As a result of poverty, child labor is widespread in southern African countries. It affects many sectors of the economy, including cobalt mining. On average, 10% of global cobalt production comes from small-scale mining in the DR Congo. An inspection of 58 small-scale mines in 2019 revealed that a total of around 2500 children were present at 29% of the sites, including many young children under the age of 10. Most of them were doing light work such as sorting ore. It is difficult to ban child labor that is driven out of poverty.[22] On the other hand, these numbers mean that approximately 97% of the global cobalt production is not associated with child labor.

Recycling currently plays a minor role in the cobalt market, amounting to around 10%. On average, around 13,000 tons of cobalt are recycled per year. However, recycling systems can help to reduce the production costs of electric cars. The recycling sector will therefore play a much greater role in the supply of cobalt and other battery raw materials in the long term.

Due to the high price, the proportion of cobalt in lithium-ion batteries for cars is being steadily reduced. In today's modern NCA/NMC 811 cells, the cobalt content is less than 5% of the cathode material.[23] Many manufacturers are starting to offer vehicles with LFP batteries that do not contain any cobalt at all. Tesla only needs around 4.5 kg of cobalt for its Model 3 long-range (LR) battery and uses no cobalt for its standard range (SR) batteries.

In 2023, the average BEV sold globally contained 4.8 kg of cobalt, down 1% from a year prior, even though the average EV pack capacity rose by 10% over the same period.[24]

13.6.5 Nickel

In addition to lithium, many but not all batteries also contain manganese and nickel. Notable exceptions are LFP and sodium–ion batteries.

Nickel is mainly used to produce stainless steel and other alloys. The proportion of nickel used in battery production is low (less than 10%).

Nickel is mainly mined in Indonesia and the Philippines but is also available in large quantities in Australia and Brazil. More than half of the production goes into the manufacture of stainless steels. The demand for nickel for battery production has played a minor role so far.

The production volume of nickel worldwide was around 3.37 million tons per year in 2022. The known reserves amount to more than 100 million tons. In addition, there are a further 300 million tons of known resources as well as extensive nickel deposits in manganese crusts and nodules on the seabed in unknown quantities.[101]

A particular advantage of nickel is its good recyclability. Around 57% of the total nickel mined worldwide is still in use today.

Nickel is therefore not currently considered a critical raw material.

---

[22] Source: BGR (2019): *Mapping of the artisanal copper-cobalt mining sector in the provinces of Haut-Katanga and Lualaba in the Democratic Republic of the Congo*. Bundesanstalt für Geowissenschaften und Rohstoffe, Germany, ISBN 978-3-943566-68-9.

[23] The other parts of a battery, namely, the anode, separator, electrolyte, current collector foils, and housing, do not contain cobalt.

[24] Source: Adams Intelligence, *Electric carmakers continue to lighten their cobalt load*, October 3, 2023.

13.6.6 Manganese

Like nickel and cobalt, manganese is also used as a cathode material in some variants of lithium-ion batteries.

Manganese is mainly mined in South Africa (36% market share), Gabun (23%), and Australia (16%). The annual consumption in 2022 was around 20 million tons. The reserves amount to 1700 million tons and the resources are almost unlimited due to the deposits of manganese nodules on the seabed.[101]

Manganese is primarily used in the iron and steel industry as an alloying element to increase strength. In addition to its use in batteries, it is used as an oxidizing agent, as a fertilizer and is required to produce chemicals.

Like Nickel, manganese is not currently considered a critical raw material.

13.6.7 Phosphate

Alongside potash and nitrogen, phosphate is one of the most important nutrients in agriculture. The most important phosphate producers are China, Morocco, and the USA. Two-thirds of all fertilizers are consumed in China, India, and Brazil.

Phosphate is available in sufficient quantities and a shortage is not expected soon. Annual production amounted to around 210 million tons in 2012. This compares to 67,000 million tons of known reserves.

Although phosphate is used in LFP batteries, the proportion of phosphate used for battery production is so small that it plays virtually no role.

13.6.8 Summary Raw Materials

As things stand today, there are no critical raw materials that would hinder the widespread introduction of electromobility.

Obviously, the resources of many raw materials will become scarce in the next 100–200 years. However, this will also affect conventional cars and many other products. That is why recycling is so important: humanity must move towards a circular economy and reuse raw materials. This applies in general and independently of electromobility. In a few hundred years, our current landfills will be coveted treasure troves.

13.7 Second Life (Re-Use)

Battery storage systems in vehicles are subject to high-performance requirements: they must charge quickly (up to 3 C is common) and discharge even faster when driving at full throttle (up to 5 C is common for BEVs, and in some cases significantly more for hybrid vehicles). In addition, batteries in cars are operated at very low (at least down to −20 °C) and high (at least up to +40 °C) ambient temperatures and must not lose too much of their capacity so that the range of the vehicle does not decrease significantly.

The effects of battery aging reduce both the capacity (SOH) and the maximum power available during acceleration and braking energy recovery (SOP). An important aging mechanism is lithium plating, i.e. the deposition of metallic lithium on the graphite anode. The capacity is reduced as part of the volume of the anode is no longer accessible and can therefore neither be charged nor discharged. In addition, the internal resistance of the cell increases, which reduces the maximum output power.

Second-life batteries may have reached the end of their automotive life but they still have a residual capacity of about 70–80%. When an EV battery reaches the end of its first life, there are basically three options[25]:

- **Repurposing** Suitable modules of the battery are selected and newly combined based on residual state, capacity, etc. They are then used directly in other applications.
- **Refurbishment** Battery modules are disassembled, single cells are reconditioned and repacked in new modules. Just as before, they can then be used in other applications.
- **Recycling (Third Life):** Valuable metals are extracted from the battery and re-used.

In many applications outside the automotive sector, for example, in home storage systems, the technical requirements for a battery are significantly lower. In the basement of a house, temperatures are usually between 15 and 20 °C all year round. The required output and charging power are low as the battery is intended to absorb solar energy and supply power to the household over a period of several hours. Typical C rates are only between ½ and ¼. A certain loss of capacity is acceptable because the weight of a stationary system is not as relevant as in a car.

There are more potential applications for a second life in local grid storage systems (B4B = Batteries for Buildings) in private homes, office, and industrial buildings and as temporary storage for the energy supply of entire streets (in local proximity to the low-voltage transformer) or districts.

Batteries from electric vehicles that have reached the end of their service life therefore do not have to be scrapped and recycled but can continue to be used in stationary applications for many years to come. Corresponding projects and feasibility studies are being carried out by several manufacturers and research institutes, including Renault, Nissan, and General Motors.

[25] Source: *What is a second life battery?*, enel x, 2024.

One recent report estimated the market volume for second-life batteries at 1.2 billion US$ in 2022 and 92 billion US$ in 2030.[26] The applications are EV chargers, commercial and industrial energy storage, grid charging, residential energy storage, and off-grid applications—all with an almost even market share.

Example 13.9 Second life applications[27]
A soccer stadium in Amsterdam uses nearly 150 new and old Nissan EV batteries to power an energy storage system.

Skoda started providing energy storage systems in 2021, each of which uses 20 plug-in hybrid batteries or five all-electric batteries, to power charging stations at up to 4000 dealerships in Europe.

In March 2022, European utility Enel SpA started operating a project with Nissan that uses 48 old EV batteries with 30 new ones in an energy storage system at a power plant that provides electricity for Melilla, Spain. The system can provide 15 min of backup power.

Enel is using second-life batteries to build an energy storage facility for a 30-megawatt solar park at Rome Fiumicino Airport.

Renault has rented out more than 250,000 batteries that will come back for repurposing or recycling in the coming years. Renault has several energy storage sites in France that combine old and new EV batteries. ◀

13.8 Recycling

Lithium-ion batteries contain many materials some of which are worth recovering (recycling) at the end of the battery's lifetime. These include lithium, cobalt, manganese, nickel, aluminum, copper, steel, and plastic.

The recycling rate (material recovery rate, recycling efficiency) is the ratio of the mass of the raw materials extracted by recycling (output) by the mass of the raw materials used in the production (input). A recycling rate of more than 90% (not taking oxygen O_2 into account) seems to be achievable in the foreseeable future. Costs are currently still proving to be a stumbling block, as recycling batteries is very energy intensive.

There are three different recycling process: low temperature (cryogenic), high temperature, and physical separation of components.

[26] Source: *Global Second-life EV Batteries Market 2023–2032*, CM Custom Market Insights, October 2022.

[27] Source: *Old Electric-Vehicle Batteries Are Getting a Second Life*, Wallstreet Journal Pro, Dieter Holger, Giulia Petroni, June 13, 2022.

The high-temperature process is the most common and involves the following steps[28]:

- Collecting and sorting suitable battery types.
- Burning off the electrolyte.
- Neutralization of the internal chemistry.
- Melting and separating the metallic components (cobalt, nickel, copper).
- Separating and cleaning the high-grade materials.
- Removal of non-recyclable waste (including lithium and aluminum).

In the cryogenic process, the battery is first frozen and then shredded. A series of filters, vibrators, and other separation processes are used to separate the components. The cryogenic process is less energy intensive, but the purity and quality of the recovered raw materials is lower than the high-temperature process.

The third option is the physical separation of the components. This process requires little energy. The batteries must first be sorted according to their internal chemistry (LFP, LCO, LMO, etc.). They are then manually disassembled into their main components (housing, cathode, anode, and separator), which can be further processed. The physical separation is relatively slow and incurs higher costs. However, the recycling rate and the purity of the recovered materials are higher than with the other two processes.

Example 13.10 Recycling capacity worldwide

The recycling capacity for lithium-ion batteries is being rapidly expanded all over the world.

In 2021, the largest electric vehicle battery recycling capacity was installed in China (231,000 tons per year t/a) followed by South Korea (24,000 t/a), the United States (20,000 t/a), Belgium (7000 t/a), Australia (3000 t/a), and Germany (1500 t/a).[29]

It is expected that Europe will reach a capacity of 1,200,000 t/a by the year 2035.[30] This is enough to recycle around 3 million car batteries each year. For reference, in 2022, there were 9.3 million new passenger cars sold across the European Union. ◀

By mid of 2023, the new Battery Regulation (2023/1542) of the EU has entered into force. Starting from 2025, the Regulation will gradually introduce declaration requirements, performance classes, and maximum limits on the carbon footprint of

[28] Source: Warner, *The Handbook of Lithium-Ion-Battery Pack Design*, Elsevier Science & Technology, 2015, Amsterdam, Netherlands.

[29] Source: *Battery recycling capacity worldwide as of June 2021*, statista 2024.

[30] Source: *Europe expands recycling of lithium-ion batteries: Focus on capacity development, demand analysis and market players*, Michael Saatkamp, Fraunhofer ISI, July 26, 2023.

batteries for electric vehicles, light means of transport (such as e-bikes and scooters), and all other types of rechargeable industrial batteries. Targets for recycling efficiency, material recovery, and recycled content will be introduced. Battery must be electronically registered with a battery passport carrying a QR code. From 2027, EV batteries must be easily removable and replaceable by independent professionals. The following key dates are of particular importance for EV batteries[31]:

- December 2025—minimum recycling efficiency is 65% lithium-based batteries.
- December 2027—critical mineral recovery rates from waste batteries set (lithium 50%, nickel 90%, cobalt 90%, copper 90%).
- February 2028—impact assessment of battery life cycle to ensure it meets the maximum carbon footprint threshold.
- August 2028—technical documentation detailing the percentages of cobalt, lead, lithium, and nickel which came from recovered sources.
- December 2030—minimum recycling efficiency of 70% lithium-based batteries.

[31] Source: *EU battery passport regulation requirements*, Chris Stretton, Curcularise, December 11, 2023.

Production and Distribution of Energy

14

Abstract

This chapter highlights various technical options for supplying mobile vehicles with energy. The energy chains from the provision of primary energy to the energy carriers at the filling or charging stations are presented. First, the most important primary energy sources and energy carriers are presented with their chemical properties. In the case of finite primary energies, the currently available reserves and the achievable ranges are presented. For renewable raw materials, the land requirements are estimated accordingly. An extensive section is devoted to electricity production from conventional and renewable energy sources, including the necessary distribution grids and their stability. The chapter concludes with a section on hydrogen. The various options for producing hydrogen and distributing it at filling stations are briefly presented. A further section highlights the many ways in which hydrogen can be used in vehicles, for grid stabilization and in industrial processes.

14.1 Introduction

Primary energies can be substances or natural energy sources. A distinction is made between finite (crude oil, natural gas, coal, uranium) and renewable (regenerative) primary energies (wind, water, sun, biomass).

Primary energies are used to produce energy carriers that are then used in the drivetrain of vehicles to generate mechanical energy (propulsion) (Fig. 14.1).

Even though there are large hydrogen gas clouds in our galaxy and hydrogen is also the fuel of all stars, including our sun, hydrogen does not occur naturally on earth. To use hydrogen, we first must produce it from primary energies at great

M. Doppelbauer, *Introduction to Electromobility*,
https://doi.org/10.1007/978-3-658-45482-1_14

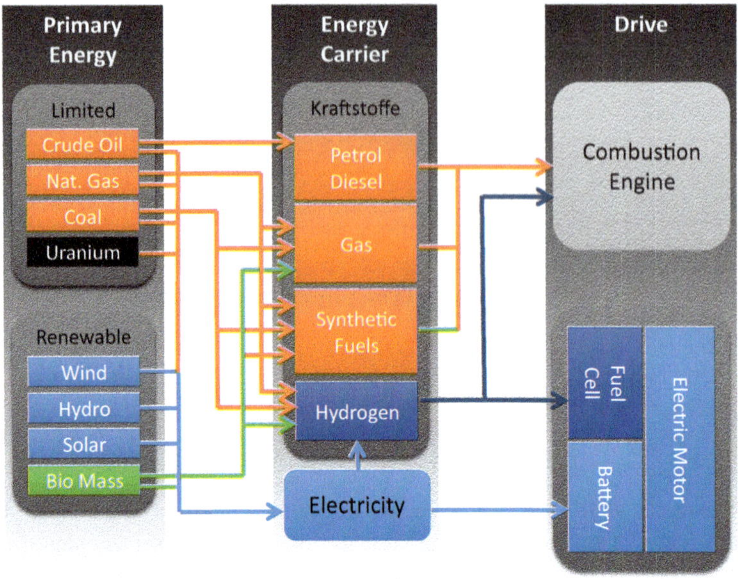

Fig. 14.1 Diversification of primary energy and energy carriers

expense, for example, from natural gas or by electrolysis from electrical energy and pure drinking water. Hydrogen is therefore an energy carrier, not a primary energy source.

The following sections describe how the various energy sources are obtained, processed, and transferred to consumers (infrastructure). The issue of the range (resource availability) of finite primary energies is also considered.

14.2 Crude Oil- and Natural Gas-Based Fuels

14.2.1 Properties

Table 14.1 summarizes the most important properties of liquid and gaseous fuels, namely, petrol and diesel. Some substances (especially natural gas) are natural products whose quality and chemical composition can vary within certain limits depending on the source. The figures given should therefore only be regarded as guidelines.

Around 2% of the energy originally stored in crude oil is lost during its extraction. In addition, 1% is required for transportation to the refinery. The production of petrol or diesel fuels causes a further 6–10% losses and transportation to the filling station another 1%. In total, the production efficiency of conventional liquid

Table 14.1 Technical properties of fuels

	Calorific value	Density at 15 °C	CO_2 emissions (direct and total[1])		
Regular petrol	11.2 kWh/kg	0.750 kg/l	2.32 kg_{CO2}/l	0.276 kg_{CO2}/kWh	2.71 kg_{CO2}/l
Premium petrol	12.0 kWh/kg	0.760 kg/l	2.38 kg_{CO2}/l	0.261 kg_{CO2}/kWh	2.78 kg_{CO2}/l
Diesel	11.8 kWh/kg	0.830 kg/l	2.65 kg_{CO2}/l	0.271 kg_{CO2}/kWh	3.25 kg_{CO2}/l
LPG	12.8 kWh/kg	0.540 kg/l	1.64 kg_{CO2}/l	0.237 kg_{CO2}/kWh	1.9 kg_{CO2}/l
Natural gas L-gas CNG	11.3 kWh/kg	180 kg/m^3	2.85 kg_{CO2}/kg	0.252 kg_{CO2}/kWh	2.9 kg_{CO2}/kg
Natural gas H-gas CNG 1 bar	13.0 kWh/kg	0.810 kg/m^3 180 kg/m^3	2.01 kg_{CO2}/m^3 2.48 kg_{CO2}/kg	0.191 kg_{CO2}/kWh	2.54 kg_{CO2}/m^3
Synthetic methane	15.4 kWh/kg	0.671 kg/m^3			

fuels is stated at around 80–90%,[2] depending on the study. Petrol is at the lower end of the spectrum and diesel at the upper end.

In the case of fuels made from gas, there is no refining, but there are losses due to compression (approx. 2.5–5% when compressing CNG to 200 bar) or liquefaction (approx. 10–25% in the case of LPG).

The figures in the very last column of the table are the total GHG emissions including the upstream chain in $CO_{2\text{-equiv}}$. They are based on typical assumptions for energy consumption and energy supply in the refining process. They provide a more realistic estimate of fuel emissions than the figures in the preceding columns but are subject to fluctuations depending on the refinery's energy source.

14.2.2 Crude Oil Production

Since Russia's attack on the Ukraine, Europe has reduced oil imports from Russia to almost zero. Previously, Russia accounted for by far the largest share of total imports. Imports of crude oil are now relatively evenly supplied from deposits in the North Sea (UK, Norway), Kazakhstan, the USA, and increasingly also from the United Arab Emirates.

[1] Sources: *Berechnung von Treibhausgas (THG)-Emissionen verschiedener Energieträger*, Umweltbundesamt Österreich, December 2023 and own calculations.

[2] The 1.6 kWh of electrical energy consumption per liter of fuel allegedly consumed in the refinery, which is often quoted on the Internet, is based on an incorrect interpretation of information from the US Department of Energy. This number refers to the total losses of a refinery, some of which also result from process heat and the manufacture of other products.

Table 14.2 Total crude oil potential, as of 2021[4]

Crude oil	Cumulative production volume in Gt	Reserves in Gt	Resources in Gt	Non-conventional resources in Gt
Arabian Peninsula	60	114	31	6
North America	51	35	26	157
South America	19	51	26	52
Russia	32	20	72	30
Africa	19	17	28	11
Asia	17	6	25	30
Europe	11	2	5	5
Worldwide	**209**	**245**	502	

Global crude oil production has remained constant at around 4.1–4.3 Gt (gigatons) per year for around 10 years.[3] It only fell slightly during the 2 coronavirus years, 2020 and 2021. The largest production areas in the world are the Middle East, followed by the USA and the CIS states.

If the cumulative production volume since the beginning of industrialization until the year 2021 (209 Gt) and the current annual production (4.1–4.3 Gt) is compared with the known reserves of 245 Gt and the resources of 502 Gt, it becomes clear that crude oil will be available for at least 100 years if current consumption continues. As new reserves are still being found, probably even for 150–200 years.

Peak oil is the time of maximum production. It can be specified on a country-specific and global basis. Predictions of peak oil have had to be repeatedly postponed in recent decades due to new oil discoveries.

Today, however, we consume around four times as much oil as new reserves are being discovered. It is assumed that global peak oil occurred around 2005–2010 and that the total production will fall by around 3% per year in the future (Tables 14.2 and 14.3).

14.2.3 Gas

A distinction is made between liquefied petroleum gas (LPG), a mixture of mainly propane and butane, and natural gas (CNG/LNG), which consists of over 80% methane.

---

[3] Source: *Globales Energie und Klimastatistik Jahrbuch 2023*, Enerdata.

[4] Source: Bundesamt für Geowissenschaften und Rohstoffe BGR, Germany, 2021.

Table 14.3 Number of vehicles with gas engines worldwide, 2017

	Natural gas (CNG and LNG) in thousand cars	Petroleum gas (autogas or LPG) in thousand cars
Asia–Pacific region	6800 (5000 in China)	3300 (2000 in Korea)
Turkey		4700
Iran	4000	
India	3000	
Pakistan	3000	
Russia		3000
Ukraine		2600
Europe	1500 (1000 in Italy)	6200 (2600 in Italy, 3300 in Poland)
South America	4700	
USA	114	147
Worldwide	**24,500**	**27,100**

LPG (also known as liquefied petroleum gas or autogas) is a by-product of crude oil distillation and is stored in liquid form in a tank at a pressure of 2–8 bar. Due to its high energy density, the tank required is comparatively small.

Apart from its density, LPG hardly differs from petrol in terms of its physical properties and can therefore be used in modified petrol engines.

Compared to petrol, LPG causes around 10% less CO_2 emissions in relation to the same energy content.

Natural gas is naturally occurring and merely dried and compressed. It can be stored in compressed form (CNG = Compressed Natural Gas) or liquefied (LNG = Liquefied Natural Gas).

CNG is stored in gaseous form at room temperature and a pressure of 200–250 bar. This requires a stable high-pressure system in the vehicle, which takes up a comparatively large amount of space.

A distinction is made between CNG H-gas (high caloric gas), usually from the states of the former Soviet Union (CIS) or the North Sea with 87–99% methane content by volume, and CNG L-Gas (low caloric gas), usually from the Netherlands or northern Germany with 80–87% methane content by volume.

Today, H-gas is much more common and used in most countries. The energy content of 1 kg of H-gas corresponds to that of around 1.5 L of premium petrol or 1.3 L of diesel, while that of L-gas is only around 1.2 L of premium petrol or 1.1 L of diesel.

Liquefied natural gas (LNG) consists of natural gas, mainly methane, like CNG. Unlike LPG, it only remains liquid at room temperature under high com-

Table 14.4 Total natural gas potential, as of 2021[5]

Natural gas	Cumulative production volume in trillion m^3	Reserves in trillion m^3	Resources in trillion m^3	Non-conventional resources in trillion m^3
CIS	36	68	131	41
North America	49	15	41	112
Arabian Peninsula	13	83	48	14
Asia	14	16	43	79
Africa	6	16	38	42
South America	5	7	21	41
Europe	14	2	8	14
Worldwide	**137**	**207**	**675**	

pression and is therefore stored in cryogenic tanks at low temperatures (−120 to −170 °C). The pressure is then only around 8 bar.

Compared to petrol, natural gas causes around a third fewer CO_2 emissions in relation to the same energy content.

14.2.4 Natural Gas Production

The annual production of natural gas amounted to around 4.2 trillion m^3 in 2021.

The known reserves are 207 trillion m^3 of which around a third is offshore. A further 675 trillion m^3 are available as resources. The supply of natural gas is secured for over 200 years if current consumption is extrapolated (Table 14.4).

14.3 Synthetic Fuels

Synthetic fuels (XtL = X-to-Liquid) are liquid fuels that are not based on natural crude oil but rather on other energy sources.

While conventional fuels are obtained by separating the crude oil into its individual components (distillation, rectification), synthetic fuels are produced in complex chemical processes. The Fischer–Tropsch synthesis for the liquefaction of energy-rich raw materials (coal, natural gas, biogas, wood, household waste, etc.) is well known since almost 100 years.

The synthetic fuels bioethanol (C_2H_6O, also written as C_2H_5OH) and biodiesel are produced by such plants.

[5] Source: *Energiestudie 2021 – Reserven, Ressourcen und Verfügbarkeit von Energierohstoffen*, Bundesanstalt für Geowissenschaften und Rohstoffe (BGR), Hannover.

Bioethanol is mainly produced from maize and wheat.

Biodiesel is mainly produced from rapeseed and palm oil. It can be used as a pure fuel (B100) or as an admixture, for example, B7 indicates 7% biodiesel.

All these synthetic fuels have similar properties to petrol or diesel. In principle, they can be used in normal combustion engines and can be distributed via the existing filling station infrastructure. However, biodiesel and ethanol attack hoses and seals and can cause damage even in small mixtures if the vehicle is not equipped accordingly.

The use of alternative fuels generated from electrical energy (e-fuels) is also being investigated. The starting point is always hydrogen obtained from electrolysis, which is then further processed. E-fuels include hydrogen (H_2), methane (CH_4), methanol (CH_4O), and ammonia (NH_3). The use of these synthetic fuels requires complex modifications to the combustion engine and exhaust gas after treatment. They are currently all experimental and are not used on a large scale to power motor vehicles.

14.3.1 Non-sustainable Synthetic Fuels

These synthetic fuels require a synthesis gas, which in turn requires coal, crude oil, or natural gas as a base material.

Processes for producing gas and diesel fuels from coal (CtL = coal-to-liquid) have been known since 1913 (Bergius–Pier process) and were used on a large scale during the Second World War. They have the disadvantage of being quite energy intensive and cost intensive. They also produce large quantities of CO_2. For this reason, coal liquefaction and other similar processes do not currently play a role and will not do so in the foreseeable future.

Another possibility is the conversion of natural gas (GtL = gas-to-liquid). Natural gas from secondary oil production is often flared today due to high transportation costs. By converting it into GtL, the energy density can be increased so that selling it may become economically viable.

14.3.2 Sustainable Synthetic Biofuels

14.3.2.1 First-Generation Biofuels

First-generation BtL (biomass-to-liquid) fuels are based on renewable raw materials (plants) that are grown agriculturally. However, biofuels can only be described as sustainable at first glance, as the high consumption of land (agriculture) and water creates new potential for conflict (hunger and thirst versus fuel production).

There are basically three types of biofuels: bioethanol (E) is a substitute or a component in petrol fuel, renewable diesel (R) and biodiesel (B) are substitutes or components in diesel fuels.

Table 14.5 Properties of biofuels

	Calorific value	Density at 15 °C	CO_2 emissions Direct and total[6]		
Bioethanol (E100)	8.3 kWh/kg	0.79 kg/l			
Biodiesel (B100)	9.9–10.6 kWh/kg	0.86 to 0.90 kg/l	0.92 kg CO_2/l	0.102 kg CO_2/kWh	1.08 kg CO_2/l
Renewable diesel (R100)	10.0–10.8 kWh/kg	0.76–0.86 kg/l			

Their production processes, chemical compositions, and compatibility with existing engines differ. Renewable diesel is more like petrodiesel and can be used in existing engines without significant modifications. Bioethanol and biodiesel may require some engine adjustments or blending with conventional petrol or diesel.

The quantity of biofuel blended with the conventional fuel is written as a number after the code letter. For example, E10 means that 10% bioethanol is blended into petrol. B100 is pure biodiesel and E100 pure bioethanol.

Renewable diesel has a higher energy density and better cold-weather performance compared to biodiesel. Biodiesel has different properties than petrodiesel, including higher viscosity, lower energy density, and poorer cold-weather performance. It may require engine modifications or blending with traditional diesel in certain proportions to ensure compatibility. Biodiesel has a calorific value that corresponds to around 87% of the calorific value of diesel.

Diesel biofuels are produced from oil-bearing plants, predominantly rapeseed.

Bioethanol can be produced from various plant-based raw materials. Production can be based on plants containing starch (e.g. grain, potatoes, maize) or sugar (e.g. sugar beet, sugar cane). German bioethanol producers mainly use grain and sugar beet.

In Brazil, Ethanol is produced by the extraction of sugars from sugar cane. The country has a large fleet of over 30 million flexible-fuel vehicles, which can use any mixture of 20% (E20) up to 100% (E100) bioethanol with regular petrol fuel. In other countries including Europe, bioethanol is blended into regular petrol in small quantities of less than 10% (E10).

Bioethanol has a higher octane rating (109–113) than petrol (87–98). But it also has a calorific value that is around a third lower. This means that a vehicle running on ethanol may experience a decrease in fuel efficiency compared to running on gasoline. Ethanol attacks plastics and therefore requires engine modifications for higher blends (Table 14.5).

[6] Source: *Berechnung von Treibhausgas (THG)-Emissionen verschiedener Energieträger*, Umweltbundesamt Österreich, December 2023.

Example 14.1 Biofuels in Germany

In Germany, ethanol is only used in small amounts to avoid damaging the vehicles. The proportion has remained constant for years at around 6 to 7% by volume (E10).

Almost 90% of the bioethanol comes from the agricultural production of feed grain (mainly maize and wheat) and the rest from sugar beet. The agricultural land required for this amounted to around 215,000 hectares in 2022, that is, 1.9% of the total agricultural land in Germany. The production volume was around 715.5 thousand tons of bioethanol, which is around 4% of the total domestic sales of petrol fuels (17 million tons).

In 2022, cultivation of rapeseed for biodiesel in Germany occupied around 665,000 hectares, or 5.7% of the agricultural land. The production volume was around 3.89 million tons of biodiesel, which is around 10% of the total domestic sales of diesel fuels (35 million tons).

Pure biodiesel (B100) currently only accounts for around 2% of the market share of biodiesel in Germany. The remaining 98% is blended with conventional diesel. ◀

14.3.2.2 Second-Generation Biofuels

For second-generation BtL fuels (bioliq process), almost all plant-based biomass can be processed into substitutes for kerosene, diesel, and petrol. Particularly suitable are cellulose-rich, dry residual biomasses such as straw, residual wood from agriculture, forestry, and horticulture.

The thermochemical decomposition process of biomass produces hydrogen and carbon monoxide, which can be converted into fuels using chemical catalysts.

1 kg of fuel is produced from 8 to 10 kg of biomass. Only around a third of the energy contained in the biomass can be converted into liquid fuels, the rest is needed to operate the plants.

Second-generation biofuels are therefore significantly more expensive to produce than biodiesel but have the advantage that they can be produced from many types of residual materials and do not require their own cultivation areas.

Due to the high costs, developments in the field of second-generation biofuels have now largely been discontinued.

Third-generation biofuels are biofuels made from algae, although this is purely a research topic and it is not yet foreseeable whether and when these fuels will reach market maturity.

14.3.3 Electricity-Generated Fuels (e-fuels)

Electrolysis converts water into hydrogen. By adding CO_2, gases (methane) or liquid fuels (methanol, ethanol, or ammonia) are produced. These plants are also referred to as PtX = Power-to-X and the fuels produced as e-fuels.

In the past, research focused on the production of oxymethylene ether (OME), which, like other liquid fuels (petrol, diesel), can be easily stored and transported and has a high energy density. OME stands for a group of chemical molecules consisting of hydrogen, carbon, and oxygen. OME can be produced based on various processes. One of these process routes requires only renewable energy or electricity, CO_2 and water as raw materials. If the required CO_2 is separated from the ambient air, this results in a closed CO_2 cycle together with the conversion of the fuel in the combustion engine.

There are no direct carbon–carbon bonds in the OME molecule and each molecule has a high proportion of oxygen. Due to these chemical properties, OME can burn soot free. OME can be in conventional combustion engines without modifications because it has similar operating parameters to diesel fuel.

However, due to the costly production of OME, other synthetic fuels are now in the foreground, in particular, hydrogen, methane, and ammonia. To be able to use these fuels in combustion engines, extensive changes must be made to engine management; exhaust gas purification; and, in some cases, injection.

The simplest fuel produced from electric energy is synthetic methane gas:

$$CO_2(44\ \text{g/mol}) + 4\ H_2(4 \cdot 2\ \text{g/mol}) \rightarrow CH_4(16\ \text{g/mol}) + 2\ H_2O(2 \cdot 18\ \text{g/mol}) \qquad (14.1)$$

This reaction requires large quantities of CO_2, which is later released again during combustion. The entire cycle is therefore CO_2 neutral, at least, when ignoring the CO_2 emissions from the construction of the production plant.

The procurement of CO_2 gas is proving to be a challenge, at least if large quantities of methane gas are to be produced synthetically. There are some natural and artificial sources of CO_2 but for obtaining large quantities sustainably, the carbon dioxide must be extracted from the ambient air (DAC = Direct Air Capture). However, the volume fraction of CO_2 in the air is low, only about 0.04% or 0.79 g/m^3. For the extraction of one cubic meter of CO_2 gas, that is, a mass of 2 kg, 4–5 kWh of electricity are needed and 2500 m^3 of air must be blown through huge filter towers.

When the energy for the extraction of CO_2 is considered in the energy balance, e-fuels achieve about 15% overall efficiency from electrical to mechanical energy. Battery electric vehicles are about five times better with a typical efficiency of around 70% (85% motor and gearbox, 95% AC/DC converter, 95% battery, 90% charging losses = 70% total).

Synthetic ammonia (NH_3) does not require carbon dioxide (CO_2) in the production process. It only produces water and nitrogen (N) or nitrogen oxides (NO) during combustion, which must be removed from the exhaust gas by a catalytic converter. However, ammonia is highly dangerous for humans, animals, and plants.

Due to the high consumption of electricity and the complex plant technology, the production of e-fuels is generally very expensive, around four to five times more expensive than petrol and diesel and will hardly become competitive even in the long term.

Example 14.2 Production of synthetic methane gas CH_4

To generate 3 kWh of mechanical energy, an internal combustion engine with a typical efficiency of 30% must burn around 10 kWh of methane (CH_4), or 0.65 kg. 10 kWh is roughly equivalent to the energy of 1 L of petrol.

The synthetic production of 0.65 kg methane requires 0.325 kg (10.8 kWh) of hydrogen (H_2) and 1.8 kg of carbon dioxide (CO_2).

The electrolysis of the hydrogen requires 15.4 kWh (efficiency of current electrolyzers around 70%). An additional 4 kWh of electrical energy is required to extract the carbon dioxide from the ambient air.

The total electricity requirement is therefore around 20 kWh,[7] which is an efficiency of around 15%.

In comparison, a battery electric vehicle would only require around 4.5 kWh of electrical energy to provide 3 kWh of mechanical energy, considering the losses in the battery, power electronics, electric motor, and charging losses. Therefore, **a BEV is about four to five times more efficient than e-fuels**.

If all 47 billion German cars were to run exclusively on e-fuels, theoretically around $8.8 \cdot 10^{13}$ m^3 of air would have to be blown through filters each year to obtain the required CO_2. That is the entire volume of air over the surface of Germany up to a height of 250 m. This air filtration alone would consume the energy of around one-third of Germany's current electricity production.

All in all, around twice the amount of electricity currently generated in Germany would be needed to supply all German cars with e-fuels. ◀

Example 14.3

In Patagonia (South America), the "Haru Oni" pilot plant, which uses wind power to generate e-fuels, went into operation in summer 2023.

It is assumed that electricity generation will cost around 1.5 ct/kWh due to the strong winds. By comparison, the cheapest wind power in Europe is around 4 ct/kWh.

First, hydrogen is produced from water by electrolysis, which is then processed into methanol in a second step using carbon dioxide from the air. For this purpose, a plant is to extract around 150 kg of CO_2 from the air in 60 minutes. Initially, the carbon dioxide was obtained from a brewery in a non-climate-neutral way. Three liters of water are needed to produce one liter of e-fuels.

The pilot plant will produce around 130,000 liters of e-fuels per year, enough for around 130 cars.

---

[7] See also: *Der Weg zum CO2-neutralen Verbrenner?* from Auto-Motor-Sport, published on 13.01.2023. In this paper, an energy requirement of 27 kWh of electricity per liter of e-fuels was calculated.

The construction of a large commercial plant a short distance from the test facility is planned for 2026. This should then produce 66 million liters of e-fuels per year.

For comparison: current fuel consumption (petrol and diesel fuels) in Germany is around 70 billion liters. With the planned plant, emissions from road traffic can therefore be improved by 0.1%, i.e. from an average of 7.4 l/100km to 7.39 l/100km. ◀

The example shows that a large fleet of millions of cars cannot be supplied with e-fuels, regardless of the country in which the fuel is produced.

It should also be emphasized that e-fuels are only CO_2 neutral in the global balance. Unlike battery electric vehicles, e-fuels do not improve local emissions (noise, particulate matter and pollutants). The same applies to biofuels, of course.[8]

Note

Leaks and unburned residues in the exhaust gas are a major risk when burning methane gas, as CH_4 has around 27.5 times more of an impact on the climate than CO_2. When burning methane (CNG or e-methane) in a piston engine, 2–3% residue is not unusual.[9] This alone produces a similar climate impact as if conventional diesel fuel had been used—even when the methane gas is produced from regenerative sources. All in all, the climate benefit of e-fuels is highly questionable.

However, e-fuels can be a sensible alternative to crude oil-based fuels wherever the direct use of electrical energy is impossible. These include heavy-duty applications, large ships, and aircraft.

14.4 Electrical Energy

14.4.1 Grids

Electricity has been produced and distributed commercially for over 100 years. The individual plants and consumers are connected to each other via a complex system of three-phase AC electricity grids.

[8] See also: *E-Fuels sind nicht sinnvoll für den großflächigen Einsatz bei Pkw und Lkw*, Martin Wietschel, Jacob Leidenberger, Fraunhofer ISI, April 04, 2023.

[9] See also: *Verbrennungssteuerung eines CNG-DI-Motors mittels Hochlast-Abgasrückführung*, Dissertationsschrift Johannes Oder, Universität Magdeburg, 2021.

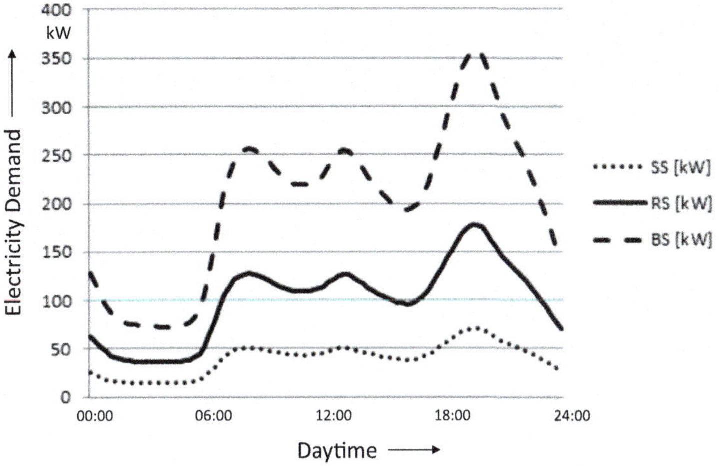

Fig. 14.2 Typical load profile of low-voltage grids[10]

14.4.2 High-Voltage Grids

At the highest level, the European interconnected grid (transport grid) connects the individual grid operators with voltage levels between 380 and 400 kV. In Germany, for example, the length of this extra high voltage grid is around 35 thousand kilometers. Regional electricity suppliers (coal, gas, and nuclear power plants) and some very large industrial plants are connected directly to this grid.

Note
If different AC grids are to be interconnected over long distances, high-voltage direct current (HVDC) connections are often used. However, these connections are not grids in the conventional sense, but rather point-to-point connections.
HVDC connections require complex power electronics at both ends to realize the transition from three-phase AC to DC. The advantage of DC connections is that they can interconnect AC grids of different frequencies, voltages, and phases. In addition, DC connections only transmit pure active power, so there are no losses due to reactive power.

[10] Source: Warweg, Schaller, Ritter, Bretschneider, *Technical and Regulatory Challenges for the Integration of Electric Vehicles into the Energy System*, 2nd IEEE PES International Conference and Exhibition on Innovative Smart Grid Technologies (ISGT Europe), 2011.

Table 14.6 Net installed capacity and net public electricity generation in Germany in 2022[12]

Energy carrier	Million kW	Billion kWh
Wind	58.01 (onshore) 8.15 (offshore)	98.61 (onshore) 24.74 (offshore)
Photovoltaic (without own consumption)	67.55	52.6
Gas (without industrial own use)	34.25	47.1
Stone coal	19.01	55.5
Brown coal	18.67	106.9
Nuclear power	4.05	32.8
Biomass	8.91	42.2
Hydro (running water)	4.94	15.0
Total	**224**	**492**

On the next lower level are various regional and national high-voltage electricity grids, which are used to connect larger industrial plants, large- and medium-sized wind farms and rail transportation. The usual voltage levels are from 50 to 220 kV (typically 110 and 220 kV). The total length of the high-voltage grids in Germany is around 80 thousand kilometers.[11]

Medium-voltage grids connect medium-sized industrial plants, larger commercial operations, larger residential complexes, office and department stores, hospitals, larger photovoltaic systems, etc. These grids operate at voltages between 6 and 60 kV. They cover around 510 thousand kilometers in Germany.

Most end consumers are connected via low-voltage grids with a nominal voltage of 480 V in the US, 415 V in India, 400 V in Europe, 380 V in China, and 200/210 V in Japan. The connection is either single phase with two wires, in which case the voltage is reduced by 42%, or three phase. Some countries use a three-wire split-phase system. For example, in the United States, two voltages of 120 V that are out of phase by 180° and have a common ground are supplied over three wires. These voltages can be used individually or can be combined to form a single phase with 240 V.[13]

[11] Source: *Overview of transmission tariffs in Europe: Synthesis 2014*, ENTSO-E Working Group Economic Framework, June 2014.

[12] Source: Fraunhofer ISE, AG Energiebilanzen, https://www.ag-energiebilanzen.de; https://www.energy-charts.de/power_inst_de.htm.

[13] The split-phase system is nevertheless single phase, i.e. neither two phase nor three phase. It is not possible to generate a rotating field in an electric motor with a split-phase connection without additional aids for phase shifting (capacitor, coil).

Table 14.7 SAIDI in selected countries in 2020[15]

Country	SAIDI 2020 in minutes/year
Slovenia	6
Finland	12
Switzerland	12
Germany	18
United Kingdom	18
France	24
Spain	30
Austria	36
Poland	66
Italy	78
USA	78
Romania	186
Croatia	200

For example, the total line length of the low-voltage grids in Germany is around 1.2 million km.

The many different electricity grids are interconnected by transformers. In Germany alone there are about 550,000 transformers.

14.4.2.1 Low-Voltage Grids

The low-voltage electricity distribution is separated in block structures, row structures, and settlement structures.[14]

Block structures (BS) are mainly found in city centers and comprise around half of all underground 400 V cables. They are typically connected to the medium-voltage grid with a 630 kVA transformer. One transformer supplies an average of eight larger buildings, each containing many apartments or stores.

Row structures (RS) are mainly found on the outskirts of cities. They comprise around a quarter of the underground low-voltage cables and are used to supply many apartments. Just like block structures, they typically have a 630 kVA transformer.

Settlement structures (SS) are used to supply village areas. They comprise the remaining quarter of the LV underground cables. The number of buildings supplied is higher (typically around 50), but the number of people supplied is lower

[14] Source: Warweg, Schaller, Ritter, Bretschneider, *Technical and Regulatory Challenges for the Integration of Electric Vehicles into the Energy System*, 2nd IEEE PES International Conference and Exhibition on Innovative Smart Grid Technologies (ISGT Europe), 2011.

[15] Data source: https://en.wikipedia.org/wiki/SAIDI.

Table 14.8 Average CO_2 emissions from electricity power plants with plausibility ranges[16]

	Emission factor in g/kWh$_{el}$ without installation	Emission factor in g/kWh$_{el}$ including installation
Brown coal power station	1000 (850–1200)	
Hard coal power station	930 (750–1100)	
Natural gas power station	480 (400–550)	
Photovoltaic	0	75 (50–100)
Hydro power station	0	25 (10–40)
Wind turbine	0	25 (10–40)
Nuclear power station (without disposal of radioactive waste)	20 (10–30)	
Biomass power station	–580 bis 150 Depending on the utilization variant (electricity, combined heat and power) and power plant technology	

(on average 2 apartments per building). Accordingly, the typical connected load is also lower at 400 kVA.

The daily load profiles of these three typical low-voltage grids are similar, but with a different power demand (Fig. 14.2).

A closer analysis of the additional electricity requirements for the installation of charging points reveals that block structures (inner cities, larger residential blocks) are particularly critical. Typically, the number of connected households is large (over 500) and the available transformer capacity and the capacity of the buried copper cables are already utilized.

In such low-voltage grids, the options for installing charging points are extremely limited. According to current estimates, typically only 10–15% of these households can be supplied with electricity for electric vehicles. There must be a significant increase in capacity in the future.

Row structures and especially settlement structures are less critical because the free capacity is higher and the number of connected households is lower. Depending on the charging speed, such grids can already supply 30–80% of all households with sufficient charging current for electric cars.

[16] Source: Wagner, Koch, Burkhardt, Böckmann, Feck, Kruse, *CO_2-Emissionen der Stromerzeugung – Ein ganzheitlicher Vergleich*, BWK Bd. 59 (2007) Nr. 10, Seite 44–52, VDI-Verlag, Düsseldorf.

14.4.2.2 Grid Stability and Security of Supply

Security of supply is a much-discussed topic, as electricity from renewable sources can only be planned statistically and production can fluctuate greatly on individual days. The question arises as to how the electricity demand can be generated at peak load times. Large-scale wind lulls at night, when neither wind energy nor PV energy can be produced, are particularly feared.

The security of supply can be achieved through a combination of several measures, see also the following chapters.

The installed capacity of wind and solar power plants must be significantly higher than the average demand and higher than the peak demand. For example, already in 2023, all power generators installed in Germany were able to produce around four times more electrical energy at full load compared than the average demand. The installed capacity of photovoltaic (PV) and wind power was approximately 130 GW, dwarfing the capacity of conventional coal-fired and nuclear power plants at around 42 GW, see Table 14.6. Despite this stark contrast, renewable sources produced about 180 TWh in the whole year, while conventional producers contributed about 200 TWh.

Storage systems must be installed to store solar energy at short notice, that is, for the evening and night time. Short-term storage can be pumped storage power plants or battery storage systems. Battery storage systems can be installed on a large industrial scale or as home storage systems in residential buildings. Parked electric cars are also suitable as short-term intermediate storage (V2G = vehicle-to-grid).

In the future, seasonal long-term storage systems for electrical energy will also be required. At the moment, there are no viable, cost-effective solutions for long-term energy storage. It is conceivable that hydrogen could play a role in the storage of electrical energy in the future. However, the efficiency of hydrogen storage systems is very poor, see Sect. 14.5.3.2, and green hydrogen is needed in large quantities for industrial use. It will therefore hardly be available for seasonal storage in the foreseeable future. One practicable solution appears to be the construction of natural gas-fired power plants (CCGT), which can take over part of the supply in an emergency.

Grid stability is measured internationally in the unit SAIDI (System Average Interruption Duration Index). SAIDI refers to the average, unplanned interruptions of electricity supply per customer in minutes per year. In most industrialized countries, the SAIDI is very low, typically less than 1 or 2 h per year (Tables 14.7 und 14.8).

The introduction of fluctuating energy sources, particularly wind and photovoltaics, and the dismantling of nuclear power plant capacities have not fundamentally changed the good stability of the electricity grids to date. However, the so-called “n − 1” rule, according to which every power grid must be able to cope with at least one failure of a line or power generator at any time without a power outage occurring, has been violated more frequently in recent years.

Active interventions to stabilize the grid are required to compensate for fluctuating loads, for example, when larger consumers are switched on or off, and in the event of a sudden loss of generation capacity, for example, in the event of line or power plant outages. Such interventions normally run automatically. But the number of manual control interventions, which used to be necessary only a few times a year to prevent a shutdown, has increased significantly.

The grid is regulated based on the grid frequency, which has the same value everywhere within an interconnected grid at any time. In the European interconnected grid (UCTE), this applies to all voltage levels from 230 to 380 kV and for all connected countries from North Africa (Morocco) to Denmark. In Europe, there is also the interconnected grid of the Northern European countries (NORDEL = Iceland, Norway, Sweden, Finland) and the grid in Great Britain (UKTSOA = England, Wales, and Scotland). These grids are connected with a high-voltage direct current (HVDC) transmission system.

If the frequency rises above the upper limit of the so-called dead band of 49.99–50.01 Hz, more power is generated than the consumers consume. The power plants have to reduce their production accordingly. However, if the frequency falls below the limit value, too little power is generated and the power plants have to ramp up their production.

The grid voltage, on the other hand, is different at every node and at every socket. It is regulated locally by variable transformers.

The power is adjusted in various stages, depending on the urgency (frequency deviation):

- instantaneous reserve,
- primary control power (time horizon a few seconds to half a minute),
- secondary reserve (time horizon half a minute to around 5–15 min), and
- minute reserve (time horizon up to 60 min).

The instantaneous reserve is provided by the flywheel masses of the large rotating power plant generators. These are mainly generators not only from coal and nuclear power plants, but also from gas and liquid fuel plants. When more and more wind turbines or PV systems are connected to the grid using power electronics, methods will be needed to ensure that they behave in a similar way to synchronous generators with their large moment of inertia. Several research and development activities and legal requirements are already underway.

Grid stabilization in the minute range is initially provided by intermediate storage facilities, for example, pumped storage power plants, and in the longer term via the readjustment of large thermal power plants. Grid operators pay fees for the provision of primary and secondary control power that are far higher than average electricity prices.

The Hornsdale Power Reserve, see Sect. 14.4.7, is a current example of a commercially successful primary control storage system based on lithium-ion batteries. Hydrogen systems are hardly suitable for primary and secondary control due to the high power-specific costs.

The following sections describe a series of measures that can be used to ensure grid stability despite fluctuating renewable electricity generation.

14.4.2.3 Management of Wind Power Generation

If individual sections of the distribution or transmission grid are overloaded, electricity from renewable energy sources may be throttled (feed-in management). A grid operator may only take this measure if all other options for bypassing the bottleneck have been exhausted.

The power that can potentially be generated but is not fed into the grid is referred to as outage energy. It has fluctuated greatly in recent years, although a slight upward trend can be observed overall.

The total outage energy in Germany amounted to around 8.1 billion kWh in 2022. This was around 3.2% of the production of renewable electrical energy or 1.4% of the total electricity production.

14.4.2.4 Network-Dependent Tariffs

A relatively new idea is the so-called grid-supportive dynamic electricity tariffs.

Cars are often driven during the day, e.g. to the shops, to work or to school, and parked overnight. However, the car does not have to be fully charged again until the next morning, the charging time is flexible. The overall load on the network can be reduced by distributing the charging processes to individual vehicles. To a certain extent, even weather conditions, such as periods of low wind, can be considered.

A first, simple step is an electricity tariff with night-time electricity, which generally applies between 10 pm and 6 am. Electricity tariffs that vary depending on the time of day and load are already common in some countries, e.g. in the USA. Similar developments can also be expected in Europe.

The next step is then to consider the current production of wind and PV electricity in the continuously adjusted tariff, e.g. on an hourly basis. The technical basis for this is provided by intelligent electricity meters (smart meters). They communicate with the electricity provider via the Internet and dynamically adjust the electricity price to the status of the electricity grid. An electric car can control its charging depending on the tariff.

Situation-dependent tariffs can even make home storage systems interesting for balancing out consumption peaks, even if no PV system is available for local energy generation. In times of cheap electricity, the home battery is charged, and the energy can later be used in times of high prices.

14.4.3 Micro-Grids

By using the grid frequency as an information carrier, it is possible to go one step further and charge consumers tariffs adapted to the respective load situation at 10-s intervals, for example. This works particularly well in island grids, which have a greater fluctuation range in grid frequency than the European interconnected grid. In Germany, too, the grid structure could be divided into smaller, DC-coupled grids (micro-grids), which could then also be regulated autonomously.[17]

The consumers in such micro-grids must react sensitively to changes in frequency to adapt their output to the current electricity balance and thus minimize their electricity costs. In millions of households, there are many appliances with high thermal time constants, which can be utilized in such scenarios. These include, for example, refrigerators and freezers; air conditioning systems; hot water heaters; and, of course, wallboxes for electric cars.

Such devices can automatically turn on or off at times that are beneficial to the grid.

Obviously, some boundary conditions must be considered, for example, the temperature in a refrigerator must not rise too much, or an electric car must be sufficiently charged in the next morning. Within these limits, the power consumption is then automatically adjusted to the respective grid frequency by the control system. In this way, millions of small consumers can work together and save a significant proportion of the primary and secondary control reserve.

14.4.3.1 Throttling of EV Chargers

Remote-controlled home chargers are common in some countries and sometimes even required by law, for example, in Luxemburg. During periods of high grid load, a signal is sent to the home chargers by the electricity grid operator. The charging current is then reduced to 6 A and the power is reduced from 11 to 4.2 kW accordingly.

In Germany, remote-controlled, time-limited electricity rationing for heat pumps and electric car charging stations was put in place in 2024. Public charging stations or charging parks are not affected by throttling. As an incentive measure, the grid operator gives an annual discount on the grid fee when a technically suitable, remote-controllable wallbox is installed.

Many household heat pumps are already connected to their own meter and a cheaper tariff, which may not be supplied by the grid operator at peak times.

14.4.4 Smart Charging

The term smart charging summarizes the various concepts and technologies used to control the bidirectional charging and discharging of electric vehicles. There must be communication between the electric vehicle, the charging station, and the power grid in order to control the flow of electricity.

For example, the charging process can be started automatically when a particularly large amount of renewable electrical energy is produced, e.g. during periods of strong wind or high solar radiation. Intelligent chargers can, for example, communicate with a private PV system and only charge the surplus solar power.

[17] Source: Easy Smart Grid GmbH, Karlsruhe, Germany.

Electric cars with their large batteries can also serve as intermediate storage media for wind and solar energy and stabilize the power grids. A battery in a typical electric car can supply an average household with electricity for a whole week. If energy becomes scarce at peak times, it can be fed back into the public grid from the cars.

To enable these technologies, the charging systems must be bidirectional: V2X (vehicle-to-everything), i.e. the transfer of energy from the car via the plug to various applications:

- V2L (vehicle-to-load) enables the connection of single-phase household appliances, e.g. hand tools or a refrigerator, to the car. Such systems are already offered in several cars.
- With V2H (vehicle-to-home), the car becomes an intermediate storage of excess solar power. During the dark evening hours and at night, the home is then supplied with the previously produced electricity. The car thus replaces a home storage unit.
- V2B (vehicle-to-building) is like V2H but may involve larger buildings with several residential units. There are additional requirements for billing the electricity supplied.
- V2G (vehicle-to-grid) is the most comprehensive technology. It enables the feeding of the temporarily stored power back into the public grid. The electricity is then remunerated by the grid operator.

The first smart charging systems were based on the CHAdeMO plug, which had already integrated a V2L protocol in 2014.

The technical requirements for smart charging are defined on a broad, international basis with the international standard ISO 15118-20 (second-generation network layer and application layer requirements) in 2022. However, the latest version 2.0.1 of the communication protocol OCPP (open charge point protocol), which was defined by the OpenChargeAlliance in 2021, is still not fully capable of V2X. This will only be implemented with the new version 2.1, which is due to be released in the next few years.

Example 14.4 Vehicle-to-grid

In the UK, Nissan launched a large-scale trial for vehicle-to-grid (V2G) systems together with energy supplier Enel in 2016. 100 Nissan Leaf and e-NV200 vehicles are connected to the electricity grid via bidirectional charging points. The vehicle owners receive a corresponding payment for providing a small part of their battery capacity as control power.

Similar concepts are being tested as prototypes in many other countries (including South Africa, Japan, the Netherlands, France, and Denmark) by a number of manufacturers (including Tesla Motors and Renault). ◀

The main difficulty in implementing V2G is currently still the bureaucratic hurdles, as feeding electricity into the public grid makes the end user a legal reseller, which entails an enormous administrative burden. However, the legal framework is gradually being improved.

So far (as of 2024), there are only a few cars on sale that are V2X-capable. And they only have the V2L function activated, i.e. they support the operation of normal household appliances such as refrigerators or hand tools on a regular domestic socket (230 V 50 Hz in Europe).

14.4.5 Local Electricity Production by PV

Local electricity generation by PV systems can contribute to grid stability because local load peaks, which occur when charging electric cars, for example, can be reduced.

Some 10 years ago, it was common practice to feed privately generated PV electricity directly into the grid with high feed-in tariffs. Feed-in tariffs have been gradually reduced over the past few years.

Example 14.5 PV feed-in tariffs in Germany

In Germany, new PV systems intended for full feed-in to the grid, e.g. on commercial rooftops, will receive a feed-in tariff of 7.1 ct/kWh from 2024.

Private PV systems that are intended for self-supply and only feed the unused surplus into the grid will receive 8.1 ct/kWh.

On the other hand, the purchase price for privately used electricity is between 30 and 40 ct/kWh. It has therefore become financially attractive to use PV systems primarily for self-consumption. ◀

Depending on the available roof area, its orientation and inclination, peak outputs between 7.5 and 15 kWp can be achieved on the roofs of single-family homes.[18] As a rule of thumb, it can be assumed that around 1000 kWh of energy is generated per kWp of installed output of a well-aligned system in Central Europa per year. A 10 kWp system therefore produces 10,000 kWh/a. This is far more than a typical four-person household needs and is a good incentive to purchase an electric car.

The ratio of self-generated PV electricity to a household's total electricity consumption is referred to as the degree of self-sufficiency. Typical private systems achieve self-sufficiency levels of around 30–40%.

[18] The peak output is the maximum output of the PV system that can be delivered when the sun is at its optimum position. During the year, however, the peak power is only reached on a few days.

14.4.6 Home Battery Storage

One disadvantage of PV systems is that a large proportion of the self-generated electricity cannot be used. The main production occurs at mid-day, when often no one is at home, while production decreases or even stops in the morning and evening hours when consumption is high.

Local intermediate storage in home storage systems can improve the use of self-produced electricity and thus significantly increase the degree of self-sufficiency. The annual average degree of self-sufficiency including the winter months can reach an average of 50–70%.

Home storage systems are battery storage units that are charged during periods of strong sunlight and then supply electricity for the evening and early morning hours. Electric cars can also be charged from home storage systems.

By reducing local consumption and capping consumption peaks, home storage systems also make a lasting contribution to stabilizing the grid.

In addition, electricity providers can form virtual electricity pools by remotely controlling many home storage units, which can take energy from the electricity grid at short notice when required and release it again later. The provision of this balancing energy is remunerated to the private operator and reduces the electricity price for the households involved. Corresponding business models are now available on the market from several providers.

Home storage systems consist of lithium-ion batteries and power electronics for the connection to the electricity grid. Typical home storage systems in Central Europe have a capacity of between 6 and 16 kWh and enable a typical two- to four-person household to be mostly self-sufficient during the sunny months from March to September.

Example 14.6 Distribution of home storage systems in Germany

In 2022, 220,000 new home storage systems were installed in Germany. Almost 240,000 further installations were added in the first half of 2023. At the beginning of 2024, almost 1.2 million home storage systems with a total capacity of 6.7 million kWh were installed in Germany.

For comparison: Germany's largest hydro pumped storage power plant in Goldisthal, south of Erfurt, has a storage capacity of 8.5 million kWh. ◀

Example 14.7 Amortization of a private PV system (new build 2017)

A 9.3 kWh peak PV system with a home storage system with a capacity of 10 kWh is installed in a residential home.

The system costs around 26,000 € net, of which around half is for the installation of the PV system and half for the home storage system.

The annual electricity requirement of the four-person household is just under 6000 kWh plus around 3000 kWh for operating an electric car (15,000 km driven annually).

Table 14.9 Average CO_2 emissions from electricity power plants in Germany[19]

	Emission factor in g/kWh$_{el}$ including installation
Photovoltaic	57
Hydro power station	2.7
Hydro transfer storage power station	25
Wind turbine (onshore)	17.7
Wind turbine (offshore)	9.7
Biomass solid	68
Biomass liquid	544
Biogas	335 (Biogas) 293 (Biomethane) 110 (Sewage gas) 144 (Landfill gas)
Geothermal power station	166

The electricity production of the system can be estimated at around 9000 kWh per year (location in southern Germany), of which around 55% can be used for self-consumption and the remaining 45% is fed into the grid. The feed-in tariff is 12.3 ct/kWh.

The household's annual electricity costs have so far amounted to around 3150 € (35 ct/kWh for 9000 kWh).

The operation of the system covers 5000 kWh of the household's own electricity requirements, while 4000 kWh must still be purchased (at a cost of around 1400 € per year). At the same time, 4000 kWh of electricity is sold (for around 500 € per year), thereby reducing the annual electricity bill from 3150 to 1100 €.

The system costs are thus amortized in around 13 years without taking interest and service costs into account. ◀

14.4.7 Short-Time Electricity Storage Systems

For intermediate storage of large amounts of electric energy on a grid level, large lithium-ion battery storage systems are being built since several years.

The cycle efficiency (round-trip efficiency) is about 80%: (electricity → power electronics [95%] → battery [90%] → power electronics [95%] → electricity).

---

[19] Source: Climate Change 49/2023, *Emissionsbilanz erneuerbarer Energieträger*, Umweltbundesamt, December 2023.

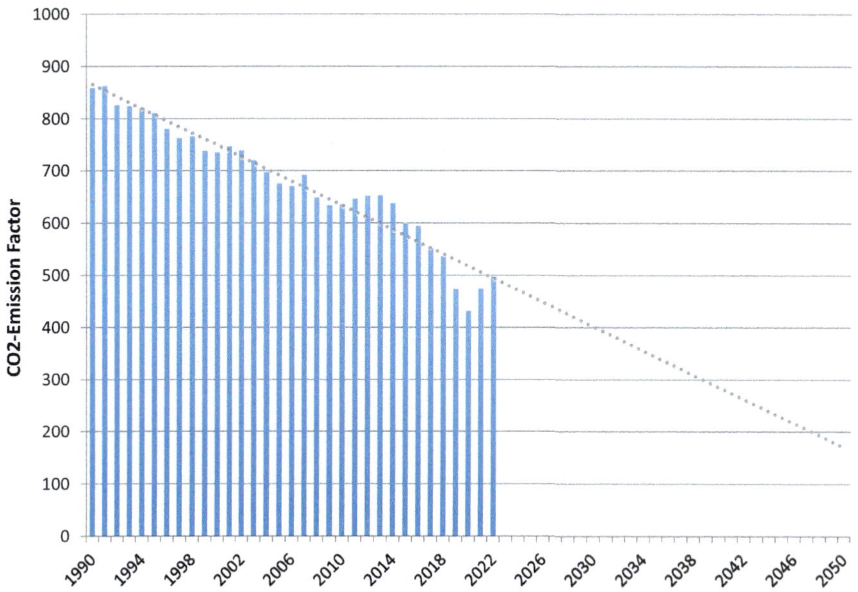

Fig. 14.3 Development and forecast of the total German CO_2 emission factor of electric energy including the upstream chain of primary energy to the power plants in g/kWh[20]

These systems store energy as a primary control power. Primary control is used to compensate for imbalances between physical power supply and demand with the aim of restoring a stable grid frequency. Primary control energy must be supplied between 30 s and 5 min after an incident. Grid operators pay relatively high prices for the provision of primary control power, which means that large battery storage systems are quite economical.

Example 14.8 Large battery storage systems

The Hornsdale Power Reserve in South Australia was the world's first big battery in 2017. Initially, the battery had a peak power of 100 MW and a capacity of 129 MWh. The operation of the storage facility was so economical that an extension was installed 3 years later. Since 2020, the system has a power of 150 MW and a capacity of 193 MWh.

However, as of 2024, the world's largest system is the Moss Landing Energy Storage Facility in Monterey County, California. The lithium-ion battery has a peak power of 750 MW and a capacity of 3000 MWh.

[20] Source: *Entwicklung der spezifischen Treibhausgas-Emissionen des deutschen Strommix in den Jahren 1990–2022*, Umweltbundesamt, Climate Change 20/2023, THG-Emissionsfaktor mit Vorketten.

Table 14.10 Production volumes of hydrogen in Germany and the world as of 2000[21]

Hydrogen	Germany in Billion Nm^3	Worldwide in Billion Nm^3
Steam reforming (natural gas, naphtha)	6	190
Partial oxidation (heavy fuel oil)	3	120
Petrochemistry (petrol reforming)	2.5	90
Petrochemistry (ethylene reforming)	3.6	33
Other chemical industry applications	0.9	7
Chlorine-alkali electrolysis (chlorine production)	0.9	10
Coal gasification (coke-oven gas)	2.1	50
Total	**19**	**500**

Other large projects in development include the Waratah Super Battery in Australia which will be at least 850 MW/1680 MWh. Construction is currently getting underway. Near Manchester, England, the UK company Carlton Power recently got planning consent for a 1040 MW/2080 MWh project. ◀

14.4.8 Seasonal Electricity Storage Systems

A high proportion of renewable electricity production, which is mainly generated by highly fluctuating wind and photovoltaic systems, can only be achieved if the available generation capacity is significantly higher than the average electricity demand. For this reason, there will always be phases of excess electricity production in which, for example, hydrogen gas can be produced from electrolysis.

In the dark winter months, daily PV electricity production is reduced to only around 10% compared to the peak times in summer. If there is a lull lasting several days and wind power production also collapses, stored energy must be used to maintain the power supply.

Battery storage systems are too expensive and technically complex to store the energy required for several days, weeks, or even months. It is therefore to be expected that power-to-gas systems for the intermediate storage of renewable electricity will play an important role in the future, see also Sect. 14.5.3.2.

[21] Source: Deutscher Wasserstoff- und Brennstoffzellen-Verband DWV, 2000.

14.4.9 Emission Factors of Power Generation Plants

Renewable electricity generation from wind turbines and PV plants is often counted with an emission factor of zero. The same is assumed for synthetic fuels and sometimes even nuclear power.

However, the construction of these plants requires raw materials and energy that leave a CO_2 footprint. The plants have a limited lifespan and must then be dismantled and hopefully recycled. This also consumes energy and leads to more CO_2 emissions. In the same way, the CO_2 emissions required for the expansion of the infrastructure for energy distribution, storage, and refueling must be considered.

Only in the very distant future can we hope that all energy will be produced CO_2 free and that all raw materials will be recycled. For the time being, the emissions from the wind turbine and PV plant construction, installation, and recycling must be written off against the amount of energy produced during their lifetime.

It is therefore wrong to assume that wind energy or PV or green hydrogen or e-fuels are CO_2 neutral!

It is obviously difficult to give exact figures for the proportionate CO_2 emissions of wind turbines or PV systems because the manufacturing processes and the consumed energy vary greatly depending on the country of production, the production process, and the manufacturer. The following tables are only an estimate based on European plants. They should provide an insight into the order of magnitude.

The German Federal Environment Agency (Umweltbundesamt) gives the following emissions for the different power plants (Table 14.9).

Using Germany as an example, the graph below shows the development of the CO_2 emission factor over the last few decades. This is the total factor, which includes all consumption, exports and imports, power plant own consumption, and the upstream chain of primary energy raw materials (Fig. 14.3).

14.5 Hydrogen

Unlike natural gas or crude oil, hydrogen (H_2) is not a primary energy source, but an energy carrier, like ions in a battery. Hydrogen must therefore first be produced from primary energy sources.

Four main processes are known:

$$\text{Steam methane reforming (SMR)} \quad CH_4 + H_2O \rightarrow CO + 3\,H_2 \tag{14.2}$$

$$\text{Dry methane reforming (DMR)} \quad CH_4 + CO_2 \rightarrow 2\,CO + 2\,H_2 \tag{14.3}$$

$$\text{Chlor - alkali electrolysis} \quad 2\,NaCl + 2\,H_2O \rightarrow 2\,NaOH + Cl_2 + H_2 \tag{14.4}$$

$$\text{Electrolysis of water} \quad 2\,H_2O \rightarrow 2\,H_2 + O_2 \tag{14.5}$$

Table 14.11 Average yearly yield of a photovoltaic system in kWh per year per installed peak power in kW

Location	Yield in annual kWh per installed kWp PV panels
Northern Europe	800–1000
Central Europe	1000–1200
Southern Europe	1500
Best locations in the world (North Africa, Central America)	Up to 1800

About half of the hydrogen produced today is obtained by steam reformation of natural gas. Less than one-third is produced from petroleum products, in particular, heavy oil or gasoline. These processes generate large amounts of CO_2.

In 2015, electrolysis only accounted for less than 4% of the total hydrogen production.[22] This proportion is expected to increase significantly in the coming years. When electricity from renewable sources is used, the hydrogen is produced from pure water and mostly CO_2 neutral.

14.5.1 Hydrogen Production

In total, around 20 billion Nm^3 of hydrogen is currently produced annually in Germany for industrial applications. By comparison, around 100 billion Nm^3 of hydrogen would be needed to run the entire German passenger car fleet.[23]

14.5.1.1 Hydrogen from Secondary Production

Some of this hydrogen, around 8 billion Nm^3, is produced as a by-product of various industrial processes,[24] e.g. in chlorine production and in the petrochemical industry (petrol reforming and ethylene production). A large proportion of this by-product is used for other chemical processes or as heating gas and is therefore not available to power fuel cell cars.

---

[22] Source: Dincer, Ibrahim; Acar, Canan (2015–09-14). "*Review and evaluation of hydrogen production methods for better sustainability*". International Journal of Hydrogen Energy.

[23] The average hydrogen consumption in passenger cars of 1.2 kg_{H2}/100 km results in 17 $Nm^3{}_{H2}$/100 km, considering 12% losses during compression and 2% transportation losses. This means that around 100 billion Nm^3 of H_2 is required for 620 billion car kilometers per year assuming that the hydrogen can be produced locally and transported in gaseous form. In the case of centralized production (wind farms), the hydrogen would have to be liquefied for the long transport, which causes significant additional losses and increases the amount of required energy by about 50%.

[24] The data on the secondary production of hydrogen is very uncertain and varies by a factor of more than 2 depending on the source.

Only a small proportion of the hydrogen from by-products is released into the atmosphere unused (one source estimates this proportion at 5%). However, the purity of this secondary hydrogen is not sufficient for direct further use in fuel cells. It would first have to be processed in suitable plants (Table 14.10).

14.5.1.2 Electrolysis (Power-To-Gas)—"Green" Hydrogen

In so-called water electrolysis (PtG = power-to-gas), water (H_2O) is broken down into hydrogen (H_2) and oxygen (O) using an electric current. Electrolysis consists of two partial reactions, which take place in the anode and the cathode chamber.

By using electricity from renewable sources, electrolyzers can produce nearly CO_2-free hydrogen, albeit with high losses. This is referred to as green hydrogen.

Renewable electricity sources include wind turbines, hydroelectric power stations, or photovoltaic (PV) systems. As an alternative, solar thermal power plants (CSP = Concentrated Solar Power) could also be used. In such plants, the light is concentrated via many reflectors, usually parabolic mirrors, and directed onto a focal point, which is usually located on a tower. The heat is then used to generate steam, which can be used to produce electricity using turbines and rotating generators. However, due to the fall in the price of solar modules, PV systems now have lower electricity generation costs than CSP systems, so they are not relevant anymore.

It should also be noted that quite a large amount of clean fresh water is required for electrolysis, about 10 kg or 10 L of water per kg of hydrogen.[25]

As the electrolyte concentration and the temperature of an electrolyte solution have a major influence on the cell resistance and therefore on the energy costs, a 25–30% potassium hydroxide solution is used in modern electrolyzers. The operating temperature is approx. 70–90 °C, the current density is approx. 0.15–0.5 A/cm^2, and the voltage is approx. 1.90 V.

To produce 1 Nm^3 of hydrogen, a typical large pressure electrolyzer requires a power of 3.5 MW (stack power) or 4.5 MW (input power of the entire system) to produce 760 Nm^3/h of hydrogen. This corresponds to an electrical energy of 5.9 kWh per 1 Nm^3 of hydrogen and therefore an efficiency of 47%. The best and most modern PEM systems (PEM = polymer electrolyte membrane or proton exchange membrane) achieve efficiencies of about 60-70%.

The theoretical physical best efficiency of the cell reaction is even around 82%. However, this is far from being achieved due to the energy consumption of the required auxiliary units, in particular, compressors, filters, and heat exchangers.

---

[25] It is often (wrongly) argued that lithium production for battery electric vehicles requires a lot of water. The fact that green hydrogen production requires even more water over the lifetime of a car does not seem to be remarkable. See also Sect. 13.6.1.

Example 14.9 Energy requirement to produce hydrogen with electrolyzers

The German Energiepark Mainz currently operates one of the world's largest and most technically advanced plants for producing hydrogen from electricity. The electrolyzers are based on PEM technology, which is more efficient than alkaline electrolysis and can react more quickly to load fluctuations.

Three units, each with a continuous output of 1.3 MW, produce 245 m^3 of hydrogen per hour. The peak output is 2 MW each with a production of 345 m^3 each. The plant delivers 35 bar operating pressure. The investment amounted to 17 M€.[26]

At 35 bar operating pressure, the compression factor is 1.02. The plant therefore has an electricity requirement of 5.2 kWh per Nm^3 H_2 or 62 kWh/kg. With a calorific value of 33.3 kWh/kg, this results in a plant efficiency of 54% at the best point at partial load and 49% at continuous operation at full capacity. ◀

Example 14.10 Cost of green hydrogen production

The world's largest electrolysis plant by its own admission was built in 2020 at the Rheinland refinery in Wesseling near Aachen, Germany.

The plant has an annual production capacity of 1300 t_{H2} and the cost was 16 M€. The capacity corresponds to the average annual consumption of 7700 cars. Around 6200 of these plants would therefore be needed to supply all 48 million cars in Germany. Each one of these plants would be able to supply just two to three average filling stations. ◀

There is an ongoing discussion about producing green hydrogen from PV electricity in countries with high levels of solar radiation and then importing it to Europe.

Table 14.11 shows the expected yields.

The world's most favorable PV location in Australia, Alice Springs, is located right in the middle of the country and has an annual average of almost 10 h of sunshine per day. PV systems achieve a yield of 1800 kWh/kWp. In the south of Australia in Melbourne, only 1300 kWh/kWp is achieved.[27] Middle East and North African countries do not achieve higher yields either: a recent Fraunhofer study[28] cites 1790 kWh/kWp.

The main costs of PV systems are caused by the purchase of components, especially PV panels and power electronics. Nowadays, these components mainly come from China and some from Western countries. The investment costs of PV systems in Australia, Africa, or South America are therefore almost the same as in Central Europe.

[26] Source: Lukas Weber, *In Mainz wird Wind zu Gas*, Frankfurter Allgemeine Zeitung, 08.07.2015.

[27] Source: https://www.solarcitizens.org.au/how_much_power_does_a_solar_panel_generate.

[28] Source: *Stromgestehungskosten erneuerbarer Energien*, Fraunhofer ISE, Freiburg, März 2018.

Table 14.12 Equivalent CO_2 emissions from the production of hydrogen with the current energy system[29]

Hydrogen	Production in Germany g_{CO2}/kWh	Production at the point of natural gas extraction g_{CO2}/kWh
Gray	460	483
Turquoise	390	346
Blue	391	370
Green produced in wind farms	45	180

The costs of PV electricity generation scale roughly in line with the electricity yield. In countries close to the equator, they therefore fall to just under half of the production costs in Germany at best.

For 2018, the Fraunhofer study estimates electricity generation costs of around 4 to 7 ct/kWh for large PV systems in Central Europe and around 3–4 ct/kWh for large systems in countries with high levels of solar radiation. However, the study assumes a depreciation period of 25 years. It must be questioned whether this long period is reasonable for countries, which are often politically unstable. This PV-generated electric energy can only be transported to Europe or other continents by means of liquid hydrogen, which reduces the energy yield to around less than half. Ultimately, the price of liquid hydrogen from distant countries will be many times higher than local production, taking into account conversion losses, expensive electrolyzers, liquefaction, and transportation in cryogenic tanks.

For comparison: The cost of onshore wind production is stated in the same study at around 4–8 ct/kWh and that of offshore wind production at around 8–14 ct/kWh. These estimates clearly show that an import of hydrogen from PV systems or wind turbines from distant countries is not worthwhile at all. Each region must produce their own PV and wind energy to stay competitive.

It is also proposed to convert the electrical PV energy directly into methane gas or e-fuels such as methanol on site. This considerably simplifies storage, transportation, and the infrastructure for refueling. On the other hand, there are additional conversion costs. The efficiency of such plants and the efficiency of the combustion engines that ultimately use these fuels are poor. The amount of electricity required and therefore the costs will hardly be lower than for a liquid hydrogen transportation.

The political situation in many of the geographically relevant countries must also be considered. The question arises as to whether we want to make ourselves dependent on politically unstable regions again.

[29] Source: *Welche Treibhausgasemissionen verursacht die Wasserstoffproduktion?*, Umweltbundesamt, Deutschland, November 30, 2022.

Table 14.13 Equivalent CO_2 emissions from the production of hydrogen in a distant future with an entirely renewable energy system[146]

Hydrogen	Production in Germany g_{CO2}/kWh	Production at the point of natural gas extraction g_{CO2}/kWh
Gray	194	171
Turquoise	60	9.0
Blue	55	31
Green	0	6.1

14.5.1.3 Steam Reforming—"Gray" Hydrogen

Most of the hydrogen produced industrially today (around 90% of direct hydrogen production or around 50% of total hydrogen production including by-products) comes from steam reforming plants.

In steam reforming, hydrogen is produced from hydrocarbons, usually natural gas (methane) CH_4, in two-process steps. In the first step, the hydrocarbon is reacted with water in the reformer at a temperature of 800–900 °C and a pressure of around 20–30 bar. This produces synthesis gas, a mixture of carbon monoxide and hydrogen:

$$CH_4 + H_2O \rightarrow CO + 3\,H_2 \tag{14.6}$$

The carbon monoxide (CO) is now separated and mixed with water vapor in the second step, known as water–gas shift reaction (WGSR). This produces another portion of hydrogen:

$$CO + H_2O \rightarrow CO_2 + H_2 \tag{14.7}$$

In addition to natural gas, biomass, crude oil, or coal can also be used as raw materials. Steam reforming releases the same amount of CO_2 as if the methane gas were burned.

The steam reformation process is obviously not sustainable if crude oil, gas, or coal is required, nor is it climate neutral, as large quantities of CO_2 are produced. The hydrogen produced is therefore also referred to as gray hydrogen.

Sustainable and climate-neutral production can be made possible in principle by using biogas.

Depending on the size of the plant, around 0.45 Nm^3 of natural gas is required to produce 1 Nm^3 of hydrogen. Natural gas produces 0.191 kg_{CO2}/kWh. With a calorific value of 13.0 kWh/kg and a density of 0.81 kg/m^3, this results in an emission factor of 0.905 kg_{CO2}/Nm^3 for hydrogen from steam reformation. This corresponds to 10.8 kg_{CO2}/kg or 323 g_{CO2}/kWh direct emissions. The efficiency of the process is about 60%. In addition, there are further emissions due to leaks and other losses.

Steam reforming is an established industrial process and is implemented in plants with capacities of up to 100,000 m^3/h.

14.5.1.4 Carbon Capture—"Blue" Hydrogen

If the CO_2 produced during the production of hydrogen from natural gas is filtered out and stored separately (CCS = carbon capture and storage), the result is so-called blue hydrogen.

There are essentially two ways to produce blue hydrogen. Both are based on steam reforming, in which methane and water are split using high heat, high pressure, and steam to produce hydrogen and carbon dioxide. In both processes, the carbon dioxide from the steam reforming is separated and stored. The difference between the two processes is whether the carbon dioxide is captured from the generators that drive the steam reforming and carbon dioxide capture processes or not.

The problem with the production of blue hydrogen is the use of natural gas, i.e. methane gas, in the production process. Methane gas has a much more harmful effect on the climate than CO_2. A certain amount of leakage is unavoidable when operating large plants and when extracting and transporting the gas.

It is also impossible to capture all CO_2 that is generated during the production of hydrogen. Capture rates of between 60 and 95% can be achieved, depending on the technology, with the remaining CO_2 escaping into the atmosphere.

A study calculates that the climate impact of blue hydrogen is only marginally better than that of gray hydrogen and even more harmful than the combustion of diesel or coal provided a CCS process is also used[30]: Gray H_2 production releases total emissions of 551 g_{CO2eq}/kWh, blue H_2 production without flue-gas capture releases 500 g_{CO2eq}/kWh, and blue $_{H2}$ production with flue-gas capture is at 486 g_{CO2eq}/kWh. These numbers include upstream CO_2 emissions and fugitive CH_4 emissions.

14.5.1.5 Methane Pyrolysis—"Turquoise" Hydrogen

In methane pyrolysis, methane (natural gas or biomethane) is split directly into its components hydrogen and carbon. This requires high temperatures of over 1000 °C. Instead of gaseous CO_2, solid carbon is produced, which can be stored without any impact on the climate.

Also with this process, a sustainable and climate-neutral production is only theoretically possible if biogas is used. In practice, just as with "blue" hydrogen, methane emissions during extraction and transportation have a harmful effect on the climate. Turquoise hydrogen is therefore not entirely climate neutral.

In addition, the hydrogen obtained in this process is contaminated with other gases and must be treated and purified in further chemical processes before it goes into the usual PEM fuel cells.

Kværner Process

A plasma burner completely breaks down hydrocarbons (methane, natural gas, biogas) at 1600 °C into carbon (pure activated carbon) and hydrogen. A pilot plant

[30] Source: *How green is blue hydrogen?*, Robert W. Howarth, Mark Z. Jacobson, Energy Science & Engineering, Volume 9, Issue October 10, 2021.

built in Canada in 1992 achieved an efficiency of 48% in terms of hydrogen production, plus 40% activated carbon and 10% superheated steam. The Kværner process is not currently used for hydrogen production, but BASF has announced its intention to introduce the process on an industrial scale by 2030.

Partial Oxidation

In partial oxidation, the raw material, usually natural gas or a heavy hydrocarbon (heating oil), is converted sub-stoichiometrically—i.e. with a lack of oxygen—in an exothermic process.

In coal-rich countries such as China or South Africa, coal can also be used as a starting material for this process, which is ground up beforehand and mixed with water to form a suspension. The main reaction products are hydrogen and carbon monoxide. Partial oxidation is considered technically mature. Again, it is only sustainable and climate neutral if biogas is used.

14.5.1.6 Summary Hydrogen Production

A study by the German Umweltbundesamt (Federal Environment Ministry) comparing hydrogen production processes concludes that the equivalent CO_2 emissions of gray, turquoise, and blue hydrogen are roughly similar. Only green hydrogen performs significantly better (Table 14.12).

These results were determined based on the CO_2 emissions of the current energy production, however assuming direct wind energy for green hydrogen production. If the energy for hydrogen electrolysis is taken from the grid, for example, when hydrogen is produced in or near filling stations, the emissions are about 700 g_{CO2}/kWh based on the electricity production in Germany in 2022.

By 2050, an entirely renewable and CO_2-neutral energy production should be in place. In this case, the production of hydrogen at the point of natural gas extraction results in lower emissions, as the fugitive methane emissions during transportation are avoided.

However, this scenario (Table 14.13) also assumes that all of the equipment needed to produce the hydrogen is made from recycled raw materials and that all the energy needed to produce the raw materials and to build the plants is also entirely CO_2 neutral, which may not be achievable by 2050.

14.5.2 Supply of Hydrogen Filling Stations

A major challenge for the supply of a large fleet of hydrogen vehicles is the supply of hydrogen to the filling stations. In principle, there are five options:

- Tank trucks filled with gaseous hydrogen (CH_2).

- Tank trucks filled with liquid hydrogen (LH_2).
- Pumping hydrogen through the existing natural gas pipelines.
- Construction of a new pipeline network for hydrogen only.
- Local production of hydrogen by electrolysis at the filling stations.

Of course, combinations of these measures are also conceivable.

Example 14.11 Hydrogen tank trucks

A conventional 40-ton tank truck can transport around 35,000 L of petrol fuel. This is enough to refuel around 1100 cars, assuming a refueling range of 400 km and an average consumption of 8.0 l/100 km per vehicle.

A similarly sized liquid hydrogen tank truck with a 45,000-L LH_2 tank can transport around 3.0 t of hydrogen. This is enough to refuel 630 FCEVs with a range of 400 km and an average consumption of 1.2 kg_{H2}/100 km per vehicle.

A similarly sized CH_2 tank truck with 200 bar gas cylinders only transports around 0.35 t of hydrogen due to the heavy weight of the gas cylinders. This is just enough to refuel 73 FCEV cars. ◀

The example calculation clearly shows that tank trucks with CH_2, which are used exclusively today, are not an option for supplying a larger fleet of vehicles and filling stations.

The transmission of hydrogen through existing natural gas pipelines is also being discussed. Gas grids with their above-ground and underground storage facilities are an efficient long-term storage facility for hydrogen and synthetic natural gas. However, hydrogen attacks steel and copper and can lead to porosity (hydrogen sickness). Therefore, only a small volume share of 5–7% H_2 can be permitted in older gas networks and up to 10% in newer, more modern natural gas pipelines. Before using this hydrogen in filling stations and fuel cell vehicles, it must first be filtered out of the natural gas and locally purified, which would require the installation of expensive and energy-intensive chemical systems at each filling station. This, again, is hardly a viable option.

There are political discussions about setting up dedicated pipeline networks for hydrogen. However, such networks are only feasible for the long-distance distribution of large quantities of hydrogen, not for final distribution to thousands of filling stations.

The last option, local electrolysis, is equally unrealistic, as a sample calculation easily shows.

Example 14.12 Local hydrogen electrolysis

On an annual average, each of the 14,000 filling stations in Germany carries out around 280 car refueling operations per day assuming an average refueling range of 400 km.

If an average filling station is converted to hydrogen, around 1.3 tons H_2 are needed per day to supply the same number of cars (not counting commercial vehicles).

The production of 1.3 tons of H_2 requires 82 MWh of electric energy (including 40% losses in electrolysis and 12% charging losses). That corresponds to a continuous electric power of 3,5 MW into the electrolysis system—24 h per day in each of the 14,000 filling stations. ◀

All in all, only tank trucks carrying liquid hydrogen (LH_2) can supply a large network of thousands of H_2 filling stations. The required hydrogen could be transported from the coast to regional hubs via dedicated pipelines.

14.5.2.1 Transportation and Refueling Losses

The losses for compression to around 200–300 bar (CH_2) are approximately 5%.

Depending on the process, around 28–46% of the energy content must be used for liquefaction (LH_2), typically resulting in an efficiency of around 67%.

During the refueling process of cars, the compression to 700 bar (CH_2) results in a further loss of around 12% of the original energy content (88% efficiency).

In total, this results in an efficiency of around 84% for the CH_2 chain from production to the tank of a vehicle and around 59% for the LH2 chain. The previously estimated CO_2 emissions increase accordingly.[31]

The type of storage must also be considered. Liquid storage in car tanks results in daily evaporation losses of around 3%. Such cars are no longer in use today. In the case of industrial liquid storage (ships, tanker trucks, pipelines), it can be assumed that the evaporation losses are lower. However, no reliable empirical values are yet available.

14.5.3 Utilization of Green Hydrogen

There is little doubt that large quantities of climate-neutral hydrogen will be produced in many industrialized nations in the future. Production in large offshore wind farms will become cheaper than it is today and there will be a high demand for renewable, CO_2-neutral energy-rich gases.

[31] Liquefaction and compression are carried out by pumps, which are of course powered by electricity and do not consume hydrogen. Strictly speaking, the pumping losses would therefore have to be weighted with the emission factor of the electricity grid. Often, however, the total energy lost is related purely to the energy balance of the hydrogen and the emission factor is calculated accordingly.

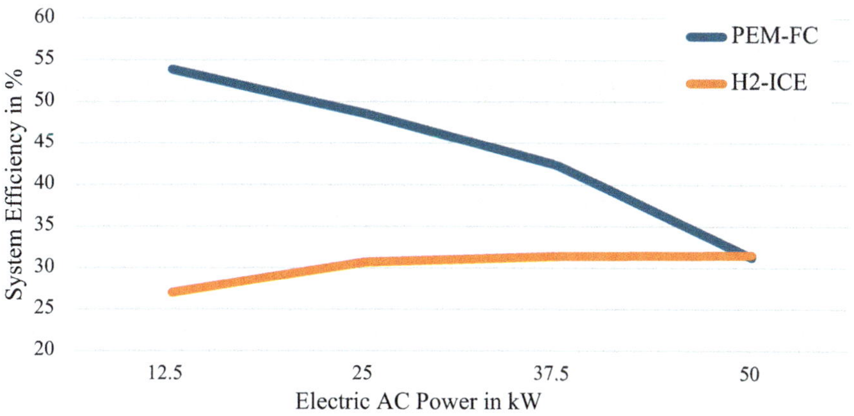

Fig. 14.4 Comparison of the efficiency from hydrogen to AC

This raises the general question of determining suitable applications for hydrogen. Numerous studies and opinions have delved into this matter.

From today's perspective, there is no alternative to green hydrogen in some areas of CO_2-neutral production. These include, for example, fertilizers, hydrogenation, methanol production, hydrocracking, desulphurization, use as a chemical raw material, crude steel production, and long-term electricity storage (reconversion in gas-fired power plants).[32]

Other possible applications with a lower probability of occurrence include mobile machinery and equipment, coastal and inland waterway shipping, remote rail transport, and decentralized methanation. Synthetic fuels based on green hydrogen are likely to be used for overseas shipping, long-haul flights, and classic cars.

However, hydrogen is uneconomical in many areas, where electrification with batteries or overhead lines would be much simpler, cheaper, and more environmentally friendly. These include, for example, subway trains and city buses, fuel cell cars, city delivery vans, two- and three-wheelers, the mass production of e-fuels, and the provision of balancing power for electricity grids.

While acknowledging the abundance of viewpoints, this book focuses on the utilization of hydrogen in vehicles. It is becoming increasingly evident, however, that hydrogen may not be optimal for this purpose.

14.5.3.1 Hydrogen as Heating Gas

Hydrogen burns almost CO_2 free, and the combustion is highly efficient, around 90%. It is therefore well suited as a fuel gas.

[32] Source: Hagedorn et al. (2021) in Anlehnung an Liebreich (2021), *Leiter der Einsatzbereiche sauberen Wasserstoffs.*

It is expected that energy-intensive industries that currently use natural gas will switch to hydrogen in the medium term. The greatest CO_2 reduction potential by hydrogen lies in the steel industry, followed by building heating and the chemical industry.

The steel industry has relatively few locations and could be supplied with hydrogen comparatively easily. A widespread supply of buildings with hydrogen for heating, on the other hand, is difficult to imagine because there is no end-user distribution network for hydrogen in any country.

Hydrogen can also be converted into synthetic methane gas. Methane can be transported in unlimited quantities in the existing natural gas grid. However, the efficiency of the conversion is only around 50–60%. In addition, the climate potential of methane is much higher than that of hydrogen, so even minor leaks would have a significant negative impact on the climate balance.

14.5.3.2 Electricity Generation from Hydrogen

During sunny days with high wind levels, much more renewable electricity is generated than is needed, especially in the summer. This electricity could be used to generate hydrogen for later reconversion into electricity. At the moment, however, such systems do not pay off, as the hydrogen production, storage, and distribution costs far exceed the income from the sale of hydrogen.

Therefore, large dedicated offshore wind farms are being planned to produce green hydrogen on a large scale with acceptable costs in the future. Producing hydrogen solely from regulated wind power at every small wind farm will probably never be worthwhile.[33]

But wherever green hydrogen is being produced in the future, there is a great need to balance out the fluctuating production of solar and wind energy with renewable and plannable hydrogen power plants.

Example 14.13 Hydrogen-capable gas-fired power plants in Germany

From 2025, new gas-fired power plants with a total capacity of around 10 GW are to be built in Germany. These power plants are intended to ensure the basic electricity supply even in times of low solar radiation and wind.

Until now, this task has been performed by coal-fired power plants, but these are to be shut down by 2030. This means that around 28 GW of secured grid capacity will be lost.

The new power plants are to be operated with natural gas or hydrogen. From 2035, the aim is to operate exclusively with climate-neutral hydrogen. ◀

[33] It is currently much cheaper to simply switch off a wind turbine in times of overproduction than to invest in an electrolysis plant and produce hydrogen that can then be sold. Converting this hydrogen back into electricity in a fuel cell is even less economical.

Hydrogen can be burned in gas turbines, which in turn drive electricity generators. The achievable efficiency of gas turbines is around 30% at best. If electrolysis and the generator are also considered, the cycle efficiency is less than 20% (electricity → electrolysis [70%] → H_2 → compression/storage [88%] → gas turbine [30%] → generator [95%] → electricity).

Alternatively, the hydrogen could be used in large, stationary fuel cells to generate electricity directly, which could then be fed into the grid as required. Efficiencies of around 70% can be achieved in the fuel cell under best circumstances. In the cycle (electricity → electrolysis [70%] → H_2 → compression/storage [88%] → fuel cell [70%] → electricity), efficiencies of around 40% should be possible.

As a third option, hydrogen can also be burned in piston engines, which then produce electricity via an electric generator. This option is described in more detail in the next section. Cycle efficiencies around 15–20% can be achieved.

However, if the surplus electrical energy is temporarily stored in lithium-ion batteries, cycle efficiencies of more than 80% are possible, see Sect. 14.4.7.

14.5.3.3 Hydrogen in Piston Engines

Hydrogen can be used instead of petrol or diesel in combustion engines to drive cars and trucks.

The combustion of hydrogen in piston engines requires significant technical modifications in the fuel supply system and in the exhaust gas after treatment but most other components are unaffected.

The efficiency of hydrogen combustion in internal combustion engines is like that of conventional diesel and petrol engines. The combustion produces nitrogen oxides (NOx), which can be removed by catalytic converters.

The advantage of hydrogen combustion is the reduced complexity compared to a fuel cell system. Also, there is a lot of know-how and experience in combustion engines and there are many component suppliers. Furthermore, a contamination of hydrogen with sulfur or other gases is not a problem. The lifetime expectancy of H_2 piston engines is therefore presumably greater than that of fuel cell systems.

The efficiency of an internal combustion engine is at its best near full load operation, which often occurs with heavy-duty drives. With hydrogen, values up to 45% can be achieved. The efficiency of a fuel cell system is best at low load, up to 60% appears possible. Under full load, however, the efficiency drops considerably. If the losses in the electric motor and power electronics are considered, an electric drive with a fuel cell is hardly better than a hydrogen engine under full load.

For these reasons, the development of hydrogen combustion engines is currently (as of 2024) being pursued by several manufacturers, primarily for use in segments of the transport sector that are difficult to electrify, such as shipping or road-based heavy-duty transport.

Combustion engines can also be used to drive electric generators and thus reconvert hydrogen into electricity.

The following figure shows a comparison of the achievable efficiency on the path from hydrogen input to AC electric energy output based on detailed models with real-world data. The system was designed for 50 kW continuous electric output power. The modeled fuel cell system includes all losses including cathode air compression, cell cooling, preparation of H_2 and air, and DC/DC and DC/AC converter. The efficiency of the path through the combustion engine includes the cooling system and an electric generator (Fig. 14.4).

At part load, the efficiency of the fuel cell is much better than the combination of combustion engine and power generator. At full load, however, the efficiency levels are similar. In addition, the fuel cell system is much more expensive and complex than the combustion engine with generator.

To determine the round-trip efficiency of the entire cycle (starting with the electrical energy, the production of hydrogen, the storage, and finally the reconversion to electricity), as discussed in the previous chapter, the hydrogen production must be considered in addition to the efficiencies shown in the figure. The cycle efficiency would then be around 20%.

14.5.3.4 Hydrogen in Fuel Cell Electric Vehicles

Fuel cell electric vehicles (FCEVs) suffer from three major technical disadvantages compared to battery electric vehicles:

- The chain efficiency from the electrical energy via electrolysis to the car and the reconversion using fuel cells is very poor. The electric energy consumption of fuel cell-based cars is four to five times higher compared to battery electric vehicles, see also Sect. 15.2. Due to the many losses, high investments in wind power and PV systems are required.
- The production, transport, storage, and refueling of hydrogen are technically very complex and therefore expensive due to its unfavorable chemical properties. Handling hydrogen is also dangerous and should only be carried out by professionals.
- The technology required in the vehicle is complex (large space requirement, high costs) and fuel cells are sensitive to hydrogen impurities. The lifetime of the fuel cell is critical.

Regardless of the production location of hydrogen in the world, there is a considerable cost disadvantage of hydrogen vehicles compared to battery electric vehicles. As this disadvantage is systemic, it will not be offset by future developments.

Therefore, fuel cells are no longer seen as a viable alternative energy source for passenger cars today. Battery electric cars are at least equivalent in all respects (weight, range) and far superior in terms of cost, installation space, service life, and safety. Even in terms of environmental impact, the battery is on a similar level as a fuel cell system, because the latter also requires critical raw materials (platinum, PFAS) and batteries can be very well recycled, unlike the complex fuel cell systems and tanks made of carbon fibers.

14.6 Summary

A distinction is made between finite and renewable primary energies. They all can be processed into a variety of different energy sources to power a vehicle.

Even though oil production peaked a few years ago (peak oil), there are still enough reserves for the next 100 years or more. The situation is similar for natural gas and LPG.

The production of biofuels is well established but requires large amounts of agricultural land. The production of synthetic fuels from electricity (e-fuels) suffers from poor efficiency and high costs.

The proportion of electricity generated from renewable sources is constantly increasing. In many countries, it is already higher than electricity generated from fossil fuels. The distribution grids will be increasingly burdened by the future increase in electricity demand. The low-voltage grids are particularly critical if many charging stations are to be connected to a local grid. Local power generation and power storage can make a balancing contribution here.

Synthetic fuels and hydrogen cannot compete with battery electric cars on the mass market because their production is far too complex and requires too much energy. In the case of hydrogen, there are also the high infrastructure costs and safety concerns. However, the use of these fuels in energy-intensive applications (aircraft, ships, possibly also road-based heavy goods vehicles) may become interesting in the future.

Life-Cycle Assessment 15

Abstract

This chapter discusses the direct and indirect harmful emissions from cars and how these can be assessed and compared. Building on the findings of the previous chapters, examples are then given of how emissions from the operation of cars with different drive technologies can be calculated. However, such considerations are not sufficient for a comprehensive assessment of the footprint of a car technology. Rather, the costs of production, the entire use phase, and possible recycling must also be considered. Such life-cycle analyses are introduced and presented using average cars as examples.

15.1 Pollutants and Environmental Impacts

A distinction is made between greenhouse gases (GHG), i.e. substances that have a significant influence on climate change, and actual pollutants, which primarily have an impact on health.

Greenhouse Gases (GHG)	**Pollutant Emissions**
• Carbon dioxide (CO_2) • Methane (CH_4) • Nitric oxide (N_2O) • Hydrogen (H_2) • (indirect effect)	• Particles (particulate matter) • Sulfur dioxide (SO_2) • Nitrogen dioxide (NO_2) • Nitric oxide (NO) • Hydrocarbon (HC) • Carbon monoxide (CO)

In addition to the impact on the climate, the environmental impact must also consider the cumulative energy input, soil acidification (lowering of the pH value),

M. Doppelbauer, *Introduction to Electromobility*,
https://doi.org/10.1007/978-3-658-45482-1_15

Table 15.1 Global warming potential (GWP) of individual substances regarding their environmental impact[1]

			Global Warming Potential
Climate impact	CO_2 equiv	CO_2	1
		CH_4	27.5
		N_2O	296
		H_2	4.3–16 est (100-year basis)
Acidification	SO_2 equiv	SO_2	1
		NO_x	0.7
		NH_3	1.88
		HCI	0.88
		HF	1.6
Eutrophication	PO_4 equiv	NO_x	0.13
		N_2O	0.27
		NH_3	0.35
Summer smog	C_2H_4 equiv	NMHC	1

eutrophication (excess nutrient supply) in soils and water bodies, summer smog, and particulate matter pollution.

The influence of the individual pollutants on the various environmental impacts is recorded using so-called global warming potential or GWP (Table 15.1), which relates the influence on a reference substance.

The comprehensive consideration of all environmental impacts of a vehicle is highly complex and depends on a wide range of assumptions and framework conditions. In the remainder of this book, we will concentrate on the climate impact.

Hydrogen has no immediate, direct effect on the climate when it enters the atmosphere. However, it is degraded by reacting with hydroxyl radicals (OH). These OH radicals fulfill important cleaning functions, for example, they reduce the lifetime of methane gas in the atmosphere. If the OH is missing, these purifications no longer take place. In addition, hydrogen in the atmosphere leads to a higher concentration of water vapor. This increases the formation of the ozone hole.

The long-term greenhouse gas potential (time horizon 100 years) of hydrogen is therefore estimated to be 4.3–16 times that of CO_2.[2]

[1] Source: *Wissenschaftlicher Grundlagenbericht Projekt UMBReLA*, ifeu Institut Heidelberg, October 2011.

[2] Source: *Ist Wassserstoff treibhausgasneutral*?, Umweltbundesamt Deutschland, November 30, 2022.

15.2 Well-to-Wheel Analysis

The well-to-wheel (WtW) energy chain refers to the entire path from the extraction of primary energy through production, transportation, and provision of the energy source to energy conversion in the vehicle and power transmission through wheel-track contact.

The WtW chain is divided into two parts:

- Well-to-Tank (WtT = provision of the fuel to the vehicle) and
- Tank-to-Wheel (TtW = use of the fuel by the vehicle).

Well-to-tank analyses are complex and depend on many assumed framework conditions: how exactly is the energy source produced from the primary energy? In which countries and regions does this take place? Which technologies are used for extraction and energy conversion? What CO_2 emissions are caused by the construction of the plants? What losses and other environmental impacts are caused by storage, transportation, further processing, and distribution to the end customer?

For many years, WtT analyses have been carried out in extensive studies by renowned research institutes around the world. This book can only provide an introduction and overview into the matter.

The tank-to-wheel analysis determines the consumption of the on-board energy source when driving the vehicle. Four methods have become established for tank-to-wheel (TtW) analyses:

- fleet tests,
- test bench measurements,
- measurements during real driving (real driving emissions), and
- calculation of the fuel consumption based on forces acting on the car and the drivetrain efficiencies.

Fleet tests are carried out as part of large-scale projects, usually involving several research partners and many vehicles and drivers. Fleet tests are often limited to one region, but occasionally they are carried out on a supra-regional or even international basis (e.g. Project CROME = Cross-border Mobility for Electric Vehicles, between France and Germany).

In test bench measurements, consumption and losses in the drivetrain of vehicles can be precisely recorded with the help of driving cycles, see also Sect. 4.5. For this purpose, all framework conditions (friction and air resistance, self-consumption through lighting and air conditioning, etc.) are precisely defined. The GHG emissions can be determined from the consumption by using the emission factor of the respective energy source. The advantage of driving cycle measurements is the objective comparison of different drive concepts. They can also be used to evaluate the product portfolio of manufacturers (fleet consumption).

A disadvantage of test bench measurements is the wide range of possibilities for manipulation. To counter these manipulations, consumption measurements in real road traffic, known as Real Driving Emission (RDE) measurements, have been required by law in the European Union since 2017. Test bench measurements are still needed for the type-approval of passenger cars. In addition, however, it must be proven in RDE tests that the test bench measurements do not deviate by more than 25%, i.e. that there is no manipulation.

Losses and efficiency of vehicles can also be determined by simulation software that takes the drag forces (see Sect. 4.1) and the drivetrain losses (see Sects. 2.4.4 and 3.4.2) into account. However, calculations usually have a high degree of uncertainty due to the large number of sometimes unknown or uncertain parameters.

To assess national or global climate effects of car fleets, test bench measurements are not suitable because the sometimes quite artificial driving cycles and other measurement conditions do not realistically reflect the actual energy consumption and pollutant emissions of millions of vehicles. Even RDE measurements tend to be artificial and atypical of driving behavior. Longer term effects (for example, the reduction of a country's pollutant emissions due to an increase in renewable electricity production) and seasonal effects are also difficult to capture.

For these purposes, it is better to use cumulative calculations of pollutant emissions from larger vehicle fleets (see Sect. 13.4). Statistical data on the total annual distance traveled by all vehicles in a country or region and on the average energy consumption are available with high accuracy in many countries. This can be offset against the total national or global emissions of the various energy sources.

Note

It can be observed time and again that certain process steps in electricity or hydrogen production and in the subsequent distribution to vehicles (infrastructure) are not considered in publicly available WtW studies, both in terms of energy requirements (efficiency) and CO_2 emissions.

The costs of liquefaction and the evaporation losses during transportation and storage are regularly forgotten or underestimated. Some studies even make the fabulous assumption that hydrogen produced from renewable electricity does not generate any CO_2 emissions at all.

For fueling, only the thermodynamic losses of 8% are often assumed. However, the pressure of 700 bar must be generated with a pump system which, like all pumps, has a poor efficiency. Hydrogen must also be cooled down considerably before refueling because it heats up during expansion in the tank. These processes result in actual losses of 12–15%.

The hydrogen consumption of FCEVs is often derived from theoretical considerations resulting in much lower values than the actual consumption of existing vehicles. If the efficiency of the fuel cell system in the vehicle is calculated, auxiliary units such as the DC/DC converter for voltage stabilization or air conditioning are often ignored. Occasionally only the partial load efficiency is considered and not the poorer efficiency at full load.

In the case of hydrogen electrolysis, it is popular to apply only the efficiency of the actual fuel cell (over 70%) and to forget the losses in the other system components (filters, pumps, compressors, air conditioning).
When considering the carbon footprint, the equivalent CO_2 emissions from the construction of electrolyzers and H_2 liquefaction plants must also be considered in addition to the direct emissions from the electricity production.

15.3 Well-to-Wheel Analysis Examples

This section demonstrates the basic procedures for preparing well-to-wheel analyses using some simplified example calculations.[3] Such calculations enable an initial assessment of the environmental impact of different drive technologies and energy sources.

The typical energy consumption of average cars has already been examined in Sect. 4.8. The emission factors, efficiencies, and other properties of energy carriers that were used in the following examples have all been introduced in the previous two chapters.

15.3.1 Vehicles with Piston Engines

(see Tables 15.2, 15.3, 15.4, and 15.5)

15.3.2 Fuel Cell Electric Vehicles

(see Tables 15.6 and 15.7)

It should be noted that this scenario is very forward-looking. Corresponding systems will only be available in a decades' time, if at all.

15.3.3 Battery Electric Vehicles

(see Tables 15.8, 15.9, and 15.10)

[3] Even though the examples are all based on German cars and the German energy system, the principles can be applied to the situation in any other country as well.

Table 15.2 CO_2 emissions of an average diesel car

Diesel Car		
Typical average fuel consumption	7.0	l_{diesel}/100 km
Energy demand 11.8 kWh/kg · 0.83 kg/l · 7.0 l/100 km	**68.6**	**kWh/100 km**
Direct CO_2 emissions Diesel: 2.65 kg_{CO2}/l · 7.0 l/100 km	18.6	kg_{CO2}/100 km
Total CO_2 emissions Including provision of the energy carrier Diesel: 3.25 kg_{CO2}/l · 7.0 l/100 km	**22.8**	**kg_{CO2}/100 km**

Table 15.3 CO_2 emissions of an average petrol car including hybrid cars

Petrol Car		
Typical average fuel consumption	7.7	l_{petrol}/100 km
Energy demand 12.0 kWh/kg · 0.76 kg/l · 7.7 l/100 km	**70.2**	**kWh/100 km**
Direct CO_2 emissions Premium petrol: 2.38 kg_{CO2}/l · 7.7 l/100 km	18.3	kg_{CO2}/100 km
Total CO_2 emissions Including provision of the energy carrier Premium petrol: 2.78 kg_{CO2}/l · 7.7 l/100 km	**21.4**	**kg_{CO2}/100 km**

Table 15.4 CO_2 emissions of an average LPG car

LPG Car		
Typical fuel consumption (1.5 L more than a similar petrol car)	9.6	l_{LPG}/100 km
Energy demand 12.8 kWh/kg · 0.54 kg/l · 9.6 l/100 km	**66.4**	**kWh/100 km**
Direct CO_2 emissions Autogas: 1.64 kg_{CO2}/l · 9.6 l/100 km	15.7	kg_{CO2}/100 km
Total CO_2 emissions Including provision of the energy carrier Autogas: 1.9 kg_{CO2}/l · 9.6 l/100 km	**18.2**	**kg_{CO2}/100 km**

15.3.4 Summary

Even though these examples are of course simplified, they provide an initial comparison of the different drivetrain technologies (Fig. 15.1).

Table 15.5 CO_2 emissions of an average CNG car

CNG Car		
Typical fuel consumption in ADAC EcoTest (ADAC: 4.9 kg/100 km plus 10% winter consumption)	5.4	kg_{CNG}/100 km
Energy demand 13.0 kWh/kg · 5.4 kg/100 km	**70.1**	**kWh/100 km**
Direct CO_2 emissions CNG-H: 2.48 kg_{CO2}/kg · 5.4 kg/100 km	13.4	kg_{CO2}/100 m
Total CO_2 emissions[4] Including provision of the energy carrier CNG-H: 2.54 kg_{CO2}/kg · 5.4 kg/100 km	**13.7**	**kg_{CO2}/100 km**

Table 15.6 CO_2 emissions of an average FCEV car, gray hydrogen

FCEV Car (Gray Hydrogen)		
Typical average fuel consumption	1.2	kg_{H2}/100 km
Energy demand incl. 12% fueling losses 33.3 kWh/kg · 1.2 kg/100 km/0.88	**45.4**	**kWh/100 km**
Direct CO_2 emissions H2: 0.46 kg_{CO2}/kWh · 33.3 kWh/100 km · 1.2 kg/100 km	18.4	kg_{CO2}/100 km
Total CO_2 emissions including 12% fueling losses H2 production: 0.46 kg_{CO2}/kWh · 33.3 kWh/100 km · 1.2 kg/100 km = 18.4 kg_{CO2}/100 km Fueling station: 33.3 kWh/kg · 1.2 kg/100 km · 0.12 · 0.445[5] kg_{CO2}/kWh_{el} = 2.4 kg_{CO2}/100 km	**20.8**	**kg_{CO2}/100 km**

15.4 Life-Cycle Analysis

In the previous chapter, the pollutant emissions generated during vehicle operation were investigated.

For a life-cycle analysis or life-cycle assessment, the entire life cycle of a technical product is considered, starting with the raw materials used in production, followed by the use phase and finishing with recycling, including all the energy required and the resulting emissions.

[4] Not including unburned CH_4 residues in the exhaust gas, which can have a significant additional climate-damaging effect.

[5] German CO_2 emission factor 2023 including losses in the upstream chain of primary energy to the power plants.

Table 15.7 CO_2 emissions of an average FCEV car, green hydrogen from wind farms, LH2 distribution

FCEV Car (Green Hydrogen from Offshore Wind Turbines)		
Typical average fuel consumption	1.2	kg_{H2}/100 km
Electric energy demand At fueling station: 33.3 kWh/kg · 1.2 kg/100 km · 12% fueling losses = 4.8 kWh/100 km At wind turbine: 33.3 kWh/kg · 1.2 kg/100 km/ 70% electrolysis/67% LH2 = 85.2 kWh/100 km	**90**	**kWh/100 km**
Total CO_2 emissions Including all losses[6] At fueling station: 33.3 kWh/kg · 1.2 kg/100 km · 0.12 · 0.445[5] kg_{CO2}/kWh_{el} = 2.4 kg_{CO2}/100 km At wind turbine: 33.3 kWh/100 km · 1.2 kg/100 km/0.70/0.67 · 0.010 kg_{CO2}/kWh = 0.9 kg_{CO2}/100 km	**3.3**	**kg_{CO2}/100 km**

Table 15.8 CO_2 emissions of an average BEV car, electricity from the German grid (2022)

BEV Car (Electricity from German Grid)		
Direct electricity consumption	20.0	kWh/100 km
Electric energy demand Including 10% charging losses	**22.2**	kWh/100 km
Total CO_2 emissions 0.445[5] kg_{CO2}/kWh_{el} · 20.0 kWh/100 km/0.9	**9.9**	**kg_{CO2}/100 km**

Table 15.9 CO_2 emissions of an average BEV car, electricity mix from the German grid (2022) and a home photovoltaic (PV) system

BEV Car (Electricity from 50% PV and 50% German Grid)		
Direct electricity consumption	20.0	kWh/100 km
Electric energy demand incl. 10% charging losses PV: 10.0 kWh/100 km/0.9 = 10.1 kWh/100 km grid: 10.0 kWh/100 km/0.9 = 10.1 kWh/100 km	22.2	kWh/100 km
Total CO_2 emissions PV: 0.025[7] kg_{CO2}/kWh_{el} · 10.0 kWh/100 km/0.9 = 0.6 grid: 0.445[5] kg_{CO2}/kWh_{el} · 10.0 kWh/100 km/0.9 = 5.5	5.6	kg_{CO2}/100 km

[6] It is assumed that compression to 700 bar (88% efficiency) requires electric energy from the grid, as it takes place at the filling stations. Electrolysis (60% efficiency) and liquefaction (67% efficiency) are calculated with the wind farm emission factor. The typical emission factor of an offshore wind turbine including installation is 0.010 kg_{CO2}/kWh and the emission factor of the German grid in 2022 including losses in the upstream chain of primary energy is 0.498 kg_{CO2}/kWh.

[7] Home-PV production with 25 g_{CO2}/kWh.

Table 15.10 CO_2 emissions of an average BEV car, electricity from a wind turbine

BEV Car (Electricity from Local, Onshore Wind Turbines)		
Direct electricity consumption	20.0	kWh/100 km
Electric energy demand including 10% charging losses and 5% grid losses	**23.4**	kWh/100 km
Total CO_2 emissions 0.018 kg_{CO2}/kWh_{el} · 23.4 kWh/100 km	**0.41**	**kg_{CO2}/100 km**

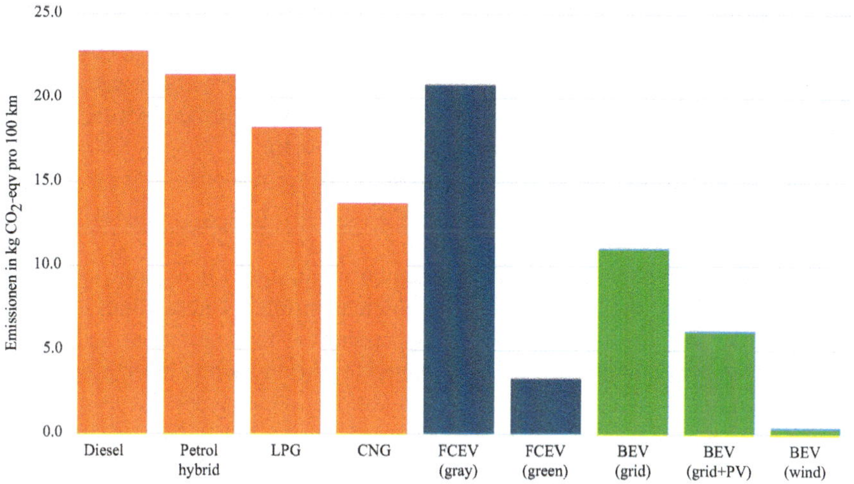

Fig. 15.1 Equivalent CO_2 emissions of average passenger cars in 2022, taking the upstream chain of primary energy into account

When calculating the life-cycle assessment of vehicles, information on their typical use is required, such as lifetime in years, kilometers driven, driving profiles, vehicle weight, recycling rates, the source of the electricity and fuel mix, average fuel and electricity consumption, and so on. This information is used to calculate not only the overall climate impact, but also the cumulative energy consumption and the environmental impact of all pollutant emissions (acidification, particulate matter, summer smog, eutrophication).

Other important environmental aspects that play a role, especially in the case of synthetic fuels and biomass, are the land and water consumption during the cultivation of the plants in agriculture and during subsequent processing.

Due to the necessary assumptions, it is difficult to make absolute statements on the environmental balance. However, different vehicle categories can be compared quite well if similar basic assumptions are made.

15.4.1 Emissions During Vehicle Production

The production of vehicles requires large amounts of energy, which generates CO_2 emissions. This applies to the production of steel and aluminum parts for the body and the combustion engine as well as the production of batteries and carbon fiber for the tanks of fuel cell vehicles. Many studies have been written in this field, some of them extensive. In the following, only the main topics will be highlighted.

There are basically three main components in the production of a vehicle: vehicle body, drivetrain, and energy storage.

15.4.1.1 Vehicle Body

There is no significant difference between the various vehicle types in terms of the environmental impact of the production of the vehicle body and chassis. On average, around two-thirds of the weight is made of steel. A further 18% is made of polymer materials and the remaining 6–8% of aluminum.

Overall, the following equivalent emissions appear reasonable to produce the vehicle body: small car 2.5 $t_{CO2\text{-}eqv}$, mid-range vehicle 3.8 $t_{CO2\text{-}eqv}$, luxury vehicle, and SUV 5 $t_{CO2\text{-}eqv}$.[8]

Additional components are the entire interior fittings, for example, the seats and steering wheel, the large number of vehicle controllers and auxiliary units, for example, the air conditioning system. Depending on the equipment and vehicle category, around 0.5–0.8 $t_{CO2\text{-}eqv}$ are estimated for the emission during production.

For electrically powered vehicles, the additional components cause somewhat higher emissions due to the power electronics and many more complex, fully electrified units. It is estimated at around 0.8–1.5 $t_{CO2\text{-}eqv}$. For fuel cell vehicles, a DC/DC converter must be added, which increases the emissions even more.

The emissions for assembling the vehicle, including painting, result in around 1.5 $t_{CO2\text{-}eqv}$. They are roughly similar for all vehicle types.

In total, around 4.5 (small cars) to 7.3 (luxury cars and SUVs) $t_{CO2\text{-}eqv}$ can be estimated for the bodywork of passenger cars.

15.4.1.2 Drivetrain

The CO_2 emissions for electric drivetrains are lower than for internal combustion engine because electric drivetrains are much smaller and contain fewer parts.

For an internal combustion engine and transmission, the emissions are estimated at around 0.4–0.8 $t_{CO2\text{-}eqv}$, depending on size and performance. The electric motor in an electric vehicle generates around 0.2–0.3 $t_{CO2\text{-}eqv}$ during production.[6]

15.4.1.3 Battery

The battery is an essential component of BEV and PHEV.

---

[8] Source: ifeu Institut für Energie- und Umweltforschung Heidelberg, *UMBReLA Umweltbilanzen Elektromobilität – Ergebnisbericht*, October 2011 sowie *Weiterentwicklung und vertiefte Analyse der Umweltbilanz*, April 2016.

Current studies assume that around half of the greenhouse gas emissions in battery production are caused by the cell production. One-third is caused using energy and transportation of the battery. The battery housing accounts for 14% and the battery management system for around 7% of the total emissions.

The lifetime of batteries has been significantly improved and a modern battery will normally last for the entire life of a vehicle, see also Sect. 6.3.5.

There are two difficulties when assessing the environmental balance of batteries:

Firstly, the energy density of cells in series production is increasing much faster than predicted in recent years. Energy densities of 360 Wh/kg at cell level and 260 Wh/kg at system level were considered utopian just a few years ago but are now announced for mass production of the new NIO ET7 in 2024. However, the manufacturing expenses and CO_2 emissions are directly linked to the energy density. Therefore, older studies on emissions from battery production are mostly outdated.

Secondly, the energy supply of the many battery manufacturing plants varies greatly and is often unclear. For example, Panasonic's production at the Tesla Gigafactory in the Nevada Desert is almost entirely powered by renewable PV electricity. Some studies make blanket calculations based on the respective national electricity mix, which is significantly worse.

The so-called Sweden study,[9] which was published in May 2017 and examines and summarizes data from 18 other studies, has become famous. The authors analyzed many older studies and found that the data varied widely. They finally gave an average value of 150–200 $kg_{CO2\text{-}eq}$ per kWh battery capacity. The energy densities used at the time were around 120 Wh/kg, that is, less than around half of current battery systems.

In its own meta-study, the ifeu Institute[155] arrives at around 140 $kg_{CO2\text{-}eq}$ per kWh, but this figure is also based on battery storage systems with 120 Wh/kg.

In November 2019, the authors of the Sweden study published an updated version[10] considering current cells and manufacturing processes. The emissions from battery production were now estimated to be only 61–106 $kg_{CO2\text{-}eq}$ per kWh.

Considering all studies, an average emission factor of around 80 $kg_{CO2\text{-}eq}$ per kWh appears to be an appropriate estimation for modern EV battery production.

The batteries of small, medium-sized, and large BEVs with 40, 80, and 100 kWh then generate equivalent emissions of 3.2, 6.4, and 8 $t_{CO2\text{-}eqv}$ during production.

---

[9] Source: Romare, M., Dhllöf, L., *The Life Cycle Energy Consumption and Greenhouse Gas Emissions from Lithium-Ion Batteries*, ivl Sweden, May 2017.

[10] Source: Emilsson, E., Dahllöff, L., *Lithium-Ion Vehicle Battery Production – Status 2019*, ivl Sweden, No. C444, November 2019.

15.4.1.4 Fuel Cell System and Hydrogen Tanks

The production of fuel cell vehicles also involves additional costs compared to conventional cars. This is mainly due to the fuel cell system, the high-pressure storage system, and finally the lithium-ion battery, which is still required for cold start and braking energy recovery.

The UMBReLA study[155] calculates around 2 $t_{CO2\text{-}eqv}$ for the fuel cell and tank and a further 0.5 $t_{CO2\text{-}eqv}$ for the additional vehicle battery required for a medium-sized FCEV.

The two tanks in a Toyota Mirai weigh around 87.5 kg in total. They are made up of a three-layered structure of carbon fiber and synthetic resin. The three tanks in a Hyundai Nexo are correspondingly heavier. Carbon fiber is produced by refining oil into acrylonitrile, which is then spun and baked. Due to the high baking temperature of 1000 °C, around 20 tons of CO_2 are emitted to produce 1 ton of carbon fiber.

Therefore, around 2.2 $t_{CO2\text{-}eqv}$ can be assumed to produce the hydrogen tanks for a medium-sized FCEV alone.

In addition to the tanks, the fuel cell system must also be manufactured. The technical effort for the various auxiliary units such as compressors, heat exchangers, etc. is considerable. For example, the Mercedes GLC F-Cell weighs around 400 kg more than the comparable vehicle with a diesel engine. Subtracting the mass of the hydrogen tanks (around 88 kg) still leaves over 300 kg extra weight for the fuel cell system. This is like the weight of a large combustion engine with gearbox. The emissions to produce the fuel cell system are therefore likely to be at a similar level and are estimated at 0.8 $t_{CO2\text{-}eqv}$.

15.4.1.5 Summary

The following table provides an overview of the approximate equivalent CO_2 emissions to be expected from vehicle production. For fuel cell vehicles, a combination of medium and large categories was used. The required space for the two to three H_2 tanks and the fuel cell system does not permit the construction of small fuel cell vehicles in the foreseeable future (Table 15.11).

15.5 Lice-Cycle Analysis of Average Passenger Cars

The relationships explained in the previous sections are now summarized in an analysis of the equivalent CO_2 emissions of average passenger cars over their lifetime.

For this purpose, the average fleet emissions in Germany are combined with the equivalent CO_2 emissions from vehicle production for a medium-sized passenger car. The results are generic CO_2 emissions that represent average values for the current German vehicle fleet and future scenarios.

According to the German Federal Motor Transport Authority, the typical lifetime of a passenger car in Germany is around 12 years. With the average yearly

Table 15.11 Equivalent CO_2 emissions from vehicle production in tons[11]

	ICE			BEV			FCEV
	Small	Medium	Large	Small 40 kWh	Medium 80 kWh	Large 100 kWh	Medium
Body	2.5	3.8	5.0	2.5	3.8	5.0	3.8
Auxiliaries	0.5	0.65	0.8	0.5	0.65	0.8	1.0
Assembly	1.5	1.5	1.5	1.5	1.5	1.5	1.5
Drivetrain	0.4	0.6	0.8	0.2	0.25	0.3	0.25
HV system	–	–	–	0.3	0.55	0.7	0.9
Battery	–	–	–	3.2	6.4	8.0	0.4
H_2 tank	–	–	–	–	–	–	2.2
Fuel cell system	–	–	–	–	–	–	0.8
Total	4.9	6.6	8.1	8.2	13.2	16.3	10.9

driving distance of 12,700 km in 2022, this results in an average total distance of 150,000 km.

An 80 kWh battery is assumed for a medium-sized electric car, which gives a range of around 400 km.

The same seven scenarios that were already used in the well-to-wheel analysis in Sect. 15.3 are analyzed. For the emission factor of the electricity grid, estimated values for the years 2025 (purchase of the car) to 2037 (end of life) according to Fig. 13.7 were used (Fig. 15.2).

Surprisingly, average diesel vehicles prove to be the most harmful to the climate due to the high costs of fuel production.

Shortly behind and roughly on a par are vehicles powered by petrol (including mild, full-, and plug-in hybrids) and fuel cell vehicles that run on gray hydrogen, which is currently the state of the art.

Battery electric vehicles save more than half of CO_2 emissions over their typical lifetime, even more if they are partially charged by a local PV system.

Hydrogen cars that are refueled with green hydrogen and battery cars that are fully charged with locally generated renewable electricity (wind or PV) perform best. They save almost two-thirds of the CO_2 compared to diesel and petrol cars. However, there is practically no supply of green hydrogen to date. The first experimental plants should go into operation soon, but their hydrogen is much more urgently needed for industrial applications than for cars.

At the end of a car's service life, even under the most favorable circumstances, emissions of at least 13 tons of CO_2 remain. They can only be avoided if the entire production chain is based on recycled materials and renewable electricity generation. We are still a long way from achieving this.

[11] Source: UMBReLA study[155] and own estimates.

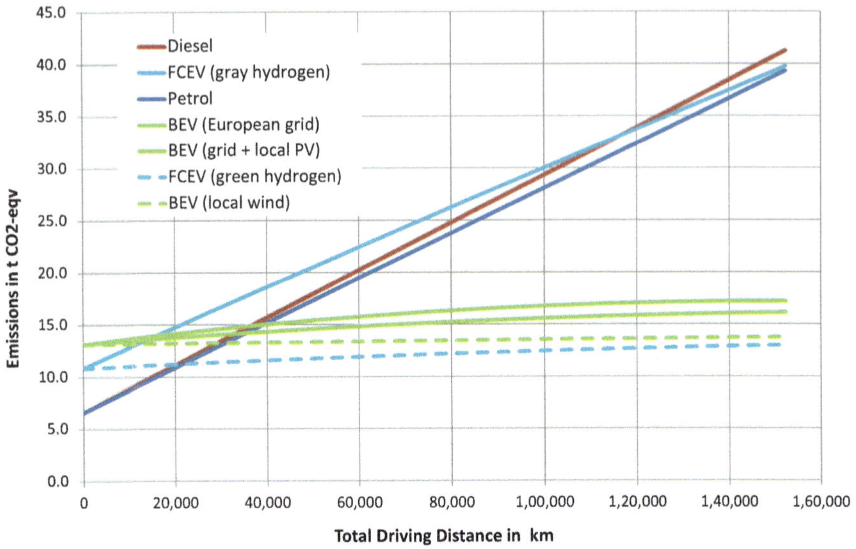

Fig. 15.2 Equivalent CO_2 emissions over the entire life cycle of an average passenger car from 2025 to 2037[12]

These estimates do not yet consider the savings from battery recycling or second life. In the best-case scenario, emissions from battery production can be halved, thereby reducing the balance by 3–4 tons.

[12] Based on the forecasted CO_2 emission factor of European electricity production plus 10% overhead to cover upstream chain emissions.

Costs of Electric Mobility 16

Abstract

This chapter estimates the costs of electric vehicles, batteries, and the necessary infrastructure. The studies on infrastructure are carried out using Germany as an example, whereby the year 2050 is assumed as the limit scenario, when hopefully all cars will be emission-free battery electric vehicles. The distribution of charging points is estimated on this basis. The estimated costs per charging point then result in an investment sum for the expansion of the infrastructure.

Like the previous chapters on energy carriers and environmental impacts, the topic of costs of electromobility can only be introduced generally in this book by using simplified examples. The aim is to develop a basic understanding of the interrelationships and the fundamental approaches. Extensive studies by various research institutions and consulting firms are available that can be consulted if necessary.

According to EU decisions, no more cars powered by conventional fuels may be sold from 2035. Assuming a typical lifetime of 10–15 years, the entire vehicle population will be converted by around 2045–2050.

Supplying all or at least a large amount of passenger cars with synthetic biofuels is impossible due to the foreseeable long-term production costs and the land requirements for agriculture, see Sect. 14.3.2. Supplying a larger fleet of passenger cars with e-fuels is also unrealistic, see Sect. 14.3.3.

Fuel cell-based hydrogen cars are also not suitable for converting a large fleet of millions of vehicles, see Sects. 12.13 and 14.5.

As a matter of fact, there is only one realistic option for realizing CO_2 neutral, individual mobility: battery electric vehicles.

M. Doppelbauer, *Introduction to Electromobility*,
https://doi.org/10.1007/978-3-658-45482-1_16

Table 16.1 Comparison of direct cost ICE vs. BEV (50 kWh battery), European compact-class vehicle[1]

	ICE 2020	BEV 2020	ICE 2030	BEV 2030
Battery	–	8000 €	–	4300 €
Engine/E-motor	3000 €	2000 €	3100 €	1800 €
Rest of drivetrain	2000 €	700 €	2100 €	600 €
Interior	2700 €	2700 €	2900 €	3000 €
Electric/electronic architecture (E/E)	1800 €	2000 €	1900 €	2000 €
Body in white (BIW) and exterior	1700 €	2100 €	1800 €	1800 €
Chassis	1200 €	1300 €	1200 €	1200 €
Assembly	1600 €	1500 €	1600 €	1300 €
Total	13,900 €	20,200 €	14,600 €	16,000 €

16.1 Cost of Vehicle and Battery Production

A study by Oliver Wyman Consultancy in 2020 compared the direct costs for a battery electric (BEV) and an internal combustion engine (ICE) European compact-class passenger car and gave an outlook to 2030. The numbers do not include indirect costs, like CAPEX, marketing, development, etc. (Table 16.1).

For 2020, the costs of battery electric vehicles were calculated to be around 45% higher than that of a conventional internal combustion engine vehicles. By 2030, the additional costs for BEV are expected to have fallen to just 9% due to better production processes and higher unit numbers.

From the author's own experience, the price of an electric drivetrain (e-motor, gearbox, power electronics) was already less than 1000 € for a mid-size electric car in 2023 (the study estimates 2400 € by 2030).

According to another study by Bloomberg/NEF, the price for EV battery cells was down to 89 $/kWh in 2023 and the price for complete EV battery packs was down to 128 $/kWh.[2] On average, the cell accounts for 78% of the total pack price (Fig. 16.1).

[1] Source: Oliver Wayman Research, 2020.

[2] Source: *Lithium-Ion Battery Pack Prices Hit Record Low*, November 26, 2023, BloombergNEF.

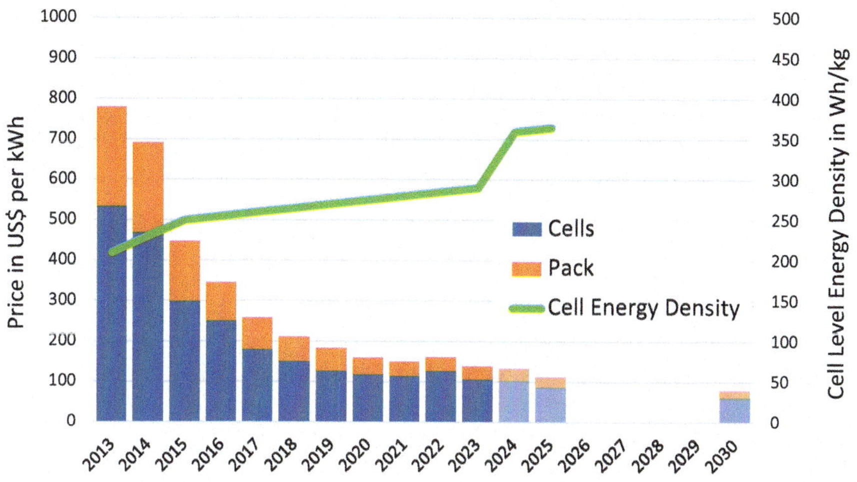

Fig. 16.1 Development of price and energy density of lithium-ion cells and packs (average across multiple end uses including cars, trucks, buses, and stationary applications)[2]

16.2 BEV Charging Points for the Whole Car Fleet of Germany

In the following sections, the costs of setting up a charging infrastructure for the total car fleet in a country are examined using Germany as an example. The analyses are based on part on the Bachelor thesis of Ms. Judith Auer, who was supervised by the Institute of Industrial Management and Industrial Production (IIP) and the author of this book at KIT in 2018.[3] It is a look ahead to 2050, when hopefully all 48 million vehicles with combustion engines will have disappeared.

There are similar considerations for other countries, e.g. the USA.[4]

The future charging infrastructure for BEVs will be divided into three areas:

- **Non-public charging points** are located in private areas (home garages, carports, underground garages of condominiums, and apartment buildings) as well as charging points provided by employers in the workplace area (employee parking lot). These are usually AC chargers with power between 3.7 and 22 kW.
- **Semi-public charging points** are located on private property with public access (e.g. privately managed parking garages, parking lots in front of supermarkets and hardware stores, recreation facilities). They are normally 22 kW or

[3] *Scope and Total Investment for a Charging Infrastructure for 100% Market Share of BEVs in Germany by 2050*; Judith Auer, Daniel Heinz, Patrick Jochem, Martin Doppelbauer; 32 Electric Vehicle Symposium EVS, Lyon, France, May 2019.

[4] See also: *Charging Up America: Assessing the Growing Need for U.S. Charging Infrastructure Through 2030*, icct White Paper, Gordon Bauer et al., July 2021.

even 50 kW DC. In addition, fast charging points are needed at service stations near busy roads and highways with typically 150 or 300 kW DC. Dedicated charging parks also fall into this category.

- **Public charging points** are located in generally accessible public spaces (e.g. parking spaces along public roads). These are mostly AC chargers with 3.7 kW up to 22 kW. They are mainly needed for charging of vehicles belonging to tenants who have not been allocated their own parking space.

16.2.1 Required Number of Charging Points

The number of charging points required to fully supply all cars can only be predicted with some uncertainty. It seems obvious that a significant proportion of the supply will be provided by non-public charging points.

16.2.1.1 Home and Workplace (Public and Non-public CP)

Private wallboxes for three-phase charging with 11 kW (16 A) or 22 kW (32 A) have fallen steadily in price in recent years and have now become quite affordable. They contain all the necessary safety devices, including the RCD. Only the power connections need to be installed. Due to the low cost of wallboxes, they will spread quickly and become standard in single and multi-family homes.

There are around 41 million private households in Germany. Around a third of these live in single-family homes, around 12% in two-family homes and the rest in residential buildings with three or more apartments.[5]

We roughly assume that 90% of all houses will be equipped with one or two private charging points. This gives charging opportunities for 18.8 million vehicles.

A further 23 million apartments are in residential buildings with 3 or more units. Around 3.5 millions of these are owner-occupied apartments, the rest are rental apartments. Some of the vehicles in these apartments will be parked at the roadside in the evening and must charge overnight at public charging points. In some cases, there will be semi-public or private parking areas or underground garages, especially for condominiums.

As a first estimate, we assume that one-third of all condominiums are equipped with private charging points as they usually have an allocated parking space. This results in charging opportunities for a further 1.2 million vehicles.

To supply all cars in Germany with sufficient charging points, public charging points must be built at roadsides for around one-third of all rental apartments. This can be done, for example, by equipping lampposts or by conventional charging columns. Each charging point can serve several cars over the course of a week. Assuming that every car without its own wallbox is parked at a roadside charging

[5] Source: Behrends, Kott: *Zuhause in Deutschland – Ausstattung und Wohnsituation privater Haushalte*, Ausgabe 2009, Statistisches Bundesamt, Wiesbaden.

Table 16.2 Estimation of the number of charging points in households, alongside streets and in employee parking lots

	Number	Rate of charging points	Number of charging points	Vehicles per charging point per week	Number of cars charging regularly at the same time
Single-family houses (non-public CP)	13.5 M	90%	12.2 M	1	12.2 M
Two-family houses (non-public CP)	5 M	90%	4.4 M	1.5	6.6 M
Condominiums (non-public CP)	3.5 M	33%	1.2 M	1	1.2 M
Rented apartments (public CP)	19 M	33%	6.3 M	2	12.6 M
Workplaces (non-public CP)	22 M	33%	7.3 M	2	14.5 M
Total			**31.3 M**		**47 M**

station about every second night, which is completely sufficient for a range of 300–500 km per week, then there are charging options for a further 12.6 million vehicles.

In addition, employers will provide large numbers of charging points in their parking lots. Of the approximately 44 million people in employment in Germany, around 60% commute to work from their home, most of them by car. If we roughly assume that employers will create charging points in one-third of their parking spaces in the long term and that workplace charging points will be used on a rotating basis (with only two different vehicles per week), this will create charging opportunities for a further 14.5 million vehicles (Table 16.2).

Based on the estimates to date, there are regular charging options for around 47 million vehicles. A charging power between 3.7 and 11 kW is completely sufficient because the vehicles are typically parked at the charging point for many hours (overnight and during working hours).

16.2.1.2 Recreational Sector (Semi-public CP)

In addition to home and workplace charging, many vehicle owners will use public and semi-public charging points during their recreational activities. This includes parking times when shopping in larger stores (supermarkets, hardware stores, etc.), in city centers (parking garages), at sports facilities (sports halls, soccer pitches, etc.), and at other destinations (restaurants, event halls, amusement parks, etc.). It can be assumed that the charging times are concentrated in the afternoon and evening hours during the week and at weekends. The typical charging power must be higher due to the shorter parking time and can range from 22 kW AC to 50 kW DC per column.

There are around 10,000 large supermarkets and around 2500 hardware stores in Germany, most of which have 100 or more parking spaces.

2.6 million parking spaces in Germany are currently subject to charges (parking garages, etc.). Around 40% of these are in city centers or shopping centers and around 30% in the event sector (halls, leisure parks, etc.).[6]

The total number of sports facilities in Germany is around 230,000, including around 66,500 uncovered sports grounds, 35,500 sports halls, 7500 swimming pools, 13,000 tennis courts, and 15,000 shooting ranges.[7] In total, there are around 140,000 larger facilities, each with many parking spaces.

In summary, there are around 5 million semi-public parking spaces in city centers, shopping malls, and the recreation sector.

As an estimate, it can perhaps be assumed that 50% of managed parking spaces in city centers or shopping centers and 25% of other parking spaces will eventually be equipped with charging points, resulting in a total of 1.4 million charging points, which will, however, be heavily frequented and used by many different vehicles.

16.2.1.3 Busy Roads (Semi-public CP)

It will also be necessary to equip the service areas and highway service stations with charging points and to build dedicated charging parks to enable fast recharging during longer journeys. There are currently around 430 service stations and 200 motorway service areas in Germany. We estimate the number of petrol pumps per service station at 10, that is, a total of around 6300 petrol pumps.

In future, these facilities will have to be equipped with DC fast charging stations of 150 kW or more so that even large batteries can be filled to at least 80% in 30 min. As the charging process is around six times slower than a diesel or petrol refueling process, which takes up to 10 min including payment, six times as many charging stations will be required, see also Sect. 12.3.

This results in an average of 60 charging points per service area to be able to handle as many electric vehicles as today's vehicles with combustion engines. Rounded up, around 40,000 fast charging points are needed alongside busy roads.

16.2.1.4 Summary of Demand for Charging Points

All these considerations result in a total of around 6.5 million charging points in public and semi-public areas and around 24 million charging points or wallboxes in private areas (private homes, employee parking lots).

---

[6] Source: Peter Sturm, *Wirtschaftliche Aspekte des Parkraummanagements*, Vortrag am 10.04.2014 im Haus am Dom, Frankfurt am Main.

[7] Source: Deutscher Olympischer Sportbund (DOSB), *Sportstätten in Deutschland – ein Überblick*, 2013.

Table 16.3 Estimated number and occupancy of charging points in Germany

	Power per CP	Number of CPs	Number of regularly served vehicles	Average CP occupancy
Non-public CP (home, work-place)	3.7 kW AC or 11 kW AC	25 M	38 M	4.5%
Public CP (residential area)	3.7 kW AC or 11 kW AC	6.3 M	12.6 M	4.3%
Semi-public CP (managed parking space in inner cities and recreation)	22 kW AC or 50 kW DC	1.4 M	–	10%
Semi-public CP (fast charger)	150/300 kW	0.08 M	–	10%
Total		**32.8 M**	**50.4 M**	**8%**

Table 16.3 gives an overview of the infrastructure, the delivered electric energy, and the average occupancy of each charging point.

The average occupancy of the charging stations with vehicles (the effective charging time used) is calculated, on the one hand, from the number of vehicles served per station and their average annual mileage (14.000 km/a with 22 kWh/100 km = 3080 kWh/a), and, on the other hand, from the theoretical charging capacity for continuous operation.

Across all charging points, a total electric energy of 138 billion kWh/a is required annually for the operation of all 48 million battery electric cars.

As can be seen from the table, the average occupancy of each charging station is rather low and only in the range of a few percent. This ensures that there are always enough free charging points available.

The VDA position paper on charging infrastructure (as of July 1, 2015) and the National Platform for Electric Mobility (NPE) recommend the installation of around 150,000 AC charging points and 10,000 DC charging points for 1 million electric cars. 10% of all parking spaces in managed parking garages and parking lots should be equipped with mandatory charging points. This corresponds to a projected 7.4 million charging points for all 48 million vehicles, which is similar to the estimates presented above (7.7 million).

16.2.2 Required Investment

The status report on the charging infrastructure of the National Platform for Electromobility lists the following costs for installation (CAPEX = investment costs for hardware, grid connection, approval, planning, assembly, construction costs)

and operation (OPEX = ongoing operating costs excluding electricity and depreciation, e.g. maintenance, communication, billing, IT):

- Non-public AC charging point (wallbox) > 3.7 kW

$$\text{CAPEX} = 1700\ €, \text{OPEX} = 500\ €/\text{a}$$

- Public and semi-public AC charging point medium-fast, 11 or 22 kW

$$\text{CAPEX} = 7500\ €, \text{OPEX} = 750\ €/\text{a}$$

- Semi-public DC charging point, 50 kW

$$\text{CAPEX} = 24,000\ €, \text{OPEX} = 1500\ €/\text{a}$$

These figures are from the year 2015. Today's prices (as of 2024) are already significantly lower in some cases. Private wallboxes with 11 kW are being sold for less than 500 € (plus installation). There are practically no operating costs apart from the energy. The target price for DC fast charging stations is around 100 € per kW (plus installation).

Public DC charging points with 50 kW currently cost around € 15,000 including installation. The installation costs of fast charging stations and charging parks vary greatly depending on the capacity of the already existing power grid.

It seems sensible to introduce a further category for simple charging points. These are non-public, medium-speed 11 kW AC charging stations that are installed in private parking lots of condominiums and employee parking lots of employers and therefore do not require a billing system. This reduces investment expenditure and, above all, running costs. We estimate these charging points with CAPEX = 5000 € and OPEX = 500 €/a.

Table 16.4 gives estimates for investment costs from today's perspective.

Based on these prices and the previously estimated quantities of charging points, the necessary investments in a charging infrastructure can be estimated as given in Table 16.5.

The investment costs for the initial installation are in the order of 100 billion €, of which around half (45 M€) would have to be borne directly by the public sector (roadside columns). In addition, investment incentives would be required for private households and commercial operators.

Added to this are the investments to expand the electricity grids on all voltage levels and the power generation (wind, PV) in the order of 20–25% of today's capacity. Assuming a full transition to electric cars by the year 2050 that translates to a 1% yearly capacity increase for the years between 2025 and 2050.

By way of comparison, the cost of building a petrol station is in the region of 700 T€ to 4 M€, depending on the location and size.[10] There are around 14,500 filling stations in Germany. The total investment costs for the filling stations

[10] Source: T*ankstelle eröffnen: Gründungsschritte, Voraussetzungen und Kosten*, Philip Zimmermann, Qonto Blog, August 30, 2023.

Table 16.4 Estimated investment (CAPEX) and operating (OPEX) costs per charging point (not including electricity)

	CAPEX	OPEX
Non-public wallbox without billing 3.7 up to 11 kW AC	500 € (wallbox) + 1000 € (installation)	200 €/a
Non-public wallbox with billing 3.7 kW up to 11 kW AC	1000 € (wallbox) + 1200 € (installation)	200 €/a
Public AC CP 22 kW AC	7500 € (fully installed)	750 €/a
Medium DC CP 50 kW DC	28,000 € (fully installed)[8]	1500 €/a
Fast DC CP 150 or 350 kW DC	75,000 € or 140,000 € (fully installed)[166]	3000 €/a

Table 16.5 Estimate of the investment costs (CAPEX) for the installation of charging points in Germany

		Number of CP	CAPEX	Operator
Family houses (non-public CP)	3.7 kW AC or 11 kW AC	16.6 M	1500 €	Private households
Condominiums (non-public CP)	3.7 kW AC or 11 kW AC	1.2 M	2200 €	Landlord
Residential area (public CP)	3.7 kW AC or 11 kW AC	6.3 M	7500 €	City Administration
Workplaces (non-public CP)	3.7 kW AC or 11 kW AC	7.3 M	2200 €	Employer
Managed parking space (semi-public CP)	22 kW AC (90%) or 50 kW DC (10%)	1.4 M	9500 € (weighted average)	Commercial Enterprise
Fast charge (semi-public)	150 kW (50%) or 300 kW (50%)	0.04 M	107,500 € (weighted average)	Commercial Enterprise
Total		**32.8 M**	**102 B€**	

therefore amount to roughly 35 B€. In addition, there is a complex infrastructure for refining the fuels and for wholesale.

16.2.3 Costs of Public Charging

Based on CAPEX and OPEX and today's electricity prices, the costs of public charging can be estimated.

[8] *DC Fast Chargers Cost $28,000–$140,000 Installed*, Property Manager Insider, June 2021.

Table 16.6 Breakdown of the total costs of a fast charging station over its lifetime[9]

Electricity cost (base price)	34%
Electricity cost (demand charges)	23%
OPEX	7%
CAPEX	24%
EBIT on total cost EBIT on OPEX and CAPEX only	12% 39%

Table 16.7 Estimate of charging price at the various charging points

		Electricity cost (2024)	CAPEX	OPEX	Target price incl. VAT
Family houses (non-public CP)	3.7 kW AC or 11 kW AC	0.35 €/kWh	0.052 €/kWh	0.072 €/kWh	0.61 €/kWh
Condominiums (non-public CP)	3.7 kW AC or 11 kW AC	0.40 €/kWh	0.101 €/kWh	0.0972 €/kWh	0.78 €/kWh
Residential area (public CP)	3.7 kW AC or 11 kW AC	0.30 €/kWh	0.172 €/kWh	0.181 €/kWh	0.90 €/kWh
Workplaces (non-public CP)	3.7 kW AC or 11 kW AC	0.35 €/kWh	0.050 €/kWh	0.048 €/kWh	0.51 €/kWh
Managed parking space (semi-public CP)	22 kW AC (90%) or 50 kW DC (10%)	0.35 €/kWh	0.037 €/kWh 0.058 €/kWh	0.039 €/kWh 0.034 €/kWh	0.53 €/kWh 0.56 €/kWh
Fast charging (semi-public CP)	150 kW (50%) or 300 kW (50%)	0.35 €/kWh	0.147 €/kWh	0.046 €/kWh	0.71 €/kWh

The total cost of a charging point breaks down as in Table 16.6.

This information can be used to calculate the target end-user prices for charging at each type of charging point. The numbers in the following table were based on an assumed 30% mark-up on OPEX and CAPEX and 19% VAT. Subsidies are not considered.

The expected end customer prices are in range of 0.50–0.90 ct/kWh, which seems reasonable compared to the actual prices as of today (2024) (Table 16.7).

[9] Data source: McKinsey Center for Future Mobility, *Can public EV fast charging stations be profitable in the United States?*, Peter Fröde et al., Oktober 5, 2023; (assuming 150 kW DC, 15% utilization, 80% charger efficiency, 0.45 US$/kWh electricity base price).

These rough estimates confirm that the development of a charging station infrastructure for the entire German car fleet is technically and economically feasible. The charging costs for a 100 km journey are between 9.50 and 17 €.

The average annual price for diesel fuel in 2023 in Germany was 1.74 €/l. With an average consumption of 7.0 l/100 km this results in costs of 12.20 €/100 km, which is on a similar level as the cost for electric cars.

Example 16.1 Electricity costs of a private PV system

A 9.3 kWh peak PV system with a home storage system with a capacity of 10 kWh is installed in a residential one-family home. For further details, see Example 14.6.

The effective electricity costs are around 30 ct/kWh or 6.60 € per 100 km of driving distance. That is half as much as an average car with a combustion engine. ◀

16.3 Service and Maintenance of Battery Electric Vehicles

Service and maintenance requirements for battery electric vehicles are lower than those for vehicles with a combustion engine because there are significantly fewer moving parts by around two orders of magnitude.

The windshield fluid, brake fluid, and coolant in electric cars also need to be checked regularly and changed if necessary. Brake fluid is generally hygroscopic, i.e. it attracts water and then loses its effectiveness.

The brake pads wear out much less in electric cars than in conventional cars because braking occurs mostly electrically by braking energy recovery without engaging the mechanical brakes. Careful drivers manage almost entirely without using the mechanical brakes. This results in the new problem that the brake disks are no longer regularly cleaned of rust. Oxidation particles can cause scoring on the disks in case hard mechanical brakes are needed from time to time. This can be prevented by regular mechanical braking, e.g. once a week in suitable traffic conditions, to remove the rust film.

Tire wear of electric cars tends to be somewhat higher due to the more powerful acceleration of an electric motor (even with nominally the same performance compared to a combustion engine) and the increased weight of the vehicle.

Battery life is a much-discussed topic. Replacing batteries within the normal life of a vehicle would lead to high costs and a poor environmental balance. Today's battery systems are typically replaced by the manufacturer when the remaining capacity falls below a limit of 70% of the new capacity. It is customary in the industry to give a warranty of at least 160,000 km or 8 years on the battery storage system. It can therefore be assumed that the typical service life projected by the manufacturer is significantly longer, otherwise they would have to expect high warranty costs, see also Sect. 6.3.5.

16.4 Summary

In the long term, the complete conversion of the German car fleet to battery electric vehicles appears to be economically feasible. Most of the charging will take place at private charging points. However, workplace charging, supermarkets, hardware stores, and recreation facilities will offer a wide range of additional charging options. Fast chargers are needed near busy roads to help cover long distances.

Considering investments and depreciation as well as the usual price increases in electricity production, electric cars can be driven at a similar cost per km as today's ICE cars. Service and maintenance costs tend to be lower due to the significantly lower number of moving parts.

Electromagnetic Fundamentals

This chapter provides a compact introduction to the fundamentals of electrical engineering. This part is especially intended for readers with little previous knowledge of the fundamentals of electrical engineering or as a refresher.

The physical–mathematical modeling of electric and magnetic circuits, including a concise summary of electric AC (alternating current) theory using complex numbers, will be presented. The laws of the generation of magnetic fields from electric current and the induction of voltage by variable magnetic fields are introduced.

A.1 Electric Current, Voltage, Power, and Energy

Ohm's law generally describes the relation between the instantaneous values of voltage u and current i across an electrical resistor R (Fig. A.1).

The following applies to all ohmic resistance R:

$$u = R \cdot i \tag{A.1}$$

Note
Analogous to a piping system, the voltage or electric potential can be imagined as the pressure with which water is forced into a pipe. The current then corresponds to the flow rate and the resistance results from the cross section and length of the pipe.

Voltage is therefore the reason of electric current flow and resistance is the property of the conducting material. Both result in the generation of a certain amount of electric current. Current corresponds to the effects of electricity, for example, heat and generated mechanical force.

The unit of voltage is Volt (V), the unit of current is Ampere (A), and the unit of resistance is Ohm (Ω).

M. Doppelbauer, *Introduction to Electromobility*,
https://doi.org/10.1007/978-3-658-45482-1

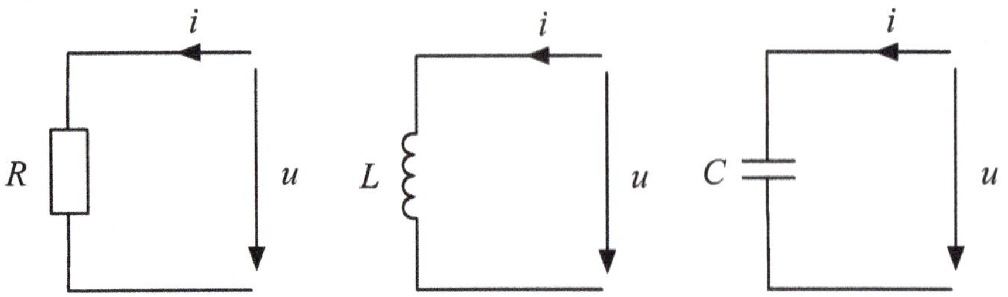

Fig. A.1 Current and voltage across resistor, coil, and capacitor

Comparable laws can also be applied to coils (***L*** = Inductance):

$$u = L \cdot \frac{\mathrm{d}i}{\mathrm{d}t} \tag{A.2}$$

and to capacitors (***C*** = Capacitance):

$$u = \frac{1}{C} \cdot \int i \cdot \mathrm{d}t. \tag{A.3}$$

or alternatively:

$$i = C \cdot \frac{\mathrm{d}u}{\mathrm{d}t} \tag{A.4}$$

The notation d/dt denotes a change of a particular quantity with time, the mathematical derivative. A changing current i that is flowing through a coil therefore causes a voltage drop u, an electric potential difference. The opposite is true for a capacitor: a changing voltage causes the flow of a current.

The integration form of Equation (A.3) describes how the capacitor absorbs a certain charge over a certain period (the electric current is ultimately a flow of charges) and then develops a proportional voltage. In principle, this simple relationship also applies to battery storage systems, although the charge curve is not as linear as with capacitors.

The unit of the capacitance is Farad (F): [C] = 1 F = 1 A·s/V.

The unit of inductanceis Henry (H): [L] = 1 H = 1 V·s/A.

Note

Variables in lower case (u) denote the instantaneous value of variables that change over time, variables in upper case (U) denote constant variables.

Effective values (rms = root mean square) are not designated further (U), mean values are written with an upper dash ($\overline{U}$), and peak values are indicated by a roof ($\hat{U}$).

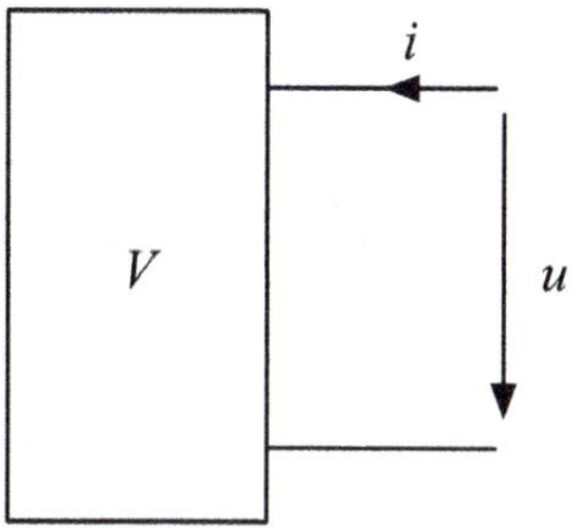

Fig. A.2 Electric power is consumed by the load *V*

Note
Curly brackets denote the numerical value of a variable, for example, $\{U\}$, square brackets denote its unit of measurement, for example, $[U]$.

Therefore: quantity = numerical value · unit, that is, $U = \{U\} \cdot [U]$. For example, $U = 400$ V; $\{U\} = 400$; $[U] = 1$ V.

Square brackets stand as an abbreviation for the text "The unit of…". Therefore, "$[U] = 1$ V" is spoken as "The unit of U is one volt".

This standardized way of writing has been defined in the ISO 80000 standard for decades. Still, incorrect usage can be found quite often, especially in the labeling of diagrams. What does "y [m]" mean? "The unit of [m] is …"? The correct axis labels in diagrams must be written in the form "y in m" or "y/m".

It should also be noted that the symbols for sizes are written in italics, while unit symbols are written in straight lines.[1]

Note
Units can be extended by prefixes. For example, V can become kV.

Often used prefixes are "k" for kilo or one thousand, "M" for mega or one million, and "m" for milli or 0.001.

At any given time, the electric power p consumed by a load is the product of voltage and current (Fig. A.2):

$$p = u \cdot i \tag{A.5}$$

[1]Sources: ISO 80000-1, Quantities and Units—Part 1: General; *Der normgerechte Umgang mit Größen, Einheiten und Gleichungen*; Firmenschrift Rhode & Schwarz GmbH & Co. KG, München, November 2016.

Note

In this book, the consumer arrow system (VZS) is always used. When the power has a positive sign, this means that power is transported from the battery to the consumer and the electric machine works as a motor.

If the sign of power is negative, the machine works as a generator, i.e. energy is fed back into the battery (braking energy recovery).

The power p is also referred to as the instantaneous value of the power. The unit is Watt: $[p] = 1$ W. As power can be calculated from voltage and current, see (A.5), the unit Watt is derived from Volts times Amperes.

If a certain amount of power flows into a consumer or is released from a battery over a certain period, this requires a certain amount of energy:

$$e = \int p \cdot \mathrm{d}t \tag{A.6}$$

The unit of energy is joule (J) or watt-second (Ws), which is the same (1 Ws = 1 J). In practical applications, kilowatt hours (kWh) are often used. The following conversion applies: 1 kWh = 3,600,000 Ws = 3,600,000 J = 3.6 MJ.

Note

It is important to have a clear understanding of the differences between capacity (given in A s), energy (given in J or W·s or kW h), and power (given in W or kW or MW).

Sometimes, the unit kW/h can be seen in non-scientific references ("kilowatt per hour"). However, this unit does not exist. It is simply nonsense.

Equally stupid is the unit kWh/h, which is used in some cars to display energy consumption at standstill. This unit is identical to kW and simply means the consumed power.

A.2 Sinusoidal Currents and Voltages

In many circuits, voltages and currents are not constant (DC) but rather fluctuate over time. When the fluctuation is sinusoidal (AC), the instantaneous value of a sinusoidal (alternating) voltage can be specified as follows:

$$u = \widehat{U} \cdot \cos(\omega t + \varphi_U) \tag{A.7}$$

And correspondingly for a sinusoidal current:

$$i = \widehat{I} \cdot \cos(\omega t + \varphi_I) \tag{A.8}$$

In these equations, ω (omega) denotes the angular frequency and φ (phi) any initial phase angle. $\hat{U}$ and $\hat{I}$ are the amplitudes of the oscillations (maximum or peak values).

The angular frequency results from the electrical frequency f to:

$$\omega = 2\pi \cdot f \tag{A.9}$$

The mean value of an oscillation is calculated from:

$$\overline{U} = \frac{1}{T} \cdot \int u \cdot \mathrm{d}t \tag{A.10}$$

T is the period of the oscillation. For sinusoidal oscillations, $T = 1/f$.

More often, however, the effective value is used, which is also called the root mean square value (RMS):

$$U = \sqrt{\frac{1}{T} \cdot \int u^2 \cdot \mathrm{d}t} = \frac{\hat{U}}{\sqrt{2}} \tag{A.11}$$

The RMS values of AC voltage and AC current correspond to the DC values of voltage and current that generate the same electric power at the same resistor R on average over time:

$$P = U \cdot I = I^2 R \tag{A.12}$$

The power P is designated here with a capital letter because it now represents the average power and not the instantaneous value.

The instantaneous power p fluctuates over time and results from the product of voltage and current:

$$\begin{aligned} p &= 2UI \cdot \cos(\omega t + \varphi_U) \cdot \cos(\omega t + \varphi_I) \\ p &= UI \cdot \cos(\varphi_U - \varphi_I) + UI \cdot \cos(2\omega t + \varphi_U + \varphi_I) \end{aligned} \tag{A.13}$$

We introduce the phase angle:

$$\varphi = \varphi_U - \varphi_I. \tag{A.14}$$

Setting $\varphi_U = 0$ results in $\varphi_I = -\varphi$.

The case $0 < \varphi < \pi$ is referred to as lagging current and the case $-\pi < \varphi < 0$ is referred to as leading current with reference to the voltage.

In inductors, the current lags and in capacitors it leads.[2] This means that the maximum of the current oscillation in inductors always occurs shortly after the maximum of the voltage oscillation, as the upper half of Fig. A.3 demonstrates.

After a short intermediate calculation, the following expression for the instantaneous power of sinusoidal currents and voltages is obtained from Equation (A.13):

$$p = UI \cdot \cos\varphi \cdot 2\cos^2(\omega t + \varphi_U) + UI \cdot \sin\varphi \cdot \sin 2(\omega t + \varphi_U). \tag{A.15}$$

[2]However, this rule only applies to the consumer arrow system (VZS). In the generator arrow system, the direction of current and therefore the meaning of leading and lagging is reversed.

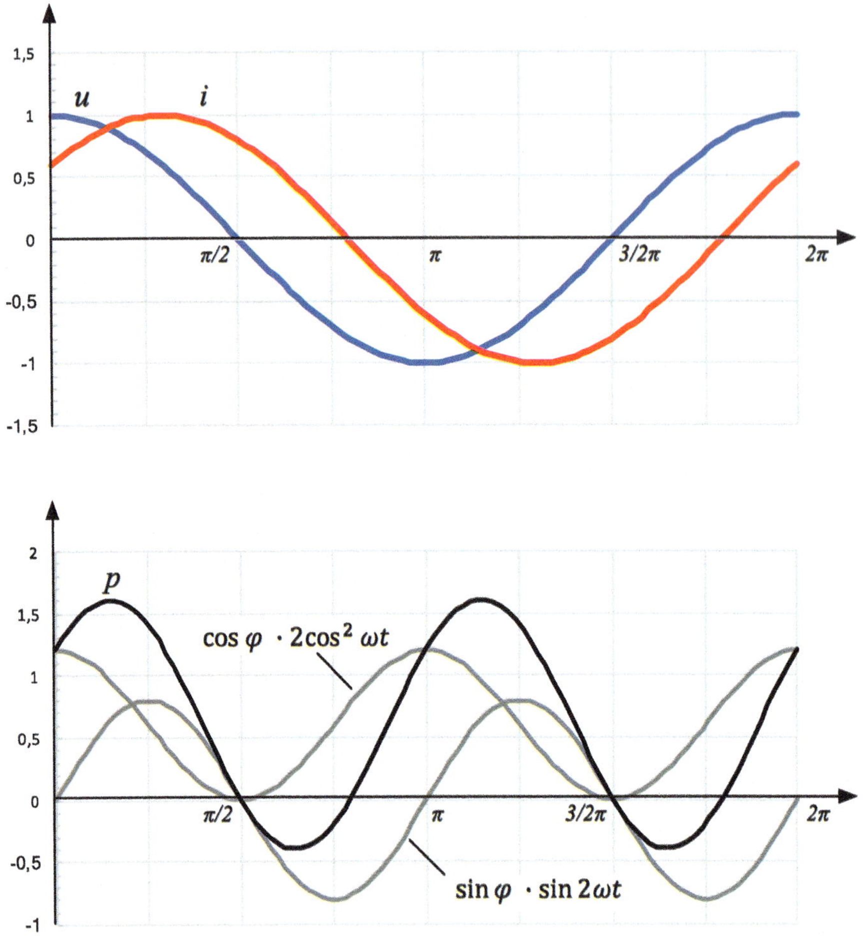

Fig. A.3 Current, voltage, and power of sinusoidal variables over time. In this example, $\varphi = 53{,}13°$, current and voltage amplitude are set to 1

The following figure illustrates the progression of the two summands and the summation function:

The first summand in (A.15) represents a sinusoidal oscillating power that has a mean value $UI \cdot \cos\varphi$. This average is called active power P and $I \cdot \cos\varphi$ is consequently called active current $\boldsymbol{I}_{\mathbf{a}}$.

The second summand also represents a sinusoidal oscillating power. However, it has an average value of zero and therefore does not contribute to the power transfer in terms of time. The part $UI \cdot \sin\varphi$ is called reactive power Q and consequently $I \cdot \sin\varphi$ is called reactive current $\boldsymbol{I}_{\mathbf{r}}$.

The sum of both parts, labeled p in the diagram, is the instantaneous power. The average value of the instantaneous power UI is called apparent power S.

Note
The concept of the different types of power in electric AC systems can be visualized with the help of a glass of beer: the liquid part corresponds to the active power P, the foam on top represents the reactive power Q. The sum of liquid and foam is the apparent power S.

Physically, the units of active, reactive, and apparent power are all Watts (W). For better differentiation, however, it has become common practice to specify the unit of reactive power Q as Volt-Ampere reactive (Var) and the unit of apparent power S as Volt-Ampere (VA). Only the unit of active power P is specified as Watts (W).

In the case of transmission lines, the active power P corresponds to the power that is transported from the generator to the consumer and can be converted into mechanical energy. The reactive power Q oscillates back and forth between the generator and consumer at twice the grid frequency. It cannot be used by the consumer to perform work.

However, the total current I and therefore the apparent power S is decisive for the thermal losses also known as heat losses of the power lines: $P_V = I^2 \cdot R$. The network companies therefore ensure that the power factor is as high as possible, i.e. that the reactive power is as low as possible.

Note
In three-phase systems, the total active power is calculated from the sum of the powers in each line, see the first part of Equation (A.15):

$$p = UI \cdot cos\varphi \cdot \Big[2cos(\omega t + \varphi_U)^2 + 2cos(\omega t + \varphi_U + 2/3\pi)^2 \\ +2cos(\omega t + \varphi_U - 2/3\pi)^2\Big]$$

The three cos-terms in the square brackets give a constant sum of the value 3 independent of the time and voltage phase angle. Therefore, in three-phase systems, the active power is three times the average active power of a single-phase system and, even more important, is it not fluctuating any more but constant the whole time.

On an ideal coil and an ideal capacitor, the phase angle is −90° and +90°, respectively. The active power is zero and the reactive power is maximum.

An ideal coil generates a magnetic field (excitation field) without the need for power and therefore without using energy. Nearly ideal coils can be built with deep-cooled superconductors whose electrical resistance is zero.

Permanent magnets also create a magnetic field without consuming power or energy. The magnetic field does not diminish over time, provided that the permanent magnet is not demagnetized by overloading, i.e. high counter fields, or by excessively high temperatures. The magnetization of a permanent magnet is carried out once by the manufacturer of the magnet or motor in a special magnetization device during the production process. The magnetic field can then be used in an electric machine to transmit power between stator and rotor.

In practical applications, coils are usually made of copper or aluminum wire. They have a certain ohmic resistance, which causes constant losses. These losses must be covered by a continuous supply of electrical energy.

The ratio of reactance ($\omega \cdot L$) and effective ohmic resistance (R) of a coil is also referred to as figure of merit (FOM) of the coil Q:

$$Q = \frac{\omega \cdot L}{R}. \tag{A.16}$$

Depending on the design, air-core coils achieve an FOM of 100–1000, coils with iron cores even far above this. The FOM of superconducting coils is infinite.

An operating point with a constant frequency f or constant angular velocity ω is often considered. In this case, inductive and capacitive reactances are used instead of inductance and capacitance. The following applies:

$$\begin{aligned} X_L &= \omega L \\ X_C &= \frac{1}{\omega C} \end{aligned} \tag{A.17}$$

The physical unit of reactances is the same as theunit of resistance, i.e. the ohm.

A.3 Complex Numbers

To understand the analytical calculation of electric machines using equivalent circuit diagrams, a basic knowledge of complex alternating current theory is required. The essential relationships are briefly summarized below.

Complex numbers are an extension of real numbers. While real numbers can be represented one-dimensionally on a number line, complex numbers consist of real and imaginary parts. They therefore lie on a two-dimensional plane, which is spanned by a real and a complex axis. In this book, complex numbers are always marked with an underscore, for example:

$$\underline{a} = \alpha + \mathrm{j}\beta \tag{A.18}$$

with the complex constant[3]:

$$\mathrm{j} := \sqrt{-1} \quad \text{or} \quad \mathrm{j} \cdot \mathrm{j} := -1 \tag{A.19}$$

[3] In mathematics, the complex constant is often denoted with the letter i. However, in electrical engineering i usually indicating an electric current and hence the letter j is used for the complex constant.

Here, α is the real part and β is the imaginary part of the complex number $\underline{a}$. Both α and β are real numbers. The complex number is only created by summing and multiplying by the complex constant *j*.

A conjugated complex number is a number whose imaginary part has a reversed sign. It is marked with a superscript star:

$$\underline{a}^* = \alpha - \mathrm{j}\beta. \tag{A.20}$$

Taking the idea of representing a number as a point in a two-dimensional plane a step further, as an alternative to specifying the real and imaginary parts, it is also possible to describe the number by an angle (phase position) and distance from the origin (amplitude), see Fig. A.4. This results in a pointer from the origin to the desired number, which represents a complex voltage or a complex current, for example.

Both representations are mathematically identical and can be converted into each other. Complex numbers can therefore also be described by using sinusoidal functions:

$$\underline{a} = |\underline{a}| \cdot (\cos\varphi + \mathrm{j} \cdot \sin\varphi). \tag{A.21}$$

The absolute value (pointer length) of a complex number is obtained from:

$$|\underline{a}| = \sqrt{\alpha^2 + \beta^2} \tag{A.22}$$

The real and imaginary parts are calculated using:

$$\alpha = \Re\{\underline{a}\} = |\underline{a}| \cdot \cos\varphi \tag{A.23}$$

$$\beta = \Im\{\underline{a}\} = |\underline{a}| \cdot \sin\varphi \tag{A.24}$$

As already mentioned, complex numbers represent points in the complex number plane. Traditionally, the real axis is drawn upwards and the imaginary axis to the left. The following figure illustrates the position of a complex number, its conjugate complex counterpart, the real and imaginary parts as well as the magnitude (pointer length) and phase angle.

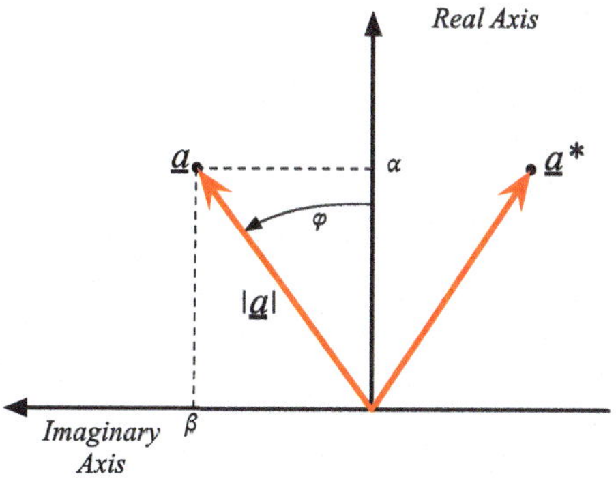

Fig. A.4 Representation of a complex numerical value $\underline{a}$ and its conjugate complex partner $\underline{a}^*$ in the complex plane

Note
Just like all other directional variables (e.g. voltages and currents), angles also have a counting arrow direction. Angles therefore have a beginning and an end, which is marked with an arrow. They are counted positively when directed counterclockwise and are counted negatively when directed clockwise.

It is technically wrong to mark angles with arrows on both ends!

The following calculation rules for complex numbers are known from mathematics with e being the Euler constant ($e = 2.71828\ldots$):

$$e^{j\varphi} = \cos\varphi + j\sin\varphi \tag{A.25}$$

$$\underline{a} = |\underline{a}| \cdot e^{j\varphi} \tag{A.26}$$

$$\underline{a} \cdot \underline{b} = |\underline{a}| \cdot |\underline{b}| \cdot \left[\cos(\varphi + \delta) + j\sin(\varphi + \delta)\right] = |\underline{a}| \cdot |\underline{b}| \cdot e^{j(\varphi+\delta)} \tag{A.27}$$

$$\frac{\underline{a}}{\underline{b}} = \frac{|\underline{a}|}{|\underline{b}|} \cdot \left[\cos(\varphi - \delta) + j\sin(\varphi - \delta)\right] = \frac{|\underline{a}|}{|\underline{b}|} \cdot e^{j(\varphi-\delta)} \tag{A.28}$$

$$\underline{a}^n = |\underline{a}|^n \cdot \left[\cos(n \cdot \varphi) + j\sin(n \cdot \varphi)\right] = |\underline{a}|^n \cdot e^{jn\varphi} \tag{A.29}$$

$$\frac{d}{dt} e^{j(\omega t+\varphi)} = j\omega \cdot e^{j(\omega t+\varphi)} \tag{A.30}$$

A.3.1 Description of Sinusoidal Quantities with Complex Numbers

With the help of complex numbers, sinusoidally changing variables (A.7) can be describes as follows:

$$u = \sqrt{2} \cdot U \cdot \cos(\omega t + \varphi_U) \tag{A.31}$$

$$\underline{u} = \sqrt{2} \cdot U \cdot e^{j(\omega t+\varphi)} = \sqrt{2} \cdot U \cdot \left[\cos(\omega t + \varphi_U) + j\sin(\omega t + \varphi_U)\right] \tag{A.32}$$

$$u = \Re\{\underline{u}\} \tag{A.33}$$

This representation applies to all sinusoidal variables, for example, currents, magnetic fluxes, and flux densities.

In AC systems, complex numbers can be visualized as pointers in the complex plane. The angle progresses with time and all pointers rotate. The actual voltage u, which can be measured at any time t, is the real part of the complex voltage $\underline{u}$. This corresponds graphically to the projection of the pointer onto the real axis.

This representation may seem complicated at first, but it has a decisive advantage: the individual alternating voltages that drop across the many elements (resistances, inductances, and capacitances) in a complicated circuit diagram are all out of phase with each other. With the help of complex numbers, the voltages are all represented by a correspondingly oriented pointer in the complex plane. To obtain the total voltage across a series connection of many such components, the individual pointers are simply graphically added (geometrically summed). The resulting pointer is the correct result in terms of both magnitude and phase angle. We will further elaborate this concept in later sections of the book.

The complex pointer of the RMS value is by definition:

$$\underline{U} = U \cdot e^{j\varphi_u} \tag{A.34}$$

This pointer $\underline{U}$ is used to draw pointer diagrams. The instantaneous value of the voltage u is obtained from $\underline{U}$ using:

$$u = \Re\{\sqrt{2} \cdot \underline{U} \cdot e^{j\omega t}\} \tag{A.35}$$

The phase angle can also be determined from the complex pointer:

$$\varphi_U = \arctan \frac{\Im\{\underline{U}\}}{\Re\{\underline{U}\}} \pm \pi \tag{A.36}$$

Note that the complex pointer $\underline{U}$ no longer contains the angular phase velocity. It therefore does not rotate.

A disadvantage of the arctangent function is the case differentiation, which is mathematically necessary. When implementing (A.36) in a programming language the `atan2(y,x)` function should be used whenever possible instead of the `atan(y/x)` function. Atan2 computes the correct arctangent value over the complete range of angles from $-\pi \ldots 0 \ldots +\pi$ without the need for a case differentiation.

The representation of sinusoidal time functions by complex numbers also elegantly solves the problem of the differential equations (A.2) and (A.4) when calculating the current at the coil and capacitor:

$$\frac{\mathrm{d}}{\mathrm{d}t} u = \Re\{\sqrt{2} \cdot j\omega \underline{U} \cdot e^{j\omega t}\} \tag{A.37}$$

$$\frac{\mathrm{d}}{\mathrm{d}t} = j\omega \underline{U} \tag{A.38}$$

The representation of currents and voltages by complex numbers therefore has the further advantage that the differential equations for determining voltage drops across capacitances and inductances now become simple algebraic equations. This enables an analytical calculation of complicated circuit diagrams with alternating voltages and currents.

It should not be forgotten that the complex representation of alternating quantities requires all quantities to be a purely sinusoidal curve. However, if the curve is complicated, for example, the audio signal of an mp3 player, then the complex representation can no longer be used. Instead, instantaneous values and the original differential equations must be used to describe the signal behavior. Ultimately, the behavior of a circuit can then only be simulated using numerical methods and computer software.

Due to the advantages of the complex representation, complicated, non-sinusoidal signal curves are often broken down into their individual frequency components, which are all sinusoidal. Each of these frequency components can then be calculated using complex variables. The result is obtained by superimposing all sinusoidal functions.

This process is done by firstly performing a Fourier analysis of the signal to obtain the sinusoidal amplitudes, frequencies, and phases. Then, the complex calculations are performed on each component independently. Finally, a Fourier synthesis is performed.

The individual frequency components are referred to as fundamental and harmonic fields, voltages, and currents and the theory required for the calculation is therefore referred to as harmonic field theory. These mathematical relationships are quite advanced and will not be discussed further in this book.

The prerequisite for this procedure is compliance with the superposition principle: the individual frequency components must be independent, i.e. they must not have any feedback effects on each other. This is only the case in linear systems. In particular, the non-linear iron saturation presented later cannot be considered in this way, or at best only as a rough approximation.

A.3.2 Equivalent Circuit Elements in Complex Representation

Equations (A.1), (A.2), and (A.4) can be rewritten and simplified by using complex pointers:

$$\underline{U} = R \cdot \underline{I} \tag{A.39}$$

$$\underline{U} = L \cdot \frac{\mathrm{d}}{\mathrm{d}t}\underline{I} = L \cdot \mathrm{j}\omega\underline{I} = \mathrm{j}X_L \cdot \underline{I} \tag{A.40}$$

$$\underline{I} = C \cdot \frac{d}{\mathrm{d}t}\underline{U} = C \cdot \mathrm{j}\omega\underline{U} = \mathrm{j}X_C \cdot \underline{U} \tag{A.41}$$

$$\underline{U} = \frac{\underline{I}}{\mathrm{j}X_C} = \frac{\mathrm{j} \cdot \underline{I}}{\mathrm{j} \cdot \mathrm{j}X_C} = -\mathrm{j}\frac{1}{X_C} \cdot \underline{I} \tag{A.42}$$

This notation enables the introduction of the complex resistance $\underline{Z}$ (impedance):

$$\underline{Z} = \frac{\underline{U}}{\underline{I}} = \frac{U}{I} \cdot e^{\mathrm{j}(\varphi_U - \varphi_j)} = Z \cdot e^{\mathrm{j}\varphi} = R + \mathrm{j}X \tag{A.43}$$

The imaginary part X of the complex resistance $\underline{Z}$ is also referred to as the reactance and the real part R as the effective resistance or ohmic resistance. It should be noted that X and R are both real values (Fig. A.5).

The imaginary factor *j*, which appears with inductances and capacitances, causes a phase shift between current and voltage. It corresponds to a rotation of

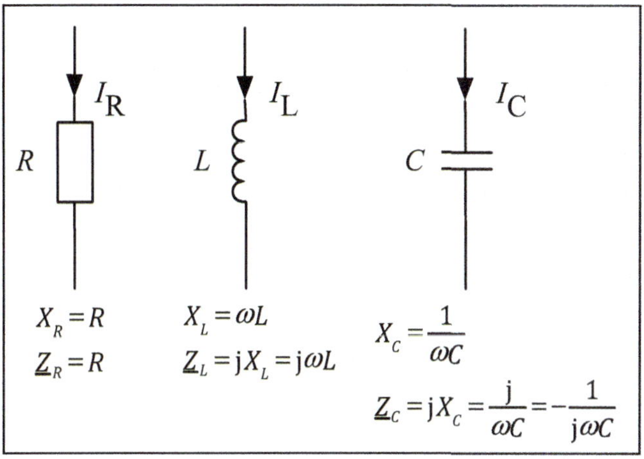

Fig. A.5 Reactive and active resistances of ohmic resistor, coil, and capacitor

the respective pointer in the complex plane by 90° in a mathematically positive sense, i.e. counterclockwise. A negative factor ($-j$) therefore rotates the pointer in a mathematically negative direction, i.e. clockwise.

Hence, the electric current I_L follows the voltage U_L by a phase shift of 90° for inductors, see (A.40): the current is multiplied by j to generate the voltage. The voltage is therefore leading by 90° and the current is lagging.[4]

The opposite is true for capacitances, see (A.42). Here, the current is multiplied by $-j$ to generate the voltage. The voltage is therefore lagging by 90° and the current is leading.

The power in an alternating current system can also be specified more easily with complex pointers. In the following figure, the voltage $\underline{U}$ has been placed randomly in the real axis as an example. P is the active power and Q is the reactive power (Fig. A.6).

The complex apparent power results from:

$$\underline{S} = \underline{U} \cdot \underline{I}^* = U \cdot I \cdot e^{j\varphi} = P + jQ \tag{A.44}$$

[4]These statements apply to the consumer arrow system. The current flowing into a machine is counted positively. Some books use the generator arrow system. In that case, the direction of the current is reversed and current coming out of an electrical machine is counted positively. This also changes the angular positions of the voltages at capacitances and inductances as well as the signs of the power. In the consumer arrow system, the absorbed electrical power (motor operation) is positive, and the output power (generator operation) is negative. In the generator arrow system, it is exactly the opposite.

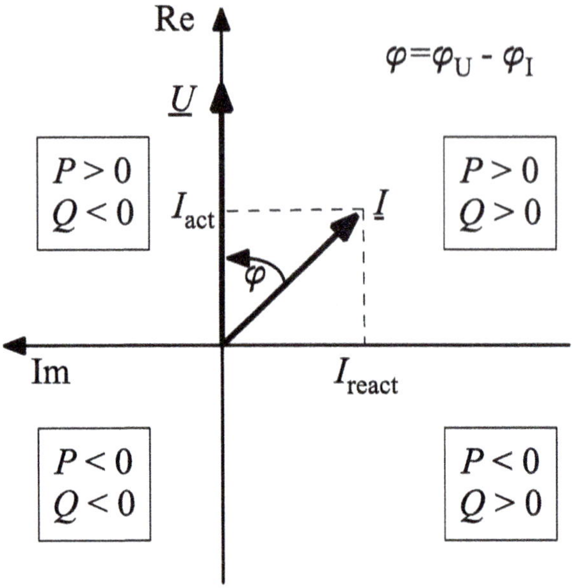

Fig. A.6 Current, voltage, and power in the complex plane

For mathematical reasons, the conjugate complex current must be used to arrive at the correct complex apparent power. The apparent power is given in volt-amperes: [S] = V · A.

The effective value of the apparent power results from:

$$S = |\underline{S}| = \sqrt{P^2 + Q^2} \tag{A.45}$$

The following applies to the effective value of the active power, specified in watts (W):

$$P = U \cdot I \cdot \cos(\varphi) \tag{A.46}$$

Accordingly, the effective value of the reactive power, specified in volt-amperes reactive (var):

$$Q = U \cdot I \cdot \sin(\varphi) \tag{A.47}$$

Purely inductive loads ($\varphi = +\pi/2$) consume reactive power ($Q > 0$), while purely capacitive loads ($\varphi = -\pi/2$) generate reactive power ($Q < 0$). Both do not influence the active power ($P = 0$).

Purely resistive loads ($\varphi = 0$), on the other hand, consume active power ($P > 0$) and do not influence the reactive power ($Q = 0$).

The part of the current that causes the active power is referred to as the active current ($I \cdot \cos\varphi$) and the part that causes the reactive power is referred to as the reactive current ($I \cdot \sin \varphi$). The geometrical sum of both is the total current I, which is also called the apparent current.

A.3.3 Three-Phase AC (Rotary Current)

Three-phase AC systems consist of three AC systems whose currents and voltages are 120° out of phase with each other. In principle, it is also possible to build multi-phase AC systems with two, four, or even more phases. Two lines are required per AC system, i.e. four lines for two phases, six lines for three phases, and so on.

A special feature of the three-phase system is that the respective return conductors can be omitted when all phases are symmetrical, i.e. when all three-phase currents have the same amplitude and hence are exactly 120° shifted in time from each other. In that case, only three lines are required to feed the motor instead of six. The remaining three lines are internally connected inside the motor.

Polyphase systems can also use two, four, and more phases. However, as already shown in the introduction, two phases have distinct disadvantages and are no longer in use today.

Phase numbers higher than three have hardly any advantages and are consequently also not used to transport electric energy.[5]

The representation of the complex pointers of current and voltage can also be applied to three-phase voltages and currents (rotary current).

The instantaneous values of the voltages in a three-phase system are phase-shifted by 120° or 2/3·π. The following applies (Fig. A.7):

$$\begin{aligned} u_1 &= \sqrt{2} \cdot U \cdot \cos(\omega t) \\ u_2 &= \sqrt{2} \cdot U \cdot \cos\left(\omega t - \frac{2}{3}\pi\right) \\ u_3 &= \sqrt{2} \cdot U \cdot \cos\left(\omega t + \frac{2}{3}\pi\right) = \sqrt{2} \cdot U \cdot \cos\left(\omega t - \frac{4}{3}\pi\right) \end{aligned} \tag{A.48}$$

This results in the complex voltage pointers:

$$\begin{aligned} \underline{U}_1 &= U \cdot e^{j0} \\ \underline{U}_2 &= U \cdot e^{-j2/3\pi} \\ \underline{U}_3 &= U \cdot e^{+j2/3\pi} \end{aligned} \tag{A.49}$$

Drawing the voltage pointers in the complex plane results in the familiar diagram:

The pointer lengths U_1, U_2, and U_3 are the effective values of the phase voltage and U_{12}, U_{23}, and U_{31} are the effective values of the line-to-line voltage.

[5]Some recent developments of power electronic circuits for cars use six-phase systems. However, these developments are based on the use of two three-phase systems in the so-called H-bridge configuration.

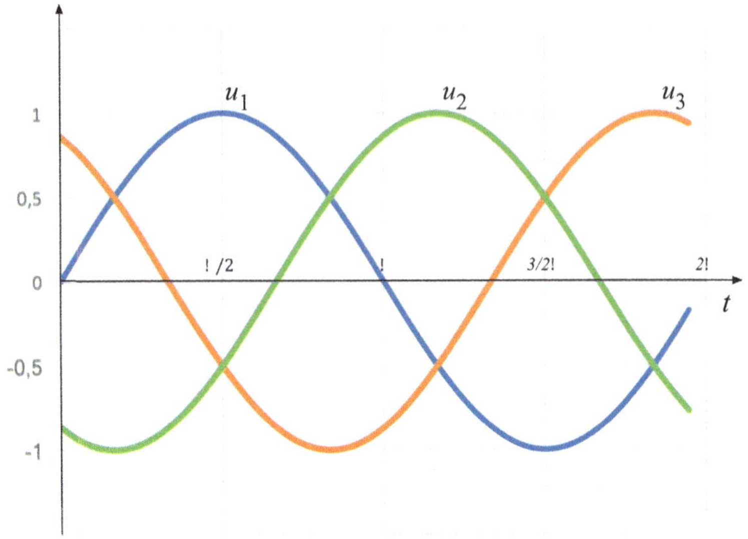

Fig. A.7 Temporal voltage curves in a three-phase system

For example, in Europe $U_1 = 230$ V and $U_{12} = 400$ V. In the U.S., $U_1 = 277$ V and $U_{12} = 480$ V.[6]

Figure A.8 shows the phase shift of 120° between the three-phase voltages ($\underline{U}_1$, $\underline{U}_2$, $\underline{U}_3$) and between the three line-to-line voltages ($\underline{U}_{12}$, $\underline{U}_{23}$, $\underline{U}_{31}$). Furthermore, there is a phase shift of 30° between the phase and line-to-line voltages, for example, between $\underline{U}_{12}$ and $\underline{U}_2$. This phase shift, however, has no practical consequences.

The RMS values of the line-to-line voltages are calculated from the RMS values of the phase voltages by

$$U_{xy} = \sqrt{3} \cdot U_x \tag{A.50}$$

The currents can be specified and calculated accordingly. Their phase angles are also shifted by 120° relative to each other. However, the phase angle between currents and voltages φ depends on the load. It is the same in all three phases both for line-to-line and for phase voltages and currents.

As explained before, the three phases in electrical equipment, i.e. in transformers and motors, are usually connected internally. In this way, instead of six external connection cables, two per phase, only three cables are required in total.

A distinction is made between star (Y) and delta (D) connections. Motors for traction applications are often star connected because the delta connection has

[6]In Europe, the nominal voltage of the public low voltage networks and the rated voltage of industrial electric motors are both 400 V. In the U.S., the nominal voltage of the public low-voltage network is 480 V but the rated voltage of low-voltage electric motors is just 460 V.

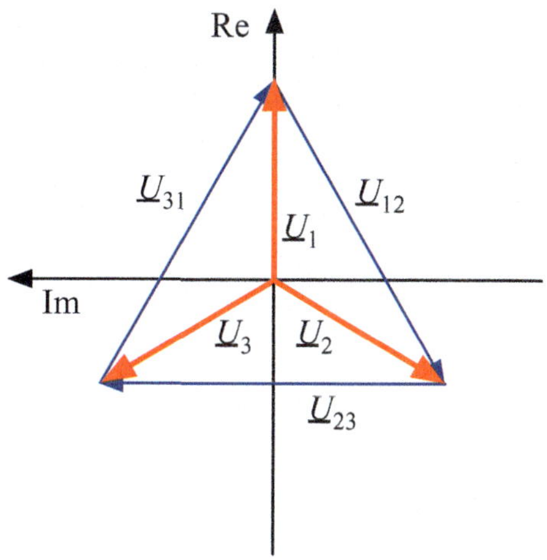

Fig. A.8 Three-phase voltage system in the complex plane

some disadvantages: circular currents can occur and lead to losses and noise. Larger motors and transformers are often delta connected but special care must be taken in the machine design to avoid excessive circular currents.

In large transformers, the three phases are divided into partial coils, which makes other circuit variants possible, for example, the zig-zag circuit.

As an example, Fig. A.9 shows the electrical equivalent circuit diagram of a permanent magnet synchronous machine, which is used in many electric cars. Each phase is represented by a voltage source, the so-called synchronous voltage, an ohmic resistor, and an inductance (coil). The ends of the three phases are internally connected to form a star point so that only three cables need to be routed to the outside.

The sum of the instantaneous values of the three currents is zero at any time:

$$i_1 + i_2 + i_3 = 0 \tag{A.51}$$

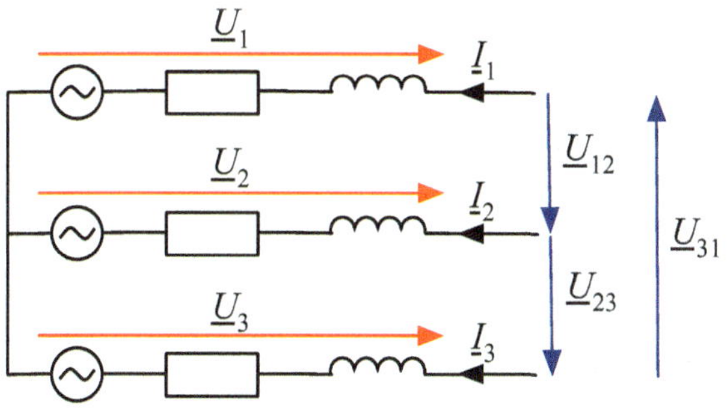

Fig. A.9 Simple electrical equivalent circuit diagram of a three-phase synchronous machine in Y-connection

The effective value of the three-phase apparent power of the machine is determined from the RMS values of the phase voltages and the RMS currents or the line-to-line RMS voltages and RMS currents:

$$S = I_1 \cdot U_1 + I_2 \cdot U_2 + I_3 \cdot U_3 = 3 \cdot I_x \cdot U_x = \sqrt{3} \cdot I_{xy} \cdot U_{xy}. \qquad (A.52)$$

In the usual case of symmetry, the RMS voltages and currents of all three phases are the same. It is therefore sufficient to calculate just one phase and multiply the result by three to determine the power of the entire system.

However, in the event of a fault, for example, a short circuit in an individual winding, the three phases will behave differently. In this case, the power of each phase must be calculated separately.

The RMS values of active and reactive power are calculated from the line-to-line values:

$$P = \sqrt{3} \cdot I_{xy} \cdot U_{xy} \cdot \cos\varphi \qquad (A.53)$$

$$Q = \sqrt{3} \cdot I_{xy} \cdot U_{xy} \cdot \sin\varphi \qquad (A.54)$$

The nameplate and the catalog data of rotating electric machines and transformers usually indicates the line-to-line values, independent from the internal star or delta connection. Formulae (A.53) and (A.54) can be applied directly.

A.4 Basics of Magnetic Circuits

The Scottish physicist James Clerk Maxwell collected and organized the existing knowledge about electromagnetism in the years 1861–1864. He recognized that light waves are only a special form of the more general electromagnetic waves. Maxwell developed a comprehensive physical theory to describe all electromagnetic phenomena.

Maxwell's original theory contained some two dozen equations. They were reduced to just four fundamental field quantities and four equations by Oliver Heaviside about 20 years later. These became known as the Maxwell–Heaviside equations and are still widely used today.

The field quantities are

- the electric field strength E (electric field strength), given in volts per meter (V/m);
- the electric displacement density D, given in ampere-seconds per square meter (As/m^2);[7]

[7]When applying the equations to electric machines, the frequencies are relatively low (quasi-stationary) and hence the rate of change of the electric displacement density D is small. dD/dt is therefore always assumed to be zero in the remainder of this book.

- the magnetic field strength H, given in amperes per meter (A/m); and
- the magnetic flux density B, given in volt-seconds per square meter (Vs/m^2) or Tesla (T) for short.

The electric field strength E multiplied by the length l corresponds to the voltage induced by a changing magnetic field in an electric conductor: $U_{ind} = E \cdot l$.

The relationship between magnetic field strength H and magnetic flux density B is fundamental for understanding electric machines. These terms can be visualized as follows:

The **magnetic field strength *H*** represents the **cause** of the magnetic flux. The magnetic field either arises from an electric current that is sent through a coil or is created by permanent magnets.

The **magnetic flux density *B*** represents the **effect** of the resulting magnetic flux. The effect is mostly the mechanical force, which then results in the torque and thus ultimately in the mechanical power produced by an electric motor. In case of generator operation, the rate of change of the flux results in the induced voltage and thus ultimately in the electric power that is fed back into the battery.

Quite often, the magnetic flux ϕ, given in volt-seconds (Vs) or Weber (Wb), is used instead of flux density B. The relationship between the two is simply:

$$B = \frac{\phi}{S} \tag{A.55}$$

The flux density B therefore results from the magnetic flux ϕ that passes through a certain cross-sectional area S.

The relationship between field strength H and flux density B is determined by the properties of the ferromagnetic material. In the simplest case of a linear material, the relationship between field strength H and flux density B is given by

$$B = \mu_0 \cdot \mu_r \cdot H \tag{A.56}$$

Here $\mu_0 = 4\pi \cdot 10^{-7}$ H/m is a physical constant, the vacuum permeability (magnetic conductivity) or simply magnetic constant. The unit H (Henry) is the already known unit of an inductance.

Similarly, the electric displacement density D and the electric field E are linked by a constant ε_0, which is called the vacuum permittivity (dielectric conductivity) or simply electric constant. Its unit is based on the unit of capacitance F (Farad): $\varepsilon_0 = 8.854 \cdot 10^{-12}$ F/m.

Magnetic and electric constants are connected by the speed of light c: $1/c^2 = \mu_0 \cdot \varepsilon_0$.

μ_r is the relative permeability, which is determined by the properties of the magnetic material. In air, μ_r is equal to 1 by definition. In ferromagnetic material, μ_r can go up to 100,000 and even more.

As already noted, formula (A.56) requires a linear behavior of the magnetic circuit. However, this is generally not the case. All ferromagnetic materials, which are mostly based on iron, become saturated under the influence of a strong magnetic field strength H. In that case, the relationship between B and H is getting non-linear and the whole magnetizing curve must be considered instead of just applying the formula.

In typical electric sheet metal for transformers and electric motors, μ_r fluctuates between around 10,000 at low field strengths and 1 at high field strengths. At typical operating points for electric machines, μ_r is in the order of 10 – 100.

Permanent magnets have a μ_r of 1.03–1.05. Their magnetic conductivity is quite surprisingly very similar to that of vacuum (air). However, the magnetization curve is very non-linear and Equation (A.56) again cannot be applied. Instead, the graphical characteristic as given in the data sheets of the manufacturer of the materials must be used.

Note

The fundamental operation of magnetic circuits can be compared with that of electric circuits. Formula (A.56) represents the ohmic law of magnetism and the expression $\mu_0\mu_r$ represents the magnetic conductivity.

The magnetic field strength H corresponds to the electric voltage U in an electric circuit. Depending on the voltage U and the ohmic resistance $R = 1/G$ (with G = conductivity) an electric current I is generated: $I = U \cdot G = U/R$. The electric current is responsible for the effects of the circuit (heat, light, etc.). This corresponds to the magnetic flux density B.

A greater μ_r indicates a lower magnetic resistance. The lower the resistance, the less field strength H (that is either current in the coils or mass of permanent magnets) is required to generate a certain amount of magnetic flux density B. As B is proportional to the generated mechanical force, the motor will become smaller and more efficient.

This is the reason why almost all electric machines use electrical sheets made of iron as their core material. This ferromagnetic steel significantly improves the magnetic conductivity, i.e. reduces the magnetic resistance, and thus makes it possible to achieve the desired force (torque, power) with much less electric current and losses.

If non-ferromagnetic materials, such as wood, aluminum, or plastic, were used instead of electrical sheets, the magnetic resistance would be much higher and much more electric current would be required to generate the same amount of force.

A.4.1 Amperes Law

Ampere's law is the fourth of Maxwell's equations. It can be expressed in integral or differential form:

$$\oint_C \vec{H} \cdot \vec{dl} = \iint_S \vec{J} \cdot d\vec{S} \tag{A.57}$$

$$\text{rot}\vec{H} = \vec{\nabla} \times \vec{H} = \vec{J} \tag{A.58}$$

where J is the current density, which results from the current I in A and the conductor cross section S in m^2 to $J = I/S$.

Ampere's law enables the calculation of the magnetic field strength (left-hand side of the equation) from the current flow in a coil (right-hand side of the equation).

Don't be put off by the seemingly complicated equations, because in practical application they are quite easy to handle, as will be shown below.

We first consider the magnetic field around a current-carrying conductor. The conductor is located far away from any ferromagnetic material.

The left-hand side of Equation (A.57) states that we must select a closed-circuit C around the conductor. The H-field is integrated along this circle.

In principle, the orbital path can be defined arbitrarily, the equation is generally valid. However, to be able to solve the integral practically, we must choose a path that exactly follows a field line. In this case, the integration is simplified to a summation.

For the problem in Fig. A.10, we select the circular path shown in blue. Neither the material (air) nor the flux density changes along this path. A constant, location-independent H is therefore to be expected and the integration, that is, the left side of the equation, provides the result: $H \cdot 2\pi r$.

The right-hand side of Equation (A.57) states that we must add up the total current included in the circulation. In the example, this is simply the conductor current. This results in the known solution:

$$H \cdot 2\pi r = I \Rightarrow H = \frac{I}{2\pi r} \tag{A.59}$$

The practical application of these rules is as follows: look for a closed path of the magnetic flux. Now follow this path and look at the ferromagnetic material and the narrowness or width of the flux tubes. As long as these two properties do not change, the H-field also remains constant. Otherwise, a new value for the H-field must be assumed.

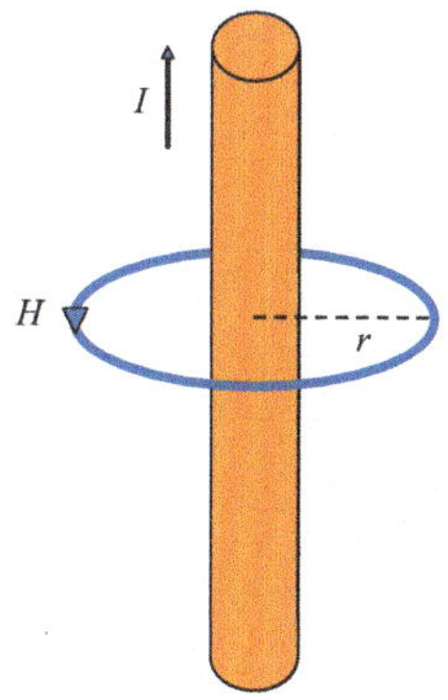

Fig. A.10 Magnetic field of a current-carrying conductor

We will now look at a more practical example, consisting of an iron core of several layered individual sheets, an air gap and a coil with six windings.

As we follow a closed path along a field line during integration, we can greatly simplify the integral (A.57):

$$\sum_i H_i \cdot l_i = w \cdot I = \Theta \tag{A.60}$$

The right-hand side of the equation, i.e. the total included current, is also known as the current linkageΘ. In this example, a coil current of 10 A and 6 windings results in a current linkage of 60 A. The current linkage is responsible for generating the magnetic field strength.

Note
Older terms for the current linkage are magnetomotive force (MMF) or ampere turns.

The left-hand part of the equation is the sum of the values of the H-field along the field line under consideration. The summands $H \cdot l$ are also called magnetic voltages V.

As mentioned, a change in H occurs when the field line passes from one material, e.g. iron, into another material, e.g. air. H also changes when the cross section changes, i.e. when the field lines "diverge", because the flux density B then changes. This is the case, for example, when the field lines pass from a narrow iron area, e.g. a tooth, into a wide iron area, e.g. the yoke. In these cases, a new, initially unknown value H_i is resulting.

In the example in Fig. A.11, a change in cross section does not occur because the iron yoke has the same cross section everywhere. However, there is a change in material when the flux flows through the air gap.

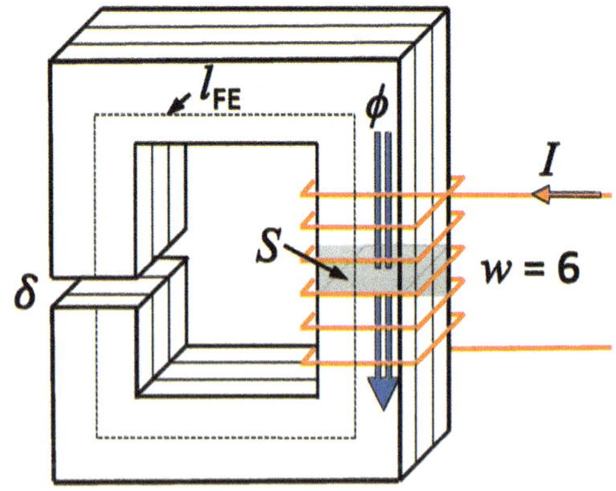

Fig. A.11 Simple magnetic circuit to explain Ampere's law

We must therefore consider two different values of the magnetic field strength: one in the iron core H_{Fe} and another one in the air gap H_{δ}.

The length l denotes the distance that the flux must travel in each case. In the air gap, l is equal to the air gap height δ. In the iron core, l is equal to the average length of the iron core l_{Fe}, for example, 20 cm.

We can therefore adapt Equation (A.60) to the example as follows:

$$H_{\delta} \cdot \delta + H_{Fe} \cdot l_{Fe} = w \cdot I \tag{A.61}$$

Note
In the analytical calculation of electric machines, it is common practice to divide the field line path into five or more sections, e.g. air gap, stator tooth, stator yoke, rotor tooth, and rotor yoke. The magnetic voltages are then assumed to be constant in each section and calculated individually.

In general, several unknown H-fields must be considered. Even this simple example has two different magnetic field strengths—one in iron and one in air. To solve for all unknown magnetic fields H_i, additional equations are therefore required.

A.4.2 Gaussian Law

The Gauss law for magnetic fields is the second Maxwell equation. It describes the solenoidality (absence of sources and sinks) of the magnetic flux density and reads in differential form:

$$\operatorname{div} \vec{B} = \vec{\nabla} \cdot \vec{B} = 0 \tag{A.62}$$

The mathematical background of the divergence function or the Nabla operator is too complicated to be detailed in this book. However, the illustrative explanation of the equation is again quite simple: the magnetic flux density B and therefore also the flux ϕ are void of sources, i.e. they are looping without a beginning or an end. Each magnetic field line runs in a closed loop and all magnetic poles always occur as pairs of north and south poles. There are no individual magnetic poles in nature.

We can now apply this knowledge to solve the example problem.

In Fig. A.11, the magnetic flux that passes through the surface S_{Fe} in the iron core, $\phi_{Fe} = B_{Fe} \cdot S_{FE}$, must be equal to the magnetic flux that passes through the air gap, $\phi_{\delta} = B_{\delta} \cdot S_{\delta}$, as no field lines can be "lost".

In the example, equal cross-sectional areas are assumed, $S_{FE} = S_{\delta}$. In that case, the flux densities must be identical: $B_{Fe} = B_{\delta}$. In electric machines, the air gap area is usually much larger than the iron cross section. Then the following generally applies:

$$B_{Fe} = \frac{S_{\delta}}{S_{Fe}} B_{\delta} \tag{A.63}$$

Equations (A.56), (A.61), and (A.63) can now be used to determine the magnetic field strengths H_{Fe} and H_{δ} and, in a further step, to calculate the flux densities.

Note
Ultimately, the calculation of the magnetic fields is like the calculation of the electric current in an electric circuit with several resistors connected in series and a voltage source.

The magnetic voltages V correspond to the electric voltage drops across the resistors. The total magnetic voltage corresponds to the applied source voltage. The magnetic flux ϕ corresponds to the electric current. The magnetic resistances are determined from:

$$R_m = \frac{V}{\phi} = \frac{H \cdot l}{B \cdot S} = \frac{l}{\mu_0 \cdot \mu_r \cdot S} \tag{A.64}$$

The current, i.e. the magnetic flux, can then be calculated from Ohm's law and the source voltage. The respective voltage drops, i.e. the individual H-fields along the path, result from the current and the individual resistances.

In this analogy, Gauss's law and Ampere's law correspond to Kirchhoff's first and second laws for electric circuits, i.e. the laws for junctions and for meshes.

A.4.3 Faraday's Law

Faraday's law, also known as law of induction, is the third and last of Maxwell's equations used in electromagnetic circuits. It establishes the connection between magnetic fields and electric voltages.

The law of induction is the inverse of Ampere's law: Ampere's law describes how an electric current generates a magnetic field (electric motor principle). The law of induction describes how a magnetic field generates an electric voltage (electric generator principle).

Faraday's law in integral and differential form is as follows:

$$\oint_C \vec{E} \cdot \vec{dl} = -\frac{d}{dt}\phi \tag{A.65}$$

$$\mathrm{rot}\,\vec{E} = \vec{\nabla} \times \vec{E} = -\frac{d}{dt}\vec{B} \tag{A.66}$$

The first equation can again be explained easily: the electric field strength E along a closed path represents the induced voltage U_{ind} at the terminals of a coil.

This voltage results from the time derivative of the magnetic flux. Put simply: if a coil is placed in a magnetic field that changes over time, a voltage is generated at its terminals.

The following illustration explains these relationships in the simplest case of a two-coil system placed in air (Fig. A.12).

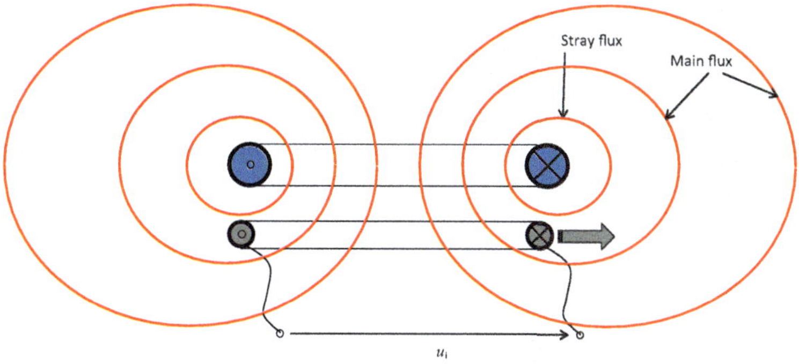

Fig. A.12 Voltage induction in the gray coil (bottom) by a magnetic field generated in the blue coil (top)

The upper coil (blue) is stationary and placed perpendicular to the plane of the book. In the left conductor (blue circle), the wires come towards the observer; in the right conductor, the wires run into the plane of the book. A sinusoidal alternating current i flows through the coil. As long as the current flows in the direction of the conductors, the magnetic field points upwards according to the right-hand rule.

The pulsating, alternating magnetic field of the blue coil produces the field curves shown in red.

Now we move a second coil (gray) that is wound in the same direction from left to right through this magnetic field. The ends of the coil are open and we measure the induced voltage:

$$u_i = -w \cdot \frac{\mathrm{d}}{\mathrm{d}t}\phi \tag{A.67}$$

where w is the number of turns of the gray (lower) coil in which the induced voltage arises and ϕ denotes the total magnetic flux received by the gray coil that was created by the upper (blue) coil.

It is important to only use that part of the flux of the blue coil in the formula that passes through the gray coil (main flux or linked flux). The inner flux lines do not reach the gray coil and therefore do not induce any voltage. This part is called leakage flux or stray flux.

Stray flux occurs in all electric machines and transformers in the order of around 2–10% of the main flux. It is useful in certain situations, in particular, in the event of a short circuit, where it limits the maximum current. However, only the main flux creates an induced voltage or transmits energy to the rotor. This is why electric motors are designed so that most of the flux is exchanged between the stator and rotor.

In general, coils will have multiple turns. In electric motors, 5–100 turns per coil are typical. The induced voltage arises equally in all turns and the turns are connected in series. Therefore, the sum of the individual voltages will be present at the terminals. Equation (A.67) takes this into account by means of the factor w.

The main flux ϕ multiplied by the number of turns w is also referred to as linked flux (or flux linkage) and the formula letter ψ is introduced for this purpose: $\psi = w \cdot \phi$.

But also the flux-producing coil, in our example, the blue coil, will generally consist of several turns, each of which produces a certain amount of magnetic flux. It is important to take both effects into consideration when calculating the induced voltage.

The change in the concatenated flux in the gray receiver coil, which is decisive for the induced voltage, can be attributed to two different physical effects:

- Variation of the current i in the energizing coil (blue): transformer voltage
- Movement of the receiving coil (gray) through the magnetic field: rotational voltage

Equation (A.67) can be rewritten to show these two causes separately:

$$u_i = -\frac{\partial \psi}{\partial i} \cdot \frac{\mathrm{d}i}{\mathrm{d}t} - \frac{\partial \psi}{\partial x} \cdot \frac{\mathrm{d}x}{\mathrm{d}t} \tag{A.68}$$

The first term represents the change in flux with current and the change in current with time ($\mathrm{d}i/\mathrm{d}t$).

The second term is the change in flux with the position and the change in position of the coil with time ($\mathrm{d}x/\mathrm{d}t$), i.e. the velocity v of the coil.

In electric transformers only the first effect is used, in rotating machines normally only the second effect. Switched reluctance machines are a special case as they also have a significant proportion of rotational induced voltage.

When all processes are sinusoidal and steady state, Equation (A.68) can be considerably simplified by using complex variables based on (A.38):

$$\underline{u}_i = -\frac{\mathrm{d}}{\mathrm{d}t}\underline{\psi} = -\mathrm{j}\omega\underline{\psi} \tag{A.69}$$

The induced voltage is proportional to the amplitude of the flux linkage and to the angular frequency. In electric machines, we can increase the induced voltage by using stronger magnets and by increasing the frequency of the supply voltage.

The equation can be rearranged to show an important relation regarding the magnetic flux that is present inside an electric machine:

$$\psi \cong \frac{U}{f}. \tag{A.70}$$

The magnetic flux remains constant when supply voltage and frequency are changed in proportion to each other (constant flux range). When the frequency is further increased while the voltage is kept constant, the flux will be reduced accordingly (field weakening range).

A.4.4 Mutual Inductances

At the beginning of the chapter, the inductance L of a coil was already used without detailed explanation.

As we have seen, the linked magnetic flux of a coil 1 $\psi_1 = w_1 \cdot \phi_1$ is generally caused by a current i_2 in a coil 2.

The ratio $L_{21} = \psi_1 / i_2$ is referred to as the mutual inductance between the two coils.

The order of the indices (21) indicates that the magnetic flux from coil 2 is considered in coil 1 ("flux from 2 to 1").

Note

The mutual inductances between two coils are the same in both directions, i.e. $L_{12} = L_{21}$.

When calculating the magnetic flux in the second coil caused by the current in the first coil ($L_{12} = \psi_2 / i_1$), the result is the same as if calculating the magnetic flux in the first coil caused by the current in the second coil ($L_{21} = \psi_1 / i_2$).

A.4.5 Self-inductance

If a pulsating magnetic field is built up in a coil by a time-variable current, then this magnetic field naturally also passes through the coil itself and consequently generates a self-induction voltage. The definition of self-inductance is the same as for mutual inductance: $L_{11} = L_1 = \psi_1 / i_1$.

An essential prerequisite for the calculation with inductances is a constant ratio of flux and current. This means that the inductance must not change with the current.

In real machines, however, the phenomenon of iron saturation results in a nonlinearity: at high currents, the magnetic flux no longer increases proportionally because the relative permeability of the electrical steel decreases. Consequently, the inductance becomes increasingly smaller at large currents and is no longer a constant.

Iron saturation can be considered approximately by determining the inductance for a specific operating point and a corresponding current. Only small variations in the current are then allowed (linearization around the operating point) and a constant inductance can be calculated.

In practice, for the precise analysis of electric machines in controllers of power electronic circuits, the use of inductances has been completely abandoned nowadays. All calculations are carried out directly with currents and fluxes. For this purpose, the corresponding characteristics $\psi = f(i)$ are determined in advance using numerical field calculation software (finite element method) and/or measurements of the electric machine. We will give examples later in the book.

A.4.6 Simplified Law of Induction

The partial derivative of the flux linkage with respect to the current, as contained in the first summand in Equation (A.68), is in the simplest case nothing other than the inductance:

$$\frac{\partial \psi}{\partial i} = const = \frac{\psi}{i} = L \tag{A.71}$$

The second summand of Equation (A.68) describes a partial derivative of the flux linkage according to the position of the coils. This can be calculated in a simplified way by assuming a constant flux density B, which passes through the coil area $S = l \cdot x$. If the coil moves in the direction x, the change in flux results in

$$\frac{\partial \psi}{\partial x} = \frac{B \cdot S}{x} = B \cdot l \tag{A.72}$$

Equation (A.68) can therefore be simplified further:

$$u_i = -L \cdot \frac{\mathrm{d}i}{\mathrm{d}t} - B \cdot l \cdot v \tag{A.73}$$

This representation directly shows the two components of the induced voltage caused by the change in current i over time and by the movement of the coils at the velocity v.

Closing Words

Due to the availability of nearly unlimited amounts of crude oil, gasoline, diesel and kerosene have been unbeatable in road, maritime, and air transport. Still, the technical effort and reliability of internal combustion engines, complex automatic transmissions, and exhaust aftertreatments leave a lot to be desired. Now, with electric propulsion (lithium-ion batteries and fuel cells) and synthetic fuels, technically viable alternatives are emerging. They all rely on renewable energy sources rather than fossil fuels and are therefore much more environmentally friendly, but they require a certain effort to be initially manufactured.

In the end, the energy form that will prevail is the one that can effectively solve a given propulsion task while demonstrating the best efficiency, meaning the lowest costs.

For passenger cars and light-to-medium-duty commercial vehicles, battery electric propulsion is undoubtedly the most energy-efficient and reliable option. Synthetic fuels, much like hydrogen, have a poor energy balance and therefore must be produced at significantly higher costs. Fuel cells suffer from low durability. Electric drives and batteries have become simpler in design, making them more reliable and durable than combustion engines required for synthetic fuels. This is visible not only in performance but also in service and maintenance costs.

Additional advantages of electric cars include reduced noise, non-existent local emissions, and decreased particulate matter pollution. Furthermore, they are comfortable and enjoyable to drive – an essential aspect for the acceptance of the technology across diverse population segments. Recharging electric cars during long vacations or business trips is typically required after several hours of driving and takes around 20–30 min at fast charging stations. This is practical and reasonable. Simply put, when buying a new car these days, the choice is between stinking, loud and bad for the climate (combustion engine) or clean, quiet and ultimately just half as bad for the climate (BEV). Ideally, one would use a bicycle, bus or train, but that is simply not possible in many life situations. Other energy sources (synthetic fuels, hydrogen) will not play a significant role in the automotive mass market even in the long term for a variety of reasons.

M. Doppelbauer, *Introduction to Electromobility*,
https://doi.org/10.1007/978-3-658-45482-1

In recent years, the production of efficient, renewable electric energy has significantly increased in all industrialized nations around the world. Particularly in China, but also in Europe, the share of photovoltaic and wind power is rapidly expanding. Already today, the electricity produced from renewable sources in Europe would be more than sufficient to power an entire car fleet based solely on battery electric vehicles.

The production of renewable electric energy is decentralized. It can be easily generated and stored by individuals using PV systems and home storage, positively impacting the environmental footprint and operating costs of electric cars. When importing large quantities of hydrogen or synthetic fuels from equatorial regions, there is a significant risk that the western countries will once again become dependent on states that disregard democratic and liberal values.

Modern battery technology has evolved rapidly since the development of lithium-ion batteries in the 1990s. Energy density and environmental performance have improved, accompanied by significant price reductions. The introduction of solid-state batteries is expected to provide a new boost to development, ensuring progress for many more years. Utilizing secondary raw materials from recycling will enable a resource-efficient and largely sustainable production of battery electric vehicles. The manufacturing costs of battery electric vehicles continue to decrease with increasing mass production and automation. It is foreseeable that electric cars with ranges of around 500 km can be produced at similar costs to today's internal combustion engine cars starting from the next decade. In China, the prices of battery-electric cars are already at and in some cases below the level of comparable combustion engine cars.

The battery integrates seamlessly with the car's design, forming a perfect symbiosis. The shallow skateboard design takes up minimal space. Additionally, the low center of gravity enhances handling and safety against rollovers in the event of an accident. The fire hazard of electric cars is about an order of magnitude lower compared to vehicles with gasoline or diesel tanks, and there are no spontaneous explosions after severe accidents. The electric drivetrain is much smaller than a conventional powertrain. Therefore, the engine compartment can be shortened, adopting a Cab-Forward Design, while simultaneously increasing the available interior and trunk space of electric cars with the same external dimensions.

Hydrogen-powered cars have, due to inherent system disadvantages, failed to gain a significant market share in their now 60-year history and have been abandoned by almost all manufacturers. They suffer from high production and service costs as well as notable drawbacks in usable vehicle interior space, owing to the complex and voluminous technology. The investment required for hydrogen infrastructure is considerably higher than for electric vehicle charging infrastructure. The efficiency chain from the electricity used to produce hydrogen to the electricity that powers the vehicle's motor remains very poor. Significant progress in this field has not been evident for a long time, and the technical possibilities seem to be exhausted. To fundamentally change this situation, a technological leap, like the impact of the lithium-ion battery on battery electric cars, would be necessary. However, such a technological leap is not foreseeable in the case of hydrogen. The weight advantage that hydrogen cars had years ago has been equalized with the latest battery generations. The weight of battery electric vehicles only plays a subordinate role in fuel consumption anyway due to regenerative braking. BEVs are now well ahead in terms of driving range. The refueling time of hydrogen cars is noticeably shorter than fast charging batteries, but that is only true for the first vehicle in the queue at the petrol pump. All subsequent cars must wait for the hydrogen to be preconditioned in the refueling station, which increases the refueling time to 20–30 minutes.

Battery-electric drivetrains are also well suited for light- and some medium-duty commercial vehicles, such as city buses, delivery vehicles, or municipal service vehicles, due to their frequent start–stop cycles, braking energy recovery, and relatively short daily driving distances.

The question of the optimal energy form remains unresolved in sectors of transportation where batteries fail to meet technical requirements. This is notably the case in heavy-duty transport, agricultural vehicles, medium- and long-haul aircraft, as well as large ships. Several studies now clearly demonstrate that hydrogen, compressed at 700 bar or even in liquid form, has significantly lower energy density than synthetic liquid fuels,[8] which can be produced regeneratively and largely CO_2 neutral. The author is convinced that hydrogen will not play a significant role in the future transportation sector at all. In all modes of transportation where battery electric is not feasible, a shift towards regenerative liquid fuels is anticipated in the long term.

Due to the unfavorable chemical properties of hydrogen, future applications for green hydrogen are primarily sought in stationary industrial sectors with high energy demands. These include, for example, the chemical industry, steel and cement manufacturing, as well as grid stabilization. Mass applications in the private sector, which typically come with low utilization rates, are not expected from today's perspective. Notably, the hazard potential of hydrogen is much more manageable in a professional industrial environment.

The main advantages of synthetic regenerative fuels are high energy density and easy distribution by an existing network of gas stations. However, the production of e-fuels is associated with an enormous consumption of energy and resources. E-fuels that are produced in distant countries should preferably be used locally for reasons of transport efficiency. In general, e-fuels should only be used where there are no more environmentally friendly alternatives. Ultimately, e-fuels are by no means climate-neutral due to the extensive chemical plant technology required for their production. Combustion engine cars powered by e-fuels are significantly less environmentally friendly than BEVs. It is unrealistic to expect a large-scale transition of many millions of vehicles to synthetic fuels. Moreover, vehicles running on synthetic fuels contribute to urban pollution just like conven-

[8] In this context, energy density includes both the fuel and the required compartment (tank). Liquid synthetic and natural fuels have approximately twice the energy density of liquid hydrogen including tanks. Gaseous hydrogen is even worse.

tional vehicles, as they are only carbon neutral in the overall balance. This has also been recognized in China, where the largest fleet of battery electric cars is now operated, especially in the city centers of major metropolitan areas.

The initial costs for transitioning the transportation sector to electric energy from renewable sources are high, and economically operating companies must focus on short- and medium-term profits. Therefore, the entire process must be accompanied and driven by policymakers. The construction of new wind turbines and power lines is urgently needed but may not be popular among the population. Public charging points are required in residential areas and the fast-charging network along highways must be expanded. The installation of wallboxes at every private parking space must become mandatory for new buildings.

Ultimately, it is crucial for policymakers to successfully support the urgently needed transition to electric cars with a thoughtful mix of incentives and regulations. Time is running out.

Glossary

Acceleration reserve The difference between the torque (current operating point) and the maximum torque at the current speed.

Air gap Distance between the inside of the stator cylinder and the outside of the rotor cylinder. Typically 0.1 … 0.8 mm depending on the machine type.

Angular displacement The angle from terminal voltage to synchronous generated voltage. Illustrative: The angle from the rotor pole to the stator field.

Anthropogenic Man-made, i.e. not natural

Balancing Charge equalization between the individual cells of a battery storage system.

Base speed range Speed range of an electric motor in which the strength of the magnetic field is kept constant. The maximum available torque is therefore also approximately constant and the power increases proportionally with the speed, usually below 40 … 60 km/h.

Bidirectional charging The vehicle's charging interface can feed energy from the vehicle battery back into the grid as required. This allows the grid to be supported at times of high demand. This function is currently not included in the CCS standard. It is only available with CHAdeMO plugs.

Boost converter Sub-function of a DC/DC converter in which the DC voltage is increased and the DC current is reduced.

Boosting The electric motor works in addition to the combustion engine. The total system output is increased.

Breakdown torque The highest torque that can be generated magnetically by the motor when a certain amount of current and voltage is available.

Buck converter Sub-function of a DC/DC converter in which the DC voltage is reduced and the DC current is increased.

M. Doppelbauer, *Introduction to Electromobility*,
https://doi.org/10.1007/978-3-658-45482-1

Calorific value The maximum amount of heat (energy content) that can be used during combustion without condensation of water vapor in the exhaust gas.

Charging point (CP) Possibility to charge an electric car (charging station, wallbox).

Charging station A charging station usually offers several charging points.

Compound-split Hybrid drive with several power split branches (two-mode).

Concentrated winding Winding in which the coil width is exactly one slot pitch (see single-tooth winding).

Constant power range See field weakening range

Constant torque range See base speed range

Continuous velocity transmission Transmission whose ratio can be changed continuously rather than in fixed gear steps.

Converter Power electronic module that converts energy between a DC and an AC voltage level—three-phase bridge circuit (DC/AC converter).

DC-machine An electric machine in which the magnetic fields of the stator and the rotor are spatially stationary.

DC/AC-converter Circuit for converting direct voltage and current into alternating voltage and current of adjustable amplitude and frequency.

DC/DC-converter Circuit for converting direct voltage and current to a higher or lower voltage and current.

Direct drive The electric motor acts directly on the wheels without a gearbox. It must therefore generate the torque required at the wheel entirely from electromagnetic force.

Distributed winding Winding in which the coil width is more than one slot pitch.

Emission factor GHG emissions from electricity generation in relation to the electricity supplied to the end consumer (socket) in g_{CO2}/kWhel.

Energy carrier Forms of energy that can be used to power a vehicle.

Feed-in management Reduction of the power output (curtailment) of renewable electricity generators when the grid capacity is overloaded.

Field weakening Speed range of an electric motor in which the strength of the magnetic field and thus the maximum available torque decrease in proportion to the speed. The power output is almost constant, usually above 40 … 60 km/h.

Flux oriented control (vector control) Control of an induction machine in a rotating coordinate system oriented to the magnetic flux of the rotor.

Free-wheeling diode Reverse diode in parallel with a semiconductor switch.

Full-hybrid Hybrid drive with similar functions as a mild hybrid, but with the addition of more powerful braking energy recovery and electric driving over shorter distances.

Gray hydrogen Hydrogen produced from natural gas or other carbon sources.

Green hydrogen Hydrogen produced from renewable energy without direct CO_2 emissions.

Gross electricity generation Actual electricity generation of the power plants, which includes own consumption (primarily by pumps).

Hole number Number of coils per pole and phase.

Hybrid vehicle Vehicle with two drive systems, usually combustion engine and electric motor.

Hybridization level Ratio of available electric traction power to total available traction power.

Input-split hybrid Power-split hybrid drive in which the power-splitting planetary gearbox is located near the combustion engine.

Laminated sheets Approx. 0.2–1.0 mm thin sheet made of a special type of steel (usually silicon iron) coated with an insulating layer of paint.

Life cycle The entire path of a product from the raw materials through production, use phase (energy requirement) to disposal (recycling).

Load-point shifting The electric motor increases (load point increase) or decreases (load point decrease) the torque of the internal combustion engine. The drive torque (system power) does not change.

Micro-hybrid Vehicle with combustion engine with automatic start/stop functionality.

Mild-hybrid Like micro-hybrid; additional load point shifting and boosting by the electric motor; no electric driving.

Multi-speed transmission Gearbox with at least two different switchable ratios.

Multi-stage Gearbox Gearbox with several gear pairs. The ratios of the individual pairs are multiplied to form the total ratio.

Nominal value Typical value of a quantity, for example nominal voltage of a network, see also rated value.

Number of poles Number of magnetic poles (north and south poles) of an electric motor.

One-mode hybrid Power-split hybrid drive with a power-splitting planetary gearbox.

One-pedal driving A relatively strong braking energy recovery can be controlled with the drive pedal, so that an experienced driver can drive almost without using the brake pedal.

Outage work Fictitious energy from renewable electricity generators that could not be generated, fed in and used due to grid overload (see also feed-in management).

Output-split hybrid Power-split hybrid drive in which the power-splitting planetary gearbox is located near the drive axle.

Parallel hybrid Direct mechanical drive from the combustion engine to the wheel. Drive jointly by ICE and electric motor.

Photovoltaics Generating electricity with solar cells from solar energy.

Plug-in hybrid Hybrid drive with the same functions as a full hybrid, but with additional battery charging capability from the socket, more powerful braking energy recovery and electric driving over longer distances.

Pole pitch Angular distance from one magnetic pole to the next.

Power-split hybrid The mechanical power is divided into a mechanical and an electrical power path. Driven jointly by ICE and electric motor.

Power-to-weight ratio The ratio of motor weight (complete weight incl. housing, shaft, flanges, connectors, etc., without cooling medium) to motor power (continuous power).

Primary energy Natural energy sources.

Pulse-width modulation (PWM) Chopping a direct voltage at high frequency to generate an alternating current of a certain frequency and amplitude.

Ragone-diagram Gravimetric power density of energy storage devices plotted against their volumetric power density.

Range-extender Standard hybrid drive in which the vehicle has a battery with a large capacity so that longer distances can also be driven purely electrically.

Rated value Design value of a variable, e.g. rated voltage of electric motors, see also nominal value.

Reluktance Attraction of ferromagnetic material, e.g. iron, in a magnetic field.

Renewable energy sources Inexhaustible energy sourcest that are self-regenerating.

Rotary current Symmetrical, 3-phase alternating current that is transmitted via three conductors.

Rotating field machine An electric machine in which the magnetic fields of the stator and rotor rotate continuously.

Rotor Inner, cylindrical part of the electric motor. The rotor carries coils, a cage or permanent magnets that are used to generate power.

Rotor-oriented control Control of a synchronous machine in a rotating coordinate system oriented to the rotor.

Sailing When the torque and power demand are small while driving, the combustion engine is automatically decoupled from the drive train and switched off to save energy.

Second life (2nd life) If a battery no longer has the energy and power density required for car operation, it can continue to be used in stationary systems where the technical requirements are lower.

Serial hybrid Drive purely via electric motor. The ICE is used exclusively to generate electrical energy using a separate generator. No mechanical through-drive to the wheel.

Single-tooth winding The coils of the single-tooth winding are each wound around a single tooth, i.e. in adjacent slots. Also referred to as concentrated winding.

Single-tooth windwing see concentrated winding

Slip The percentage deviation of the actual speed oft he rotor from the synchronous speed.

Slot pitch Angular distance from one slot to the next.

Slots Slotted holes in the stator or rotor laminated sheets, into which the copper conductors of the coils are inserted.

Squirrel-cage A special construction of the rotor of an induction machine in which the winding consists of a large number of electrically conductive bars, each of which is short-circuited at both ends by a ring.

Standstill ratio Negative ratio of the number of teeth of the ring gear and pinion of a planetary gear.

Stator Outer cylinder of an electric motor. The stator carries coils that generate the stator field.

Sustainability Use of raw materials and energies that are inexhaustible.

Synchronous generated voltage The voltage induced in the stator by the rotor of synchronous machines.

Torque vectoring If both rear wheels are driven individually, faster cornering can be achieved by individually varying the engine torque. Particularly advantageous in motorsport.

Two-mode hybrid Power-split hybrid drive with two or more power-split planetary gearboxes.

Voltage induction Generation of a voltage in a conductor that is moved through a magnetic field.

Index

M. Doppelbauer, *Introduction to Electromobility*,
https://doi.org/10.1007/978-3-658-45482-1

The manufacturer's authorised representative in the EU is Springer Nature Customer Service Centre GmbH, Europaplatz 3, 69115 Heidelberg, Germany. If you have any concerns regarding our products, please contact ProductSafety@springernature.com

Printed and bound by CPI Group (UK) Ltd, Croydon, CR0 4YY
30/09/2026
02225438-0002